Biological Treatment Technologies for Petroleum Hydrocarbon-Contaminated Environment

This book summarizes diverse biological treatment technologies employed in the remediation of hazardous petroleum hydrocarbon-contaminated sites. It explains a wide range of approaches, including bioremediation, biostimulation, bioaugmentation, bioattenuation, bioventing, biosparging, phytoremediation, mycoremediation, phycoremediation, electrobioremediation, nano-remediation, trichoremediation, and more. It describes the management of hydrocarbon waste through circular economy approaches and the successful degradation of contaminant sites using a zero-waste approach.

Key Features:

- Discusses the past, present, and future of biological remedies for the treatment of hydrocarbon pollutants.
- Introduces petroleum hydrocarbons, ecotoxicity, and green remediation approaches applicable to contaminated and wastewater streams.
- Covers the utilization of hydrocarbon waste-contaminated areas for economic growth and novel opportunities.
- Includes reports and case studies on environmental hydrocarbon contaminants and organic amendments for strategic recovery of the contaminated sites.
- Provides a clear understanding of the sustainable, greener technology approaches for the degradation of hydrocarbon contamination from different sites.

This book is aimed at researchers, professionals, and graduate students in chemical, environmental, biochemical, and bioengineering.

Biological Treatment Technologies for Petroleum Hydrocarbon-Contaminated Environment

Edited by
Ipsita Dipamitra Behera and Alok Prasad Das

CRC Press
Taylor & Francis Group
Boca Raton London New York

CRC Press is an imprint of the
Taylor & Francis Group, an **informa** business

Cover image: shutterstock

First edition published 2026
by CRC Press
2385 NW Executive Center Drive, Suite 320, Boca Raton FL 33431

and by CRC Press
4 Park Square, Milton Park, Abingdon, Oxon, OX14 4RN

CRC Press is an imprint of Taylor & Francis Group, LLC

First edition published by CRC Press, 2026

ISBN: 9781032824321 (hbk)
ISBN: 9781032832609 (pbk)
ISBN: 9781003508496 (ebk)

DOI: 10.1201/9781003508496

Typeset in Times
by codeMantra

Contents

Preface

The increasing world population, industrialization, and rapid expansion of the petroleum sector have led to severe environmental contamination due to the release of petroleum hydrocarbon pollutants into the ecosystem. Petroleum hydrocarbons are the organic compounds derived from crude oil and the refining of this produces gasoline, diesel, lubricating oil, and petrochemicals. The petroleum hydrocarbons include alkanes, which are saturated hydrocarbons, aromatics, and polycyclic aromatic hydrocarbons (PAHs). Benzene, toluene, xylene, and ethylbenzene are aromatic compounds considered highly toxic. The petroleum hydrocarbons are considered as persistent organic pollutants due to their hazardous effects on the ecological system and living beings. Water contamination, soil degradation, air pollution, and toxicity to living organisms are the major impacts of petroleum hydrocarbons. To date, a number of physicochemical methods have been implemented to treat the petroleum hydrocarbon pollutants but the generation of secondary pollutants, high energy consumption, and high chemical utilization are the major problems with these processes. Therefore, bioremediation of petroleum hydrocarbons has become a pressing concern, demanding sustainable and efficient treatment technologies. Among various approaches, biological treatment technologies have emerged as a promising solution due to their eco-friendliness, cost-effectiveness, and ability to degrade complex hydrocarbons into less toxic forms.

This book, *Biological Treatment Technologies for Petroleum Hydrocarbon-Contaminated Environment*, aims to provide a comprehensive overview of sources of petroleum hydrocarbons, and impacts on the ecosystem, marine bodies, and living organisms have been discussed. In addition to these, recent advancements in the biological degradation of petroleum hydrocarbons have been explained. It investigates the principles, mechanisms, and applications of various microbial and enzymatic processes employed for the bioremediation of petroleum-contaminated environments. This book highlights the role of bacteria, fungi, and algae in hydrocarbon degradation, elucidating their metabolic pathways and enzymatic functions. Additionally, it covers in situ and ex situ bioremediation strategies, including bioaugmentation, biostimulation, and phytoremediation, with case studies demonstrating their effectiveness. Furthermore, this book presents insights into cutting-edge research and innovative approaches, such as greener technology, fungal treatment, phytoremediation, and biosurfactant-mediated degradation. The challenges associated with biological treatment, including environmental constraints and microbial adaptability, are also discussed to provide a balanced perspective on the feasibility of large-scale applications. This book is intended for researchers, academicians, environmental professionals, and students in the fields of environmental biotechnology, biochemical engineering, and petroleum remediation. It serves as a valuable resource for understanding the scientific and technological aspects of biological treatment methodologies, fostering the development of sustainable and effective solutions for petroleum hydrocarbon pollution. We express our sincere gratitude to all contributors, researchers, and reviewers who have enriched this work with their expertise. We also acknowledge the support of our institutions and funding agencies that have facilitated our research endeavors in this field. We hope this book will inspire further research and innovations toward achieving a cleaner and healthier environment.

Editor Biographies

Ipsita Dipamitra Behera's research focuses on the development of bioremediation technology and process modernization in the area of petroleum waste management. She has published several research articles in leading and high-impact journals of international repute. She has more than 14 years of teaching experience with an area of expertise in environmental pollution control, petroleum hydrocarbon pollution, and its sustainable management.

Alok Prasad Das is presently serving as an assistant professor in the Department of Life Sciences, Rama Devi Women's University, Odisha, India. He is an academician and author of several books and journal articles. He is an editorial member of several reputed journals. He has more than 15 years of research experience. His area of expertise includes wastewater treatment, environmental microbiology, environmental pollution and management, microplastic and microfiber pollution, and bioremediation for sustainable solutions. He is the editor of several books. He has published more than 100 research and review articles in reputed international journals. He is an active lead guest editorial board member in top-rated journals. His current Google Scholar Citations is 2906, and his h-index is 30. He is also listed in the World's Top 2% Scientists for 2023, drawn up by Stanford University in the USA on the basis of subject-wise analysis.

Contributors

K Naga Sai Abhiram
Department of Biotechnology
Koneru Lakshmaiah Education Foundation (Deemed to be University)
Guntur, Andhra Pradesh, India

Kumari Aditi
Post Graduate Department of Zoology
Magadh University
Bodh Gaya, Bihar, India

Sarfaraz Ali
P.G. Department of Biotechnology
Magadh University
Bodh Gaya, Bihar, India

Maha Awjan Alreshidi
Department of Chemistry, College of Science
University of Ha'il
Ha'il, Saudi Arabia

Rajalaxmi Badra
Department of Life Sciences
Rama Devi Women's University
Bhubaneswar, Odisha, India

Kiran Bala
Department of Zoology, Deshbandhu College
University of Delhi
New Delhi, India

Nilambar Bariha
Assam Energy Institute
Sivasagar, Assam, India
Department of Fire and Safety Engineering
Sivasagar Centre of Rajiv Gandhi Institute of Petroleum Technology
Sivasagar, Assam, India

Anamika Basu
Department of Biochemistry
Gurudas College, Affiliated to University of Calcutta
Calcutta, West Bengal, India

Somanath Behera
School of Life Sciences
Sambalpur University
Burla, Odisha, India

Akalabya Bissoyi
University of Manchester
Greater Manchester, UK

Melissa Chen
Biotechnology Division Department of Biotechnology
Karunya Institute of Technology and Sciences (Deemed to be University)
Karunya Nagar, Coimbatore, India

Chiranjibi Chhotaray
Department of Medicine, Center for Emerging Pathogens, New Jersey Medical School, Rutgers
The State University of New Jersey
Newark, New Jersey, USA

Sudeshna Dey
Department of Life Sciences
Rama Devi Women's University, Vidya Vihar
Bhubaneswar, Odisha, India;
and
Centre for Biotechnology
Siksha 'O' Anusandhan (Deemed to be University)
Bhubaneswar, Odisha, India

Arushi Dogra
Faculty of Life Sciences and Biotechnology
South Asian University
Delhi, India

Priya Dubey
Department of Biosciences, Faculty of Science
Integral University
Lucknow, Uttar Pradesh, India

Amel Gacem
Department of Physics, Faculty of Sciences
University 20 Août 1955
Skikda, Algeria

Ramakrishnan Gnansekaran
Department of Biotechnology
Koneru Lakshmaiah Education Foundation (Deemed to be University)
Guntur, Andhra Pradesh, India

Aswetha Iyer
Biotechnology Division Department of Biotechnology
Karunya Institute of Technology and Sciences (Deemed to be University)
Karunya Nagar, Coimbatore, India

Suman Jagatee
Department of Life Sciences
Rama Devi Women's University
Bhubaneswar, Odisha, India

Akbar John
College of Marine Science and Aquatic Biology
University of Khorfakkan
Sharjah, UAE

Selcan Karakus
Department of Chemistry
Istanbul University-Cerrahpaşa
Istanbul, Türkiye

Pamreishang Kasar
Department of Testing
Central Institute of Petrochemical Engineering & Technology (CIPET)
Chennai, Tamil Nadu, India

Abhishek Kumar
Department of Chemical Engineering
National Institute of Technology Rourkela
Rourkela, Odisha, India

Ashish Kumar
Department of Science, Technology and Technical Education
Nalanda College of Engineering, Bihar Engineering University
Patna, Bihar, India

Raj Kumar
Department of Zoology
S.N. Sinha College
Gaya, Bihar, India

Sunny Kumar
Department of Chemical Engineering
ChBE, Georgia Institute of Technology
Atlanta, Georgia, USA

Mukta Mayee Kumbhar
Department of Life Sciences
Rama Devi Women's University
Bhubaneswar, Odisha, India

Chhagan Lal
Department of Chemistry, Swami Shraddhanand College
University of Delhi
New Delhi, India

Chandra Mohan
School of Basic & Applied Sciences
K R Mangalam University
Gurugram, Haryana, India

Jyotirmayee Mohanty
Department of Life Sciences
Rama Devi Women's University
Bhubaneswar, Odisha, India

Sevanan Murugan
Biotechnology Division Department of Biotechnology
Karunya Institute of Technology and Sciences (Deemed to be University)
Karunya Nagar, Coimbatore, India

Harjeet Nath
Department of Chemical and Polymer Engineering
Tripura Central University
Suryamaninagar, Tripura, India

Deeptimayee Pal
Department of Chemical Engineering
National Institute of Technology
Rourkela, Odisha, India

Maminee Panda
P.G. Dept. of Zoology
Samanta Chandra Sekhar Autonomous College
Puri, Odisha, India

Subhashree Parida
Department of Life Sciences
Rama Devi Women's University
Bhubaneswar, Odisha, India

Sailaja Priyadarsini
Department of Life Sciences
Rama Devi Women's University
Bhubaneswar, Odisha, India

Sujata Priyadarshini
Department of Life Sciences
Rama Devi Women's University
Bhubaneswar, Odisha, India

Smiti Kana Ray
Department of Life Sciences
Rama Devi Women's University
Bhubaneswar, Odisha, India

G. Koteswara Reddy
Department of Biotechnology
Koneru Lakshmaiah Education Foundation (Deemed to be University)
Guntur, Andhra Pradesh, India

Aditi Roy
Department of Botany (Environmental Science)
University of Lucknow
Lucknow, Uttar Pradesh, India

Priyanka Singh
Department of Biotechnology, School of Engineering
IILM University
Greater Noida, Uttar Pradesh, India

Samriddhi Singh
Department of Biotechnology, School of Engineering
IILM University
Greater Noida, Uttar Pradesh, India

Meenu Singh
Department of Biotechnology, School of Engineering
IILM University
Greater Noida, Uttar Pradesh, India

Luiyaiwoshi Shimray Songachan
Department of Botany
Banaras Hindu University (BHU)
Varanasi, Uttar Pradesh, India

Abhinay Thakur
Division of Research and Development
Lovely Professional University
Phagwara, Punjab, India

Krishna Kumar Yadav
Department of Environmental Science
Parul Institute of Applied Sciences, Parul University
Vadodara, Gujarat, India;
and
Environmental and Atmospheric Sciences Research Group, Scientific Research Center
Al-Ayen University
Nasiriyah, Thi-Qar, Iraq

Noushi Zaidi
Department of Basic Sciences, Preparatory Year
King Faisal University
Al-Ahsa, Saudi Arabia

1 Hydrocarbon Pollution

An Introductory Overview of Environmental Concerns

Deeptimayee Pal and Abhishek Kumar

1.1 INTRODUCTION

Hydrocarbons are organic molecules consisting only of hydrogen and carbon atoms, serving as the primary components of fossil fuels, including coal, crude oil, and natural gas. Hydrocarbons are the foundation of contemporary energy generation, transportation, and chemical sectors, significantly influencing the global economy (Bala et al., 2024; Dey, Bissoyi, et al., 2024). While hydrocarbons are vital for industrial and commercial advancement, their extraction, processing, and use have resulted in extensive environmental degradation, referred to as hydrocarbon pollution. Hydrocarbon contamination occurs when hydrocarbon compounds are unintentionally released into the environment through actions such as oil spills, leaks, automobile exhaust emissions, industrial discharges, and improper disposal of petroleum products (Pal & Sen, 2023a). Pollutants persist in many habitats, contaminating air, water, and soil, leading to significant consequences for human health and the natural environment. Hydrocarbon contamination has emerged as a worldwide environmental concern due to the significant dependence on fossil fuels (Behera et al., 2020, 2021, 2022). It appears in several ways, including air pollution from hydrocarbon burning and water and soil contamination due to oil spills or industrial effluents. The environmental fate of hydrocarbons is contingent upon their chemical characteristics; lighter compounds often volatilize into the atmosphere, while heavier molecules endure in soils and sediments (Dey, Dash, et al., 2024). Polycyclic aromatic hydrocarbons (PAHs) exhibit notable resistance to degradation and are recognized for their poisonous, mutagenic, and carcinogenic properties. Combustion of hydrocarbons in the atmosphere produces detrimental by-products, including volatile organic compounds (VOCs) and particulate matter, which contribute to smog production and global warming. In aquatic ecosystems, hydrocarbons may create surface slicks, disturbing marine organisms, while their deposition in sediments might result in enduring ecological consequences. Likewise, hydrocarbon pollution in soils diminishes land production and threatens groundwater quality (Zhang et al., 2012).

The extensive reach of hydrocarbon contamination significantly affects ecosystems and human health. Significant oil spills, such as the Exxon Valdez incident in 1989 and the Deepwater Horizon leak in 2010, have shown the devastating consequences of hydrocarbon pollution on marine ecosystems, resulting in habitat damage, widespread species death, and prolonged ecological deterioration (Dey, Tripathy, et al., 2023).

DOI: 10.1201/9781003508496-1

These instances underscore the susceptibility of coastal ecosystems and the challenges inherent in restoration efforts. Alongside catastrophic spills, the persistent discharge of hydrocarbons from traffic, industrial operations, and urban runoff progressively deteriorates the ecosystem in more insidious but equally detrimental ways (Kanungo et al., 2023a,b). Vehicle emissions are a significant source of air pollution in metropolitan areas, leading to respiratory illnesses and diminishing the quality of life for millions. Water pollution, often caused by industrial discharges, jeopardizes the health of aquatic ecosystems, resulting in the bioaccumulation of harmful hydrocarbons in fish and other marine organisms, which may then infiltrate the human food chain. The enduring presence of hydrocarbons in the environment presents considerable long-term hazards. PAHs, found in both air and water pollution, are recognized for their ability to bioaccumulate in organisms, leading to chronic health impacts such as cancer, reproductive disorders, and immunological suppression. The economic repercussions of hydrocarbon pollution are significant, including biodiversity loss, harm to fisheries and tourists, and the financial strain of remediation operations (Honda & Suzuki, 2020).

Mitigating hydrocarbon pollution is crucial for environmental preservation, safeguarding human health, and ensuring economic viability. The widespread occurrence of hydrocarbon pollution, along with its enduring ecological impacts, requires the formulation and execution of efficient methods to diminish emissions, manage waste, and restore polluted ecosystems. Hydrocarbon contamination severely impacts biodiversity and ecological services. Minimizing hydrocarbon emissions into the environment is crucial for safeguarding terrestrial and aquatic ecosystems, which perform key activities such as carbon sequestration, nitrogen cycling, and water purification. The detrimental health consequences linked to hydrocarbon exposure, including cancer, respiratory disorders, and cardiovascular ailments, highlight the need for rigorous restrictions and pollution mitigation strategies (Dey, Das, et al., 2023; Mohanty et al., 2024; Singh Jadaun et al., 2022). Minimizing exposure to airborne hydrocarbons and averting water pollution are essential measures for safeguarding public health, especially in heavily populated urban and industrial areas (Kumar & Gangawane, 2023b). The economic ramifications of hydrocarbon pollution are substantial, including cleaning efforts after oil spills and healthcare expenses associated with pollution-induced ailments. Shifting to greener energy sources and enhancing waste management methods may alleviate these fiscal pressures while fostering long-term economic stability and sustainability. The burning of hydrocarbons significantly contributes to global climate change. Minimizing dependence on fossil fuels and shifting to renewable energy sources is crucial for curtailing greenhouse gas emissions and alleviating the effects of climate change, including elevated temperatures, severe weather phenomena, and rising sea levels.

1.2 TYPES OF HYDROCARBONS

Hydrocarbons are classified according to their structure and the kind of bonds connecting carbon atoms. The categories include alkanes, alkenes, alkynes, and aromatic hydrocarbons, as shown in Figure 1.1. Each kind has unique chemical and physical features that influence their environmental behavior and their effects on ecosystems when they serve as pollutants.

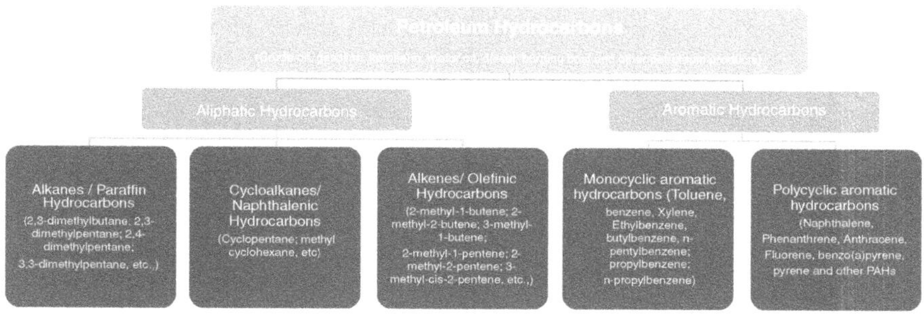

FIGURE 1.1 Types of hydrocarbons (Sharma et al., 2023).

Alkanes are hydrocarbons characterized by only single bonds between carbon atoms, indicating they are "saturated" with hydrogen atoms. Every carbon atom in an alkane establishes four single covalent connections, either with hydrogen atoms or with other carbon atoms. The chemical formula for alkanes is C_nH_{2n+2}, with "n" denoting the number of carbon atoms. Alkanes are hydrocarbons characterized exclusively by single bonds between carbon atoms, indicating that they are "saturated" with hydrogen atoms. Every carbon atom in an alkane forms four single covalent connections, either with hydrogen atoms or with other carbon atoms. The molecular formula for alkanes is C_nH_{2n+2}, with "n" denoting the number of carbon atoms. Alkanes are prevalent environmental pollutants due to their presence in crude oil and petroleum derivatives. Alkanes contribute to soil and water contamination in oil spills and leaks (Das et al., 2024; Mishra et al., 2022, 2024). Shorter-chain alkanes, such as methane and ethane, may volatilize into the atmosphere, but long-chain alkanes often accumulate in sediments and soils. The environmental persistence of alkanes is linked to their biodegradability. Certain microorganisms, including bacteria and fungi, may metabolically break down alkanes, but at a sluggish rate, especially for longer and heavily branched alkanes (Jiang et al., 2017).

Alkenes are unsaturated hydrocarbons characterized by the presence of at least one carbon–carbon double bond (C=C), differentiating them from alkanes. The existence of this double bond renders alkenes more chemically reactive than alkanes. The empirical formula for alkenes is C_nH_{2n}, signifying a reduced number of hydrogen atoms compared to the comparable alkane. The double bond in alkenes facilitates certain chemical processes, including polymerization, oxidation, and hydrogenation, which are not feasible with alkanes. Alkenes have heightened chemical reactivity due to their double bond, making them susceptible to oxidation and other chemical changes in the environment. Alkenes have a significant role in atmospheric chemistry and industrial pollutants. Ethylene is naturally produced by plants as a hormone and is also emitted during combustion processes, including biomass burning and industrial operations. Alkenes in the environment may combine with pollutants like nitrogen oxides (NO_x) to produce ground-level ozone, a detrimental air pollutant. The presence of double bonds in alkenes makes them more vulnerable to deterioration by chemical processes, such as oxidation by hydroxyl radicals in the environment. In contaminated soils and water, alkenes may endure and function as pollutants until adequately remediated (Kuppusamy et al., 2020).

Alkynes are unsaturated hydrocarbons characterized by the presence of at least one carbon–carbon triple bond (C≡C). The triple bond confers a linear configuration to the molecule, making alkynes more reactive than alkenes. The empirical formula for alkynes is C_nH_{2n-2}, indicating a reduced hydrogen content relative to alkanes and alkenes. Alkynes have significant reactivity, particularly in the presence of catalysts that may cleave the triple bond, facilitating reactions such as hydrogenation. They are less prevalent in natural sources than alkanes and alkenes. Alkynes are less common in environmental contamination compared to alkanes and alkenes, mostly due to their higher reactivity and lower abundance in fossil fuels. Nonetheless, they remain present in industrial pollutants and certain categories of trash. Alkynes, owing to their reactivity, are less likely to last in the environment, since they are more easily decomposed by chemical and photochemical processes.

Aromatic hydrocarbons, or arenes, are defined by the presence of a minimum of one aromatic ring, which is a cyclic arrangement of carbon atoms including alternating single and double bonds. This arrangement, referred to as "conjugation," imparts distinct stability and reactivity to aromatic hydrocarbons due to resonance. Benzene (C_6H_6) is the most prevalent aromatic hydrocarbon. Aromatic hydrocarbons possess distinctive chemical characteristics attributed to their ring configurations, making them generally stable but reactive under certain circumstances. They are generally non-polar and hydrophobic, making them insoluble in water but soluble in organic solvents. PAHs are a category of chemical compounds consisting of many interconnected aromatic rings formed exclusively from carbon and hydrogen atoms. These structures exhibit stability and persistence due to the resonance of alternating single and double bonds within the rings. PAHs lack heteroatoms (such as oxygen, nitrogen, or sulfur) and any substituents on the aromatic rings. The fundamental structure of PAHs consists of two or more fused benzene-like rings arranged in linear, angular, or clustered configurations. Examples include naphthalene, anthracene, phenanthrene, pyrene, benzo[a]pyrene, fluoranthene, chrysene, and benz[a]anthracene. Aromatic hydrocarbons, especially PAHs, pose significant environmental risks due to their toxicity, persistence, and capacity for bioaccumulation in living organisms (Logeshwaran et al., 2018). PAHs are generated by the incomplete combustion of fossil fuels, biomass, and other organic materials, leading to their widespread presence in urban air pollution, industrial emissions, and vehicular exhaust. Upon release into the environment, PAHs have a strong affinity for binding to soil particles and sediments, leading to their persistence. They are recognized as carcinogenic, mutagenic, and teratogenic, presenting substantial hazards to human health and animals (Ray et al., 2024). Benzene and its derivatives, including toluene and xylenes, are hazardous and often present in polluted groundwater and industrial effluents. Aromatic hydrocarbons are stringently regulated in several countries due to their extensive use in industrial processes and the associated concerns for environmental and human health (Pal & Sen, 2023b).

Straight-chain hydrocarbons are hydrocarbons characterized by a linear structure, with carbon atoms linked in a sequential manner. They are often less complex and more susceptible to microbial breakdown than branched hydrocarbons. Branched-chain hydrocarbons possess side chains or branches extending from the primary carbon chain. They often exhibit greater resistance to microbial destruction,

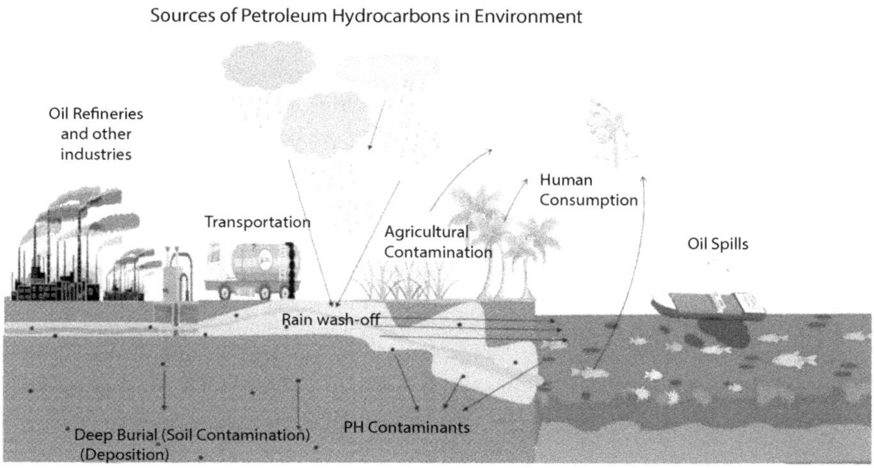

FIGURE 1.2 Sources of hydrocarbons (Sharma et al., 2023).

hence enhancing their persistence in the environment. Cyclic hydrocarbons have carbon atoms organized in ring formations, devoid of the alternating double bonds characteristic of aromatic molecules. Cyclohexane (C_6H_{12}) exemplifies a saturated cyclic hydrocarbon. These molecules are present in petroleum and natural gas, exhibiting characteristics that are intermediate between straight-chain hydrocarbons and aromatic hydrocarbons (Blanco et al., 2001).

1.3 SOURCES

Hydrocarbons are organic molecules consisting of hydrogen and carbon, serving as the primary constituents of fossil fuels, including crude oil, natural gas, and coal. Hydrocarbon pollution occurs when hydrocarbon chemicals are emitted into the environment, leading to the contamination of air, water, and soil. Environmental hydrocarbon contamination originates from a variety of sources, which may be categorized as natural or anthropogenic (human-made), as depicted in Figure 1.2. Comprehending the sources of hydrocarbon contamination is essential for formulating effective measures to alleviate its effects on ecosystems and human health.

1.4 ANTHROPOGENIC SOURCES

Anthropogenic activities are the primary sources of hydrocarbon pollution, responsible for the majority of hydrocarbon emissions into the environment. These activities include fossil fuel extraction, transportation, refining, industrial processes, and waste management.

The extraction of oil and gas is a major source of hydrocarbon contamination. Oil spills are a prominent and devastating form of hydrocarbon contamination. They occur during offshore and onshore drilling, pipeline breaches, tanker incidents, and well blowouts. Significant oil spills lead to extensive pollution of marine and terrestrial ecosystems. The environmental and economic ramifications of such spills are substantial, impacting marine life, fisheries, coastal ecosystems, and tourist sectors. In the extraction of oil and gas, substantial quantities of water, referred to as "produced water," are obtained with hydrocarbons. Produced water comprises dissolved hydrocarbons, heavy metals, salts, and chemicals used during extraction, posing a risk of environmental contamination if not controlled appropriately. The disposal of generated water, whether by discharge into aquatic systems or subsurface injection, presents problems of groundwater pollution and soil deterioration. Flaring denotes the combustion of natural gas that is extracted with oil but not used for commercial purposes. This process leads to the release of gaseous hydrocarbons, including methane and other VOCs, into the atmosphere. Venting involves the direct emission of gases without burning. Both behaviors worsen air pollution, increase greenhouse gas emissions, and contribute to the production of ground-level ozone (Ahmed & Fakhruddin, 2018).

The transportation of crude oil, petroleum derivatives, and natural gas through pipelines, vessels, trucks, and railways is a major source of hydrocarbon contamination. Incidents and spills during transit can lead to significant environmental pollution. Pipelines are extensively used for the transportation of crude oil and natural gas over considerable distances. While pipelines are often regarded as a secure transmission method, they may experience leaks and ruptures due to corrosion, mechanical failure, or external damage. When pipelines rupture, hydrocarbons may infiltrate the adjacent soil and water, polluting groundwater and surface water supplies. Underground pipelines, specifically, pose a danger of prolonged contamination that is difficult to identify and rectify. Oil tankers carry substantial quantities of crude oil across seas. Tanker incidents, while infrequent, may result in substantial oil leaks, causing catastrophic effects on marine ecosystems. Minor spills from refueling activities, bilge discharges, or inadvertent releases during loading and unloading may lead to persistent hydrocarbon contamination in ports and coastal regions. The transportation of oil by rail and road is a possible source of hydrocarbon contamination. Train derailments and truck collisions may lead to oil spills that pollute rivers, streams, and land (Kuppusamy et al., 2020).

The refining of crude oil and the manufacturing of petrochemicals release hydrocarbons into the environment through various channels, including atmospheric emissions, wastewater discharge, and accidental spills. Oil refineries and petrochemical facilities emit a range of hydrocarbons, including methane, benzene, toluene, ethylbenzene, xylene (BTEX), and PAHs, during the refining process. These emissions exacerbate air pollution, facilitate smog formation, and pose health risks to adjacent residents. VOCs are precursors of ground-level ozone, which may result in respiratory ailments and other health complications. Refinery activities produce effluent including hydrocarbons, heavy metals, and various contaminants. If inadequately cleaned before disposal, this effluent may pollute adjacent water bodies, endangering aquatic organisms, and threatening human health. Refining and petrochemical operations are prone to leaks, spills, and accidents that discharge hydrocarbons into the surrounding environment. Decrepit infrastructure, equipment malfunctions, and human mistakes may result in significant pollution of soil, groundwater, and surface water (Elijah, 2022).

Urban runoff is a significant source of hydrocarbon contamination, especially in regions characterized by high traffic density and industrial operations. Precipitation that accumulates on highways, parking lots, and other impermeable surfaces transports hydrocarbons and other contaminants into storm drains, rivers, and coastal waterways. Hydrocarbons from automobiles, including engine oil, gasoline, and lubricants, significantly contribute to urban runoff pollution. Vehicle leaks, tire degradation, and exhaust emissions contribute to the buildup of hydrocarbons on roadways and pavements. During precipitation events, these hydrocarbons are transported into stormwater systems, subsequently entering rivers, lakes, and oceans. Urban regions characterized by extensive industrial and commercial operations contribute to hydrocarbon pollution via runoff from production sites, petrol stations and storage facilities (Oyo-Ita et al., 2016). Spills, leaks, and inappropriate waste disposal from these operations may result in hydrocarbon pollution of adjacent water bodies.

Inadequate disposal of hydrocarbon-laden trash, including spent motor oil, chemicals, and plastic materials, may result in environmental pollution. Landfills, specifically, pose potential hydrocarbon contamination risks if waste management is insufficient. Landfills produce leachate, a liquid created when water infiltrates waste items. This leachate may include dissolved hydrocarbons, especially from waste oil, plastics, and other petroleum-derived items. Failure of landfill liners or inadequate treatment of leachate may result in pollution of groundwater or surface water. The unlawful disposal of hazardous materials, such as used oil and solvents, in uncontrolled locations is a prevalent issue. These pollutants may permeate soil and groundwater, resulting in prolonged hydrocarbon pollution (Ossai et al., 2020).

1.5 NATURAL SOURCES

Apart from anthropogenic sources, hydrocarbons are also emitted into the environment via natural processes. Although these sources are often of lesser magnitude than human emissions, they still contribute significantly to the global hydrocarbon cycle.

Natural oil seeps occur when crude oil or natural gas escapes from subterranean reservoirs and rises to the surface, often through fissures or cracks in the Earth's crust. These leaks release hydrocarbons directly into the environment, particularly in marine habitats. Natural oil seeps are prevalent in oceanic regions characterized by intense tectonic activity, including the Gulf of Mexico, the Caspian Sea, and the California coast. These seeps may discharge substantial quantities of oil into the water, where it ascends to the surface and creates slicks. Although a portion of the oil undergoes natural degradation via microbial activity, substantial quantities may collect and impact marine ecosystems. Oil leaks on land may pollute soil and groundwater. Natural oil seeps in the Middle East have been documented for ages and persist in discharging hydrocarbons into the environment (E. Ite et al., 2018).

Volcanic activity is an additional natural source of hydrocarbon emissions. Hydrocarbons, including methane and other VOCs, may be discharged into the atmosphere during volcanic eruptions or by volcanic gases produced from fumaroles and hot springs. Although volcanic hydrocarbon emissions are rather minor compared to human sources, they still contribute to the global hydrocarbon flow. The breakdown of organic materials, especially in wetland ecosystems, generates hydrocarbons such as methane. Wetlands, including swamps, marshes, and peatlands, are important natural sources of methane, a potent greenhouse gas. The anaerobic breakdown of organic matter by microorganisms in saturated soils produces methane, which is then released into the atmosphere. Some plants and microbes synthesize hydrocarbons as part of their metabolic activities. Certain tree species, including conifers, emit volatile hydrocarbons such as isoprene and terpenes into the atmosphere. Biogenic hydrocarbons influence atmospheric chemistry by contributing to aerosol production and the generation of secondary organic pollutants (Haider et al., 2021).

1.6 TRANSPORTATION AND FATE OF HYDROCARBONS

Hydrocarbons, released into the environment by natural processes or anthropogenic activities, traverse complex paths and undergo several changes that govern their final outcome. The environmental fate of hydrocarbons depends on their physical and chemical characteristics, prevailing environmental conditions, and the ecosystems into which they are released. Comprehending these routes is essential for evaluating the enduring effects of hydrocarbon contamination and for formulating efficient remediation approaches. Hydrocarbons are organic molecules consisting of hydrogen and carbon, existing in gaseous, liquid, and solid phases. These substances may be classified into aliphatic hydrocarbons (linear or branching chains) and aromatic hydrocarbons (cyclic structures such as benzene). Their durability, mobility, and environmental toxicity are contingent upon these molecular attributes. Upon the release of hydrocarbons into the environment, they interact with the physical, chemical, and biological processes that govern their dispersion, transformation, and final outcome (Truskewycz et al., 2019).

Hydrocarbons are introduced into the environment by natural processes and human activities. Natural hydrocarbon emissions arise from activities such as oil seepage from the oceanic substrate, volcanic activity, and the breakdown of organic materials. Biogenic hydrocarbons are generated by certain plants and microbes. Human activities are the principal sources of hydrocarbon contamination. Primary sources include oil spills (resulting from transportation, offshore drilling, and pipeline breaches), industrial effluents, fuel burning, urban runoff, and inadequate waste disposal. Oil spills in aquatic ecosystems and leaks from subterranean storage tanks are significant sources of environmental pollution. When introduced into the environment, hydrocarbons infiltrate air, water, or soil compartments, with their environmental behavior shaped by characteristics such as volatility, solubility, and molecular structure.

Hydrocarbons undergo several physical processes that influence their distribution and behavior in the environment. Volatilization is the process by which volatile hydrocarbons convert from a liquid or solid state to a gaseous one, subsequently entering the atmosphere. This process is especially crucial for lighter hydrocarbons, including methane, ethane, propane, and other VOCs. Upon entering the atmosphere, these hydrocarbons may be carried over considerable distances before undergoing degradation or being deposited back to the ground. Temperature, wind velocity, and surface area are critical determinants affecting the rate of volatilization.

Elevated temperatures and increased surface exposure enhance the volatilization rate of hydrocarbons, while wind dispersion may disseminate these hydrocarbons across extensive regions (Manzetti, 2013).

Certain hydrocarbons dissolve in aquatic settings upon contact, a phenomenon referred to as dissolution. Hydrocarbons typically have limited solubility in water; however, aromatic hydrocarbons, including benzene, toluene, ethylbenzene, and xylene (BTEX), have greater solubility than aliphatic hydrocarbons. Dissolution is a pivotal step in assessing the transport of hydrocarbons in aquatic systems, since dissolved hydrocarbons may disseminate across water currents and impact aquatic life. The solubility of hydrocarbons is contingent upon their molecular structure, with smaller, lighter hydrocarbons exhibiting greater solubility than larger, more intricate counterparts. Water temperature, salinity, and pH influence dissolution rates.

Adsorption is the mechanism by which hydrocarbons attach to the surfaces of solid particles, including soil or sediment particles (Pal & Sen, 2024b). Hydrophobic hydrocarbons, characterized by their low solubility in water, often adsorb onto organic matter or mineral surfaces in soil and sediments. Upon adsorption, hydrocarbons exhibit reduced mobility but may last in the environment for extended durations, possibly serving as a persistent source of pollution. The concentration of organic matter, clay content, and particle size of the soil or sediment substantially affect the adsorption of hydrocarbons. Elevated organic content enhances hydrocarbon adsorption owing to the hydrophobic characteristics of hydrocarbons.

Hydrocarbons discharged into aquatic ecosystems, especially denser components like tar balls or aged oil, may submerge and become entrenched in sediments. Hydrocarbons, once entrapped in sediments, may last for years, gradually degrading under anaerobic circumstances or remaining stable. Sediment pollution is most pronounced in coastal locations, estuaries, and wetlands, where hydrocarbons may collect and adversely affect benthic organisms. Hydrocarbons participate in several chemical reactions in the environment, leading to their conversion into various chemicals. Hydrocarbons in the air and water undergo oxidation processes, breaking down into smaller, less hazardous molecules. Oxidation in the atmosphere is mostly facilitated by processes involving hydroxyl radicals ($\cdot OH$), ozone (O_3), or nitrogen oxides (NO_x), resulting in the production of carbon dioxide, water, and other by-products. Oxidation in water occurs through similar routes, often aided by sunlight, through photochemical processes. Sunlight, especially ultraviolet (UV) radiation, may trigger photochemical processes that decompose hydrocarbons, particularly in surface waters and the top strata of soil or air. This mechanism is crucial for the degradation of certain aromatic compounds. Hydrolysis is the chemical interaction between hydrocarbons and water, leading to the decomposition of hydrocarbon molecules. This process is often more pronounced for chlorinated hydrocarbons than for unrefined hydrocarbons such as oil or gas. In maritime settings, hydrocarbon "weathering" denotes the amalgamation of physical and chemical processes that modify the makeup of spilled oil over time. Weathering includes the evaporation of lighter constituents, solubility in water, emulsification (creation of oil-water combinations), and oxidation. This process may result in the creation of tar balls or asphalt-like leftovers, which may persist in the environment for extended periods (Tveit et al., 2021).

Biodegradation is the principal biological process through which hydrocarbons are decomposed in the environment. Microorganisms, such as bacteria, fungi, and archaea, are essential for the breakdown of hydrocarbons, using them as a carbon and energy source. Aerobic bacteria metabolize hydrocarbons in the presence of oxygen via enzyme activities, converting them into simpler molecules such as carbon dioxide and water. Aerobic biodegradation is more efficient and rapid than anaerobic processes, occurring in surface soils, aquatic environments, and regions with ample oxygen availability. Bacteria such as Pseudomonas, Alcanivorax, and Mycobacterium are recognized for their ability to degrade aliphatic and aromatic hydrocarbons under aerobic conditions. In anoxic conditions, including deep soils, sediments, or groundwater, anaerobic degradation occurs via microbes that use alternative electron acceptors like nitrate, sulfate, or iron. Anaerobic biodegradation often occurs at a slower rate than aerobic processes and is particularly crucial for the decomposition of heavier hydrocarbons and PAHs. Hydrocarbon degradation often entails a consortium of microbes, each adept at distinct phases of the degradation process. One type of microbe may decompose complex hydrocarbons into simpler compounds, which are then further degraded by other species. This collaborative process improves the overall efficacy of hydrocarbon biodegradation. The rate of biodegradation is influenced by several parameters, including oxygen availability, nutrient concentrations (particularly nitrogen and phosphorus), temperature, and the presence of appropriate microbial populations. The structure of hydrocarbons influences degradation rates, with smaller, simpler molecules undergoing degradation more readily than complex, branching, or aromatic compounds (Whelan et al., 2015).

Upon discharge, hydrocarbons may migrate via several environmental mediums, such as air, water, and soil. Volatile hydrocarbons may be carried through the atmosphere over extensive distances, especially when they evaporate from spills or emissions. This process may result in the extensive dissemination of contaminants, particularly ozone precursors such as VOCs, which exacerbate air quality problems and pose dangers to human health. In aquatic ecosystems, hydrocarbons may be conveyed by surface water currents, groundwater movement, or soil penetration. Dissolved hydrocarbons can spread across aquatic systems, affecting habitats far from the initial pollution source. In terrestrial contexts, hydrocarbons may migrate through soil strata via leaching or lateral movement, thereby contaminating groundwater supplies. In aquatic ecosystems, sediment-associated hydrocarbons may be transported by water currents and accumulate in low-energy settings such as estuaries or wetlands. Numerous environmental variables affect the trajectories and final disposition of hydrocarbons in the ecosystem. Elevated temperatures often enhance the rates of volatilization, biodegradation, and chemical processes, thereby promoting the decomposition of hydrocarbons. The availability of oxygen significantly improves biodegradation processes, particularly aerobic degradation. In oxygen-deficient conditions, hydrocarbons may last for extended periods due to the reduced rate of anaerobic breakdown. The microbial breakdown of hydrocarbons relies on the availability of nutrients, namely nitrogen and phosphorus. Nutrient-deficient conditions often restrict the rate of biodegradation. The soil or sediment type influences the adsorption and movement of hydrocarbons. Fine-grained soils with elevated organic content exhibit enhanced hydrocarbon adsorption, hence diminishing mobility while augmenting permanence.

1.7 ADVERSE EFFECTS OF HYDROCARBONS

Hydrocarbon contamination poses considerable risks to environmental integrity and human well-being. The consequences are extensive and may last for decades, impacting ecosystems, biodiversity, air, and water quality and resulting in significant health issues for both people and animals.

1.8 ENVIRONMENTAL EFFECTS

Hydrocarbon contamination affects environmental systems through physical, chemical, and biological processes, altering ecosystems and inflicting enduring harm. These effects occur in both terrestrial and aquatic habitats, with especially grave repercussions in marine ecosystems.

Hydrocarbons that spill on land may permeate the soil, where they can attach to soil particles or migrate through the subsurface, impacting soil quality. Hydrocarbons disrupt soil porosity, water retention, and nutrient availability, hindering plant development and modifying the composition of the soil microbial community. Contamination may result in decreased agricultural output and reduced vegetation cover in impacted regions. Numerous hydrocarbons, especially PAHs and VOCs, are detrimental to plants, resulting in inhibited growth, diminished seed germination, foliar necrosis, and even mortality. These compounds disrupt photosynthesis and nutrient uptake, impairing plants' capacity to flourish. In severely polluted regions, hydrocarbons may lead to deforestation and a reduction in plant cover, intensifying soil erosion and desertification. Soil microbes are essential for nutrient cycling and the breakdown of organic materials. Hydrocarbons perturb microbial ecosystems by modifying the quantity and diversity of microbial populations. Some hydrocarbons may suppress the proliferation of advantageous bacteria and fungi, while others may facilitate the growth of hydrocarbon-degrading microorganisms. The alteration of microbial populations may adversely affect soil health and fertility, undermining ecosystem functionality (Aislabie et al., 2004).

Hydrocarbons released in or near aquatic environments, including rivers, lakes, and seas, rapidly spread over the water's surface. The formation of oil slicks hinders oxygen transfer between the atmosphere and water, adversely impacting aquatic organisms that depend on dissolved oxygen. Decreased oxygen levels may result in hypoxia, jeopardizing fish, invertebrates, and other aquatic animals. Hydrocarbons, particularly PAHs, exhibit significant toxicity to aquatic creatures, including fish, shellfish, and plankton. Exposure may result in several sub-lethal and deadly consequences, including reproductive failure, developmental abnormalities, immunological suppression, and cancer. The early life stages of fish, including eggs and larvae, exhibit heightened sensitivity to hydrocarbon exposure, resulting in population decreases. Hydrocarbons, especially PAHs, may accumulate in the tissues of marine creatures and become amplified as they ascend the food chain. Biomagnification leads to elevated amounts of hydrocarbons in apex predators, including marine animals, seabirds, and people that ingest polluted seafood. The buildup of toxic hydrocarbons may result in chronic health consequences, such as cancer, organ damage, and reproductive abnormalities in both animals and people. Denser hydrocarbon molecules, such as those present in crude oil, may descend to the ocean bottom, getting ensnared in sediments. These hydrocarbons last for decades and provide enduring hazards to benthic animals, such as crabs, mollusks, and worms. Contaminated sediments may impair nutrient cycling and deteriorate the health of coastal ecosystems, including coral reefs and mangrove forests (Speight, 2011).

Marine habitats are especially susceptible to hydrocarbon contamination due to the magnitude of oil spills and operational discharges from vessels and offshore drilling activities. The distinctive characteristics of hydrocarbons, such as their persistence, toxicity, and capacity for extensive dispersion, make marine ecosystems vulnerable to significant and enduring harm. Significant oil spills, exemplified by the Deepwater Horizon incident in 2010, discharge millions of gallons of crude oil into the ocean, causing extensive environmental destruction. The physical effects of oil slicks on marine ecosystems are significant, impacting surface-dwelling creatures, including seabirds, marine mammals, and fish, by contaminating their fur, feathers, or gills with oil (Onwuka & Adu, 2024). This results in hypothermia, drowning, or asphyxiation. Hydrocarbon contamination, especially from oil spills, significantly harms coral reefs, which rank among the most biodiverse ecosystems on the planet. Oil may suffocate coral polyps, impede photosynthesis in symbiotic algae, and disrupt coral reproduction. The degradation of coral reefs not only decreases marine biodiversity but also undermines the ecological services they provide, including coastal protection and fisheries support. Mangrove forests and coastal wetlands, serving as natural barriers against storm surges and floods, are particularly vulnerable to hydrocarbon pollution. Oil and hydrocarbons may envelop the roots and trunks of mangroves, asphyxiating the plants and exterminating the related wildlife. The deterioration of these ecosystems intensifies the impacts of climate change and sea-level rise, since wetlands and mangroves are crucial for carbon storage.

1.9 HEALTH EFFECTS

Hydrocarbon pollution adversely affects not only the environment but also poses risks to human health through both direct and indirect exposure via air, water, soil, and food. Hydrocarbons may infiltrate the human body by inhalation, ingestion, and skin exposure, resulting in both acute and chronic health issues. Figure 1.3 illustrates the various health effects of hydrocarbons on human health.

VOCs, including BTEX, are hydrocarbon constituents that readily vaporize and enter the atmosphere. Inhalation of VOCs is prevalent in proximity to refineries, petrochemical facilities, or spill locations. Acute exposure to elevated concentrations of VOCs may induce symptoms like headaches, dizziness, nausea, and breathing difficulties. Prolonged exposure to VOCs, especially benzene, is linked to severe health issues, including leukemia, other cancers, and immune system impairment. Hydrocarbons may aggravate respiratory ailments, such as asthma and chronic obstructive pulmonary disease (COPD), particularly in at-risk groups, including youngsters and the elderly. Fine particulate matter and aerosolized hydrocarbons from spills or industrial pollutants may infiltrate the lungs, leading to inflammation and diminished lung function (Abdel-Shafy & Mansour, 2016).

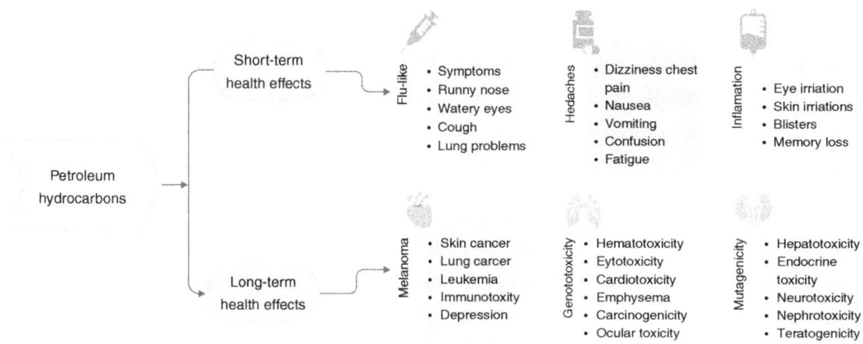

FIGURE 1.3 Effects of hydrocarbons on human health (Ambaye et al., 2022).

Hydrocarbons, such as PAHs and VOCs, may pollute drinking water sources, presenting a direct risk to human health. Consuming polluted water may result in gastrointestinal issues, hepatic and renal damage, and a heightened risk of cancer. Communities next to oil extraction plants or refineries may have heightened hazards of water pollution resulting from inadequate disposal of generated water or unintentional spills. Hydrocarbon contamination in aquatic habitats results in the bioaccumulation of hazardous substances in seafood, including fish and shellfish. Individuals who consume contaminated seafood may be exposed to elevated levels of PAHs, which are carcinogenic and can cause DNA damage, reproductive issues, and neurodevelopmental disorders in fetuses and young children. The danger is especially elevated for people who depend significantly on seafood as a dietary mainstay. Direct exposure to hydrocarbons, as encountered in oil spill remediation efforts, may result in skin irritation, rashes, and burns (Venkatraman et al., 2024). Extended exposure to hydrocarbons, especially in work environments, may result in dermatitis and a heightened risk of skin cancer due to the mutagenic and carcinogenic characteristics of certain PAHs.

Numerous hydrocarbons, especially aromatic chemicals like PAHs and benzene, are identified as established human carcinogens. Prolonged exposure, even at minimal levels, markedly elevates the chance of developing cancers of the lung, bladder, skin, liver, and hematologic malignancies (such as leukemia). Employees in the oil and gas sector, along with residents of regions adjacent to hydrocarbon extraction facilities, have an increased risk of cancer due to prolonged exposure. Certain hydrocarbons, especially certain PAHs and those produced from plastics, may act as endocrine disruptors, interfering with hormonal activity in people and animals. Endocrine imbalance may result in reproductive complications, developmental disorders, and heightened vulnerability to illnesses like diabetes and obesity. Prolonged exposure to certain hydrocarbons, particularly toluene and xylene, may result in neurotoxic consequences, such as memory deficits, cognitive dysfunction, and mood disturbances. The implications are especially alarming for employees in sectors like petrochemical processing or oil spill remediation, where extended exposure to toxic substances is prevalent (Ambaye et al., 2022).

1.10 REMEDIATION OF HYDROCARBON POLLUTION

Hydrocarbon pollution presents substantial threats to environmental and human health, requiring efficient remediation solutions to mitigate its effects. Hydrocarbons can permeate soil, water, and air, leading to persistent pollution due to their chemical stability and potential toxicity. Remediation treatments for hydrocarbon contamination have progressed over time, transitioning from traditional physical and chemical processes to more sustainable biological strategies. The choice of a remediation method is contingent upon many aspects, including the degree of pollution, the type of hydrocarbons, environmental circumstances, and economic viability.

1.11 PHYSICAL TECHNIQUES

Physical remediation methods are often used to contain or eliminate hydrocarbon pollutants from the environment without altering their chemical structure. These techniques are particularly effective for surface spills or situations where rapid elimination of contaminants is necessary.

Skimmers and booms are often used as containment techniques for oil spills on water. Skimmers are apparatuses that extract oil from the water's surface by using mechanical or suction mechanisms to separate it from water. Booms, which are buoyant barriers, are used to confine and restrict the dispersion of hydrocarbons, thus limiting their encroachment into vulnerable shorelines or ecosystems. Although useful for substantial spills, skimmers and booms are often restricted to calm weather conditions and may only extract hydrocarbons from the surface, leaving subsurface and emulsified oils unaddressed. Sorbents are substances used to absorb or adsorb hydrocarbons, often in oil spill situations. Prevalent sorbent materials include organic substances (e.g., peat, straw), inorganic substances (e.g., clay, zeolites), and synthetic substances (e.g., polypropylene) (Pal & Sen, 2024a). Sorbents are especially effective for minor spills or as supplementary containment measures. The recovery and disposal of oil-saturated sorbents present problems, as they produce supplementary waste requiring treatment or secure disposal. Soil washing is a remediation method that uses water, sometimes in conjunction with chemical additives, to physically extract hydrocarbons from polluted soil. The procedure depends on the dissolution and suspension of pollutants in the washing solution, enabling the cleaned soil to be either reinstated at the site or subjected to further treatment. Soil washing is very efficient for remediating light hydrocarbons and VOCs but may be less successful for heavy hydrocarbons and hydrophobic molecules that are heavily adsorbed to soil particles. Moreover, the method is energy-demanding and produces substantial quantities of polluted wash water, necessitating further treatment. Thermal desorption is a technique that uses heat to vaporize hydrocarbons from polluted soils or sediments. The contaminated substance is subjected to temperatures between 100°C and 550°C, depending on the specific hydrocarbons involved. The vaporized pollutants are then collected in gas-phase treatment devices, including condensers or oxidizers. Thermal desorption is efficient for a diverse array of hydrocarbon pollutants, including VOCs and semi-volatile organic compounds (SVOCs). However, the significant energy demands and the possibility of inadequate contaminant recovery are major disadvantages of this approach (Kumar & Gangawane, 2023a).

1.12 CHEMICAL TECHNIQUES

Chemical remediation approaches use chemical agents to convert or decompose hydrocarbon pollutants into less detrimental compounds. These approaches are especially effective for addressing dissolved or dispersed hydrocarbons in aqueous or terrestrial environments, as well as for degrading stubborn chemicals that resist biological decomposition. AOPs include the production of extremely reactive species, including hydroxyl radicals (OH), capable of chemically oxidizing and degrading hydrocarbons. AOPs include techniques like Fenton's reaction, ozonation, photocatalysis, and electrochemical oxidation. These reactions decompose hydrocarbons into carbon dioxide, water, and simpler organic molecules. Fenton's reaction involves the interaction of hydrogen peroxide (H_2O_2) with ferrous ions (Fe^{2+}) to produce hydroxyl radicals, capable of degrading hydrocarbons in polluted soils and water. Fenton's reaction is proficient in addressing many hydrocarbons, including PAHs, but requires careful regulation of pH and iron concentrations (Rayaroth et al., 2023). The ozonation process involves the introduction of ozone (O_3) into polluted water or air, where it combines with hydrocarbons, oxidizing them into simpler molecules. Ozonation is notably efficient for the remediation of volatile hydrocarbons and low-molecular-weight pollutants; however, it necessitates specialized apparatus and meticulous management due to the reactivity of ozone. The photocatalysis method employs light, usually with a semiconductor catalyst like titanium dioxide (TiO_2), to produce reactive species that decompose hydrocarbons. Photocatalysis is an innovative method that provides benefits in the remediation of soil and water, particularly for recalcitrant organic contaminants (Kumar et al., 2022). Nonetheless, the scalability of the technology and its reliance on external factors (e.g., sunshine) continue to pose issues. Chemical dispersants are surfactants that fragment oil spills into tiny droplets, facilitating their dispersion and destruction within the water column. Dispersants do not eliminate oil from the environment; rather, they enhance natural biodegradation by augmenting the surface area of oil droplets accessible to bacteria. Although dispersants effectively mitigate the spread of oil spills, their use is contentious due to potential harm to marine life and concerns about the long-term ecological consequences of the dispersed oil. Electrochemical oxidation involves the use of an electric current to decompose hydrocarbons in aqueous or terrestrial environments. Anodes composed of materials like boron-doped diamond (BDD) or graphite produce oxidizing species, such as hydroxyl radicals, that oxidize and decompose hydrocarbons. This technique is efficacious for the remediation of hydrocarbons in generated water, especially within the oil and gas sector, and may attain elevated removal efficiencies for chemical oxygen demand (COD) and total organic carbon (TOC). Nonetheless, the substantial energy consumption linked to electrochemical oxidation continues to be a constraint for extensive usage (Ji et al., 2024).

1.13 BIOLOGICAL TECHNIQUES

Bioremediation utilizes the metabolic functions of microorganisms or plants to convert hydrocarbons into less toxic compounds (Das & Chandran, 2011). This method is considered ecologically friendly and economically efficient, especially for addressing extensive, dispersed pollution sites.

Biodegradation is the process whereby native or foreign microorganisms, including bacteria and fungi, decompose hydrocarbons into simpler molecules via enzymatic processes. Microorganisms that degrade hydrocarbons may use them as a carbon and energy source, transforming them into carbon dioxide, water, and biomass. Biodegradation is particularly efficient for low-molecular-weight hydrocarbons, but larger hydrocarbons, including PAHs, exhibit greater resistance to microbial breakdown (Agamuthu et al., 2013). Bioaugmentation entails the introduction of specialized microbial strains with superior hydrocarbon-degrading abilities to polluted locations. This method is effective for expediting the decomposition of certain hydrocarbons that may be challenging for native microbes to metabolize. Biostimulation aims to enhance environmental parameters, including nutrient concentrations, oxygen availability, and moisture levels, to promote the proliferation and activity of hydrocarbon-degrading bacteria. Biostimulation is often used with bioaugmentation to enhance the efficacy of microbial decomposition. Phytoremediation is a plant-based remediation method that employs certain plants, termed hyperaccumulators, to absorb, decompose, or stabilize hydrocarbons in polluted soils or water. Plants may either metabolize hydrocarbons in their tissues or promote microbial breakdown in the rhizosphere, the soil area affected by plant roots. Phytoremediation is especially appealing for addressing low-to-moderate hydrocarbon pollution, as it is seen as a cost-effective, visually pleasing, and eco-friendly solution. Nonetheless, it is a protracted procedure limited by the slow growth rate of plants, making it unsuitable for severely polluted or urgent circumstances (Bramley-Alves et al., 2014).

1.14 HYBRID TECHNIQUES

Integrated remediation systems use numerous techniques, usually physical, chemical, and biological methods simultaneously, to address hydrocarbon-contaminated settings. Hydrocarbon pollution, due to its complexity and different manifestations, often necessitates several treatment methods to comprehensively mitigate contamination, particularly across varied environmental matrices such as soil, water, and air. The integration of multiple approaches can markedly improve the efficacy, efficiency, and sustainability of remediation initiatives, addressing constraints that may arise when using a singular approach. Integrated remediation strategies provide a thorough and adaptable method for addressing hydrocarbon contamination. These solutions may address the complexity of hydrocarbon contamination in various environmental matrices, such as soil, water, and sediments, by combining physical, chemical, and biological approaches. Despite specific constraints regarding cost, complexity, and compatibility, the advantages of improved efficiency, sustainability, and the ability to address a diverse array of pollutants make integrated methods an essential tool in the continuous endeavor to alleviate hydrocarbon pollution. Future advancements in technology and science will enhance these approaches, making them more accessible, efficient, and ecologically sustainable.

1.15 CHALLENGES AND FUTURE PROSPECTS

Hydrocarbon pollution poses complex and dynamic concerns due to the extensive utilization of petroleum products and the intrinsic limitations in preventing and alleviating environmental damage. Despite progress in detection and remediation technology, future management of hydrocarbon pollution must address new challenges such as climate change, escalating industrialization, and increased oil drilling operations (Lv et al., 2018).

The development of novel remediation technologies is crucial for addressing the shortcomings of existing approaches. Future research will concentrate on optimizing current technologies, including bioremediation, AOPs, and electrochemical approaches, to improve their efficiency and cost-effectiveness. Moreover, emerging technologies such as nanotechnology and genetically modified microbes (GEMs) provide exciting opportunities for enhancing hydrocarbon breakdown. Nanoparticles, owing to their extensive surface area and catalytic characteristics, may facilitate the degradation of hydrocarbon molecules (Kumar & Gangawane, 2022a). GEMs may be engineered to selectively target certain hydrocarbons with remarkable efficiency, even under harsh environmental circumstances. Nonetheless, challenges concerning the extensive use of these sophisticated technologies persist. The environmental safety of nanoparticles and the ecological concerns associated with genetically engineered microorganisms need thorough evaluation. Research must also examine the scalability of these technologies to guarantee their successful deployment in real-world situations, such as maritime oil spills or deep subterranean pollution. A comprehensive solution to hydrocarbon pollution necessitates a fundamental transition to sustainable practices in businesses dependent on hydrocarbons. The oil and gas sector must use cleaner technology to reduce spills, emissions, and waste during production, refining, and transportation. There is increasing interest in green chemistry, which aims to develop chemical processes that use less harmful substances and produce fewer by-products. Industries may engage in carbon capture and storage (CCS) systems to mitigate the emission of hydrocarbons and other pollutants into the atmosphere (Kumar & Gangawane, 2022b).

Shifting to renewable energy sources, including solar, wind, and biofuels, is a crucial measure for reducing dependence on hydrocarbons. This transformation will take time, but policies advocating for renewable energy and energy efficiency may substantially reduce hydrocarbon emissions in the long run. A significant difficulty, however, lies in the economic and political variables that shape energy regulations, as fossil fuel businesses continue to dominate in several areas due to their profitability and established infrastructure. Climate change complicates the control of hydrocarbon pollution. Elevated temperatures, increasing sea levels, and a heightened occurrence of severe weather events might intensify the dissemination and effects of hydrocarbon pollutants. Melting permafrost in Arctic locations may unleash stored hydrocarbons and other toxins into the environment. Likewise, heightened storms and hurricanes may elevate the danger of oil leaks and compromise infrastructure, including pipelines and storage facilities (Brito et al., 2015).

Future initiatives must prioritize adaptive management solutions that address the changing climate. This may include enhancing oil spill response mechanisms, conducting more stringent risk evaluations for hydrocarbon-associated infrastructure, and advancing climate-resilient cleanup technology. Regulatory frameworks are essential for overseeing hydrocarbon pollution control; however, current rules may inadequately tackle the intricacies of forthcoming difficulties. There is a need for more international collaboration on environmental legislation, especially in regions affected by transboundary hydrocarbon contamination, such as international waterways. Entities like the United Nations Environment Programme (UNEP) and the International Maritime Organization (IMO) must persist in advocating for more stringent implementation of legislation designed to prevent and mitigate oil spills and other types of hydrocarbon pollution.

Moreover, regulatory agencies must adapt to technological progress, guaranteeing that novel cleanup methods are properly executed and do not unintentionally inflict further environmental damage. Public policy should prioritize incentivizing research and development of sustainable alternatives to hydrocarbons while promoting the adoption of cleaner practices within the industry (Aziz et al., 2024).

Challenges persist, especially with political and economic interests that support the fossil fuel sector. Policies designed to mitigate hydrocarbon pollution often encounter opposition from influential industrial entities, complicating the implementation of essential reforms on a worldwide scale. The disparate allocation of resources and technical expertise across nations presents obstacles to attaining consistent advancement in hydrocarbon pollution control.

Public participation and awareness will be crucial in the future control of hydrocarbon contamination (Mishra et al., 2021; Mishra & Das, 2021). A better understanding of the environmental and health consequences of hydrocarbons may catalyze behavioral modifications and bolster support for governmental initiatives designed to diminish hydrocarbon use. Educational initiatives may motivate companies, governments, and people to embrace more sustainable practices and endorse greener technology.

Engagement of stakeholders, especially indigenous tribes and local inhabitants, is essential. Numerous hydrocarbon extraction sites are situated in proximity to at-risk people whose livelihoods rely on robust ecosystems. Involving these stakeholders in decision-making processes guarantees that their perspectives are acknowledged and that remediation strategies are customized to address the specific challenges of each location. The primary problem in this sector is sustaining public attention and engagement in environmental issues, particularly while contending with other urgent social matters. Moreover, a frequent disparity exists between public knowledge and implementable change, since several remedies require long-term investments in policy and infrastructure (Imam et al., 2021).

1.16 CONCLUSIONS

Hydrocarbon contamination is a major environmental concern, propelled by the worldwide dependence on petroleum-derived goods and the widespread use of hydrocarbons in industrial, agricultural, and energy sectors. The enduring presence of hydrocarbons in the environment and their many pathways of introduction, whether through accidental spills, industrial discharges, or natural seepage, make them a prevalent and challenging pollutant to control. The thorough analysis of hydrocarbon contamination encompasses its origins, classifications, environmental behavior, ecological consequences, and existing remediation methods, highlighting the complex problems of addressing this worldwide environmental threat. Hydrocarbons may originate from both natural and manmade sources in the environment. Natural causes include oil seeps and forest fires; however, the predominant cause of environmental hydrocarbon pollution comes from anthropogenic activities, including oil extraction, transportation, refining, and the utilization of petroleum products in urban settings. Industrial activities, vehicle emissions, and incidents like oil spills are significant anthropogenic factors that lead to the pollution of air, soil, and water environments.

Hydrocarbons may be categorized as aliphatic or aromatic. Aliphatic hydrocarbons are often less dangerous and include substances like methane and ethane, whereas aromatic hydrocarbons (e.g., benzene, toluene, and PAHs) are more perilous due to their potential carcinogenic and mutagenic properties. The ecological behavior of these chemicals varies markedly, affecting their dispersion and persistence in ecosystems. Upon discharge into the environment, hydrocarbons undergo several physicochemical and biological modifications that influence their persistence, mobility, and bioavailability. Hydrocarbons in air, soil, and water may undergo volatilization, adsorption, photodegradation, chemical oxidation, and biodegradation. The effectiveness of these processes varies according to the environmental matrix, meteorological circumstances, and the distinct chemical characteristics of the hydrocarbons involved.

In aquatic environments, hydrocarbons may either skim on the water surface, dissolve, or sink, depending on their density and solubility, leading to pollution in both water columns and sediments. Soil environments encounter difficulties due to hydrocarbon adsorption onto soil particles, which may impede microbial activity and hinder biodegradation. Volatile hydrocarbons exacerbate air pollution and may contribute to atmospheric interactions, leading to the formation of secondary pollutants. The destiny of hydrocarbons in different environments dictates their ecological consequences, which can include toxicity to plant and animal life, bioaccumulation in food webs, and long-term deterioration of soil and water quality.

The consequences of hydrocarbon contamination on the environment are varied and grave, affecting ecosystems, biodiversity, and human health. Hydrocarbons, especially PAHs and VOCs, are recognized as poisonous, carcinogenic, and mutagenic, presenting significant hazards to both animal and human populations. Aquatic ecosystems are particularly susceptible to oil spills and hydrocarbon releases, as they may contaminate water bodies and adversely affect marine organisms by coating fish, seabirds, and invertebrates, upsetting the whole food web. Hydrocarbons contribute to soil pollution, diminishing soil fertility and obstructing plant development. The emission of hydrocarbons into atmosphere may result in photochemical smog and the production of ground-level ozone, exacerbating air pollution and related health risks, including respiratory issues. Moreover, the bioaccumulation of certain hydrocarbons in organisms may result in chronic impacts on individual species and whole ecosystems, leading to prolonged ecological imbalances.

The remediation of hydrocarbon contamination encompasses several physical, chemical, and biological techniques, each with distinct benefits and limitations. The choice of a remediation technique is contingent upon the characteristics of the polluted environment, the specific hydrocarbons involved, and the extent of the contamination. Physical techniques, including skimming, booms, and sorbents, are often used to mitigate oil spills on aquatic surfaces. These procedures facilitate the prompt containment and extraction of floating hydrocarbons; nevertheless, they are often constrained by meteorological circumstances and the inadequate elimination of contaminants. Soil washing and thermal desorption are useful methods for extracting hydrocarbons from soil matrices in soil remediation; however, they may be energy-intensive and expensive. Chemical treatments, such as advanced oxidation processes (AOPs) and chemical dispersants, are used to expedite the degradation of hydrocarbons. AOPs include the production of extremely reactive radicals, such as hydroxyl radicals, which decompose hydrocarbons into less harmful substances. Nonetheless, these procedures may generate by-products that could pose additional environmental hazards. Electrochemical techniques and photocatalysis show potential for the efficient breakdown of hydrocarbons; nonetheless, issues about scalability and cost persist. Bioremediation is progressively acknowledged as a viable method for addressing hydrocarbon contamination. Indigenous or imported microorganisms, including bacteria and fungi, may degrade hydrocarbons into innocuous by-products under optimal circumstances. Methods like bioaugmentation (the introduction of specialized microbial strains) and biostimulation (the improvement of environmental conditions to foster microbial activity) have shown efficacy in both terrestrial and aquatic ecosystems. Phytoremediation, which involves using plants to eliminate, stabilize, or decompose hydrocarbons, presents a cost-effective and environmentally sustainable option; however, its effectiveness is limited by the time required for plant maturation and the certain environmental conditions needed.

Despite substantial advancements in comprehending and addressing hydrocarbon contamination, serious obstacles persist. The partial extraction of hydrocarbons, especially in intricate settings like marine ecosystems or groundwater, presents persistent hazards. The long-term ecological effects of residual hydrocarbons and their degradation products remain little known, requiring more study into the toxicological characteristics of these breakdown products. The advancement of integrated remediation strategies, which amalgamate physical, chemical, and biological techniques, offers a potential pathway for enhancing the efficiency and sustainability of pollution control. Nanotechnology, genetic engineering, and sophisticated monitoring instruments are expected to improve the identification and remediation of hydrocarbon pollutants.

Mitigating hydrocarbon contamination requires sustained multidisciplinary collaboration in scientific inquiry, technology advancement, policy development, and public involvement. The future of hydrocarbon management will depend on the enhancement of remediation methods and the widespread implementation of sustainable practices that minimize the environmental impact of hydrocarbons. Through the integration of new solutions, preventive measures, and regulatory frameworks, the global community may enhance the protection of ecosystems and human health against the widespread danger of hydrocarbon pollution.

REFERENCES

Abdel-Shafy, H. I., & Mansour, M. S. M. (2016). A review on polycyclic aromatic hydrocarbons: Source, environmental impact, effect on human health and remediation. *Egyptian Journal of Petroleum*, *25*(1), 107–123. https://doi.org/10.1016/j.ejpe.2015.03.011

Agamuthu, P., Tan, Y. S., & Fauziah, S. H. (2013). Bioremediation of hydrocarbon contaminated soil using selected organic wastes. *Procedia Environmental Sciences*, *18*, 694–702. https://doi.org/10.1016/j.proenv.2013.04.094

Ahmed, F., & Fakhruddin, A. N. M. (2018). A review on environmental contamination of petroleum hydrocarbons and its biodegradation. *International Journal of Environmental Sciences & Natural Resources*, *11*(3), 1–7.

Aislabie, J. M., Balks, M. R., Foght, J. M., & Waterhouse, E. J. (2004). Hydrocarbon spills on Antarctic soils: Effects and management. *Environmental Science & Technology*, *38*(5), 1265–1274. https://doi.org/10.1021/es0305149

Ambaye, T. G., Chebbi, A., Formicola, F., Prasad, S., Gomez, F. H., Franzetti, A., & Vaccari, M. (2022). Remediation of soil polluted with petroleum hydrocarbons and its reuse for agriculture: Recent progress, challenges, and perspectives. *Chemosphere*, *293*, 133572. https://doi.org/10.1016/j.chemosphere.2022.133572

Aziz, Z. S., Jazza, S. H., Dageem, H. N., Banoon, S. R., Balboul, B. A., & Abdelzaher, M. A. (2024). Bacterial biodegradation of oil-contaminated soil for pollutant abatement contributing to achieve sustainable development goals: A comprehensive review. *Results in Engineering*, *22*(February), 102083. https://doi.org/10.1016/j.rineng.2024.102083

Bala, K., Dogra, A., Lata, P., & Das, A. P. (2024). Discussing sources and biological-chemical remediation approaches for microplastic pollution. In: *Sustainable Microbial Technology for Synthetic and Cellulosic Microfiber Bioremediation*, 37–59. https://doi.org/10.1007/978-3-031-62600-5_3

Behera, I. D., Basak, G., Kumar, R. R., Sen, R., & Meikap, B. C. (2020). Treatment of petroleum refinery sludge by petroleum degrading bacterium *Stenotrophomonas pavanii* IRB19 as an efficient novel technology. *Journal of Environmental Science and Health, Part A*, *56*(2), 226–239. https://doi.org/10.1080/10934529.2020.1866924

Behera, I. D., Manoranjan, N., Subrata, B., Meikap, B. C., & Sen, R. (2021). Enhanced biodegradation of total petroleum hydrocarbons by implementing a novel two-step bioaugmentation strategy using indigenous bacterial consortium. *Journal of Environmental Management*, *292*, 112746, ISSN 0301-4797. https://doi.org/10.1016/j.jenvman.2021.112746

Behera, I. D., Nayak, M., Mishra, A., Meikap, B. C., and Sen, R. (2022). Strategic implementation of integrated bioaugmentation and biostimulation for efficient mitigation of petroleum hydrocarbon pollutants from terrestrial and aquatic environment. *Marine Pollution Bulletin*, *177*, 113492, ISSN 0025-326X. https://doi.org/10.1016/j.marpolbul.2022.113492

Blanco, M., Maspoch, S., Villarroya, I., Peralta, X., González, J. M., & Torres, J. (2001). Geographical origin classification of petroleum crudes from near-infrared spectra of bitumens. *Applied Spectroscopy*, *55*(7), 834–839. https://doi.org/10.1366/0003702011952857

Bramley-Alves, J., Wasley, J., King, C. K., Powell, S., & Robinson, S. A. (2014). Phytoremediation of hydrocarbon contaminants in subantarctic soils: An effective management option. *Journal of Environmental Management*, *142*, 60–69. https://doi.org/10.1016/j.jenvman.2014.04.019

Brito, E. M. S., De la Cruz Barrón, M., Caretta, C. A., Goñi-Urriza, M., Andrade, L. H., Cuevas-Rodríguez, G., Malm, O., Torres, J. P. M., Simon, M., & Guyoneaud, R. (2015). Impact of hydrocarbons, PCBs and heavy metals on bacterial communities in Lerma River, Salamanca, Mexico: Investigation of hydrocarbon degradation potential. *Science of the Total Environment*, 521–522, 1–10. https://doi.org/10.1016/j.scitotenv.2015.02.098

Das, A., Dey, S., Bala, K., Gnanasekaran, R., & Das, A. P. (2024). Microbial bioremediation technology for sustainable treatment and management of synthetic microfiber waste. In: *Sustainable Microbial Technology for Synthetic and Cellulosic Microfiber Bioremediation*, 221–238. https://doi.org/10.1007/978-3-031-62600-5_12

Das, N., & Chandran, P. (2011). Microbial degradation of petroleum hydrocarbon contaminants: An overview. *Biotechnology Research International*, *2011*, 1–13. https://doi.org/10.4061/2011/941810

Dey, S., Bissoyi, A., Koteswara Reddy, G., & Das, A. P. (2024). Synthetic microfibers: An urban emerging pollutant demanding for sustainable environmental solutions. In: *Sustainable Microbial Technology for Synthetic and Cellulosic Microfiber Bioremediation*, 107–121. https://doi.org/10.1007/978-3-031-62600-5_7

Dey, S., Das, A., Mallick, K., Sahu, A., & Das, A. P. (2023). Environmental petroleum waste: Pollution, toxicity, sustainable remediation. In: Behera, I. D., Das, A. P. (eds) *Impact of Petroleum Waste on Environmental Pollution and its Sustainable Management through Circular Economy*, 159–175. Springer, Cham. https://doi.org/10.1007/978-3-031-48220-5_7

Dey, S., Dash, N., Ray, S. K., Priyadarsini, S., Chhotaray, C., & Das, A. P. (2024). Occurrence and quantification of tire wear plastics pollutants from highway city dust of Bhubaneswar city of Odisha. In *Sustainable Microbial Technology for Synthetic and Cellulosic Microfiber Bioremediation*, 77–89. https://doi.org/10.1007/978-3-031-62600-5_5

Dey, S., Tripathy, B., Kumar, M. S., & Das, A. P. (2023). Ecotoxicological consequences of manganese mining pollutants and their biological remediation. *Environmental Chemistry and Ecotoxicology*, *5*, 55–61. https://doi.org/10.1016/J.ENCECO.2023.01.001

Elijah, A. A. (2022). A review of the petroleum hydrocarbons contamination of soil, water and air and the available remediation techniques, taking into consideration the sustainable development goals. *Earthline Journal of Chemical Sciences*, *7*(1), 97–113. https://doi.org/10.34198/ejcs.7122.97113

Haider, F. U., Ejaz, M., Cheema, S. A., Khan, M. I., Zhao, B., Liqun, C., Salim, M. A., Naveed, M., Khan, N., Núñez-Delgado, A., & Mustafa, A. (2021). Phytotoxicity of petroleum hydrocarbons: Sources, impacts and remediation strategies. *Environmental Research*, *197*(March), 111031. https://doi.org/10.1016/j.envres.2021.111031

Honda, M., & Suzuki, N. (2020). Toxicities of polycyclic aromatic hydrocarbons for aquatic animals. *International Journal of Environmental Research and Public Health*, *17*(4), 1363. https://doi.org/10.3390/ijerph17041363

Imam, A., Kanaujia, P. K., Ray, A., & Suman, S. K. (2021). Removal of petroleum contaminants through bioremediation with integrated concepts of resource recovery: A review. *Indian Journal of Microbiology*, *61*(3), 250–261. https://doi.org/10.1007/s12088-021-00928-4

Ite, E. A., Harry, T. A., Obadimu, C. O., Asuaiko, E. R., & Inim, I. J. (2018). Petroleum hydrocarbons contamination of surface water and groundwater in the Niger Delta Region of Nigeria. *Journal of Environment Pollution and Human Health*, *6*(2), 51–61. https://doi.org/10.12691/jephh-6-2-2

Ji, Z., Zhang, N., Huang, C., Duan, X., Ren, D., & Huo, Z. (2024). The degradation of polycyclic aromatic hydrocarbons (PAHs) by ozone-based advanced oxidation processes: A review. *Ozone: Science & Engineering*, *46*(1), 26–42. https://doi.org/10.1080/01919512.2023.2192751

Jiang, Z., Duan, H., Liang, C., Wu, J., Zhang, W., & Zhang, J. (2017). Classification of hydrocarbon-bearing fine-grained sedimentary rocks. *Journal of Earth Science*, *28*(6), 693–976. https://doi.org/10.1007/s12583-016-0920-0

Kanungo, J., Sahoo, T., Bal, M., & Behera, I. D. (2023a). Performance of bioremediation strategy in waste lubricating oil pollutants: A review. *Geomicrobiology Journal*, *41*(4), 360–373. https://doi.org/10.1080/01490451.2023.2245395

Kanungo, J., Sahoo, T., Swain, L. P., & Behera, I. D. (2023b). Toxicity of persistent hydrocarbon pollutants, sources and sustainable remediation process. In: Behera, I. D., Das, A. P. (eds) *Impact of Petroleum Waste on Environmental Pollution and its Sustainable Management through Circular Economy*. Environmental Science and Engineering. Springer, Cham. https://doi.org/10.1007/978-3-031-48220-5_2

Kumar, A., & Gangawane, K. M. (2022a). Effect of precipitating agents on the magnetic and structural properties of the synthesized ferrimagnetic nanoparticles by co-precipitation method. *Powder Technology*, *401*, 117298. https://doi.org/10.1016/j.powtec.2022.117298

Kumar, A., & Gangawane, K. M. (2022b). Synthesis and effect on the surface morphology & magnetic properties of ferrimagnetic nanoparticles by different wet chemical synthesis methods. *Powder Technology*, *410*, 117867. https://doi.org/10.1016/j.powtec.2022.117867

Kumar, A., & Gangawane, K. M. (2023a). Advanced computational approaches for water treatment. In: *Advanced Computational Approaches for Water Treatment*, 1–12. CRC Press. https://doi.org/10.1201/9781003325147-1

Kumar, A., & Gangawane, K. M. (2023b). Molecular dynamics for the membrane process. In: *Advanced Computational Approaches for Water Treatment*, 13–38. CRC Press. https://doi.org/10.1201/9781003325147-2

Kumar, A., Gangawane, K. M., & Ramanjaneyulu, B. (2022). Ferrofluids for waste-water treatment. In: *Advances in Chemical, Bio and Environmental Engineering*, 723–744. Springer. https://doi.org/10.1007/978-3-030-96554-9_48

Kuppusamy, S., Maddela, N. R., Megharaj, M., & Venkateswarlu, K. (2020). Total petroleum hydrocarbons. In: *Total Petroleum Hydrocarbons: Environmental Fate, Toxicity, and Remediation*. Springer International Publishing. https://doi.org/10.1007/978-3-030-24035-6

Logeshwaran, P., Megharaj, M., Chadalavada, S., Bowman, M., & Naidu, R. (2018). Petroleum hydrocarbons (PH) in groundwater aquifers: An overview of environmental fate, toxicity, microbial degradation and risk-based remediation approaches. *Environmental Technology & Innovation*, *10*, 175–193. https://doi.org/10.1016/j.eti.2018.02.001

Lv, H., Su, X., Wang, Y., Dai, Z., & Liu, M. (2018). Effectiveness and mechanism of natural attenuation at a petroleum-hydrocarbon contaminated site. *Chemosphere*, *206*, 293–301. https://doi.org/10.1016/j.chemosphere.2018.04.171

Manzetti, S. (2013). Polycyclic aromatic hydrocarbons in the environment: Environmental fate and transformation. *Polycyclic Aromatic Compounds*, *33*(4), 311–330. https://doi.org/10.1080/10406638.2013.781042

Mishra, S., & Das, A. P. (2021). Current treatment technologies for removal of microplastic and microfiber pollutants from wastewater. In: *Wastewater Treatment: Cutting-Edge Molecular Tools, Techniques and Applied Aspects*, 237–251. https://doi.org/10.1016/B978-0-12-821881-5.00011-8

Mishra, S., Dash, D., Al-Tawaha, A. R. M. S., & Das, A. P. (2022). A review on heavy metal ion adsorption on synthetic microfiber surface in aquatic environments. *Applied Biochemistry and Biotechnology*, *194*(10), 4639–4654. https://doi.org/10.1007/s12010-022-04029-w

Mishra, S., Dash, D., & Das, A. P. (2024). Isolation and molecular characterization of polyester-degrading bacterial strains from municipal wastewater and their application in microplastic degradation in contaminated wastewater. *Geomicrobiology Journal*. https://doi.org/10.1080/01490451.2024.2376101

Mishra, S., Swain, S., Sahoo, M., Mishra, S., & Das, A. P. (2021). Microbial colonization and degradation of microplastics in aquatic ecosystem: A review. *Geomicrobiology Journal*. https://doi.org/10.1080/01490451.2021.1983670

Mohanty, J., Dey, S., Bala, K., & Das, A. P. (2024). Recent advances in microplastic pollution for its sustainable management. In: *Sustainable Management of Urban Plastic Waste through Circular Economic Approaches*, 27–48. https://doi.org/10.1201/9781003465447-3

Onwuka, O. U., & Adu, A. (2024). Eco-efficient well planning: Engineering solutions for reduced environmental impact in hydrocarbon extraction. *International Journal of Scholarly Research in Multidisciplinary Studies*, *4*(1), 033–043. https://doi.org/10.56781/ijsrms.2024.4.1.0028

Ossai, I. C., Ahmed, A., Hassan, A., & Hamid, F. S. (2020). Remediation of soil and water contaminated with petroleum hydrocarbon: A review. *Environmental Technology & Innovation*, *17*, 100526. https://doi.org/10.1016/j.eti.2019.100526

Oyo-Ita, I. O., Oyo-Ita, O. E., Dosunmu, M. I., Domínguez, C., Bayona, J. M., & Albaigés, J. (2016). Distribution and sources of petroleum hydrocarbons in recent sediments of the Imo River, SE Nigeria. *Archives of Environmental Contamination and Toxicology*, *70*(2), 372–382. https://doi.org/10.1007/s00244-015-0237-5

Pal, D., & Sen, S. (2023a). Emerging petroleum pollutants and their adverse effects on the environment. In: *Impact of Petroleum Waste on Environmental Pollution and its Sustainable Management through Circular Economy*, 103–137. https://doi.org/10.1007/978-3-031-48220-5_5

Pal, D., & Sen, S. (2023b). In-depth coverage of petroleum waste sources, characteristics, environmental impact, and sustainable remediation process. In: *Impact of Petroleum Waste on Environmental Pollution and its Sustainable Management through Circular Economy*, 1–38. https://doi.org/10.1007/978-3-031-48220-5_1

Pal, D., & Sen, S. (2024a). Optimal synthesis of dolochar derived faujasite zeolite X for highly effective Cd(II) removal. *Environmental Research*, *240*, 117494. https://doi.org/10.1016/j.envres.2023.117494

Pal, D., & Sen, S. (2024b). Valorization of sponge iron industrial waste into iron-modified zeolite X for ciprofloxacin removal: a multi-parameter optimization study. *Environmental Science and Pollution Research*, *31*(40), 53381–53398. https://doi.org/10.1007/s11356-024-34730-8

Ray, S. K., Dey, S., Das, A. P., & John, A. (2024). Microplastic waste management through greener technologies: A review. In: *Sustainable Management of Urban Plastic Waste through Circular Economic Approaches*, 237–261. https://doi.org/10.1201/9781003465447-14

Rayaroth, M. P., Marchel, M., & Boczkaj, G. (2023). Advanced oxidation processes for the removal of mono and polycyclic aromatic hydrocarbons: A review. *Science of the Total Environment*, *857*, 159043. https://doi.org/10.1016/j.scitotenv.2022.159043

Sharma, S., Shaikh, S., Mohana, S., Desai, C., & Madamwar, D. (2023). Current trends in bioremediation and bio-integrated treatment of petroleum hydrocarbons. *Environmental Science and Pollution Research*, *31*(46), 57397–57416. https://doi.org/10.1007/s11356-023-30479-8

SinghJadaun, J., Bansal, S., Sonthalia, A., Rai, A. K., & Singh, S. P. (2022). Biodegradation of plastics for sustainable environment. *Bioresource Technology*, *347*, 126697. https://doi.org/10.1016/J.BIORTECH.2022.126697

Speight, J. G. (2011). Production, properties and environmental impact of hydrocarbon fuel conversion. In: *Advances in Clean Hydrocarbon Fuel Processing*, 54–82. Elsevier. https://doi.org/10.1533/9780857093783.1.54

Truskewycz, A., Gundry, T. D., Khudur, L. S., Kolobaric, A., Taha, M., Aburto-Medina, A., Ball, A. S., & Shahsavari, E. (2019). Petroleum hydrocarbon contamination in terrestrial ecosystems—fate and microbial responses. *Molecules*, *24*(18), 3400. https://doi.org/10.3390/molecules24183400

Tveit, M. R., Khalifeh, M., Nordam, T., & Saasen, A. (2021). The fate of hydrocarbon leaks from plugged and abandoned wells by means of natural seepages. *Journal of Petroleum Science and Engineering*, *196*(October 2020), 108004. https://doi.org/10.1016/j.petrol.2020.108004

Venkatraman, G., Giribabu, N., Mohan, P. S., Muttiah, B., Govindarajan, V. K., Alagiri, M., Abdul Rahman, P. S., & Karsani, S. A. (2024). Environmental impact and human health effects of polycyclic aromatic hydrocarbons and remedial strategies: A detailed review. *Chemosphere*, *351*(January), 141227. https://doi.org/10.1016/j.chemosphere.2024.141227

Whelan, M. J., Coulon, F., Hince, G., Rayner, J., McWatters, R., Spedding, T., & Snape, I. (2015). Fate and transport of petroleum hydrocarbons in engineered biopiles in polar regions. *Chemosphere*, *131*, 232–240. https://doi.org/10.1016/j.chemosphere.2014.10.088

Zhang, J., Dai, J., Du, X., Li, F., Wang, W., & Wang, R. (2012). Distribution and sources of petroleum-hydrocarbon in soil profiles of the Hunpu wastewater-irrigated area, China's northeast. *Geoderma*, *173–174*, 215–223. https://doi.org/10.1016/j.geoderma.2011.12.004

2 An Overview of the Ecotoxicity of Petroleum Hydrocarbons

Jyotirmayee Mohanty, Sudeshna Dey, Amel Gacem, G. Koteswara Reddy, Ipsita Dipamitra Behera, Kiran Bala, Chandra Mohan, Maha Awjan Alreshidi, and Alok Prasad Das

2.1 INTRODUCTION

Up to one-third of the world's energy demand is met by petroleum, making it the greatest energy source on the planet (Erdem, 2010). The ecosystem is contaminated by virtually every petrochemical and petroleum exploration activity. Petroleum spills have become more commonplace worldwide in recent years. Groundwater, the ocean, and the soil have all been contaminated to varying degrees by ship mishaps, industrial sewage leaks, and petroleum pipeline leaks. When petroleum hydrocarbons (PHs) are introduced into a pristine ecosystem, the ecosystem's composition alters right away. Numerous microbial species are killed or inhibited by the added hydrocarbons, altering the microbial community and the ecosystem's overall performance (Falciglia et al., 2011). If there are minimal primary producers and functional microbial biogeochemical networks, the compromised habitat has little capacity to support higher-order organisms (Dey, Bissoyi, et al., 2024). Weathering is the phenomenon that occurs when PHs are introduced into the environmental matrix. Various factors, including physical (dispersion), physiochemical (evaporation, dissolution, sorption), chemical (photo-oxidation, auto-oxidation), and biological (plant and microbial hydrocarbon catabolism) processes, collectively contribute to the weathering dynamics of PHs within the ecosystem. Because PHs are more persistent and biotoxic than other substances, they pose a serious hazard to the environment. Crude oil landings deplete natural habitats, squander resources, and damage the environment (Ray et al., 2024). For this reason, it is essential to remove PHs from contaminated environments and develop ways to lessen their harmful impact on living things to protect the environment and public health. Principally, four groups of oil hydrocarbons can be distinguished: asphaltenes (phenols, ketones, esters, porphyrins, and fatty acids), aromatics, resins (carbazoles, sulfoxides, pyridines, quinolines, and amides), and aliphatic. Among the diverse array of remediation methodologies, land farming, soil washing, thermal desorption, and phytoremediation are often employed strategies

 DOI: 10.1201/9781003508496-2

for the treatment of oil-contaminated soils, whereas thermal desorption and pyrolysis have demonstrated effective application in the extraction of petroleum-derived pollutants within a comparatively brief duration of treatment (Khatun et al., 2023). When determining the harm and risk to polluted soil ecosystems, plant species serve as markers of the toxicity of leachable components in the soil (Das et al., 2024; Dey, Tripathy, et al., 2023). Early life history vascular plant bioassays have several advantages for assessing soil ecotoxicity, including the ability to gauge sensitive characteristics such as seed germination and seedling growth (Liwarska-Bizukojc, 2022). The embryonic zebrafish assay has been utilized to examine the toxicological effects of soil derived from various industrial locations, including soil contaminated with polycyclic aromatic hydrocarbons (PAHs), gas-impacted soil, and Superfund soil contaminated with creosote (Chibwe et al., 2015). PHs can disrupt reproductive and developmental processes in wildlife (Behera, 2020, 2021, 2022). For instance, exposure to these substances has been linked to developmental deformities in fish and amphibians and decreased fertility in various species. Refinery byproducts worsen the physiochemical and biochemical characteristics of soils, limiting plant development and yield (Bala et al., 2024). The primary alterations in soil characteristics associated with interaction with petroleum-derived substances include deficiencies in water and oxygen, as well as in vital nutrients like nitrogen and phosphorus (Haider et al., 2021). In terrestrial environments, PHs can alter soil structure and chemistry, affecting plant growth and microbial communities. Contaminated soils can lead to decreased agricultural productivity and reduced soil health. It is generally believed that plant lipid layers accumulate PHs. It has been demonstrated that PHs cause oxidative stress, stunted growth, and distorted leaves, resulting in the necrosis of plant tissues and cells (Mukome et al., 2020). PHs can interfere with the structural integrity of chloroplasts, leading to disturbances in electron transport and photosynthesis. Furthermore, they can bind covalently to proteins, DNA, and RNA within cells, disrupting enzymatic function, growth, metabolism, and hydrophilicity while boosting reactivity and electrophilicity (Bao et al., 2020). Aromatic hydrocarbons, such as PAHs, are known to disrupt cellular processes (Das, Dey, Bala, et al., 2024). They can interfere with the normal function of enzymes, affect DNA integrity, and alter cell signaling pathways, leading to mutagenic and carcinogenic effects. The most hazardous contaminants found in commercial PCBs, which amount to hundreds of thousands of metric tons and persist in aquatic sediments, are thought to be polychlorinated biphenyls (PCBs) because of their carcinogenicity, toxicity, and sluggish biodegradation in the environment. PCBs are extensively employed in transformers, hydraulic fluids, plasticizers, adhesives and lubricants, and dielectric fluids, as well as in flame retardants. In response to both biotic and abiotic degradation processes, polychlorinated dibenzo-p-dioxins and dibenzofurans (PCDD/Fs), which are the most prominent contaminants found in nature, are frequently designed to be resistant. The use of motor oil has increased due to the availability of numerous types of automobiles and machinery vehicles, putting pressure on PHs and other energy sources as modern society has grown (Behera et al., 2020; Dey, Das, et al., 2023; Lahiri et al., 2021). In particular, pH compounds in soil environments can combine naturally occurring organic chemicals with cycling processes, altering the physicochemical properties of soils and eventually affecting plant growth. Aromatic molecules are among the most stable of these and can linger in the environment for extended periods. Continuous exposures can lead to subcellular effects such as changed metabolism, alterations in cell structure and function, or an increase in chromosome mutations. This series of biological effects linked to long-term pollution from frequent, smaller spills is often considered to be more dangerous than acute exposure from tanker accidents. Whether from natural or artificial sources, continuous or severe, oil pollution in the ocean is a significant environmental concern (Alzahrani & Rajendran, 2019). As PAHs are derived from crude oil their byproducts pose significant environmental and health risks due to their widespread use and potential for accidental spills. These compounds, which include a variety of alkanes, cycloalkanes, and aromatic hydrocarbons, are known for their persistence in the environment and their ability to cause severe ecological damage. Their toxic effects can disrupt aquatic and terrestrial ecosystems, leading to contamination of water sources, soil, and air. Understanding the ecotoxicity of PHs is crucial for assessing their impact on wildlife, plants, and human health, as well as for developing effective remediation strategies to mitigate their adverse effects (Sahu & Das, 2024). This exploration of PH ecotoxicity highlights the importance of addressing these pollutants to safeguard environmental and public health.

The purpose of this chapter is to give a thorough analysis of the ecotoxicity resulting from the pollution of PHs. The chapter outlines a summary of the prominent sources of the pollution of PHs; in addition, it describes the impacts on the environment and various proactive strategies for the mitigation of PH pollution.

2.2 SOURCES OF PETROLEUM HYDROCARBON

Petroleum predominantly consists of carbon (83%–87%) and hydrogen (12%–14%), encompassing a sophisticated mixture of hydrocarbons that includes gaseous hydrocarbons, paraffins, naphthenes, and aromatic hydrocarbons. There are three main types of PH origins: biological, petrogenic, and pyrogenic (Odukoya et al., 2019). Examples of PAHs with human origins from wide-point sources impacted by incomplete combustion exist. Oils are erratic mixtures of hydrocarbons that vary not just between fuel types but also within each fuel type according to manufacturer, region, and traditional use. The results of these emissions include a wide variety of hydrocarbon mixtures. Only a small number of these blends have toxicological information accessible. This makes it difficult to determine which prosperity risk is represented by oil hydrocarbons. Scattered sources include smaller point sources such as emissions from automobiles, smoke from wood stoves, jet airplane exhaust, tobacco and cigarette smoke (Dey, Dash, Ray, et al., 2024). Environmental contamination caused by petroleum oil spills, sewage sludge, and waste items containing tarry or creosote are other significant causes of PHs. Natural seepages of crude oil occur where oil reservoirs are exposed at the Earth's surface. These seepages can be found in regions like asphalt lakes and oil seeps in various parts of the world. Such natural leaks contribute hydrocarbons to the environment, albeit on a smaller scale compared to human activities. The extraction of crude oil from underground reserves involves drilling and pumping oil to the surface. This process often results in the release of hydrocarbons into the environment, particularly through routine operations and accidental spills (Paul et al., 2023; Pradhan et al., 2023). Crude oil refining separates crude oil into various products like gasoline, diesel, and kerosene. During refining, some hydrocarbons can be released into the environment, particularly if proper controls are not in place. A large portion of these pollutants originate from nearby outlets, such as automobile emissions from nearby roads (Bhattacharjee et al., 2021; Das & Ghosh, 2022). Some helium photons originate from sources farther away and travel varying distances in the atmosphere. PHs can be added to soils that have infill materials in them; after they have accumulated on the surface of the earth, they will become mobile when conditions permit. The main processes that produce and release PHs are the burning of fossil fuels, waste incineration, wood processing, local heating, forest fires, and other related activities. These processes are highly dependent on the PHs' molecular hydrophobicity. Typically, PHs molecules enlarge and add more aromatic rings to make them more hydrophobic and environmentally stable. Alongside many other prominent sources, natural gas, emissions from mud volcanoes, oil sand, synthetic fuels, oil spills, and urban and industrial runoffs contribute to PHs pollution (Mishra et al., 2019). Crude oil is the most significant natural source of PHs. It is extracted from underground reservoirs and consists of a complex mixture of hydrocarbons. Natural gas is composed mainly of methane and often accompanies crude oil deposits, containing hydrocarbons in gaseous form. Oil sands or tar sands are deposits of sand or sandstone saturated with a thick, tar-like form of crude oil known as bitumen. Apart from the above processes, many industrial and refining processes, including synthetic production such as coal liquefaction and gas-to-liquid technologies, are the primary sources of PH pollution. Industries that use or process petroleum products often discharge wastewater containing hydrocarbons into water bodies. This can occur through direct discharge or runoff from industrial sites. Volcanic eruptions can release hydrocarbons as part of the gas phase of magma. This release includes various organic compounds, including simple hydrocarbons. While this is a less common source, it still contributes to natural hydrocarbon emissions. Some agricultural chemicals, including certain pesticides and fertilizers, contain petroleum-based hydrocarbons. Runoff from agricultural fields can carry these compounds into water bodies, contributing to environmental pollution (Khatun et al., 2023). Some of the major sources of PHs are summarized in Figure 2.1.

2.3 IMPACTS OF PETROLEUM HYDROCARBON ON THE LIVING ORGANISMS

2.3.1 Impact on Terrestrial Animals

Although PH is a major source of energy and is typically necessary for advancement in the economy, it has also resulted in a vast area of contamination and negative effects. Worldwide, PH-related environmental contamination is regarded as a major problem. It has terrible and harmful effects on many biotic elements in the environment in addition to the health of humans (Ghosh & Das, 2018). Animals are surrounded by petroleum in a variety of ways, both directly and indirectly. During petroleum refining and processing, some byproducts are utilized for making other highly dangerous compounds. Petroleum exposure affects animals in a variety of

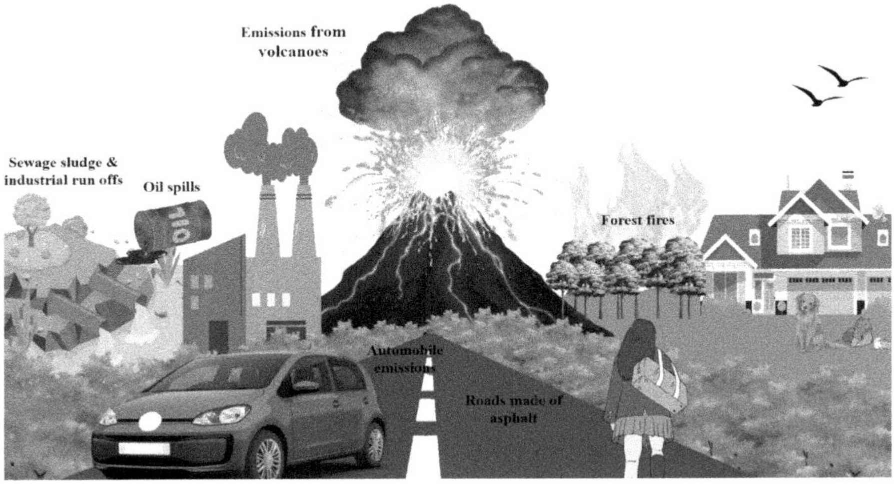

FIGURE 2.1 Represents potential sources of petroleum hydrocarbons.

direct and indirect ways. During the refining and processing of petroleum, certain byproducts are produced that are used to make other highly dangerous compounds (Das, Dey, & Das, 2024). These hazardous substances are routinely unintentionally released into the environment, and if the impact of these releases is linked to the global consequences of unintentional crude oil spills, then the combined sources of unrestrained hydrocarbons are the primary contributors to environmental contamination (Dey, Das, et al., 2023; Singh et al., 2020). Petroleum products contain a lot of hydrocarbons; however, only a tiny percentage of these molecules have thorough toxicity profiles. All things taken into account, oil metabolites must enter the body, usually by ingestion, inward breath, or retention, to have a direct biological impact on terrestrial vertebrates. The primary way that most living organisms are exposed to polyaromatic hydrocarbons is by ingesting contaminated soils, waste products, and food items (Das & Ghosh, 2018; Ghosh & Das, 2020). In light of this, organisms that ingest a lot of sediment-related invertebrates will often be more vulnerable to PAH exposure than higher-order consumers (Brooks et al., 2009). PAHs seldom demonstrate sustenance web biological accumulation and biomagnification; their capacity to disrupt the natural order of things is restricted. This mostly refers to vertebrates' enhanced ability to utilize, and so remove, PAH deposits.

2.3.2 Impact on Humans

PHs, which are complex mixtures of organic compounds derived from crude oil, can have a range of impacts on human health. These impacts depend on the specific hydrocarbons involved, the level and duration of exposure, and individual susceptibility. The known components of petroleum and petroleum-determined materials are polyaromatic hydrocarbons. Since PAHs are oncogenic, they are essential ecological toxins. These mixtures are regularly chosen for drinking water, groundwater, and modern wastewater. Certain hydrocarbons, such as benzene and PAHs, are known carcinogens. Long-term exposure to these substances can increase the risk of developing cancers, particularly leukemia and lung cancer. Long-term exposure can lead to more serious neurological effects, including cognitive impairments and persistent mood changes. Some hydrocarbons have been linked to reproductive and developmental problems, potentially affecting fertility and fetal development. Even very low quantities of PHs exposed over an extended period may affect behavior and memory. An evaluation of this product's impact on the neurological system has resulted from this assertion. Almost any hydrocarbon solvent vapor can elicit a widespread CNS depression with a single relatively high concentration exposure that, at large dosages, can result in unconsciousness. In severe cases, exposure to high concentrations of PHs can lead to chemical pneumonitis, a condition where the lungs become inflamed due to inhalation of irritant chemicals (Mishra et al., 2021). This can cause symptoms like chest pain, fever, and difficulty breathing. Long-term exposure to low levels of PHs can lead to chronic bronchitis and asthma, characterized by persistent cough, mucus production, and airflow obstruction. Benzene is a known respiratory irritant and has been associated with respiratory tract cancers. It is also associated with inducing oxidative stress, leading to cellular damage in the lungs and contributing to chronic respiratory conditions. Chronic exposure can cause more serious respiratory difficulties, such as chronic obstructive pulmonary disease (COPD) (Mishra et al., 2024). Diesel fume particles have also been connected to a reduction in male reproductive ability. Previous research has indicated that DEP exposure aggravated spermatogenesis, resulting in lower daily sperm quantity and motility, higher morphological sperm deviations from the norm, and ultrastructural changes in mouse Leydig cells.

Unfavorable pregnancy outcomes may be linked to long-term exposure to PAHs. Toxic PAH metabolites can upset the hormonal and redox balance, which can prevent a pregnancy from progressing and lead to a miscarriage. In a study, there has been an evaluation of the relationship between dietary exposure to contaminated mussels containing PAHs and changes in reproductive hormones, oxidative stress indicators, and PAH metabolites. The findings indicate a possible association of chronic PAH accumulation with recurrent pregnancy loss in women (El-Sikaily et al., 2023). Various studies have shown decreased creatinine in the urine, hematuria due to exposure to xylene, and enlarged, sharp kidney tubules. The effects of xylene on the kidneys were dependent on focus and portion, which resulted in the accumulation of m-xylene in the kidney's peripheral fat. Histopathological analysis revealed a negligible chronic kidney disease (John & Das, 2024). Despite the fact that urine was frequently produced, the main negative effects were increasing hyaline bead alterations in male rats and kidney damage in female rats, which resulted in cell toxicity (Yuan et al., 2020). Workers in the oil and gas industry, as well as those involved in refining, transporting, and handling petroleum products, are at a higher risk of exposure to harmful hydrocarbons. Chronic exposure in these occupations can lead to long-term health issues, including respiratory diseases, cancer, and neurological disorders. Gas station employees who frequently handle fuels may experience health effects from inhaling gasoline vapors or having skin contact with hydrocarbons. They may be at risk for respiratory issues and skin conditions. Also, workers involved in cleaning up oil spills can be exposed to high concentrations of hydrocarbons. This exposure can lead to acute health effects, such as respiratory and skin issues, as well as long-term health risks if proper protective measures are not taken.

2.3.3 Impact on Marine Organisms

Exposure of marine organisms to PHs can occur through various pathways and can have extensive impacts. The detrimental effects of petroleum contamination on such an essential component of marine biological systems may be extremely noteworthy. Since unsaturated hydrocarbons, naphthenic acids, and other substances comprising aromatic groups and nitrogen cause oil to be toxic to biota, the amount of non-volatile material that has broken down is a strong indicator of the actual deleterious effects of the oil. Marine organisms come into direct contact with PHs through surface oil spills or contaminated sediments (Paul et al., 2023; Priyadarshini & Das, 2024a). This can affect organisms such as fish, birds, and invertebrates that are in or near the oil slick. They ingest PHs either directly by consuming contaminated water or indirectly by eating contaminated prey. For example, filter-feeding organisms like mussels and oysters can accumulate hydrocarbons from contaminated water. For species that breathe oxygen from the water, such as fish, hydrocarbons dissolved in the water can be taken up through the gills. Volatile compounds in petroleum are also inhaled when they are present in the air, affecting marine mammals and seabirds. PHs can accumulate in the tissues of marine organisms over time, a process known as bioaccumulation. Biomagnification is the process by which pollutants become more concentrated as they move up the food chain. Predators at the top of the food chain, including larger fish, marine mammals, and seabirds, are particularly vulnerable to oil spills. PHs can settle into sediments on the ocean floor (Guo et al., 2022). Organisms that live in or on these sediments, such as benthic invertebrates, can be exposed to these contaminants over long periods. Short-term exposure to high levels of PHs has been observed to cause immediate health issues such as respiratory distress, neurological effects, and death. Symptoms might include reduced feeding behavior, lethargy, and disorientation. Long-term or repeated exposure to lower concentrations of hydrocarbons also leads to chronic health issues such as liver damage, reproductive failures, and immune system suppression. Chronic exposure can also lead to developmental abnormalities in offspring. Hydrocarbons interfere with the reproductive systems of marine organisms, resulting in reduced fertility, abnormal embryo development, and lower survival rates of juveniles. For example, fish embryos exposed to hydrocarbons may exhibit malformations and increased mortality. Exposure to hydrocarbons impairs the immune systems of marine organisms, making them more susceptible to infections, parasites, and diseases, leading to higher mortality rates and disruptions in population dynamics. The effects on individual organisms cascade through the ecosystem.

2.3.4 Impact on Plants

PHs can be directly toxic to plants. Compounds like PAHs and alkanes can interfere with physiological processes. For instance, they may inhibit seed germination, root development, and overall plant growth. When PHs contaminate the soil, they can be absorbed by plant roots, leading to toxic effects. These can include stunted growth, chlorosis, and necrosis (Dey, Dash, & Das, 2024). PHs alter soil properties such as texture, structure, and nutrient availability, which in turn can affect plant growth indirectly by making the soil less hospitable for roots and reducing nutrient and water uptake. PHs alter soil microbial populations, which are essential for nutrient cycling and soil health. Microbial activity changes can affect plant nutrient availability and overall soil fertility (Priyadarshini et al., 2024a). Exposure to PHs can slow down plant growth rates, leading to smaller plants with reduced biomass. They can cause physical damage to plant roots, impairing their ability to absorb water and nutrients, which can lead to wilting, reduced growth, and increased susceptibility to drought stress. PHs can affect plant reproductive processes, leading to reduced seed production and viability. This can result in fewer seedlings and reduced population growth (Sahoo et al., 2011). The development of seedlings can be hindered by exposure to PHs, leading to higher mortality rates and poor establishment of new plants. Hydrocarbons can impair photosynthetic processes by affecting chlorophyll content and function, leading to reduced photosynthesis and overall plant productivity. To date, not much research has been done on how petroleum contamination affects PHC plant absorption and growth in agricultural settings. The majority of experiments on plants growing in soil or nutrient solutions containing high quantities of PAHs have been carried out in laboratory conditions. The literature reports very few field research studies. In a study, it was comprehensively shown that since the physical and chemical characteristics of soil were adversely impacted by oil spills, *Zea mays* plant development was inhibited by PHs present in the soil. *Zea mays* growing in both polluted and controlled soils had the same amount of PHCs. The findings pointed to a biogenic genesis for PHCs in plants compared to an anthropogenic one (Grifoni et al., 2020). Some of the major impacts exhibited by the plants and animals due to overexposure to deadly sources of PHs are shown in Table 2.1.

2.3.5 Impact on Soil Flora and Fauna

PH leaches into the groundwater from the soil when it is released into the environment by accidents, industry releases, byproducts of commercial or private uses, spills, or leaks. PH-contaminated soil is a concern for several reasons, as will be discussed below. First, when PHCs are released into the soil, their volatility can pose a risk of fire or even explosion, especially when the vapor enters confined places. Degradation of the land can result from contaminants interfering with the flow of nutrients and water. For years, petroleum residues can cling to soil particles and remain there. The majority of soil biota remain poisonous to pollutants, even if they may benefit from carbon sources as degraders (Dey et al., 2024). When PHCs provide an unpleasant taste, smell, or appearance in environmental media, they might ruin the aesthetic. Finally, PHCs may be harmful to the ecosystem. PHs are a significant source of soil contamination due to their widespread use and frequent accidental spills. These contaminants impact soil health and the organisms that inhabit it, including plants, microorganisms, and invertebrates. The impact on soil flora and fauna is multifaceted, affecting plant growth, soil structure, microbial communities, and invertebrate populations.

TABLE 2.1
Demonstrates the Impacts on Various Organisms due to Excessive Exposure to Petroleum Hydrocarbons and Their Sources

Sl No	Adverse Effects	Plants/Animals Studied	References
1.	Reduced levels of chlorophyll in the leaves and an increase in the shoot's total proteins and amino acids	Jojoba (*Simmodsia chinensis*)	Shukry et al. (2013)
2.	Lowered heights and girths, as well as hindered germination	Okra (*Abelmoschus esculentus*)	Oyedeji et al. (2012)
3.	Increased lipid peroxidation activity generally in the ovary, hemolymph, and hepatopancreas	Giant Mud Crab (*Scylla serrata*)	Karam and Al-Wazzan (2021)
4.	Disruption of the embryonic axis through the Wnt/ß-catenin signaling pathway	Zebrafish	Cherr et al. (2017)
5.	A substantial decline in seedling height, leaf length, and leaf count	Soybean (*Glycine max*)	Ogunsola et al. (2023)
7.	Retardation of both longitudinal development and germination	(*Festuca arundinacea*)	Borowik et al. (2019)
8.	Root growth is diminished.	Corn (*Zea mays*) and Red beans (*Phaseolus nipponesis*)	Martínez-Alonso et al. (2022)
9.	Reduced production of the aboveground and root sections of yellow lupines	*Lupinus arboreus*	Arukalam et al. (2021)
10.	High levels of pollution greatly inhibited growth and germination.	*Sorghum vulgare*	Luhach and Chaudhry (2249)

These effects can have far-reaching implications for soil health and ecosystem function. Soil microorganisms, including bacteria, fungi, and protozoa, are highly sensitive to PHs (Bal et al., 2018). Hydrocarbons inhibit the activity of soil microbes responsible for decomposing organic matter. This leads to reduced nutrient cycling and slower organic matter decomposition. Contamination can shift microbial community composition, favoring hydrocarbon-degrading microbes while suppressing other beneficial organisms (Moharana et al., 2024; Pradhan et al., 2024). Many plants rely on symbiotic relationships with soil microbes (e.g., mycorrhizal fungi) for nutrient uptake. Petroleum contamination disrupts these relationships, further affecting plant health (Jia et al., 2023). PHs lead to a reduction in microbial diversity. Sensitive species get eliminated or reduced in abundance, while certain hydrocarbon-degrading microbes become more dominant. This shift can decrease the overall resilience of microbial communities and their ability to perform essential soil functions. Contaminated soils often experience shifts in microbial community composition. Hydrocarbon pollution can favor the growth of specific bacteria and fungi that are capable of degrading hydrocarbons, while other beneficial microbes may be suppressed. For example, bacteria like *Pseudomonas* and *Mycobacterium* are known to degrade hydrocarbons, whereas soil fungi and other bacteria involved in nutrient cycling may be negatively affected. PHs inhibit microbial activity, leading to reduced rates of organic matter decomposition. Microorganisms responsible for breaking down plant residues and other organic materials may become less active or die off. This reduction in decomposition led to the accumulation of undecomposed organic matter and a decrease in soil fertility. Microbial respiration, a key indicator of microbial activity, is altered by various environmental pollutants (Jagatee et al., 2024; Sahu et al., 2024). Contamination leads to reduced soil respiration rates as microbes struggle to metabolize hydrocarbons or suffer from toxicity. Lower respiration rates indicate reduced microbial activity and impact soil health and plant growth. Hydrocarbon pollution impairs the nutrient cycling processes by affecting the microbes responsible for nitrification, denitrification, and other nutrient transformations (Mishra et al., 2023; Sahoo et al., 2022). Higher PH concentrations also enhance the carbon to nitrogen ratio in soil, which enhances competition among microbes for nitrogen sources and favors the rapid development of some nitrogen-fixing bacteria, resulting in a loss in the diversity of the microbial population. On the contrary, low levels of PHs may enhance microbial diversity, in line with the ecological theories' intermediate disturbance hypothesis (IDH), which predicts that an ecosystem's species diversity will peak at mild disturbance (Ma et al., 2021). This happens because low PH concentrations give soil-dwelling bacteria access to carbon supplies and energy as well as encourage the growth of microbes that can break down PHs. In a study, it is inferred that high PH concentration stress increased the modularity coefficient of microbial networks, indicating that a more deterministic approach to microbial community building is induced by higher PH concentrations (Jia et al., 2023). The original keystone taxa's niche breadth in the soil gradually shrunk as the concentration of PHs increased because they were unable to endure the toxic effects of high PH concentrations. In contrast, some resilient microbes' niche breadth gradually widened as a result of their improved ability to adapt to high PH concentrations. This illustrates how the concentration of PHs increases and how different soil microbial communities evolve. The keystone genus's enhanced niche overlap was caused by high PH concentrations. This is because when the concentration of PHs increased, resources became more concentrated, new keystone microorganisms in the community developed comparable resource requirements, and interspecies competition increased.

2.4 MITIGATION OF PETROLEUM HYDROCARBON POLLUTION

Total Petroleum Hydrocarbon (TPH) released into the environment, according to the Environmental Protection Agency (USEPA), will endanger public health and safety by contaminating drinking water; causing explosions and fires; lowering air and water quality; jeopardizing agriculture; destroying recreational areas, habitats, and food; and wasting non-renewable resources. It could take several years for the soil to recover after PHs have contaminated it. Therefore, mitigation of PH pollution has become a crucial aspect of environmental management, aimed at reducing the adverse impacts of oil and its derivatives on ecosystems and human health. Currently, TPH is analyzed using a broad range of specialized and non-specific techniques (Dey, Das, et al., 2023). Gravimetric determination and infrared spectrometry are two key conventional non-specific techniques, as are field screening gas chromatography with flame ionization (GC-FID) or photo ionization detection (GC-PID), thin-layer chromatography (TLC), supercritical fluid chromatography (SFC), high-performance liquid chromatography (HPLC), and fiber optic IR sensors for identifying various petroleum samples. Over time, bioremediation, which breaks down toxins using bacteria or other biological means, has become one of the most advanced approaches for treating systems damaged by petroleum (Ghosh & Das, 2018). Using living organisms to break down and remove contaminants is a practical and efficient treatment method since the natural environment is home to a wide variety of microorganisms and plants (Biswal et al., 2015, 2021, Nag et al., 2023). Typically, PHs experience a two-step biodegradation in polluted areas. The rate of elimination is high during the first stage and is mostly constrained by the native microbial community's microbial degradation kinetics (Lahiri et al., 2021). During the second phase, the rate of PH removal decreases and is mostly determined by the contaminant's availability followed by slow desorption from the soil's mineral and organic matter fractions (Ławniczak et al., 2020). A wide range of organic amendments, including crop residues, green wastes, manures, biosolids, and biochar, can be employed in the context of remediating soil PH pollution because of their favorable impacts on soil health and plant growth (Kanungo et al., 2023). PH-contaminated soil can be cleaned up by limiting its bioavailability in the soil through chemical and biological immobilization, which lowers PH exposure to receptors. On the other hand, chemical washing and phytoextraction can be used to mobilize the contaminants to treat localized contamination in urban settings. Organic amendments are being implemented to improve the impacted soil's surface area, surface charge density, and number of exchange sites, which improve surface adsorption and complexation and lower the soil's PH mobility. In TPH-contaminated soils, a range of organic wastes, including food scraps, leaves, animal dung, and even sewage sludge, can be composted and applied (Behera et al., 2020). The creation of metal-humic complexes is a common cause of changed TPH availability following the addition of composts. Composts can raise soil organic matter (SOM) content, which lowers the bioavailability of TPH in polluted soil. Moreover, humic materials in composts are thought to have a cation exchange capacity that is comparable to clay minerals in soil, which helps the soil hold onto vital nutrients. Rhizoremediation is an innovative and eco-friendly technique for cleaning up contaminated soil and water, leveraging the natural processes associated with plant roots. This method uses the interactions between plant roots and microorganisms in the rhizosphere (the soil region around plant roots) to degrade or detoxify pollutants, including PHs. Compared to other remediation technologies, rhizoremediation often requires less maintenance and can be more economical. It utilizes natural processes and can reduce the need for chemical treatments or extensive physical interventions. The method has been proved to be effective against a range of contaminants, including heavy metals, pesticides, and PHs. Hence, it has been used for having a promising and sustainable approach to environmental cleanup, harnessing the natural capabilities of plants and microbes to address contamination issues (Singha & Pandey, 2021). Numerous studies on bioaugmentation-based soil remediation have been carried out. Utilizing a biopile technology system and an indigenous strain of *Bacillus sonorensis* for bioaugmentation with optimal nutrients, weathered oil hydrocarbons were investigated for their biodegradation. The ideal conditions for the introduction of *Bacillus sonorensis* were as follows: moisture (10%), temperature (37°C), carbon/nitrogen/phosphorus (100/10/1). When ammonium nitrate was used as the nitrogen source, the diesel-range organic and PAH removal efficiencies of weathered oil-contaminated soil after 160 days were 39.2% and 32.4%, respectively (Oualha et al., 2019). In another investigation, colonies of bacterial strains resistant to heavy metals were isolated from sediments and cultivated in Luria Bertani (LB) broth that had been modified with 1% metal solution for testing on the remediation of crude oil and phenanthrene substrates. Ultimately, after 28 days, the bacilli stain revealed that phenanthrene and crude oil had degraded by 94% and 85%, respectively. The most popular technique for increasing the rate at which pH-polluted soil detoxifies is biostimulation, which involves adding nutrients including carbon, nitrogen, phosphorus, and oxygen. However, since it could lead to a nutrient imbalance in the microbiota, the concentration of additional nutrients needs to be regulated. By providing nutrients to heavily contaminated areas, biostimulation of PHs encourages microorganisms to break down hazardous and toxic pollutants quickly (Das, Dutta, et al., 2023; Mohanty et al., 2024). The most common nutrients that encourage microbial development and highly effective HC-degrading microorganisms can be added in situ to speed up biostimulation tactics for PH mineralization in the soil (Guarino et al., 2017). Since the 1980s, fungi have been known to effectively remove contaminants and advance environmental biotechnology. Fungi come in millions of species, are widely distributed in many types of settings, and have been used in a wide range of contexts. By distributing spores and secreting different enzymes, fungi are able to thrive under a wide range of environmental conditions. As a result, a great deal of study has been done on the basic and applied aspects of mycoremediation in soil and water. As a result, mycoremediation has established itself as a reliable technique for remediating soil without running the risk of contaminants such as PHs spreading or bioaccumulating in food chains. Effective mitigation of PH pollution requires a multifaceted approach that integrates prevention, containment, treatment, and ongoing monitoring (Das & Ghosh, 2023f;). By employing strategies such as spill prevention measures, advanced recovery technologies, and innovative treatment methods like bioremediation, we can significantly reduce the environmental and health impacts of petroleum contaminants. Emphasizing preventive measures and leveraging natural processes, such as rhizoremediation, not only aids in addressing current pollution but also helps to build resilience against future incidents. Collaborative efforts between industries, regulatory bodies, and communities, alongside continuous research and adaptation of best practices, are essential to ensure the long-term success of these mitigation strategies (Behera et al., 2020; Mallick et al., 2023). Ultimately, a proactive and comprehensive approach will help protect our ecosystems, safeguard public health, and foster sustainable management of our natural resources.

2.5 CONCLUSION

Ecotoxicity of PHs represents a profound and multifaceted challenge to environmental health. These substances, which include a range of complex chemicals from simple alkanes to more toxic PAHs, have been shown to exert detrimental effects on various ecosystems. The persistence and bioaccumulation of these hydrocarbons further complicate their environmental impact. Many petroleum compounds resist natural degradation processes, leading to prolonged exposure and accumulation in the food chain. This bioaccumulation can result in chronic health effects for both wildlife and humans, including cancer and reproductive impairments. This chapter delves into the sources and impacts on the living beings, along with certain mitigation strategies of PHs. The multifaceted nature of these impacts necessitates a comprehensive approach to managing petroleum pollution. Effective response strategies must include rigorous environmental monitoring to detect and quantify contamination, innovative cleanup technologies to address spills and leaks, and robust regulations to prevent and manage pollution incidents. Ultimately, addressing the ecotoxicity of PHs requires a concerted effort from scientists, policymakers, industry stakeholders, and the public. By fostering collaboration and investing in both preventative and remedial measures, the adverse effects on our environment can be mitigated, stepping toward a more sustainable future.

REFERENCES

Alzahrani, M. A., & Rajendran, P. (2019). Petroleum Hydrocarbon and Living Organisms. In: *Hydrocarbon Pollution and Its Effect on the Environment*. IntechOpen. https://doi.org/10.5772/intechopen.86948

Arukalam, I. O., Madu, I. O., & Ishidi, E. Y. (2021). High performance characteristics of Lupinus arboreus gum extract as self-healing and corrosion inhibition agent in epoxy-based coating. *Progress in Organic Coatings*, *151*. https://doi.org/10.1016/j.porgcoat.2020.106095

Bal, B., Ghosh, S., & Das, A. P. (2018). Microbial recovery and recycling of manganese waste and their future application: a review. *Geomicrobiological Journal*, *36*(1), 85–96. https://doi.org/10.1080/01490451.2018.1497731

Bala, K., Dogra, A., Lata, P., & Das, A. P. (2024). Discussing Sources and Biological-Chemical Remediation Approaches for Microplastic Pollution. In: *Sustainable Microbial Technology for Synthetic and Cellulosic Microfiber Bioremediation* (pp. 37–59). https://doi.org/10.1007/978-3-031-62600-5_3

Bao, H., Wang, J., Zhang, H., Li, J., Li, H., & Wu, F. (2020). Effects of biochar and organic substrates on biodegradation of polycyclic aromatic hydrocarbons and microbial community structure in PAHs-contaminated soils. *Journal of Hazardous Materials*, *385*. https://doi.org/10.1016/j.jhazmat.2019.121595

Behera, I. D., Basak, G., Kumar, R. R., Sen, R., & Meikap, B. C. (2020). Treatment of petroleum refinery sludge by petroleum degrading bacterium Stenotrophomonas pavanii IRB19 as an efficient novel technology. *Journal of Environmental Science and Health - Part A Toxic/Hazardous Substances and Environmental Engineering*, *56*(2), 226–239. https://doi.org/10.1080/10934529.2020.1866924

Behera, I. D., Nayak, M., Biswas, S., Meikap, B. C., & Sen, R. (2021). Enhanced biodegradation of total petroleum hydrocarbons by implementing a novel two-step bioaugmentation strategy using indigenous bacterial consortium. *Journal of Environmental Management*, *292*, 112746.

Bhattacharjee, J., Mishra, S., & Das, A. P. (2021). Recent advances in sensor-based detection of toxic dyes for bioremediation application: a review. *Applied Biochemistry and Biotechnology*, *2021*, 1–20. https://doi.org/10.1007/S12010-021-03767-7

Biswal, P., Pal, A., & Das, A. P. (2015). A new light of therapy for non-alcoholic fatty liver disease: symbiotic. *International Journal of Pharmaceutical Sciences Review and Research*, *33*(1), 326–336.

Biswal, P., Pal, A., & Das, A. P. (2021). Screening for probiotic potential of Lactobacillus rhamnosus strain CRD4. *Biointerface Research in Applied Chemistry*, *11*, 10147–10184. https://doi.org/10.33263/BRIAC113.1017410184

Borowik, A., Wyszkowska, J., Gałązka, A., & Kucharski, J. (2019). Role of Festuca rubra and Festuca arundinacea in determinig the functional and genetic diversity of microorganisms and of the enzymatic activity in the soil polluted with diesel oil. *Environmental Science and Pollution Research*, *26*(27), 27738–27751. https://doi.org/10.1007/s11356-019-05888-3

Brooks, A. C., Gaskell, P. N., & Maltby, L. L. (2009). Importance of prey and predator feeding behaviors for trophic transfer and secondary poisoning. *Environmental Science and Technology*, *43*(20), 7916–7923. https://doi.org/10.1021/es900747n

Cherr, G. N., Fairbairn, E., & Whitehead, A. (2017). Impacts of petroleum-derived pollutants on fish development. *Annual Review of Animal Biosciences*, *5*, 185–203. https://doi.org/10.1146/annurev-animal-022516-022928

Chibwe, L., Geier, M. C., Nakamura, J., Tanguay, R. L., Aitken, M. D., & Simonich, S. L. M. (2015). Aerobic bioremediation of PAH contaminated soil results in increased genotoxicity and developmental toxicity. *Environmental Science and Technology*, *49*(23), 13889–13898. https://doi.org/10.1021/acs.est.5b00499

Das, A., Dey, S., Bala, K., Gnanasekaran, R., & Das, A. P. (2024). Microbial Bioremediation Technology for Sustainable Treatment and Management of Synthetic Microfiber Waste. In: *Sustainable Microbial Technology for Synthetic and Cellulosic Microfiber Bioremediation* (pp. 221–238). https://doi.org/10.1007/978-3-031-62600-5_12

Das, A., Dey, S., & Das, A. P. (2024). Ecotoxicological impacts of synthetic microfiber pollutants and development of sustainable mitigation strategies. *Environmental Chemistry and Ecotoxicology*. https://doi.org/10.1016/J.ENCECO.2024.12.003

Das, A., & Ghosh, S. (2022). Role of microorganisms in extenuation of mining and industrial wastes. *Geomicrobiology Journal*, *39*, 173–175. https://doi.org/10.1080/01490451.2022.2038953

Das, A. P., Dutta, K., Khatun, R., Behera, I. D., Singh, S., & Mishra, S. (2023). Microfiber pollution and its microbial mitigation: a review on current trends and future prospects. *Journal of the Taiwan Institute of Chemical Engineers*, 105104. https://doi.org/10.1016/J.JTICE.2023.105104

Das, A. P., & Ghosh, S. (2018). Bioleaching of manganese from mining waste materials. *Materials Today: Proceedings*, *5*(1), 2381–2390. https://doi.org/10.1016/J.MATPR.2017.11.459

Dey, S., Bissoyi, A., Koteswara Reddy, G., & Das, A. P. (2024). Synthetic microfibers: An urban emerging pollutant demanding for sustainable environmental solutions. *Sustainable Microbial Technology for Synthetic and Cellulosic Microfiber Bioremediation*, 107–121.

Dey, S., Das, A., Mallick, K., Sahu, A., & Das, A. P. (2023). Environmental Petroleum Waste: Pollution, Toxicity, Sustainable Remediation. In: *Impact of Petroleum Waste on Environmental Pollution and its Sustainable Management through Circular Economy* (pp. 159–175). https://doi.org/10.1007/978-3-031-48220-5_7

Dey, S., Dash, A., & Das, A. P. (2024). Bioremediation as a Sustainable and Circular Solution for Mining Waste Management and Bioenergy Generation. In: *Sustainable Management of Mining Waste and Tailings: A Circular Economy Approach* (pp. 107–119). CRC Press.

Dey, S., Dash, N., Ray, S. K., Priyadarsini, S., Chhotaray, C., & Das, A. P. (2024). Occurrence and Quantification of Tire Wear Plastics Pollutants from Highway City Dust of Bhubaneswar City of Odisha. In: *Sustainable Microbial Technology for Synthetic and Cellulosic Microfiber Bioremediation* (pp. 77–89). https://doi.org/10.1007/978-3-031-62600-5_5

Dey, S., Tripathy, B., Kumar, M. S., & Das, A. P. (2023). Ecotoxicological consequences of manganese mining pollutants and their biological remediation. *Environmental Chemistry and Ecotoxicology*. https://doi.org/10.1016/J.ENCECO.2023.01.001

Erdem, Z. B. (2010). The contribution of renewable resources in meeting Turkey's energy-related challenges. Renewable and Sustainable Energy Reviews, 14(9), 2710-2722

El-Sikaily, A., Helal, M., Nsonwu-Anyanwu, A. C., Azab, H., Abd ElMoneim, N., Farahat, E. O. S., & Saad, A. (2023). Impacts of PAH accumulation on reproductive hormones, indices of oxidative stress and BPDE-albumin adduct in women with recurrent pregnancy loss. *Toxicological Research*, *39*(3), 517–531. https://doi.org/10.1007/s43188-023-00181-5

Falciglia, P. P., Giustra, M. G., & Vagliasindi, F. G. A. (2011). Low-temperature thermal desorption of diesel polluted soil: influence of temperature and soil texture on contaminant removal kinetics. *Journal of Hazardous Materials*, *185*(1), 392–400. https://doi.org/10.1016/j.jhazmat.2010.09.046

Ghosh, S., & Das, A. P. (2018). Metagenomic insights into the microbial diversity in manganese-contaminated mine tailings and their role in biogeochemical cycling of manganese. *Scientific Reports*, *8*(1), 1–12. https://doi.org/10.1038/s41598-018-26311-w

Ghosh, S., & Das, A. P. (2020). Microbial Metagenomics: Current Advances in Investigating Microbial Ecology and Population Dynamics. In: S. K. Nayak & B. B. Mishra (Eds.), *Frontiers in Soil and Environmental Microbiology* (1st ed., pp. 247–254). CRC Press.

Grifoni, M., Rosellini, I., Angelini, P., Petruzzelli, G., & Pezzarossa, B. (2020). The effect of residual hydrocarbons in soil following oil spillages on the growth of Zea mays plants. *Environmental Pollution*, *265*. https://doi.org/10.1016/j.envpol.2020.114950

Guarino, C., Spada, V., & Sciarrillo, R. (2017). Assessment of three approaches of bioremediation (Natural Attenuation, Landfarming and Bioagumentation – Assistited Landfarming) for a petroleum hydrocarbons contaminated soil. *Chemosphere*, *170*, 10–16. https://doi.org/10.1016/j.chemosphere.2016.11.165

Guo, W., Wang, X., Liu, S., Kong, X., Wang, P., & Xu, T. (2022). Long-term petroleum hydrocarbons pollution after a coastal oil spill. *Journal of Marine Science and Engineering*, *10*(10). https://doi.org/10.3390/jmse10101380

Haider, F. U., Ejaz, M., Cheema, S. A., Khan, M. I., Zhao, B., Liqun, C., Salim, M. A., Naveed, M., Khan, N., Núñez-Delgado, A., & Mustafa, A. (2021). Phytotoxicity of petroleum hydrocarbons: sources, impacts and remediation strategies. *Environmental Research*, *197*. https://doi.org/10.1016/j.envres.2021.111031

Jagatee, S., Priyadarshini, S., Rath, C. C., & Das, A. P. (2024). Synthetic Microfiber: An Enduring Environmental Problem Linked to Sustainable Development. In: *Environmental Science and Engineering*, *Part F2333* (pp. 93–112). https://doi.org/10.1007/978-3-031-51792-1_6

Jia, W., Cheng, L., Tan, Q., Liu, Y., Dou, J., Yang, K., Yang, Q., Wang, S., Li, J., Niu, G., Zheng, L., & Ding, A. (2023). Response of the soil microbial community to petroleum hydrocarbon stress shows a threshold effect: research on aged realistic contaminated fields. *Frontiers in Microbiology*, *14*. https://doi.org/10.3389/fmicb.2023.1188229

John, A., & Das, A. P. (2024). Government Strategies for Achieving Circular Economy for Plastic Waste Management in Southeast Asia. In: *Sustainable Management of Urban Plastic Waste through Circular Economic Approaches* (pp. 72–82). CRC Press. https://www.taylorfrancis.com/chapters/edit/10.1201/9781003465447-6/government-strategies-achieving-circular-economy-plastic-waste-management-southeast-asia-akbar-john-alok-prasad-das

Kanungo, J., Sahoo, T., Bal, M., & Behera, I. D. (2023). Performance of bioremediation strategy in waste lubricating oil pollutants: a review. *Geomicrobiology Journal*. https://doi.org/10.1080/01490451.2023.2245395

Karam, Q., & Al-Wazzan, Z. (2021). Toxicity of petroleum hydrocarbons to Brachyuran crabs: a review of deleterious effects of oil-related xenobiotics on life stages. *Journal of the Marine Biological Association of the United Kingdom*, 101(3), 561–576. Cambridge University Press. https://doi.org/10.1017/S0025315421000370

Khatun, R., Singh, S., Dubey, N. K., & Prasad Das, A. (2023). A Review on Marine-Based Phytochemicals and Their Application in Biomedical Research. In: *Recent Frontiers of Phytochemicals: Applications in Food, Pharmacy, Cosmetics, and Biotechnology* (pp. 383–395). https://doi.org/10.1016/B978-0-443-19143-5.00013-X

Lahiri, D., Nag, M., Dey, A., Sarkar, T., Joshi, S., Pandit, S., Das, A. P., Pati, S., Pattanaik, S., Tilak, V. K., & Ray, R. R. (2021). Biofilm mediated degradation of petroleum products. *Geomicrobiology Journal*, *39*, 1–10. https://doi.org/10.1080/01490451.2021.1968979

wniczak, Ł., Woźniak-Karczewska, M., Loibner, A. P., Heipieper, H. J., & Chrzanowski, Ł. (2020). Microbial degradation of hydrocarbons: basic principles for bioremediation: a review. *Molecules*, *25*(4). https://doi.org/10.3390/molecules25040856

Liwarska-Bizukojc, E. (2022). Phytotoxicity assessment of biodegradable and non-biodegradable plastics using seed germination and early growth tests. *Chemosphere*, *289*. https://doi.org/10.1016/j.chemosphere.2021.133132

Luhach, J., & Chaudhry, S. (2012). *Effect of Diesel Fuel Contamination on Seed Germination and Growth of Four Agricultural Crops*. www.environmentaljournal.org

Ma, Y., Zhao, H., Shan, Q., Xu, Y., Yu, M., Cui, J., Liu, T., Qiao, L., & He, X. (2021). K-Strategy species plays a pivotal role in the natural attenuation of petroleum hydrocarbon pollution in aquifers. *Journal of Hazardous Materials*, *420*. https://doi.org/10.1016/j.jhazmat.2021.126559

Mallick, K., Sahu, A., Dubey, N. K., & Das, A. P. (2023). Harvesting marine plastic pollutants-derived renewable energy: a comprehensive review on applied energy and sustainable approach. *Journal of Environmental Management*, *348*, 119371. https://doi.org/10.1016/J.JENVMAN.2023.119371

Martínez-Alonso, C., Taroncher, M., Castaldo, L., Izzo, L., Rodríguez-Carrasco, Y., Ritieni, A., & Ruiz, M.-J. (2022). *Effect of Phenolic Extract from Red Beans (Phaseolus vulgaris L.) on T-2 Toxin-Induced Cytotoxicity in HepG2 Cells*. https://doi.org/10.3390/foods

Mishra, S., Dash, D., & Das, A. P. (2023). Aquatic microbial diversity on plastisphere: colonization and potential role in microplastic biodegradation. *Geomicrobiology Journal*, 1–12. https://doi.org/10.1080/01490451.2023.2209750

Mishra, S., Dash, D., & Das, A. P. (2024). Isolation and molecular characterization of polyester-degrading bacterial strains from municipal wastewater and their application in microplastic degradation in contaminated wastewater. *Geomicrobiology Journal*. https://doi.org/10.1080/01490451.2024.2376101

Mishra, S., Rath, C. C., & Das, A. P. (2019). Marine microfiber pollution: a review on present status and future challenges. *Marine Pollution Bulletin*, *140*, 188–197. https://doi.org/10.1016/J.MARPOLBUL.2019.01.039

Mishra, S., Rout, P. K., & Das, A. P. (2021). Emerging Microfiber Pollution and Its Remediation. In: R. Prasad (Ed.), *Environmental and Microbial Biotechnology* (pp. 247–266). Springer, Singapore. https://doi.org/10.1007/978-981-15-5499-5_9

Mohanty, M., Mohanty, J., Dey, S., Dutta, K., Shah, M. P., & Das, A. P. (2024). The face mask: a tale from protection to pollution and demanding sustainable solution. *Emerging Contaminants*, *10*(2), 100298. https://doi.org/10.1016/J.EMCON.2023.100298

Moharana, S., Dey, S., Priyadarsini, S., Kumar, M. S., & Das, A. P. (2024). Identification and Characterization of Microplastic Pollutants from the Marine Sediments of Paradeep Coast of Bay of Bengal, India for their Sustainable Management. *Environmental Science and Engineering*, *Part F2333* (pp. 57–73). https://doi.org/10.1007/978-3-031-51792-1_4

Mukome, F. N. D., Buelow, M. C., Shang, J., Peng, J., Rodriguez, M., Mackay, D. M., Pignatello, J. J., Sihota, N., Hoelen, T. P., & Parikh, S. J. (2020). Biochar amendment as a remediation strategy for surface soils impacted by crude oil. *Environmental Pollution*, *265*. https://doi.org/10.1016/j.envpol.2020.115006

Nag, M., Lahiri, D., Ghosh, S., Sarkar, T., Pati, S., Das, A. P., Ram, D. K., Bhattacharya, D., & Ray, R. R. (2023). Application of microorganisms in biotransformation and bioremediation of environmental contaminant: a review. *Geomicrobiology Journal*. https://doi.org/10.1080/01490451.2023.2261443

Odukoya, J., Lambert, R., & Sakrabani, R. (2019). Understanding the impacts of crude oil and its induced abiotic stresses on agrifood production: a review. In *Horticulturae*, *5*(2). MDPI Multidisciplinary Digital Publishing Institute. https://doi.org/10.3390/horticulturae5020047

Ogunsola, K. E., Odesola, K. A., Eniola, B. F., & Majebi, O. E. (2023). Growth and mitotic chromosomes of soybean (*Glycine max L. Merril*) plants as affected by spent engine oil in sandy-loam soils. *Agro-Science*, *21*(4), 49–55. https://doi.org/10.4314/as.v21i4.8

Oualha, M., Al-Kaabi, N., Al-Ghouti, M., & Zouari, N. (2019). Identification and overcome of limitations of weathered oil hydrocarbons bioremediation by an adapted Bacillus sorensis strain. *Journal of Environmental Management*, *250*. https://doi.org/10.1016/j.jenvman.2019.109455

Oyedeji, A. A., Adebiyi, A. O., Omotoyinbo, M. A., & Ogunkunle, C. O. (2012). Effect of crude oil-contaminated soil on germination and growth performance of *Abelmoschus esculentus* L. moench: a widely cultivated vegetable crop in Nigeria. *American Journal of Plant Sciences*, *03*(10), 1451–1454. https://doi.org/10.4236/ajps.2012.310174

Paul, A., Dey, S., Ram, D. K., & Das, A. P. (2023). Hexavalent chromium pollution and its sustainable management through bioremediation. *Geomicrobiology Journal*, 1–11. https://doi.org/10.1080/01490451.2023.2218377

Pradhan, G., Dey, S., Priyadarsini, S., Kumar, M. S., & Das, A. P. (2024). Characterization and Quantification of Microplastics Pollutants in Sediment Samples from Daya River of Odisha State in India for their Appropriate Management. *Environmental Science and Engineering*, *Part F2333* (pp. 75–92). https://doi.org/10.1007/978-3-031-51792-1_5

Pradhan, G., Tripathy, B., Ram, D. K., Digal, A. K., & Das, A. P. (2023). Bauxite mining waste pollution and its sustainable management through bioremediation. *Geomicrobiology Journal*, 1–10. https://doi.org/10.1080/01490451.2023.2235353

Priyadarshini, B., & Das, A. P. (2024a). The Adverse Impact of Synthetic Microplastics Released from Disposable Face Masks Pollutants on Aquatic Microbiota, Aquatic Species and Humans: A Review. *Environmental Science and Engineering*, *Part F2827* (pp. 67–81). https://doi.org/10.1007/978-3-031-50840-0_5

Priyadarshini, B., Santhosh Kumar, M., & Das, A. P. (2024a). Microbial Degradation of Synthetic Plastics Released from COVID-19 Generated Single Use Disposable Face Masks. *Environmental Science and Engineering*, *Part F2827* (pp. 1–16). https://doi.org/10.1007/978-3-031-50840-0_1

Ray, S., Dey, S., Digal, A. K., Yadav, K. K., & Das, A. P. (2025). Sanitary waste and microplastic pollution in the Euro-Mediterranean region: Challenges and solutions. *Euro-Mediterranean Journal for Environmental Integration*, *10*(6), 5417–5440.

Sahoo, P. P., Singh, S., Rout, P. K., Mishra, S., & Das, A. P. (2022). Microbial remediation of plastic pollutants generated from discarded and abandoned marine fishing nets. *Biotechnology and Genetic Engineering Reviews*. https://doi.org/10.1080/02648725.2022.2152629

Sahu, A., Mallick, K., & Das, A. P. (2024). Environmental Impact of Single-Use Synthetic Face Mask and Its Recycling: A Sustainable Approach. (pp. 197–212). https://doi.org/10.1007/978-3-031-50840-0_11

Sahu, S., & Das, A. P. (2024). Microplastics Degradation Mechanism by Aquatic Plastiomes: A Review. (pp. 213–226). https://doi.org/10.1007/978-3-031-50840-0_12

Shukry, W., Al-Hawas, G., Al-Moaik, R., & El-Bendary, M. (2013). Effect of petroleum crude oil on mineral nutrient elements and soil properties of jojoba plant (Simmondsia chinensis). *Acta Botanica Hungarica*, *55*(1–2), 117–133. https://doi.org/10.1556/ABot.55.2013.1-2.8

Singh, R. P., Mishra, S., & Das, A. P. (2020). Synthetic microfibers: pollution toxicity and remediation. *Chemosphere*, *257*, 127199. https://doi.org/10.1016/j.chemosphere.2020.127199

Singha, L. P., & Pandey, P. (2021). Rhizosphere assisted bioengineering approaches for the mitigation of petroleum hydrocarbons contamination in soil. *Critical Reviews in Biotechnology*, *41*(5), 749–766. Taylor and Francis Ltd. https://doi.org/10.1080/07388551.2021.1888066

Yuan, T. H., Ke, D. Y., Wang, J. E. H., & Chan, C. C. (2020). Associations between renal functions and exposure of arsenic and polycyclic aromatic hydrocarbon in adults living near a petrochemical complex. *Environmental Pollution*, *256*. https://doi.org/10.1016/j.envpol.2019.113457

3 Application of Advanced and Greener Technology for Sustainable Management of Hydrocarbon Pollutants

Pamreishang Kasar and
Luiyaiwoshi Shimray Songachan

3.1 INTRODUCTION

Toxic chemicals from organic sources, pesticides, and petroleum products are polluting the environment with hydrocarbons. Petroleum hydrocarbons are especially concerning due to their harmful effects on plants and animal species. Unprocessed oil pollution is widespread because of its wide-ranging use and the risks of discharge and leaks. Petroleum is a mixture of various hydrocarbons, both low and high molecular weight, including branched and saturated alkanes, alkenes, and heterocyclic compounds like naphthenes. It also contains aromatic compounds with sulfur, nitrogen, and oxygen, as well as functional groups such as carboxylic acids and ethers. Additionally, it includes high-molecular-weight aromatics like resins, naptheno-aromatics, and asphaltenes.

Heavy-element pollutants in crude oil are linked to porphyrins with a pyloric structure. Crude oil is processed into components such as light oil, lube oil waxes, naphtha, asphaltenes, kerosene, and diesel. "Light ends" are fractions distilled under normal atmospheric conditions, while "heavy ends" include asphaltenes and lube oil. Light ends have fewer aromatic chemicals and light hydrocarbons. Heavy ends have higher molecular weight hydrocarbons, high-mass molecules, aromatic compounds, and organometallic substances. This portion of petroleum is comparatively rich in metals and compounds containing sulfur, nitrogen, and oxygen (Xu et al., 2018). Human activities are leading to an increase in the accumulation of heavy metals in the soil. This rise in heavy metal and metalloid levels is significantly impacting certain areas (Prince et al., 2013).

The main sources of hydrocarbons include petroleum and its combustion. These releases are classified as natural (from plants), petroleum-based, and combustion-related. Knowing where they come from is essential for recognizing the type of pollutant and how it spreads (Brooks, Gaskell, and Maltby, 2009). Hydrocarbons

 DOI: 10.1201/9781003508496-3

are introduced into the environment through processes such as dispersion, evaporation, dissolution, adsorption, and other mechanisms associated with both petroleum and its combustion (Christine, Jill, Stefan, Philip, and Sabrina, 2014; Jones, Magnin-Bissel, and Holladay, 2009). Sources originating from petroleum typically lead to groundwater contamination and pose environmental risks when substances like leaks of lubricants and fuels from tanks enter the surroundings (Oris and Roberts, 2013).

The US Environmental Protection Agency (USEPA) identified 16 important polycyclic aromatic hydrocarbons (PAHs) originating from petroleum, including series such as alkylated naphthalenes, dibenzothiophenes, fluorenes, phenanthrenes, and chrysenes (Alonso-Alvarez, Munilla, and López-Alonso, and Velando, 2007). Pyrogenic PAHs are generated through fuel combustion under high-temperature anaerobic conditions. Other significant sources of these hydrocarbons involve the pyrolysis of fats and inefficient combustion, particularly in power plants (Wu et al., 2019). Hydrocarbons and their by-products pose a substantial environmental concern due to their widespread use, contaminated methods of production, and high prevalence in aquatic environments (Correa Pabón, Souza Filho, and de Oliveira, 2019; Jones, 1990). Industrial activities and chemical plants are major sources of PAHs, categorized as both petrogenic and natural sources of these compounds. The impact caused by microfibers and microplastics is considerable (Kasar and Songachan, 2024). Through fat pyrolysis and inadequate burning methods, PAHs are anthropogenically emitted into the environment (Fowzia and Fakhruddin, 2018; Jyoti and Smita, 2012). PAH sources are classified as industrial, natural, household, farming, and transportation (Riihimäki and Savolainen, 1980). Hydrocarbons come from various sources such as forest fires, oil leaks, volcanic eruptions, and other natural events. They primarily result from chemical conversions of organic materials within microorganisms, fungi, plants, and sediments (Altomare et al., 2013; Peng et al., 2010).

Oil spills represent a significant source of hydrocarbon pollution, primarily resulting from accidents on oil platforms and ships during transportation. Additionally, spills can occur from the discharge of water used to clean tankers into the ocean (Abha and Singh, 2012). Crude oil and petroleum products create a water-resistant layer on the surface of water bodies, inhibiting oxygen exchange and posing risks to plants, animals, and humans (Dey, Tripathy, et al., 2023; Pradhan, Dey, Priyadarsini, Kumar, and Das, 2024). Remediation approaches are tailored based on the oil's composition and the specific physical and chemical characteristics of the affected area. Cleaning up environments contaminated with oil is challenging and contingent upon factors including the chemical and physical characteristics as well as the pH of the affected water or soil (Kanungo, Sahoo, Bal, and Behera, 2023). Management of oil spills typically involves chemical and biological methods, with chemical methods being more prevalent historically, although bioremediation is increasingly recognized worldwide. Recent advances in environmental clean-up have introduced various techniques, including mechanical, chemical, and biochemical methods for remediation.

Methods such as incineration, pyrolysis, and solvent extraction, which fall under physiochemical approaches, are often inefficient, impractical, and costly. Moreover, their use can lead to unpredictable changes in ecosystems due to residual chemicals, as well as transitional products and co-products, posing possible risks. In contrast, bioremediation emerges as a promising and sustainable solution for eliminating hydrocarbon pollutants, aligning well with the additional criteria for sustainable development (Liu, Sun, Cui, and Ding, 2019).

Bioremediation stands out as one of the most advanced and eco-friendly sustainable methods for responsibly managing hydrocarbon waste. This method is both eco-friendly and highly effective in breaking down a wide range of hydrocarbon pollutants (Mohanta, Pradhan, and Behera, 2023). Microbial activities play a crucial role in transforming these pollutants directly or indirectly into simpler molecules, thereby facilitating their remediation in the environment (Pradhan, Tripathy, Ram, Digal, and Das, 2023). Bioremediation, employing microbes to disintegrate oil and hydrocarbon pollutants, presents an eco-conscious and cost-efficient solution to mitigate these issues. The choice of geographical area is crucial for the success of in situ bioremediation. Bioaugmentation and biostimulation are key bioremediation methods. Biostimulation involves enriching the site with abundant nutrients, creating an optimal aerobic atmosphere, adjusting acidity, and temperature to boost microbial populations for improved biodegradation. Conversely, bioaugmentation entails introducing external microorganisms into the environment to accelerate biodegradation rates (Lange, Kotiukov, and Lebedeva, 2023).

Another approach is integrated treatment, combining biostimulation and bioaugmentation, which enhances functionality and applicability in ex situ treatment methods. Bioremediation shows promise as a technology for remediating soil polluted by PAHs (Adams, Fufeyin, Okoro, and Ehinomen, 2015; Blenis, Hue, Maaz, and Kantar, 2023). The treatment method demonstrated versatility in degrading diverse natural hydrocarbon contaminants such as petroleum pollutants, explosive compounds, agrochemicals, chlorophenols, and PAHs. Despite numerous biological solution approaches for diverse organic compounds, an in-depth review of cutting-edge technologies specifically targeting information on the most harmful petroleum hydrocarbons is scarce in the literature. Field investigations often provide insufficient indication of impactful long-term bioremediation, especially in challenging scenarios like superfund sites. Therefore, there is a focus on exploring the operational and eco-friendly features of various bioremediation techniques.

The investigation offers a detailed systematic comparison of the productivity of various remediation approaches to propose a comprehensive approach suitable for both in situ and ex situ conditions. Additionally, the chapter underscores the implication of fine-tuning microbial essentials for a successful and sustainable bioremediation technique.

3.2 HYDROCARBON REMEDIATION TECHNIQUES

Petroleum contamination is widespread within the ecosystem, affecting both surface and underground water (Balasubramaniam, Boyle, and Voulvoulis, 2007). Oil exploration, storage tank failures, and the manufacturing and delivery of oil all impact the environment (Nadim, Hoag, Liu, Carley, and Zack, 2000). Contamination threatens public safety and health concerns and can pollute surface and groundwater (Balasubramaniam et al., 2007).

Both national and international efforts aim to remediate hydrocarbon pollution, which poses ecological and public health risks. Three methods are used to clean up such contaminated sites (Abha and Singh, 2012; Dave and Ghaly, 2011). They are as follows:

1. Phytoremedial treatment
2. Biological remediation
3. Chemical-based remediation

3.3 PHYTOREMEDIAL TREATMENT

Phytoremedial treatment, which is also known as phytoremediation, is a method that employs plants to break down, extract, and eliminate contaminants from the water, air, and soil. It encompasses several mechanisms that can result in the breakdown of contaminants, their dispersion, immobilization, and accumulation (Kamath, Rentz, Schnoor, and Alvarez, 2007).

3.3.1 Mechanisms of Phytoremedial Treatment

Using plants for remediation is often more practical for cleaning contaminated land and water, as they involve various pollutant suppression methods compared to physical and chemical decontamination strategies (Huang, El-Alawi, Gurska, Glick, and Greenberg, 2005). Plants, owing to their stationary nature, have developed diverse abilities to manage hazardous compounds. They function as solar-powered pumping and filtration systems by absorbing pollutants from the soil via their roots. These pollutants are then dispatched to various plant parts, where they can be released, processed, or stored by plant tissues (Abhilash, Jamil, and Singh, 2009; Gerhardt, Huang, Glick, and Greenberg, 2009). Plants employ various mechanisms to remove pollutants from the soil, including both chemical and biophysical procedures. These mechanisms involve uptake, translocation, and transport, including mineralization and transformation through plant enzymes. Plants degrade halogenated substances such as TCE through oxidative degradation pathways, which involve plant-specific dehalogenases. Even after the plant dies, the activity of these dehalogenases continues to be maintained (Nzengung, Wolfe, Rennels, McCutcheon, and Wang, 1999). Plants contain a variety of contaminant-degrading enzymes, including laccases, P450 monooxygenases, peroxidases, dioxygenases, nitroreductases, phosphatases, nitrilases, and dehalogenases (Chaudhry, Blom-Zandstra, Gupta, and Joner, 2005; Susarla, Medina, and McCutcheon, 2002). Phytoremedial treatment in higher plants and associated microorganisms involves fundamental physiological mechanisms such as nutrient uptake, photosynthesis, water loss, and metabolic functions. The plant roots absorb both organic and inorganic constituents from the soil and can effectively bind and stabilize substances on their external surfaces through interactions with microorganisms in the rhizosphere. Additionally, molecules are taken up or released through gas exchange between the aerial parts of the plant and the atmosphere. To tackle various contaminants in different substrates, the Interstate Technology and Regulatory Cooperation has identified six phytotechnologies:

1. Phytotransformation is perfect for organic impurities across all types of substrata.
2. Rhizosphere bioremediation is pragmatic to soil with organic pollutants.
3. Phytostabilization is employed in soil to manage both inorganic and organic contaminants.
4. Phytoextraction is effective for substrates that contain inorganic pollutants.
5. Phytovolatilization is employed to address volatile matters.
6. Evapotranspiration can be used to regulate hydraulic flow in contaminated environments.

3.4 BIOLOGICAL REMEDIATION

Biological remediation, also known as bioremediation, is an economical technique for treating contaminated soil using native microbes. Effective land treatment involves managing appropriate moisture levels, adding nutrient fertilizers, optimizing microbial degradation through aeration and mixing, and adjusting the pH (Salanitro et al., 1997). Enzymes break down certain inorganic and the majority of organic substances via biological interactions. Biological remediation is a technique that utilizes these natural degradation processes to remove or neutralize pollutants from the environment (Ghosh et al., 2023).

Oil spills lead to soil contamination, which is a major global issue. Petroleum pollution has severe effects on human health, groundwater quality, agricultural productivity, and can result in economic and ecological problems. The toxicity of petroleum hydrocarbons impacts plants, animals, microorganisms, and humans. These spills and accidents often occur during the transportation of crude oil, typically via tankers on water or pipelines on land. The problem is exacerbated because countries that produce oil are not necessarily the top consumers, requiring the transportation of petroleum to other regions. Certain microorganisms can degrade petroleum hydrocarbons, using them as a basis of energy and carbon for their growth and maintenance. Soil contamination can be addressed through various methods, featuring both physicochemical and biological methods.

Many researchers have conducted out experiments to treat hydrocarbon-contaminated soil and have analyzed oily sludge (OS) obtained from refineries (Atagana, 2008), using modified techniques to address the pollutant sulfamethoxazole through wetland remediation. These efforts are essential for understanding how bioremediation affects the decontamination of hydrocarbon pollutants.

The biological remediation experiments were carried out on OMW sludge obtained from seven well-established evaporation ponds that were heavily contaminated with complex organic compounds (Martínez-Gallardo et al., 2020). The treatment method showed flexibility in breaking down various hydrocarbon-based organic contaminants, including explosives, chlorophenols, pesticides, and PAHs. Recent research utilizing aged field soil contaminated with PAHs from a gas production facility and the soil from a former diamond mining site has confirmed the viability of bioremediation (Atagana, 2008; Vasudevan and Natarajan, 2022).

Microorganisms are essential in the biodegradation of hydrocarbons, a complex process where microbial activities break down hydrocarbons into simpler forms. Both aerobic and anaerobic microorganisms degrade hydrocarbons mainly through enzyme activation and subsequent catalysis under optimized circumstances. The strain KA2 of *Acinetobacter radioresistens*, obtained from oily waste sludge, was used in a two-stage process that successfully reduced total petroleum hydrocarbons (TPHs) by 80% within 16 weeks, effectively remediating crude oil (Poorsoleiman, Hosseini, Etminan, Abtahi, and Koolivand, 2020) over 3 months achieved an 84% reduction in TPH within 16 weeks. Additionally, degradation of the contaminant from an OS was achieved using a culture-based medium containing *Enterobacter hormaechei* strain KA6. This in-vessel trial, carried out over 4 months, demonstrated a TPH removal rate of up to 80% (Atagana, 2008).

The performance criteria are affected by a range of abiotic and biotic factors, including microbial communities, aeration levels, moisture content, and temperature, among others. Furthermore, selecting the right method is essential for successful bioremediation. Various sequencing techniques are now available that facilitate the identification of novel microbes from unique and extreme environments. (Hussain et al. 2018). Advances in genome sequencing have enabled the swift identification and characterization of microbial strains (Nzila, 2018). It is important to recognize that chemometric techniques can enhance conditions for bioremediation, thus increasing the effectiveness of the breakdown process (Pandya and Kumar, 2021).

Commonly used chemometric techniques for optimizing the conditions of bioremediation include response surface methodology (RSM), principal component analysis (PCA), design of experiments (DoEs), artificial neural networks (ANNs), and genetic algorithms (GAs) (Oliveira et al. 2021; Wang et al., 2014).

Biological techniques are generally more cost-effective, environmentally friendly, sustainable and efficient compared to physicochemical methods often ensuring the full mineralization of the intended environmental contaminants. The speed at which petroleum products break down can be improved by developing various remediation strategies. Bioremediation using microorganisms is regarded as the most effective biological method, though the degradation of large molecular mass hydrocarbons is hindered by their low adsorption and solubility, limiting their availability to microorganisms (Dey, Bissoyi, Koteswara Reddy, and Das, 2024; Ray, Dey, Das, and John, 2024).

The eco-friendly bioremediation process is essential for tackling petroleum contaminants, hydrocarbon waste, and leakage. To enhance these efforts, scaling up next-generation techniques, such as enzyme and microbial modifications, along with microbial adsorption methods, is necessary for practical use. Improvements in binding processes and the application of restrained enzymes are anticipated to greatly improve biodegradation. Furthermore, the use of advanced recombinant tools and strategies can accelerate the bioremediation process. A deeper understanding of microbial mechanisms and their modes of action could reveal new target sites and enzymatic processes to further boost degradation rates. Additionally, combining biological and non-biological methods for hydrocarbon pollutant treatment should be pursued with a focus on cost efficiency.

3.4.1 Principle of Bioremediation

Crude oil is a complex blend of several chemical substances. Microbes identify oil and its components using bio emulsifiers and biosurfactants, which help them bind to the hydrocarbons and utilize them as a source of energy and carbon. However, high-molecular-weight hydrocarbons are less reachable to microorganisms because of their low adsorption and solubility. The addition of biosurfactants enhances the degradation rates of oil by increasing the solubility and removal of these pollutants. Oil components differ in their susceptibility, volatility, and biodegradability. While some substances degrade easily, others are resistant or nonbiodegradable. Different microbial species target various compounds, leading to varying rates of petroleum degradation. Microorganisms yield enzymes that assault hydrocarbon molecules when the source of carbon is presence. These enzymes and inculcation routes are crucial for degrading hydrocarbons, and a lack of suitable enzymes can hinder hydrocarbon degradation (Thapa, Kumar, and Ghimire, 2012). The bioremediation process involves using natural microorganisms to clean and restore contaminated environments. This procedure transforms impurities into harmless or beneficial matters using naturally occurring microbes such as bacteria, fungi, and yeast. This process also includes microorganisms improving environmental quality by decomposing and metabolizing chemical substances (Dave and Ghaly, 2011) (Table 3.1).

3.4.2 Microbes

3.4.2.1 Bacteria

Microbial species possess effective hydrocarbon breakdown capabilities in natural environment. Different microorganisms have been secluded from severely contaminated coastal regions, oil spills, and petroleum-contaminated soil. These isolates were selected based on their capability to break down various carbon sources, including

TABLE 3.1
List of Key Microorganisms Involved in Hydrocarbon Remediation

Yeast and Fungi	Bacteria
Aspergillus	*Achromobacter*
Rhodotorula	*Bacillus*
Fusarium	*Flavobacterium*
Cladosporium	*Alcaligenes*
Trichoderma	*Corynebacterium*
Penicillium	*Arthrobacter*
Sporobolomyces	*Brevibacterium*
Trichoderma	*Nocardia,*
Candida	*Pseudomonas*
	Vibrio
	Acinetobacter

aliphatic and aromatic complexes and their chlorinated by-products. Improvement culture techniques were employed to obtain these microorganisms, with selection criteria based on achieving the highest final cell density or specific growth rate. The range of microorganisms involved in degrading petroleum hydrocarbons includes fungi, microalgae, bacteria, and yeast. Among these, bacteria are particularly crucial for hydrocarbon degradation. Studies have reported that bacteria can rapidly degrade low molecular weight alkanes. The capacity of microorganisms to use hydrocarbons for cellular growth and energy requirements is a key factor driving petroleum biodegradation. In general, mixed cultures are more effective at biodegrading petroleum compared to pure cultures (Ghazali, Rahman, Salleh, and Basri, 2004). Many ecosystems have a sufficient indigenous microbial community capable of biodegrading oil, but effective oil degradation requires favorable environmental conditions. Indigenous microorganisms offer several advantages over introducing new microorganisms for hydrocarbon degradation. Sporadic addition of microbial consortia consisting of *Bacillus firmus*, *Shewanalla chilikensis*, and *Halomonas hamiltonii* can degrade the oil sludge hydrocarbons present in short duration of time by generating and sustaining the biosurfactant and biocatalysts in persistent level during the degradation process (Suganthi, Murshid, Sriram, and Ramani, 2018).

3.4.2.2 Fungi

In the decomposition of hydrocarbons in soils, fungi play a more critical role than bacteria. Filamentous fungi, commonly present in aquatic environments, are often associated with external part of the sediments and films. These fungi use enzymatic procedures similar to those found in mammals to break down PAHs.

Two primary types of cytochrome P450 monooxygenases are explicitly defined in yeasts and filamentous fungi. Many fungi can oxidize PAHs into phenols, dihydrodiols, and other metabolites, but only certain fungi, like *Phanerochaete chrysosporium*, can fully catabolize these compounds to CO_2. *Phanerochaete chrysosporium* is known for its complete mineralization of PAHs.

3.4.2.3 Yeast

The biodegradability of different yeasts follows this order: n-alkanes > branched alkanes > low molecular weight aromatic hydrocarbons > cycloalkanes > high-molecular-weight aromatic and polar compounds. The bioremediation method, which utilizes the diverse metabolic abilities of microorganisms, is an emerging method for removing contaminants from the environment and managing petroleum industry by-products (Medina-Bellver et al., 2005). Bioremediation is regarded as a cost-effective and non-invasive technique for removing petroleum and other hydrocarbon impurities from the environment. Microorganisms are the primary agents in this process, offering a more affordable alternative to other remediation technologies (Dey, Tripathy, et al., 2023; Nag et al., 2023). Essential for this method are microorganisms with suitable metabolic capabilities. Alkylaromatic breakdown is performed by several microorganisms, including *Arthrobacter, Sphingomonas, Mycobacterium, Burkholderia, Pseudomonas*, and *Rhodococcus*. Fungi, yeast, and bacteria all play key roles in the biological decomposition of hydrocarbons. For soil fungi, 6% (Jones, Knight, and Byrom, 1970) to 82% (Pinholt, Struwe, and Kjøller, 1979) of efficiency of biodegradation, 0.13% (Jones, Knight, and Byrom, 1970) to 50% (Pinholt, Struwe, and Kjøller, 1979) and 0.003%–100% (Mulkins-Phillips and Stewart, 1974) for marine bacteria for soil bacteria. Complex hydrocarbon mixtures, such as crude oil, in aquatic environments, freshwater, and soil, are broken down by diverse microbial communities with broad enzymatic capabilities.

Bioremediation consists of two main processes:

i. Bioaugmentation
ii. Biostimulation

3.4.2.3.1 Bioaugmentation

Bioaugmentation involves adding microorganisms to degrade harmful hydrocarbons, aiming to reduce pollutants (Sharma, 2012). It also involves injecting polluted water with microorganisms that are able to degrade hydrocarbons (Dave and Ghaly, 2011). This process may sometimes include the biodegradation of hydrocarbon contaminants by introducing genetically modified microorganisms into the contaminated water (Gentry, Rensing, and Pepper, 2004). Bioaugmentation is not frequently used for hydrocarbon breakdown due to microorganisms that can naturally break down hydrocarbons already present in the environment. Additionally, this method is often less efficient at oil spill remediation sites, particularly when involving non-native microorganisms (Dey, Das, Mallick, Sahu, and Das, 2023; Lahiri et al., 2021). Bioaugmentation, especially the application of high-density microbial consortia, holds significant potential for remediation and offers a more effective approach to environmental protection. Consequently, suggesting implementing bioaugmentation on a large scale in areas where environmental degradation is prevalent. In this process, organisms are systematically isolated from a particular contaminated site (such as soil) and then activated by culturing them under controlled conditions. Therefore, knowing the types of microbial species and contaminant levels is necessary (Mrozik and Piotrowska–Seget, 2010). This process has been considered by investigators as the most efficient intervention for freshly spilled hydrocarbon.

3.4.2.3.2 Biostimulation

Biostimulation involves adding nutrients to support indigenous hydrocarbon-degrading microbes, enhancing their ability to degrade harmful compounds (Sharma, 2012). The growth of microorganisms accountable for oil breakdown during spills is stimulated by increased carbon levels. Adding appropriate concentrations of supplemental nutrients boosts the microorganisms' ability to break down hydrocarbons (Behera et al., 2020, 2021, 2022). This leads to optimal growth rates and, consequently, the highest rates of contaminant removal (Boufadel, Suidan, and Venosa, 2006). Fullest biostimulation is obtained by providing the optimal nutrient intensity for the best microbial development and upholding this intensity for as long as feasible (Lee, Lee, and Kim, 2007).

3.5 CHEMICAL-BASED REMEDIATION

Chemical-based remediation, which is also known as chemical remediation process, involves using chemicals to treat contaminants, which can alter their chemical and physical properties (Vergetis, 2002). Chemical remediation is grouped into three categories: dispersants, solidifiers, and chemical oxidants (Abha and Singh, 2012; Dave and Ghaly, 2011).

3.5.1 Dispersants

Dispersants contain surfactants that shatter oil slicks into finer particles. These droplets then spread rapidly in the water, facilitating easier and faster degradation (Lessard and DeMarco, 2000). Chemical dispersants boost the surface exposure of oil droplets, boosting the rate of natural breakdown. This action reduces the oil's stickiness to surfaces, slows the formation of oil–water suspension, and enables faster intervention. This method reduces the harm of oil spills to marine life and living organisms by fragmenting oil slicks into droplets that bacteria can more easily degrade (Abha and Singh, 2012; Lessard and DeMarco, 2000). Examples of chemical dispersants include Nokomis 3-F4, Slickgone NS, Finasol OSR 52, SPC 1000™, Neon AB3000, ZI-400, Corexit 9500, Corexit 8667, and Saf-Ron Gold (USEPA, 2011).

3.5.2 Solidifiers

In this process, oil is physically removed by using dry granular materials that transform the oil from a liquid to a rubberlike solid. Solidifiers come in various forms, such as balls, pucks, and sponges. The solidification process can be improved in seawater due to the mixing energy it provides. However, solidifiers are problematic to recuperate after they have solidified the oil and are less efficient, which are significant drawbacks to their use (Fingas et al., 1995).

3.5.3 Chemical Oxidation

This methods involves using chemical substance that can oxidize organic pollutants. Chemical agents are introduced using agitation devices and injections into water or soil at polluted sites. The effectiveness of this process depends on the quality of the oxidant, the efficiency of contact among the contaminant and oxidant, geological circumstances, and the oxidant's dwelling time (Karpenko, Lubenets, Karpenko, and Novikov, 2008). This process is advantageous due to its speed and suitability for all weather conditions. Table 3.2 provides further details on its additional benefits and drawbacks.

3.6 MECHANICAL AND CHEMICAL REMEDIATION METHODS

3.6.1 Oil Discharged onto the Sea

There are several methods for removing oil from the sea surface and preventing it from reaching the coastline. The most commonly used techniques are mechanical retrieval and the application of emulsifier. Since crude oil has a lower density than water, it spreads quickly on the surface of the sea, forming a very thin film. The rate at which the oil spreads depends on factors such as oil type, atmospheric temperature, tides, water temperature, and wind. In the event of an oil trip, skimmers and booms can be used to contain the spill within a small area, allowing the oil to

TABLE 3.2
Merits and Demerits of Chemical Treatment

Chemical Treatment	Merits	Demerits
Dispersants	It is suitable for all weather conditions and effective for a wide range of oils. It accelerates oil degradation through natural processes, with advanced formulations addressing previous toxicity concerns. It requires less manpower and is more cost-effective than mechanical methods.	It does not facilitate oil recovery and is ineffective on highly viscous, non-spreading, or waxy oils. A localized and temporary rise in oil concentration in water could impact the surrounding aquatic life. If dispersion is not realized, the effectiveness of other response methods may be diminished due to the less dispersed oil.
Solidifiers	Quick in all weather conditions	Limited practical application, requires large quantities, depends on specific oil type, not effective, and offers no oil recovery.

be collected into a container. Biosurfactants, which are generally environmentally friendly, can be applied to solubilize the oil. For oil discharge mitigation in situ burning is sometimes employed as an alternative technique, but it is effective only when the oil is fresh, floating on the water surface, or contained within a limited space by booms (Kanungo et al., 2023). However, this method has the disadvantage of polluting the aquatic system with by-products and smoke from the burning oil. The effectiveness of clean-up methods depends on weather, tides, and ocean currents. If the oil reaches the coast, diverse techniques are used to clean gravel and sand. Oil sorbents, similar to sponges, can sometimes absorb the oil, and water washing is used to remove oil from vegetation, though severely damaged plants must be completely removed.

When the volume of polluted water is relatively small, an ex situ remediation technique is employed, where the impaired water is pumped to a production facility for treatment. Similarly, shore sand and gravel are extracted and purified at a separate location away from the polluted area.

3.6.2 Oil Spilled on Soil

Soil effluence can result from leaks in tubing and wellheads during offshore oil operations and drilling, underground leaks petroleum storage tanks, overflows from collection centers, petroleum by-products, and improper disposal of petroleum waste. This method poses significant risks during the mining, transport, and management of polluted materials. Finding new landfill sites for the final dumping of these substances is particularly challenging. Continuous monitoring and maintenance of separation barriers are required, as pollutants remain on-site, making the cap-and-containment technique a short-term solution. The following are methods for treating soil contamination:

1. In situ
2. Ex situ

3.6.3 In Situ Method

This process includes physicochemical methods such as air sprinkling, soil vapor extraction, or a combination of both applied to the polluted site. In situ treatment utilizes vertical and horizontal drilling equipment typically used for fossil fuels. This technique is more efficient on sandy soils compared to clay soils. Air sparging, also known as soil venting, is used to remove soil pollutants and promotes the development of aerobic bacteria by providing oxygen. Air sparging can also be conducted beneath the water table if contamination reaches groundwater, using extraction wells or through surface gravity sorting. Oil can be extracted from oil-saturated groundwater or partly saturated soil using a method known as slurping. Additionally, steam injection into contaminated soil can extract volatile substance confined within the soil.

3.6.4 Ex Situ Method

This approach entails removing and transporting contaminated soil to a remote remediation facility. Several methods are utilized for ex situ remediation, including:

1. **Land Farming**: In this process, oil-contaminated soil is exhumed and spread across a bed, where it is regularly turned or mixed to enhance the degradation of contaminants. Usually, a soil layer between 15 and 35 cm thick is treated using this method.
2. **Composting**: This technique enhances microbial growth by combining contaminated soil with inoffensive organic constituents, thereby accelerating the cessation of pollutants.
3. **Bioreactors**: These are used for the bioprocessing of polluted sediment, water and soil, by incessant fraternization of the soil, liquid, and gas segments to upsurge the degree of biodegradation. Prior to loading into bioreactors, the soil is pre-treated, and pollutants endure chemical reactions that change detrimental compounds into harmless ones. UV light or dechlorination is frequently used to catalyze oxidation reactions.

However, these methods have restrictions, such as high costs because of the complexity of the processes. In contrast, bioremediation, which utilizes natural biological activity, offers a more cost-effective solution for breaking down hydrocarbon contaminants.

3.7 RELEVANCE OF BIOREMEDIATION

1. The process is environmentally friendly and natural; the population of microorganisms rises in the presence of contaminants and decreases naturally as the pollutants break down. The by-products of this biological treatment, such as water, fatty acids, and carbon dioxide, are generally non-toxic, with the CO_2 produced being usable by plants for photosynthesis.
2. Bioremediation works by eliminating the targeted chemicals rather than merely moving the contaminants from one location to another.
3. Other methods for cleaning up hazardous waste are more expensive than bioremediation. For instance, during the Exxon Valdez leakage clean-up, the cost of one day of physical cleaning exceeded the cost of bioremediating 120 km of coastline.
4. Bioremediation involves in situ treatment, avoiding the necessity to transfer large quantities of polluted waste off-site, thereby eliminating the risks associated with movement.
5. The effectiveness of microbes in the bioremediation process can be improved by using nutrient formulations.
6. The by-products, such as CO_2, fatty acids, and water, produced through biological dealing are typically non-hazardous.

This technique is more cost-effective compared to other methods applied to clean up noxious waste. Hydrocarbons, because of their lower polarity compared to polar compounds like methanol, have reduced solubility. However, the degradation of hydrocarbons isn't solely determined by solubility. Various microorganisms, such as *Bacillus licheniformis, Pseudomonas aeruginosa, Pseudomonas putida, Bacillus laterosporus, Bacillus cereus, Bacillus subtilis*, and secrete emulsifiers that surge the external area of the substrate. Microorganisms also facilitate the absorption of hydrophobic substances by altering their cell surfaces. The physicochemical properties of hydrocarbons can be modified, which influences the behavior of both individual hydrocarbons and mixtures.

3.8 CONCLUSION

Hydrocarbon pollutants significantly impact land, aquatic, and atmospheric ecosystems. This issue has persisted since the advent of fossil fuel use and the industrial revolution. Rapid population growth, frequent oil spills, pipeline leaks, and widespread pesticide use have all contributed to a substantial rise in pollution. These factors collectively endanger the lives of animals and native microbial populations in the air, land, and water, both on the surface and below. Therefore, environmental remediation is crucial for human survival. This chapter emphasizes the application of advanced yet eco-friendly technologies for the sustainable management of hydrocarbon pollutants. It outlines various bioremediation techniques employed in pollution correction, including (1) phytoremediation, (2) bioremediation, and (3) chemical remediation. We can also conclude that numerous technologies are available for the clean-up or remediation of hydrocarbon-contaminated sites. However, each technique has its own advantages and disadvantages, and no single method provides a complete solution for the remedial needs of a specific type of pollution or location. As a result, each problem will likely require a combination of technologies to achieve full remediation.

REFERENCES

Abha S, Singh CS. (2012). Hydrocarbon pollution: Effects on living organisms, remediation of contaminated environments and effects of heavy metals co-contamination on bioremediation. In R-Z Laura (Ed.), *Introduction to Enhanced Oil on Recovery (EOR) Processes and Bioremediation of Oil Contaminated Sites*, 186–206. InTech Publisher.

Abhilash PC, Jamil S, Singh N. (2009). Transgenic plants for enhanced biodegradation and phytoremediation of organic xenobiotics. *Biotechnology Advances*, *27*, 474–488. https://doi.org/10.1016/j.biotechadv.2009.04.002

Adams GO, Fufeyin PT, Okoro SE, Ehinomen I. (2015). Bioremediation, biostimulation and bioaugmentation: A review. *International Journal of Environmental Bioremediation*, *3*, 28–39.

Alonso-Alvarez C, Munilla I, López-Alonso M, Velando A. (2007). Sublethal toxicity of the prestige oil spill on yellow-legged gulls. *Environment International*, *33*, 773–781. https://doi.org/10.1016/j.envint.2007.02.012

Altomare DF, Di Lena M, Porcelli F, Trizio L, Travaglio E, Tutino M, et al. (2013). Exhaled volatile organic compounds identify patients with colorectal cancer. *The British Journal of Surgery*, *100*(1), 144–150. https://doi.org/10.1002/bjs.8942

Atagana HI. (2008). Compost bioremediation of hydrocarbon-contaminated soil inoculated with organic manure. *African Journal of Biotechnology*, *7*, 1516–1525.

Balasubramaniam A, Boyle AR, Voulvoulis N. (2007). Improving petroleum contaminated land remediation decision making through the MCA weighting process. *Chemosphere*, *66*, 791–798. https://doi.org/10.1016/j.chemosphere.2006.06.039

Behera ID, Nayak M, Biswas S, Meikap BC, Sen R. (2021). Enhanced biodegradation of total petroleum hydrocarbons by implementing a novel two-step bioaugmentation strategy using indigenous bacterial consortium. *Journal of Environmental Management*, *292*, 112746.

Behera ID, Nayak M, Mishra A, Meikap BC, Sen R. (2022). Strategic implementation of integrated bioaugmentation and biostimulation for efficient mitigation of petroleum hydrocarbon pollutants from terrestrial and aquatic environment. *Marine Pollution Bulletin*, *177*, 113492.

Behera ID, Das AP. (2023). *Impact of petroleum waste on environmental pollution and its sustainable management through circular economy*. Springer.

Blenis N, Hue N, Maaz TM, Kantar M. (2023). Biochar production, modification, and its uses in soil remediation: A review. *Sustainability*, *15*, 3442.

Boufadel MC, Suidan MT, Venosa AD. (2006). Tracer studies in laboratory beach simulating tidal influences. *Journal of Environmental Engineering*, *132*, 616–623.

Brooks AC, Gaskell PN, Maltby LL. (2009). Importance of prey and predator feeding behaviours for trophic transfer and secondary poisoning. *Environmental Science & Technology*, *43*, 7919–7923. https://doi.org/10.1021/es900747n

Chaudhry Q, Blom-Zandstra M, Gupta SK, Joner E. (2005). Utilizing the synergy between plants and rhizosphere microorganisms to enhance breakdown of organic pollutants in the environment. *Environmental Science and Pollution Research*, *12*, 34–38. https://doi.org/10.1065/espr2004.08.213

Christine MBB, Jill AO, Stefan W, Philip CS, Sabrina S.T. (2014). Effects of oil on terrestrial vertebrates: Predicting impacts of the Macondo blowout. *Bioscience*, *64*, 820–828.

Correa Pabón RE, Souza Filho CR, de Oliveira WJ. (2019). Reflectance and imaging spectroscopy applied to detection of petroleum hydrocarbon pollution in bare soils. *Science of the Total Environment*, *649*, 1224–1236. https://doi.org/10.1016/j.scitotenv.2018.08.231

Dave D, Ghaly AE. (2011). Remediation technologies for marine oil spills: A critical review and comparative analysis. *American Journal of Environmental Sciences*, *7*, 423–440.

Dey S, Tripathy B, Kumar MS, Das AP. (2023). Ecotoxicological consequences of manganese mining pollutants and their biological remediation. *Environmental Chemistry and Ecotoxicology*, *5*, 55–61.

Dey S, Bissoyi A, Koteswara Reddy G, Das AP. (2024). Synthetic microfibers: An urban emerging pollutant demanding for sustainable environmental solutions. In *Sustainable Microbial Technology for Synthetic and Cellulosic Microfiber Bioremediation*, 107–121. https://doi.org/10.1007/978-3-031-62600-5_7

Dey S, Das A, Mallick K, Sahu A, Das AP. (2023). Environmental Petroleum Waste: Pollution, Toxicity, Sustainable Remediation. 159–175. https://doi.org/10.1007/978-3-031-48220-5_7

Fingas MF, Kyle DA, Larouche N, Fieldhouse B, Sergy G, Stoodley G. (1995). Effectiveness testing of oil spill-treating agents. In P Lane (Ed.), *The Use of Chemicals in Oil Spill Response*, 286–298. ASTM International.

Fowzia A, Fakhruddin ANM. (2018). A review on environmental contamination of petroleum hydrocarbons and its biodegradation. *International Journal of Environmental Sciences & Natural Resources*, *11*(3), 555811.

Gentry TJ, Rensing C, Pepper IL. (2004). New approaches for bioaugmentation as a remediation technology. *Critical Reviews in Environmental Science and Technology*, *34*, 447–494.

Gerhardt KE, Huang XD, Glick BR, Greenberg BM. (2009). Phytoremediation and rhizoremediation of organic soil contaminants: Potential and challenges. *Plant Science*, *176*, 20–30. https://doi.org/10.1016/j.plantsci.2008.09.014

Ghazali MF, Rahman RNZA, Salleh AB, Basri M. (2004). Biodegradation of hydrocarbons in soil by microbial consortium. *International Biodeterioration and Biodegradation*, *54*, 61–67. https://doi.org/10.1016/j.ibiod.2004.02.002

Ghosh S, Sinha JK, Ghosh S, Vashisth K, Han S, Bhaskar R. (2023). Microplastics as an emerging threat to the global environment and human health. *Sustainability, 15*(14), 10821. https://doi.org/10.3390/SU151410821

Huang XD, El-Alawi YS, Gurska J, Glick BR, Greenberg B. (2005). A multiprocess phytoremediation system for decontamination of persistent total petroleum hydrocarbons (TPHs) from soils. *Microchemical Journal*, *81*, 139–147. https://doi.org/10.1016/j.microc.2005.01.009

Hussain F, Hussain I, Khan AHA, Muhammad YS, Iqbal M, Soja G, Reichenauer TG, Yousaf S. (2018). Combined application of biochar, compost, and bacterial consortia with Italian ryegrass enhanced phytoremediation of petroleum hydrocarbon contaminated soil. *Environmental and Experimental Botany*, *153*, 80–88.

Jones DE, Magnin-Bissel G, Holladay SD. (2009). Detection of polycyclic aromatic hydrocarbons in the shed skins of corn snakes (Elaphe guttata). *Ecotoxicology and Environmental Safety*, *72*, 2033–2035. https://doi.org/10.1016/j.ecoenv.2008.11.001

Jones JG, Knight M, Byrom JA. (1970). Effect of gross population by kerosene hydrocarbons on the microflora of a moorland soil. *Nature*, *227*, 1166.

Jones RD. (1990). Xylene/amitraz: A pharmacologic review and profile. *Veterinary and Human Toxicology*, *32*(5), 446–448.

Jyoti L, Smita C. (2012). Effect of diesel fuel contamination on seed germination and growth of four agricultural crops. *Universal Journal of Environmental Research and Technology*, *2*(4), 311–317.

Kamath R, Rentz JA, Schnoor JL, Alvarez PJJ. (2007). Phytoremediation of hydrocarbon contaminated soils: Principles and applications. In Q-RR Vazquez-Duhalt (Ed.), *Petroleum Biotechnology: Developments and Perspectives (Studies in Surface Science and Catalysis)* (151st ed.), 447–478. Elsevier.

Kanungo J, Sahoo T, Bal M, Behera ID. (2023). Performance of bioremediation strategy in waste lubricating oil pollutants: A review. *Geomicrobiology Journal*, *41*(4), 360–373. https://doi.org/10.1080/01490451.2023.2245395

Karpenko O, Lubenets V, Karpenko E, Novikov V. (2008). Chemical oxidants for remediation of contaminated soil and water. A review. *Chemistry & Chemical Technology*, *3*, 41–45.

Kasar P, Songachan LS. (2024). Microfiber waste management and recycling with zero waste adaptation technology. In NP Das, AP Behera, ID Das (Eds.), *Renewable Energy Generation and Value Addition from Environmental Microfiber Pollution through Advanced Greener Solution. Environmental Science and Engineering*, 231–259. Springer. https://doi.org/10.1007/978-3-031-51792-1_12

Lahiri D, Nag M, Dey A, Sarkar T, Joshi S, Pandit S, Das AP, Pati S, Pattanaik S, Tilak VK, Ray RR. (2021). Biofilm mediated degradation of petroleum products. *Geomicrobiology Journal*, 1–10. https://doi.org/10.1080/01490451.2021.1968979

Lange I, Kotiukov P, Lebedeva Y. (2023). Analyzing physical-mechanical and hydrophysical properties of sandy soils exposed to long-term hydrocarbon contamination. *Sustainability*, *15*, 3599.

Lee SH, Lee S, Kim DY, Kim J-G. (2007). Degradation characteristics of waste lubricants under different nutrient conditions. *Journal of Hazardous Materials*, *143*, 65–72.

Lessard RR, DeMarco G. (2000). The significance of oil spill dispersants. *Spill Science and Technology Bulletin*, *6*, 59–68.

Liu S, Sun S, Cui P, Ding Y. (2019). Molecular modification of fluoroquinolone-biodegrading enzymes based on molecular docking and homology modelling. *International Journal of Environmental Research and Public* Health, *16*, 3407.

Martínez -Gallardo MR, López MJ, Jurado MM, Suárez-Estrella F, López-González JA, Sáez JA, Moral R, Moreno J. (2020). Bioremediation of Olive Mill Wastewater sediments in evaporation ponds through in situ composting assisted by bioaugmentation. *Science of the Total Environment*, *703*, 135537.

Medina -Bellver JI, Marin P, Delgado A, Rodríguez-Sánchez A, Reyes E, Ramos JL, Marqués S. (2005). Evidence for in situ crude oil biodegradation after the prestige oil spill. *Environmental Microbiology*, *7*, 773–779. https://doi.org/10.1111/j.1462-2920.2005.00742.x

Mohanta S, Pradhan B, Behera ID. (2023). Impact and remediation of petroleum hydrocarbon pollutants on agricultural land: A review. *Geomicrobiology Journal*, *41*(4), 345–359. https://doi.org/10.1080/01490451.2023.2243925

Mrozik AM, Piotrowska–Seget Z. (2010). Bioaugmentation as a strategy for cleaning up of soils contaminated with aromatic compounds. *Microbiological Research*, *165*, 363–375.

Mulkins -Phillips GJ, Stewart J. (1974). Distribution of hydrocarbon utilizing bacteria in Northwestern Atlantic waters and coastal sediments. *Canadian Journal of Microbiology*, 955–962. https://doi.org/10.1139/m74-147

Nadim F, Hoag GE, Liu S, Carley RJ, Zack P. (2000). Detection and remediation of soil aquifer systems contaminated with petroleum products: An overview. *Journal of Petroleum Science and Engineering*, *26*, 169–178. https://doi.org/10.1016/S0920-4105(00)00031-0

Nag M, Lahiri D, Ghosh S, Sarkar T, Pati S, Das AP, Ram DK, Bhattacharya D, Ray RR. (2023). Application of microorganisms in biotransformation and bioremediation of environmental contaminant: A review. *Geomicrobiology Journal*. https://doi.org/10.1080/01490451.2023.2261443

Nzengung VA, Wolfe LN, Rennels DE, McCutcheon SC, Wang C. (1999). Use of aquatic plants and algae for decontamination of waters polluted with chlorinated alkanes. *International Journal of Phytoremediation*, *1*, 153–226. https://doi.org/10.1080/15226519908500016

Nzila A. (2018). Current status of the degradation of aliphatic and aromatic petroleum hydrocarbons by thermophilic microbes and future perspectives. *International Journal of Environmental Research and Public Health*, *15*, 2782.

Oliveira LG, Araújo KC, Barreto MC, Bastos MEP, Lemos SG, Fragoso WD. (2021). Applications of chemometrics in oil spill studies. *Microchemical Journal*, *166*, 106216.

Oris JT, Roberts AP. (2013). Cytochrome P450 1A (CYP1A) as a biomarker. In JA Wiens (Ed.), *Oil Spill Assessments*, 201–219. Cambridge University Press.

Pandya DK, Kumar MA. (2021). Chemo-metric engineering designs for deciphering the biodegradation of polycyclic aromatic hydrocarbons. *Journal of Hazardous Materials*, *411*, 125154.

Peng G, Hakim M, Broza YY, Billan S, Abdah-Bortnyak R, Kuten A, Tisch U, Haick H. (2010). Detection of lung, breast, colorectal, and prostate cancers from exhaled breath using a single array of nanosensors. *British Journal of Cancer*, *103*(4), 542–551. https://doi.org/10.1038/ sj.bjc.6605810

Pinholt Y, Struwe S, Kjøller A. (1979). Microbial changes during oil decomposition in soil. *Holarctic Ecology*, *2*, 195–200. https://doi.org/10.1111/j.1600-0587.1979.tb00701.x

Poorsoleiman MS, Hosseini SA, Etminan A, Abtahi H, Koolivand A. (2020). Effect of two-step bioaugmentation of an indigenous bacterial strain isolated from oily waste sludge on petroleum hydrocarbons biodegradation: Scaling-up from a liquid mineral medium to a two-stage composting process. *Environmental Technology & Innovation*, *17*, 100558.

Pradhan G, Dey S, Priyadarsini S, Kumar MS, Das AP. (2024). Characterization and quantification of microplastics pollutants in sediment samples from Daya River of Odisha State in India for their appropriate management. In *Environmental Science and Engineering, Part F2333*, 75–92. https://doi.org/10.1007/978-3-031-51792-1_5

Pradhan G, Tripathy B, Ram DK, Digal AK, Das AP. (2023). Bauxite mining waste pollution and its sustainable management through bioremediation. *Geomicrobiology Journal*, 1–10. https://doi.org/10.1080/01490451.2023.2235353

Prince RC, McFarlin KM, Butler JD, Febbo EJ, Wang FC, Nedwed TJ. (2013). The primary biodegradation of dispersed crude oil in the sea. *Chemosphere*, *90*(2), 521–526. https://doi.org/10.1016/j.chemosphere.2012.08.020

Ray SK, Dey S, Das AP, John A. (2024). Microplastic waste management through greener technologies: A review. In *Sustainable Management of Urban Plastic Waste Through Circular Economic Approaches*, 237–261. https://doi.org/10.1201/9781003465447-14

Riihimäki V, Savolainen K. (1980). Human exposure to m-xylene. Kinetics and acute effects on the central nervous system. *The Annals of Occupational Hygiene*, *23*(4), 411–422.

Suganthi SH, Murshid S, Sriram S, Ramani K. (2018). Enhanced biodegradation of hydrocarbons in petroleum tank bottom oil sludge and characterization of biocatalysts and biosurfactants. *Journal of Environmental Management*, *220*(December 2017), 87–95. https://doi.org/10.1016/j.jenvman.2018.04.120

Susarla S, Medina VF, McCutcheon SC. (2002). Phytoremediation: An ecological solution to organic chemical contamination. *Ecological Engineering*, *18*, 647–658. https://doi.org/10.1016/S0925-8574(02)00026-5

Thapa B , Kumar KCA, Ghimire A. (2012). A review on bioremediation of petroleum hydrocarbon contaminants in soil. *Kathmandu University Journal of Science, Engineering and Technology*, *8*, 164–170. https://doi.org/10.3126/kuset.v8i1.6056

USEPA. (2011). *National Contingency Plan Product Schedule*. Washington, DC: US Environmental Protection Agency.

Vasudevan M, Natarajan N. (2025). Towards achieving sustainable bioplastics production and nutrient recovery from wastewater—A comprehensive overview on polyhydroxybutyrate. *Biomass Conversion and Biorefinery*, *15*(2), 1729–1748.

Vergetis E. (2002). *Oil Pollution in Greek Seas and Spill Confrontation Means Methods*.

Wang C, Chen S, Wang Z. (2014). Electrophysiological follow-up of patients with chronic peripheral neuropathy induced by occupational intoxication with n-hexane. *Cell Biochemistry and Biophysics*, *70*(1), 579–585. https://doi.org/10.1007/s12013-014-9959-7

Wu Q, Liu Z, Liang J, Kuo DTF, Chen S, Hu X, Deng M, Zhang H, Lu Y. (2019). Assessing pollution and risk of polycyclic aromatic hydrocarbons in sewage sludge from wastewater treatment plants in China's top coal-producing region. *Environmental Monitoring and Assessment*, *191*(2), 102. https://doi.org/10.1007/s10661-019-7225-6

Xu X, Liu W, Tian S, Wang W, Qi Q, Jiang P, Gao X, Li F, Li H, Yu H. (2018). Petroleum hydrocarbon degrading bacteria for the remediation of oil pollution under aerobic conditions. *Frontiers in Microbiology*, *9*(2885), 1–11. https://doi.org/10.3389/fmicb.2018.02885

4 Advanced Biological Technologies for the Transformation and Detoxification of Petroleum Hydrocarbon Waste

Nilambar Bariha and Sunny Kumar

4.1 INTRODUCTION

Exploration of crude oil has increased the disposal of hydrocarbon waste, and the treatment of sludge has been one of the major challenges in recent years. Currently, various methods are used in the treatment, such as stabilization/solidification, oxidation, electrocoagulation, biodegradation, and bioremediation technologies used for the recovery of oil sludge (Johnson & Affam, 2019; Dey et al., 2023; Lahiri et al., 2021). The large quantities of oil sludge are composed of rebellious compounds that may have a severe impact on the environment and build up technology. India's rapidly growing petroleum industry generates over 28,000 tons of oil sludge annually (Roy et al., 2018), making it a significant environmental challenge. This waste is primarily a byproduct of the refining process. The petroleum industry generates wastewater, which consists of a range of pollutants such as petroleum hydrocarbons, mercaptans, oil and grease, phenol, ammonia, sulfide, and some organic compounds. These compounds, present in complex forms within the industry's discharged water, pose direct and indirect threats to the environment (Das, Dey, et al., 2024; Mishra et al., 2022; Mishra & Das, 2021). Several techniques currently used to treat oily waste include membrane technology, photocatalytic degradation, advanced oxidation processes, and electrochemical catalysis, among others (Varjani et al., 2020).

In this chapter, our aim is to explore different scenarios for applying various treatment technologies to manage petroleum industry waste, incorporating physical, chemical, and biological processes.

Other advanced modified operations improve the treatment and disposal of petroleum sludge, such as solvent extraction, centrifugation, surfactant-enhanced oil recovery, freeze/thaw, pyrolysis, microwave irradiation, electro-kinetic techniques, ultrasonic irradiation, and froth flotation (Da Silva et al., 2012).

 DOI: 10.1201/9781003508496-4

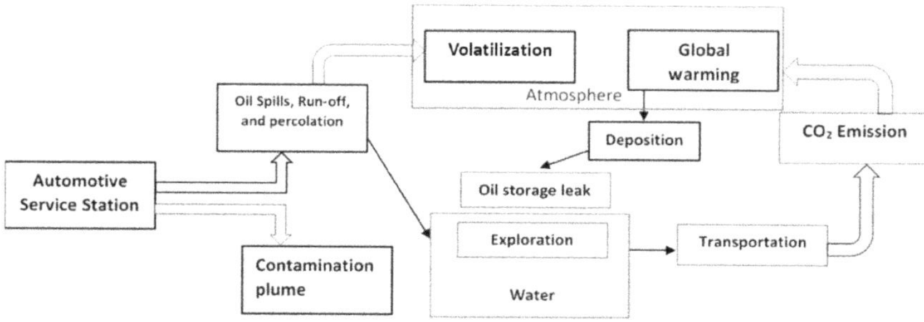

FIGURE 4.1 Sources of different types of waste generated and their complexity in petroleum industries.

Crude oil wastes, such as spills from pipelines, trains, trucks, tanks, and other sources, are considered hazardous due to their high content of benzene, related organic compounds, and heavy metals like arsenic, cadmium, chromium, lead, mercury, and selenium (Priyadarsini & Das, 2024, 2025; Sattar et al., 2022). Hydrocarbons and their waste aggregate in various states, such as emulsions, solids, and slurries. The exploration, processing, and transportation of these substances can harm the environment, aquatic and terrestrial life, and human health (Das et al., 2024a).

The automotive service station generates oil spill, runoff, percolation in water, and contamination plumes in the environment. The oil spill can generate volatile vapors (through volatilization) and can impact the deposition or leakage of storage tanks. This volatilization and waste generation can increase the emission of carbon dioxide or other hazardous gases, which are a consequence of global warming, as shown in Figure 4.1. The major composition of petroleum wastes primarily consists of water (55.2%), light hydrocarbons (23.2%), waxes (10.5%), clay (9%), and asphaltenes (1.9%). However, the exact composition can vary depending on the region's geological and biological factors (Sattar et al., 2022). For example, the Deepwater Horizon spill of 2010 released over 600 million L of oil into the Gulf of Mexico, causing significant damage to marine life (Jabbar et al., 2022; Kleindienst et al., 2015). Recovery efforts took months

4.2 CHALLENGES TO USING DIFFERENT TECHNOLOGIES FOR THE TREATMENT OF PETROLEUM HYDROCARBON WASTE

The treatment of petroleum hydrocarbon waste presents several challenges, including the complexity of waste composition, the high cost of treatment technologies, the limited availability of appropriate technologies, the potential for secondary waste generation, regulatory compliance, and public perception. Petroleum hydrocarbon waste is complex and contains a wide range of contaminants that require specialized treatment technologies. The composition of the waste can vary widely depending on the source and can include a mix of hazardous and non-hazardous waste (Das, Ray, et al., 2024; Mohanty et al., 2018; Tripathy et al., 2024). The cost of treatment technologies for petroleum hydrocarbon waste can be high, particularly for advanced technologies such as bioremediation, thermal treatment, and chemical oxidation.

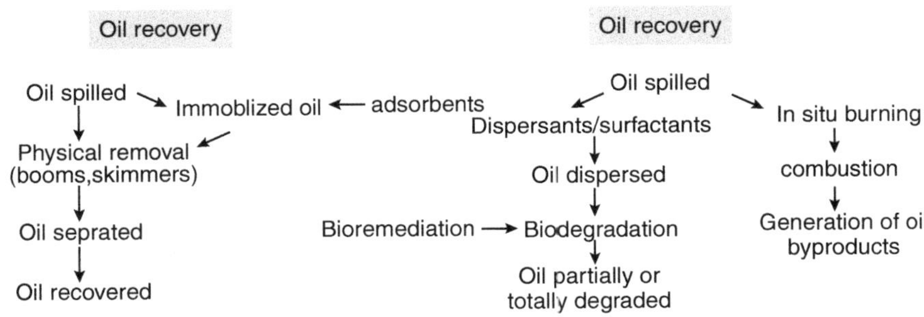

FIGURE 4.2 Oil spill treatment methods based on oil recovery or degradation (Silva et al., 2022).

The high cost can be a barrier to the adoption of these technologies, particularly in developing countries. There are limited technologies available that are appropriate for treating certain types of petroleum hydrocarbon waste. For example, some technologies may be effective for treating contaminated soil but may not be suitable for treating contaminated groundwater. Some treatment technologies can generate secondary waste, such as ash or sludge, which may require further treatment or disposal. This can increase the overall cost and complexity of the waste management process.

Oil recovery is a complicated task because it requires liquid–liquid separation. The spilled oil requires physical removal methods such as booms, skimmers, and adsorbents (Gaaseidnes & Turbeville, 1999). After that, separate the oil from the physical systems and recover it. Another method for oil recovery is using dispersants or surfactants and in situ burning. The surfactants can help lower the surface tension of dispersed oil and the bioremediation method can partly or totally degrade the oil (Ramirez & Collins, 2018). In situ burning is a process of full or partial combustion that generates oil byproducts as part of the oil recovery treatment, as illustrated in Figure 4.2.

4.2.1 Bioremediation Technologies

Bioremediation is a promising technology for the treatment of petroleum waste. It involves the use of microorganisms to break down and degrade contaminants in the waste, converting them into harmless substances. There are various bioremediation technologies such as in situ, ex situ, biopile, and composting (Jabbar et al., 2022; Ossai et al., 2022) as represented in Figure 4.3. In situ bioremediation involves the treatment of petroleum waste at the site where it is located, without the need for excavation or removal (Das et al., 2015; Dey, Dash, & Das, 2024; Nag et al., 2023; Paul et al., 2023). This method involves the introduction of microbial populations that can break down and degrade petroleum contaminants in the soil or groundwater. The in situ method also involves attenuation, bioaugmentation, biostimulation, and phytoremediation. Ex situ bioremediation involves the removal of petroleum waste from the site and its treatment in a separate facility. This method involves the use of bioreactors, which are containers that provide a controlled environment for microbial activity to take place. Landfarming is a type of ex situ bioremediation that involves the spreading of petroleum waste on

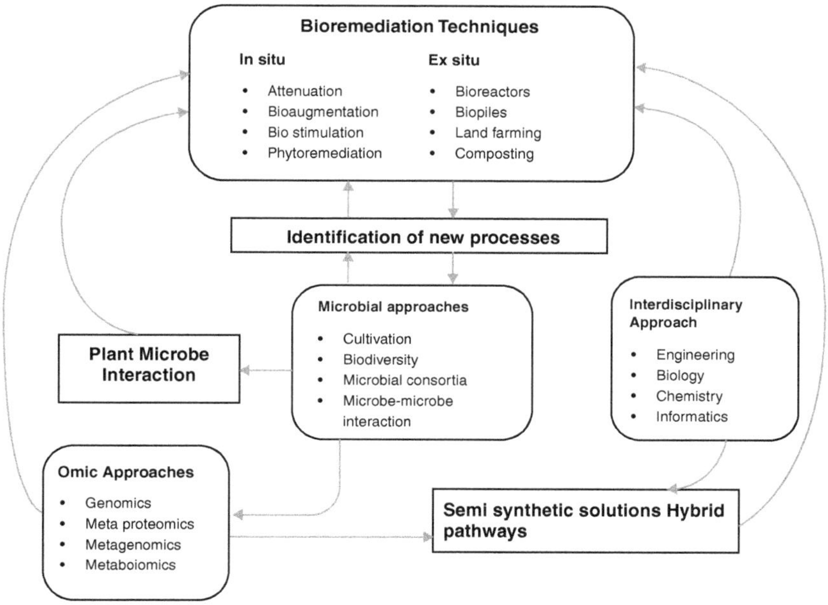

FIGURE 4.3 Bioremediation technique for petroleum waste (Singh et al., 2016).

the surface of the soil and the addition of nutrients and microorganisms to promote the degradation of the contaminants (Bala et al., 2024; Bhattacharjee et al., 2021; Dey, Bissoyi, et al., 2024; Dey, Dash, Ray, et al., 2024). The treated soil can then be returned to its original location. A biopile is a type of ex-situ bioremediation that involves the excavation of contaminated soil and its placement in a lined containment area. Nutrients and microorganisms are added to the soil to promote degradation, and the treated soil is then returned to its original location. Composting is a type of ex situ bioremediation that involves the mixing of petroleum waste with organic materials, such as yard waste or food waste, to create a compost mixture. The compost mixture is allowed to degrade over time, with the help of microorganisms, into a nutrient-rich soil amendment. Microbial cultures developed from petroleum waste can be consumed by these microorganisms (Das, Nayak, et al., 2024; Das & Ghosh, 2023b; Satapathy et al., 2024). This will be converted into plant nutrition and utilized by the plants for their growth. There is a challenge in that it is very difficult to identify which molecules are converted into nutrition. There are various factors that affect the efficiency of the process to convert petroleum waste. Further treatment is done by an omics approach, which involves the genomics, metaproteomics, metagenomics, and metabolomics (Das & Ghosh, 2023c, 2023a).

4.2.2 Electrocoagulation

Electrocoagulation is an electrochemical process that can be used for the treatment of petroleum waste (Moneer et al., 2023). This process involves the use of an electrical current to cause the formation of metal hydroxide flocs that can trap and remove contaminants from the waste stream (Kadhim & Abdulredha, 2021). It is a promising

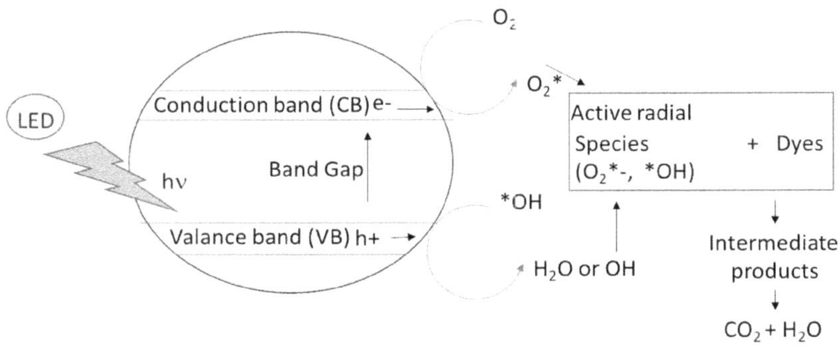

FIGURE 4.4 Mechanism of photocatalytic degradation.

technology for the treatment of petroleum waste, particularly in situations where other treatment methods may not be effective or cost-efficient (Akkaya, 2016).

4.3 PHOTODEGRADATION TECHNIQUE FOR PETROLEUM WASTE

Photodegradation is a treatment technique that uses ultraviolet (UV) radiation to break down petroleum waste into simpler, less harmful compounds. The petroleum waste is exposed to UV radiation, which emits high-energy light that can break down the chemical bonds in petroleum compounds (Behera et al., 2020, 2021, 2022). It causes the formation of reactive oxygen species (ROS) such as hydroxyl radicals, as shown in Figure 4.4. These ROS react with petroleum compounds such as aliphatic and phenolic (Tetteh et al., 2020), breaking them down into simpler compounds such as water and carbon dioxide that are less harmful to the environment. TiO_2 is a widely used catalyst for the photodegradation of petroleum waste in various reactors. When exposed to UV radiation, TiO_2 can generate ROS, which can break down petroleum compounds into simpler, less harmful compounds (Rabahi et al., 2019). ZnO is another catalyst that can enhance the photodegradation of petroleum waste (Singh, 2016). Similar to TiO, ZnO can generate ROS when exposed to UV radiation, which can improve the degradation of petroleum compounds. Fe_2O_3 can act as a catalyst for the photodegradation of petroleum waste by generating ROS and enhancing the degradation of petroleum compounds. Fe_2O_3 can also be used in combination with other catalysts, such as TiO_2, to improve the efficiency of photodegradation. Carbon-based materials, such as activated carbon, graphene, and carbon nanotubes, can also be used as catalysts for the photodegradation of petroleum waste. These materials can absorb petroleum compounds and enhance the formation of ROS, leading to more efficient degradation of the waste.

4.4 OPPORTUNITY FOR ECONOMIC BENEFIT AND SUSTAINABLE ENVIRONMENT

The clean environment can reduce health regulations and attract investment. It can be used for sustainable development through integrated environmental, health, safety, risk, and social consulting and sustainability-related services (Azuazu et al., 2023).

Petroleum waste presents significant environmental challenges, but it also offers opportunities for economic benefits and a sustainable environment. Through various approaches such as recycling, energy recovery, bioremediation, carbon capture, and the circular economy, petroleum waste can be managed effectively to reduce environmental impacts and create economic opportunities (Mohanty et al., 2024; Ray et al., 2024).

Petroleum waste can be recycled and re-refined to produce base oils that can be used as lubricants. This process not only reduces waste but also helps to conserve natural resources, and it can create job opportunities in the recycling and re-refining industries (Gozbenko et al., 2019). Petroleum waste can also be used to generate energy through incineration or gasification. This process reduces waste and produces energy that can be used to power homes and businesses. It can reduce greenhouse gas emissions by displacing fossil fuels. Petroleum waste can be treated through bioremediation, which uses microorganisms to break down contaminants (Kanungo et al., 2023a, b). This process can remediate contaminated soil and water, and it can also create opportunities for biotechnology and environmental consulting firms. Petroleum waste can also be used to capture carbon dioxide emissions from industrial processes. The captured carbon dioxide can then be used for enhanced oil recovery, which increases the production of oil and gas while reducing greenhouse gas emissions (Wienchol et al., 2020). The circular economy approach promotes the reduction, reuse, and recycling of waste. By adopting this approach, petroleum waste can be converted into new products, which can generate revenue and create a sustainable environment.

4.5 WASTE MANAGEMENT, HANDLING, AND UTILIZATION FOR USEFUL PRODUCTS

There are millions of tons of petroleum waste generated every year. The handling and management of the waste are significant problems. Waste management involves several steps, including collection, treatment, disposal, and utilization for beneficial products (Qin et al., 2009). The waste can be collected at an individual level and gathered in one place. Furthermore, it should be treated and transported within a week. Following treatment, beneficial products are used for various purposes. In addition to disposal, there are various ways to utilize and extract useful products from crude oil waste, which can help reduce the environmental impact and provide valuable resources. It is important to choose the appropriate method of waste management based on the type and nature of the waste, as well as the potential environmental impact. Recycling is one of the most effective ways of managing crude oil waste and gives rise to the idea of a waste refinery, as shown in Figure 4.5. Used oil can be re-refined, which removes contaminants and produces high-quality base oil that can be used for lubrication purposes. The re-refined oil can be used in engines, hydraulic systems, and other machinery (Asim et al., 2021). There is another method of incineration for managing crude oil waste in which the waste is burned at high temperatures, which destroys the contaminants and converts the waste into ash. Incineration can be an effective way of managing hazardous waste, but it can also release toxic gases into the air. Landfill disposal involves burying the waste at a designated landfill site.

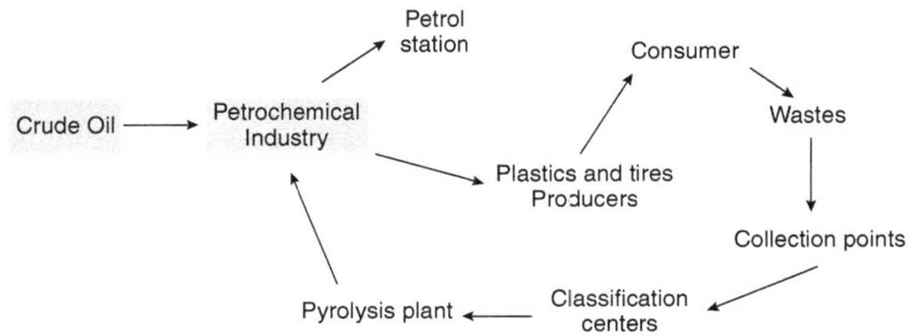

FIGURE 4.5 Waste generation and utilization from the crude oil refinery (Palos et al., 2021).

This method can be effective for non-hazardous waste, but it can also lead to contamination of the soil and groundwater. Bioremediation is a process that involves the use of microorganisms to break down contaminants, as discussed in the technology section (Das, Dey, et al., 2024). There are various useful products that can be extracted from crude oil waste, such as asphalt, bitumen, fuel oils, and waxes. These products can be extracted from waste and used for road construction.

4.6 REGULATORY CHALLENGES FOR PETROLEUM WASTE MANAGEMENT

Petroleum waste management presents significant regulatory challenges and public perceptions due to its potential environmental impact and increased demand. The key regulatory challenges include a lack of uniform standards, limited enforcement, limited liability, and high costs. Public perception needs to change regarding environmental impacts, health risks, and a lack of trust. There are various regulatory bodies working at both national and international levels. There are three major challenges include reducing operational costs, improving environmental impact, and optimizing waste management related to industrial products (Ahmad, 2018). The oil industry's regulatory compliance can modify laws, rules, and directives to ensure a sustainable environment. Government and regulatory bodies such as the Occupational Safety and Health Administration (OSHA), the Petroleum and Natural Gas Regulatory Board (PNGRB), and the National Oil Spill Detection and Response Agency (NOSDRA) can enforce these regulations with the participation of local, state, federal, and international laws (Sam et al., 2017). A significant obstacle to improving petroleum waste management systems in developing countries is the lack of qualified engineers and environmental professionals with the necessary experience in technology (Kumar et al., 2017). Comprehensive and well-coordinated strategies can improve the minimization of system costs, control waste flow directly to landfills, and reduce energy during treatment. Further, training and skill development can help overcome challenge and reduce the shortage of manpower.

4.7 CONCLUSIONS

We have concluded that petroleum waste is a significant problem for creating sustainable environments. The role of petroleum waste treatment technologies, including bioremediation, electrocoagulation, and photodegradation, are discussed along with their advantages and limitations. These technologies may be more or less suitable depending on the specific type of waste being handled. The challenges of petroleum waste management are also discussed, such as regulatory compliance, public perceptions, and engaging with stakeholders to build trust and understanding around waste management practices, considering waste complexity. However, effective waste management and its strategies provide opportunities for economic benefit and sustainable environmental practices. It includes waste reduction, reuse, and recycling, as well as careful monitoring of waste management streams to ensure compliance with environmental regulations. Overall, effective petroleum waste management requires a multifaceted approach that takes into account the complex nature of waste as well as the regulatory, economic, and social contexts, leveraging opportunities to achieve sustainable environmental and economic benefits.

REFERENCES

Ahmad, T. (2018). Regulatory Challenges to Oil & Gas Industry and Environment Protection: A Critical Analysis of Asia. https://papers.ssrn.com/abstract=3177399 (accessed August 24, 2023).

Akkaya, E. (2016). ANFIS based prediction model for biomass heating value using proximate analysis components. *Fuel*, 180, 687–693. https://doi.org/10.1016/j.fuel.2016.04.112

Asim, N., Badiei, M., Torkashvand, M., Mohammad, M., Alghoul, M.A., Gasaymeh, S.S., & Sopian, K. (2021). Wastes from the petroleum industries as sustainable resource materials in construction sectors: opportunities, limitations, and directions. *Journal of Cleaner Production*, 284, 125459. https://doi.org/10.1016/j.jclepro.2020.125459

Azuazu, I.N., Sam, K., Campo, P., & Coulon, F. (2023). Challenges and opportunities for low-carbon remediation in the Niger Delta: towards sustainable environmental management. *Science of the Total Environment*, 900, 165739. https://doi.org/10.1016/j.scitotenv.2023.165739

Bala, K., Dogra, A., Lata, P., & Das, A.P. (2024). Discussing sources and biological-chemical remediation approaches for microplastic pollution. In *Sustainable Microbial Technology for Synthetic and Cellulosic Microfiber Bioremediation*, 37–59. https://doi.org/10.1007/978-3-031-62600-5_3

Behera, I.D., Basak, G., Kumar, R.R., Sen, R., & Meikap, B.C. (2020). Treatment of petroleum refinery sludge by petroleum degrading bacterium Stenotrophomonas pavanii IRB19 as an efficient novel technology. *Journal of Environmental Science and Health, Part A*, 56(2), 226–239. https://doi.org/10.1080/10934529.2020.1866924.

Behera, I.D., Manoranjan, N., Subrata, B., Meikap, B.C., & Sen, R. (2021). Enhanced biodegradation of total petroleum hydrocarbons by implementing a novel two-step bioaugmentation strategy using indigenous bacterial consortium. *Journal of Environmental Management*, 292, 112746. ISSN 0301–4797, https://doi.org/10.1016/j.jenvman.2021.112746

Behera, I.D., Nayak, M., Mishra, A., Meikap, B.C., & Sen, R. (2022). Strategic implementation of integrated bioaugmentation and biostimulation for efficient mitigation of petroleum hydrocarbon pollutants from terrestrial and aquatic environment. *Marine Pollution Bulletin*, 177, 113492. ISSN 0025-326X, https://doi.org/10.1016/j.marpolbul.2022.113492

Bhattacharjee, J., Mishra, S., & Das, A.P. (2021). Recent advances in sensor-based detection of toxic dyes for bioremediation application: a review. *Applied Biochemistry and Biotechnology*, 1–20. https://doi.org/10.1007/S12010-021-03767-7

Da Silva, L.J., Alves, F.C., & de França, F.P. (2012). A review of the technological solutions for the treatment of oily sludges from petroleum refineries. *Waste Management & Research*, 30, 1016–1030. https://doi.org/10.1177/0734242X12448517

Das, A., Dey, S., Bala, K., Gnanasekaran, R., & Das, A.P. (2024). Microbial bioremediation technology for sustainable treatment and management of synthetic microfiber waste. In *Sustainable Microbial Technology for Synthetic and Cellulosic Microfiber Bioremediation*, 221–238. https://doi.org/10.1007/978-3-031-62600-5_12

Das, A., Ray, S.K., Mohanty, M., Mohanty, J., Dey, S., & Das, A.P. (2024). Ecotoxicity of microplastic wastes and their sustainable management: a review. *Environmental Chemistry and Ecotoxicology*, 6, 144–152. https://doi.org/10.1016/J.ENCECO.2024.05.003

Das, A.P., & Ghosh, S. (2023a). Metagenomic insights into unculturable microbial diversity. In *Manganese Mining Microorganisms*, 63–76. https://doi.org/10.1016/B978-0-12-822147-1.00002-9

Das, A.P., & Ghosh, S. (2023b). Microorganisms in manganese biomining processes. In *Manganese Mining Microorganisms*, 55–61. https://doi.org/10.1016/B978-0-12-822147-1.00013-3

Das, A.P., & Ghosh, S. (2023c). Role of metabolites and metabolomics for understanding of manganese bioleaching. In *Manganese Mining Microorganisms*, 117–127. https://doi.org/10.1016/B978-0-12-822147-1.00011-X

Das, A.P., Ghosh, S., Mohanty, S., & Sukla, L.B. (2015). Advances in manganese pollution and its bioremediation. In L.B. Sukla, N. Pradhan, S. Panda, & B.K. Mishra (Eds.), *Environmental Microbe Biotechnology* (1st ed.), 313–328. Springer, Cham. https://doi.org/10.1007/978-3-319-19018-1_16

Das, A.P., Nayak, B., & Behera, I.D. (2024). Applications of microorganisms in advanced geomicrobiological and biotechnological exploration. *Geomicrobiology Journal*, 41(4), 309–311. https://doi.org/10.1080/01490451.2024.2327059

Dey, S., Bissoyi, A., Koteswara Reddy, G., & Das, A.P. (2024). Synthetic microfibers: an urban emerging pollutant demanding for sustainable environmental solutions. In *Sustainable Microbial Technology for Synthetic and Cellulosic Microfiber Bioremediation*, 107–121. https://doi.org/10.1007/978-3-031-62600-5_7

Dey, S., Das, A., Mallick, K., Sahu, A., & Das, A.P. (2023). Environmental petroleum waste: pollution, toxicity, sustainable remediation. 159–175. https://doi.org/10.1007/978-3-031-48220-5_7

Dey, S., Dash, A., & Das, A. P. (2024). Bioremediation as a sustainable and circular solution for mining waste management and bioenergy generation. In *Sustainable Management of Mining Waste and Tailings: A Circular Economy Approach*, 107–119. https://doi.org/10.1201/9781003442455-6

Dey, S., Dash, N., Ray, S.K., Priyadarsini, S., Chhotaray, C., & Das, A.P. (2024). Occurrence and quantification of tire wear plastics pollutants from highway city dust of Bhubaneswar City of Odisha. In *Sustainable Microbial Technology for Synthetic and Cellulosic Microfiber Bioremediation*, 77–89. https://doi.org/10.1007/978-3-031-62600-5_5

Gaaseidnes, K., & Turbeville, J. (1999). Separation of oil and water in oil spill recovery operations. *Pure and Applied Chemistry*, 71, 95–101. https://doi.org/10.1351/pac199971010095

Ghosh, S., & Das, A.P. (2020). Microbial metagenomics: current advances in investigating microbial ecology and population dynamics. In S.K. Nayak & B.B. Mishra (Eds.), *Frontiers in Soil and Environmental Microbiology* (1st ed.), 247–254. CRC Press, Boca Raton, FL.

Gozbenko, V.E., Kargapoltsev, S.K., & Karlina, A.I. (2019). Environmental benefits of new industrial waste-based lubricant compositions. *IOP Conference Series: Earth and Environmental Science*, 229, 012020. https://doi.org/10.1088/1755-1315/229/1/012020

Jabbar, N.M., Alardhi, S.M., Mohammed, A.K., Salih, I.K., & Albayati, T.M. (2022). Challenges in the implementation of bioremediation processes in petroleum-contaminated soils: a review. *Environmental Nanotechnology, Monitoring and Management*, 18, 100694. https://doi.org/10.1016/j.enmm.2022.100694

Johnson, O.A., & Affam, A.C. (2019). Petroleum sludge treatment and disposal: a review. *Environmental Engineering Research*, 24, 191–201. https://doi.org/10.4491/eer.2018.134
Kadhim, N.R., & Abdulredha, M. (2021). Petroleum effluent treatment using electrocoagulation: a case study of phenols removal. *IOP Conference Series: Materials Science and Engineering*, 1184, 012009. https://doi.org/10.1088/1757-899X/1184/1/012009
Kanungo, J., Sahoo, T., Bal, M., & Behera, I. D. (2023a). Performance of bioremediation strategy in waste lubricating oil pollutants: a review. *Geomicrobiology Journal*, 41(4), 360–373. https://doi.org/10.1080/01490451.2023.2245395
Kanungo, J., Sahoo, T., Swain, L.P., & Behera, I.D. (2023b). Toxicity of persistent hydrocarbon pollutants, sources and sustainable remediation process. In I.D. Behera & A.P. Das (Eds.), *Impact of Petroleum Waste on Environmental Pollution and Its Sustainable Management through Circular Economy*. Environmental Science and Engineering. Springer, Cham. https://doi.org/10.1007/978-3-031-48220-5_2
Kleindienst, S., Paul, J.H., & Joye, S.B. (2015). Using dispersants after oil spills: impacts on the composition and activity of microbial communities. *Nature Reviews Microbiology*, 13, 388–396. https://doi.org/10.1038/nrmicro3452
Kumar, S., Smith, S.R., Fowler, G., Velis, C., Kumar, S.J., Arya, S., Rena, Kumar, R., & Cheeseman, C. (2017). Challenges and opportunities associated with waste management in India. *Royal Society Open Science*, 4, 160764. https://doi.org/10.1098/rsos.160764
Lahiri, D., Nag, M., Dey, A., Sarkar, T., Joshi, S., Pandit, S., Das, A.P., Pati, S., Pattanaik, S., Tilak, V.K., & Ray, R.R. (2021). Biofilm mediated degradation of petroleum products. *Geomicrobiology Journal*, 1–10. https://doi.org/10.1080/01490451.2021.1968979
Mishra, S., & Das, A.P. (2021). Current treatment technologies for removal of microplastic and microfiber pollutants from wastewater. In *Wastewater Treatment: Cutting-Edge Molecular Tools, Techniques and Applied Aspects*, 237–251. https://doi.org/10.1016/B978-0-12-821881-5.00011-8
Mishra, S., Singh, R.P., Rout, P.K., & Das, A.P. (2022). Membrane bioreactor (MBR) as an advanced wastewater treatment technology for removal of synthetic microplastics. In *Development in Wastewater Treatment Research and Processes*, 45–60. https://doi.org/10.1016/B978-0-323-85583-9.00022-3
Mohanty, J., Dey, S., Bala, K., & Das, A.P. (2024). Recent advances in microplastic pollution for its sustainable management. In *Sustainable Management of Urban Plastic Waste through Circular Economic Approaches*, 27–48. https://doi.org/10.1201/9781003465447-3
Mohanty, S., Ghosh, S., Bal, B., & Das, A.P. (2018). A review of biotechnology processes applied for manganese recovery from wastes. *Reviews in Environmental Science and Bio/Technology*, 17(4), 791–811. https://doi.org/10.1007/S11157-018-9482-1
Moneer, A.A., Thabet, W.M., Khedawy. M., El-Sadaawy, M.M., & Shaaban, N.A. (2023). Electrocoagulation process for oily wastewater treatment and optimization using response surface methodology. *International Journal of Environmental Science and Technology*. https://doi.org/10.1007/s13762-023-05003-7
Nag, M., Lahiri, D., Ghosh, S., Sarkar, T., Pati, S., Das, A.P., Ram, D.K., Bhattacharya, D., & Ray, R.R. (2023). Application of microorganisms in biotransformation and bioremediation of environmental contaminant: a review. *Geomicrobiology Journal*. https://doi.org/10.1080/01490451.2023.2261443
Ossai, I.C., Hamid, F.S., Hassan, A., Ossai, I.C., Hamid, F.S., & Hassan, A. (2022). Biological treatments for petroleum hydrocarbon pollutions: the eco-friendly technologies. In *Hazardous Waste Management*. IntechOpen. https://doi.org/10.5772/intechopen.102053
Palos, R., Gutiérrez, A., Vela, F.J., Olazar, M., Arandes, J.M., & Bilbao, J. (2021). Waste refinery: the valorization of waste plastics and end-of-life tires in refinery units. A review. *Energy Fuels*, 35, 3529–3557. https://doi.org/10.1021/acs.energyfuels.0c03918

Paul, A., Dey, S., Ram, D.K., & Das, A.P. (2023). Hexavalent chromium pollution and its sustainable management through bioremediation. *Geomicrobiology Journal*, 1–11. https://doi.org/10.1080/01490451.2023.2218377

Priyadarsini, S., & Das, A.P. (2024). Analytical and structural characterization of waste lithium-ion batteries for their effective recycling strategy. In *Environmental Chemistry and Ecotoxicology*. https://doi.org/10.1016/j.enceco.2024.12.004

Priyadarsini, S., & Das, A.P. (2025). Lithium bioleaching: a review on microbial-assisted sustainable technology for lithium bio-circularity. *Journal of Water Process Engineering*, 69, 106744. https://doi.org/10.1016/j.jwpe.2024.106744

Qin, X.S., Huang, G.H., & He, L. (2009). Simulation and optimization technologies for petroleum waste management and remediation process control. *Journal of Environmental Management*, 90, 54–76. https://doi.org/10.1016/j.jenvman.2008.07.002

Rabahi, A., Assadi, A.A., Nasrallah, N., Bouzaza, A., Maachi, R., & Wolbert, D. (2019). Photocatalytic treatment of petroleum industry wastewater using recirculating annular reactor: comparison of experimental and modeling. *Environmental Science and Pollution Research*, 26, 19035–19046. https://doi.org/10.1007/s11356-018-2954-6

Ramirez, D., & Collins, C.D. (2018). Maximisation of oil recovery from an oil-water separator sludge: influence of type, concentration, and application ratio of surfactants. *Waste Management*, 82, 100–110. https://doi.org/10.1016/j.wasman.2018.10.016

Ray, S.K., Dey, S., Das, A.P., & John, A. (2024). Microplastic waste management through greener technologies: a review. In *Sustainable Management of Urban Plastic Waste Through Circular Economic Approaches*, 237–261. https://doi.org/10.1201/9781003465447-14

Roy, A., Sar, P., Sarkar, J., Dutta, A., Sarkar, P., Gupta, A., Mohapatra, B., Pal, S., & Kazy, S.K. (2018). Petroleum hydrocarbon rich oil refinery sludge of North-East India harbours anaerobic, fermentative, sulfate-reducing, syntrophic and methanogenic microbial populations. *BMC Microbiology*, 18, 151. https://doi.org/10.1186/s12866-018-1275-8

Sam, K., Coulon, F., & Prpich, G. (2017). Management of petroleum hydrocarbon contaminated sites in Nigeria: current challenges and future direction. *Land Use Policy*, 64, 133–144. https://doi.org/10.1016/j.landusepol.2017.01.051

Satapathy, S., Kar, D., Priyadarsini, S., Das, A.P., & Kuanar, A. (2024). Application of microorganisms in emerging waste management strategies and remediation of synthetic and cellulosic microfibers. In *Sustainable Microbial Technology for Synthetic and Cellulosic Microfiber Bioremediation*, 183–206. https://doi.org/10.1007/978-3-031-62600-5_10

Sattar, S., Hussain, R., Shah, S.M., Bibi, S., Ahmad, S.R., Shahzad, A., Zamir, A., Rauf, Z., Noshad, A., & Ahmad, L. (2022). Composition, impacts, and removal of liquid petroleum waste through bioremediation as an alternative clean-up technology: a review. *Heliyon*, 8, e11101.

Silva, I.A., Almeida, F.C.G., Souza, T.C., Bezerra, K.G.O., Durval, I.J.B., Converti, A., & Sarubbo, L.A. (2022). Oil spills: impacts and perspectives of treatment technologies with focus on the use of green surfactants. *Environmental Monitoring and Assessment*, 194, 143. https://doi.org/10.1007/s10661-022-09813-z

Singh, J. (2016). A roadmap for production of sustainable, consistent and reliable electric power from agricultural biomass: an Indian perspective. *Energy Policy*, 92, 246–254. https://doi.org/10.1016/j.enpol.2016.02.013

Singh, P., Ojha, A., Borthakur, A., Singh, R., Lahiry, D., Tiwary, D., & Mishra, P.K. (2016). Emerging trends in photodegradation of petrochemical wastes: a review. *Environmental Science and Pollution Research*, 23, 22340–22364. https://doi.org/10.1007/s11356-016-7373-y

Tetteh, E.K., Rathilal, S., & Naidoo, D.B. (2020). Photocatalytic degradation of oily waste and phenol from a local South Africa oil refinery wastewater using response methodology. *Scientific Reports*, 10, 8850. https://doi.org/10.1038/s41598-020-65480-5

Tripathy, B., Dey, S., Dubey, N.K., Lo, W.-C., & Das, A.P. (2024). An Omics-Based Approach for Managing the Risk of COVID-19-Generated Wastes. 51–66. https://doi.org/10.1007/978-3-031-50840-0_4

Varjani, S., Joshi, R., Srivastava, V.K., Ngo, H.H., & Guo, W. (2020). Treatment of wastewater from petroleum industry: current practices and perspectives. *Environmental Science and Pollution Research*, 27, 27172–27180. https://doi.org/10.1007/s11356-019-04725-x

Wienchol, P., Szlęk, A., & Ditaranto, M. (2020). Waste-to-energy technology integrated with carbon capture: challenges and opportunities. *Energy*, 198, 117352. https://doi.org/10.1016/j.energy.2020.117352

5 Sustainable Remediation of Hydrocarbon Pollutants Using Bioremediation, Biosurfactants, and Advanced Oxidation Processes

Deeptimayee Pal and Abhishek Kumar

5.1 INTRODUCTION

Hydrocarbon pollution is a substantial ecological and public health concern, especially from sources like pesticides, petroleum, or other hazardous organic substances. Petroleum-induced environmental degradation is particularly concerning owing to the poisonous characteristics of petroleum hydrocarbons, which endanger all living forms. Crude oil contamination is prevalent due to its extensive use, with spills and inappropriate disposal practices being significant factors in this problem. Petroleum denotes an intricate amalgamation of several hydrocarbons with differing molecular weights. This composition comprises saturated alkanes, branching alkanes, alkenes, naphthenes (both homocyclic and heterocyclic), and aromatic compounds, some of which include heteroatoms like sulfur, oxygen, and nitrogen. Furthermore, it encompasses substantial aromatic compounds, including resins and asphaltenes, alongside hydrocarbons with functional groups such as carboxylic acids and ethers. Additionally, crude oil contains heavy metals, many of which are associated with porphyrin structures, particularly pyrrolic compounds. These compounds are classified into light (low-molecular-weight) and heavy (high-molecular-weight) ends based on their molecular weights and distillation temperature.

Produced water (PW) is the predominant wastewater outflow generated during oil and gas exploration and production operations. This brine is extracted from hydrocarbon-bearing formations and consists of a blend of PW and either seawater or freshwater, confined for long periods beneath hydrocarbons within reservoir rock, along with injection water, minor quantities of condensed water from gas production,

 DOI: 10.1201/9781003508496-5

and remnants of treatment chemicals introduced during hydro-fracture operations to improve extraction efficiency. In gas fields, where water injection is often absent, PW mainly consists of formation and condensed water, resulting in a somewhat reduced amount of PW compared to oilfields. Furthermore, PW from gas fields often exhibits more excellent acidity than oilfields, a trait ascribed to dissolved acidic gases, including carbon dioxide (CO_2) and hydrogen sulfide (H_2S). Oil and gas exploration and production operations use diverse solvents and chemicals, with reports indicating that over one-third of these substances are released into PW. The water cut, defined as the percentage of water generated to total fluid production, often increases with the well's age. Production capacity may decrease markedly in depleting fields, leading to water cuts of up to 98% (Fakhru'l-Razi et al., 2009). The constituents of PW are illustrated in Figure 5.1.

Exposure to petroleum and its derivatives occurs via several direct and indirect pathways, impacting all living creatures. Numerous byproducts generated during petroleum refining and processing, used in producing various industrial goods, are very hazardous. These hazardous substances are perpetually and often unintentionally discharged into the environment. The environmental damage from unintentional crude oil spills significantly contributes to global hydrocarbon pollution when combined with these emissions. Although petroleum contains many hydrocarbons, few have been well characterized in terms of their toxicological characteristics.

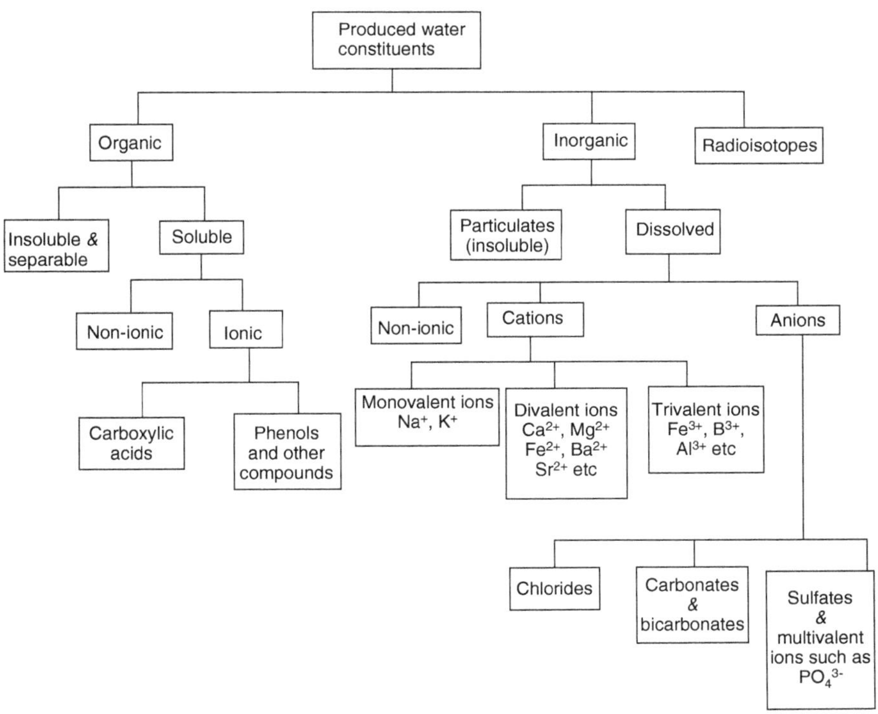

FIGURE 5.1 Constituents of produced water (Olajire, 2020).

Petroleum hydrocarbons display diverse molecular weights and boiling temperatures, influencing their differing degrees of environmental toxicity (Behera et al., 2020, 2021, 2022). Their chemical composition and physical characteristics significantly affect the toxicity and bioavailability of these hydrocarbons for microbial breakdown. Petroleum is intrinsically hazardous and can be fatal to both people and animals, depending on the precise proportion, exposure route, and length of exposure (Dey, Bissoyi, et al., 2024). Besides hydrocarbons, the compounds and dispersants included in crude oil substantially enhance the toxicity associated with oil spills. These compounds may have various health impacts on humans and animals, depending on the level of exposure and individual sensitivity. Crude oil contains highly hazardous compounds that may impact almost all major human organ systems, including the neurological, respiratory, circulatory, immunological, reproductive, sensory, and endocrine systems. Moreover, damage to the kidneys and liver may occur, resulting in various illnesses and disorders among affected people (Pal & Sen, 2023).

Efficient treatment of oil and gas PW is crucial to complying with the quality criteria established by regulatory authorities for disposal or reuse. The improper disposal of PW, often including hazardous materials, threatens environmental sustainability (Mohanta et al., 2023). Consequently, it is essential to comprehend PW's chemical and physical composition to reduce the concentration of deleterious elements to acceptable levels before disposal or reuse. Effective management solutions are necessary to guarantee the safe disposal or reuse of oil and gas PW without detrimental environmental effects (Bala et al., 2024; Das, Dey, et al., 2024). These initiatives are essential for advancing environmental sustainability and ensuring the enduring profitability of the energy industry. The appropriate treatment and management of PW can reduce the environmental risks linked to hydrocarbon extraction while promoting the prudent and sustainable use of energy resources.

5.2 ADVANCED AND GREENER TECHNOLOGIES FOR THE MANAGEMENT OF HYDROCARBON POLLUTANTS

Hydrocarbons, upon environmental discharge, provide substantial risks to ecosystems, human health, and economic stability. Hydrocarbons, particularly in the form of oil spills and petroleum leaks, may inflict significant harm on both terrestrial and aquatic ecosystems. Hydrocarbon pollution in rivers, lakes, and aquifers renders water unsuitable for consumption, agriculture, and industrial use (Dey, Bissoyi, et al., 2024; Dey, Dash, et al., 2024). Therefore, it is essential for the management of hydrocarbon contamination. Conventional techniques such as physical extraction, chemical applications, and burning have been extensively used. However, they often provide considerable constraints. Innovative and sustainable approaches for hydrocarbon pollutant management, including bioremediation, biosurfactants (BioSs), and other advanced oxidation processes (AOPs), have several benefits compared to conventional treatment methods.

5.2.1 Bioremediation

Petroleum pollutants are among the most hazardous contaminants worldwide, necessitating efficient management to protect marine organisms and ecosystems. A key strategy for regulating these contaminants has been assessing their degradability in indigenous microbial ecosystems. The seriousness of the issue is evident from the discovery of drilling mud waste contaminated with hydrocarbons in numerous tanks and the sludge generated as waste in oil processing facilities. Comprehending the mechanics and functions of bacteria may provide insights into improved bioremediation methods. This treatment approach has shown efficacy in degrading various organic hydrocarbon contaminants, such as explosives, pesticides, chlorophenols, and polycyclic aromatic hydrocarbons (PAHs), as depicted in Figure 5.1. Studies on PAH-contaminated soil from a gas production facility and a diamond mining site have shown the viability of bioremediation (Paul et al., 2024). Researchers have undertaken studies to repair hydrocarbon-contaminated soil and oily sludge from refineries, investigating amendment strategies for pollutants such as sulfamethoxazole during wetland rehabilitation. This research is essential for comprehending the effects of bioremediation on hydrocarbon contaminants and strategizing large-scale implementations (Chang et al., 2008).

Hydrocarbon degradation occurs through complex microbial processes that convert hydrocarbons into simpler molecules. Aerobic and anaerobic microbes are essential, usually functioning via enzyme-mediated processes tailored for particular experimental circumstances. Acinetobacter radioresistens strain KA2, isolated from oily waste sludge, was used in a two-stage procedure, resulting in an 80% reduction in total petroleum hydrocarbons (TPH) over 16 weeks. In a separate investigation, the A. radioresistens strain KA5 and Enterobacter hormaechei strain KA6, both isolated from petroleum waste sludge, attained a TPH removal rate of 84% during a comparable 16-week period using a two-stage bioremediation strategy. Experiments using the E. hormaechei strain KA6 in a culture-based medium over four months resulted in an 80% decrease in TPH) from oily sludge pollutants. Similarly, Staphylococcus equorum strain KA4 and E. hormaechei strain KA3, extracted from heavy oil sludge, were examined in a bioreactor configuration, resulting in a TPH removal efficiency of 89% over 16 weeks (Parhamfar et al., 2020).

Besides bacteria, fungi have considerable potential for bioremediation. Fungal species have shown the capacity to mitigate DDT pollution in soil. Aspergillus ochraceus H2 and Scedosporium apiospermum H16, extracted from olive mill waste, efficiently decomposed PAHs. Bioremediation tests using a mix of cow bile and bacterial species have shown a TPH clearance rate of up to 90%. These results emphasize the significance of microbial activity in the bioremediation of hydrocarbon contaminants while indicating the need for more research on optimizing physical factors. Microbial populations, aeration, moisture content, temperature, and pH greatly influence bioremediation efficacy (Kumar et al., 2022). Advancements in genome sequencing technology have facilitated the rapid discovery of new microorganisms from severe habitats, contributing to the formulation of more efficient bioremediation techniques. The efficacy of the microbial population relies on sustaining ideal conditions: a moisture content of 50%–55%, a near-neutral pH, and a carbon-to-nitrogen (C/N) ratio under 50 for expedited biodegradation (Li & Yang, 2021).

The use of chemometric techniques has enhanced the optimization of bioremediation settings. Chemometrics facilitates the identification of parameters such as pH, nutrient content, and temperature by modeling the link between input and output variables, therefore optimizing the biodegradation process (Das et al., 2011). These approaches enable the observation of microbial population dynamics and metabolite synthesis, yielding essential insights into the degradation process. The biodegradation process generally transpires in two phases: the maturation period, including the mesophilic (298.15–318.15 K) and thermophilic (>318.15 K) stages, and the curing phase, referred to as the secondary mesophilic phase. The mixing ratio is crucial since inadequate mixing might impede microbial activity. Two-stage methodologies are often used for the remediation of petroleum pollutants (Dey et al., 2023; Lahiri et al., 2021). An open-state biodegradation method was used to treat soil polluted by diamond mining activities. Bioreactors have been used in laboratory research for the bioremediation of petroleum sludge, using finished compost as the bulking agent. Ideal parameters for these processes include a moisture content of 12%–25%, a neutral pH, and a temperature range commensurate with biodegradation efficacy.

Supplementary bioremediation methods include in situ remediation of OMW-contaminated soil using biowaste, animal waste, and vermicomposting procedures. This approach has shown versatility in decomposing phenolic compounds. A comparable bioremediation experiment used a microbial-fungal consortium extracted from OMW to address evaporation pond pollution. A 60-day in-vessel bioremediation method for pyrene-contaminated soil was extended by 14 days to break down PAHs efficiently. Diverse inoculation methods have been examined depending on substrate, pollutants, and biogeochemical conditions. Researchers injected strains of A. radioresistens and E. hormaechei into cylindrical bioreactors. In a separate work, used a 1.5×10^8 CFU/g inoculum to facilitate the biodegradation of petroleum hydrocarbons in bioreactors. Olive mill waste sludge was subjected to treatment for OMW biodegradation using an inoculum of 7×10^7 CFU/vol. The amalgamation of microbial consortia has shown superior efficacy in hydrocarbon degradation relative to mono-species inoculations (Koolivand et al., 2020).

5.2.2 Biosurfactants

BioSs are synthesized by several microorganisms, each demonstrating considerable structural diversity and unique chemical and surface-active characteristics. These microorganisms, adept at synthesizing many forms of BioSs, include both low molecular weight and large molecular weight molecules. Moreover, recent research revealed that Paenibacillus sp. D9 is a proficient BioS generator, augmenting the variety of bacteria that create BioSs. BioSs may be categorized into many categories according to their molecular structure. Glycolipids (xylolipids, rhamnolipids, and mannosylerythritol) and lipopeptides (subtilisin, viscosin, serrawetin, surfactin, polymyxin, and iturin) are the most frequently synthesized low molecular weight surfactants (Martins & Martins, 2018).

Besides BioSs, bioemulsifiers constitute another category of surface-active biomolecules. In contrast to BioSs, bioemulsifiers do not markedly decrease surface tension but are essential for stabilizing emulsions, especially between hydrocarbons and aqueous phases. Bioemulsifiers are often significant molecular-weight substances with intricate combinations of lipoproteins, heteropolysaccharides, proteins, and

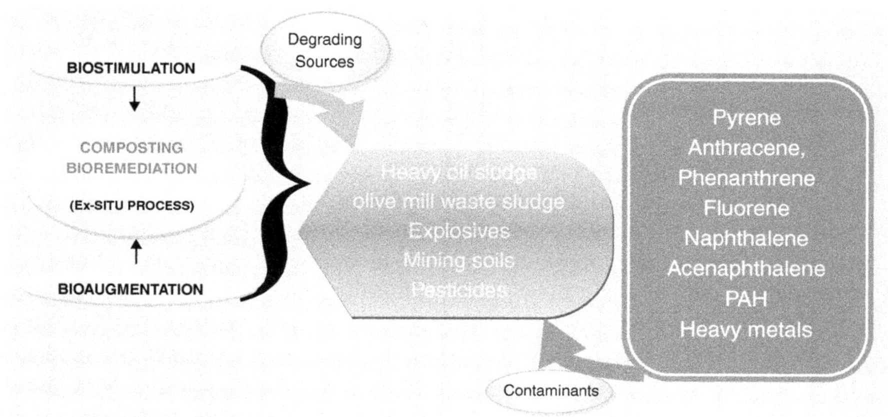

FIGURE 5.2 Bioremediation for hydrocarbon management (Radhakrishnan et al., 2023).

lipopolysaccharides. Notwithstanding this difference, "BioSs" and "bioemulsifiers" are sometimes used interchangeably in the literature. BioSs are economically desirable because of their eco-friendly characteristics, providing benefits over chemical surfactants in several industrial applications. Their biodegradability, low toxicity, and efficacy under harsh circumstances offer them a viable alternative to synthetic surfactants, notwithstanding some problems associated with manufacturing costs and scalability, as shown in Figure 5.2.

The increasing environmental issues linked to synthetic surfactants, known for their considerable toxicity, have prompted the investigation of environmentally friendlier alternatives. A primary benefit of BioSs is their biodegradable and non-toxic characteristics, which are crucial for ecologically sustainable product formulations (Das, Ray, et al., 2024; Dey, Dash, et al., 2024). Research suggests that anionic and cationic BioSs possess markedly reduced toxicity relative to synthetic surfactants, rendering them more environmentally and industrially benign. The biodegradability of BioSs renders them suitable for diverse environmental applications. These processes depend on the capacity of BioSs to diminish surface and interfacial tension, promoting the disintegration and elimination of pollutants (Mishra et al., 2024). Furthermore, the stability and effectiveness of BioSs have shown advantages in petroleum industry applications, including oil sludge removal, microbial enhanced oil recovery (MEOR), storage tank cleaning, and immobilization of oil residues (Jha et al., 2016).

BioSs (emulsifiers/de-emulsifiers) are contingent upon their interaction with hydrophobic contaminants. The process of dispersing an immiscible liquid into droplets forms an emulsion, a heterogeneous system, usually >0.01 cm in diameter. Two principal categories of emulsions are often documented: oil-in-water (o/w) and water-in-oil (w/o) emulsions. Comparative studies indicate that BioSs and bioemulsifiers provide equivalent or enhanced emulsifying efficacy relative to synthetic surfactants. Moreover, owing to their microbial origin, they exhibit enhanced ecological compatibility. A significant benefit of BioSs is their capacity to serve as anti-adhesive and anti-biofilm agents. BioSs may alter surface hydrophobicity, affecting microbial contaminant adherence and inhibiting biofilm development. Microbial surface interactions, enabled by synthesizing surface-active BioSs, are essential for biofilm formation, whereby an organized community of bacteria attaches to a solid substrate. This characteristic is crucial for inhibiting biofouling and improving microbial efficacy in biodegradation processes (Sharma & Saharan, 2016).

The synthesis of BioS is a crucial approach for augmenting the solubilization of hydrophobic contaminants, enhancing their bioavailability, and promoting their degradation. Hydrophobic organic molecules, including hydrocarbons, often resist biodegradation due to their poor water solubility. The bioavailability of these chemicals may be enhanced by pseudo-solubilization, a method that facilitates their accessibility for microbial breakdown. Research indicates that BioSs generated by Paenibacillus sp. D9 markedly improve the solubilization of 10% diesel fuel, exhibiting substantial emulsification capacity and stability under hazardous hydrophobic conditions. The lipopeptide BioSs synthesized by Paenibacillus sp. D9 can emulsify a range of hydrophobic substances. The emulsifying capability of BioSs renders it highly appropriate for various applications, including oil tank cleanup, emulsified oil transportation, biodegradation of PAHs, and bioremediation (Gudiña et al., 2015).

BioSs significantly enhance the biodegradation of hydrophobic contaminants, mainly diesel fuel and hydrocarbon substrates. Studies indicate that incorporating natural BioSs significantly enhances the biodegradation rates of diesel fuel. The introduction of 500 ppm of Paenibacillus sp. D9BioSs enhanced the diesel fuel biodegradation rate by 65.1% relative to cultures devoid of BioSs, underscoring its potential for bioremediation. The BioSs generated by Bacillus mojavensis A21 improved diesel fuel degradation by 20%, 27%, and 40%, with a peak degradation rate of 58% at a concentration of 2 g/L of BioSs. The mechanism underlying this enhanced degradability pertains to the capacity of BioSs to augment the bioavailability of hydrophobic compounds by facilitating their pseudo-solubilization and transport to the aqueous phase, thereby enabling emulsification and improved biodegradation. BioSs provide a safer and more ecologically sustainable alternative to synthetic surfactants, which are toxic and less biodegradable, particularly for applications like the remediation of oil-contaminated soil. Research has repeatedly shown that BioSs are superior to chemical surfactants in removing hydrocarbon mixtures.

BioSs generated by Bacillus subtilis CN2 eliminated 84.6% of oil-contaminated soil, while those from Paenibacillus dendritiformis removed 78%–81% of motor oil from similarly contaminated sand. A BioS generated by Serratia marcescens UCP 1549 eliminated 94% of oil-contaminated soil, while distilled water alone achieved a 63% clearance rate. The results underscore the need for efficient techniques to assess and enhance BioS levels during bioremediation procedures. BioSs have shown efficacy in eliminating diesel fuel adsorbed on polluted sand. Research has shown that Bacillus subtilis BioSs facilitated 78.5%–81.8% diesel removal at critical micelle concentration (CMC) and three times the CMC, whereas distilled water produced only 37.9% oil removal. Rhamnolipid BioSs and saponins have shown efficacy in enhancing the extraction of polychlorinated biphenyls (PCBs) from sediments, with removal rates escalating from 30% to 55%–60% (Lászlová et al., 2018).

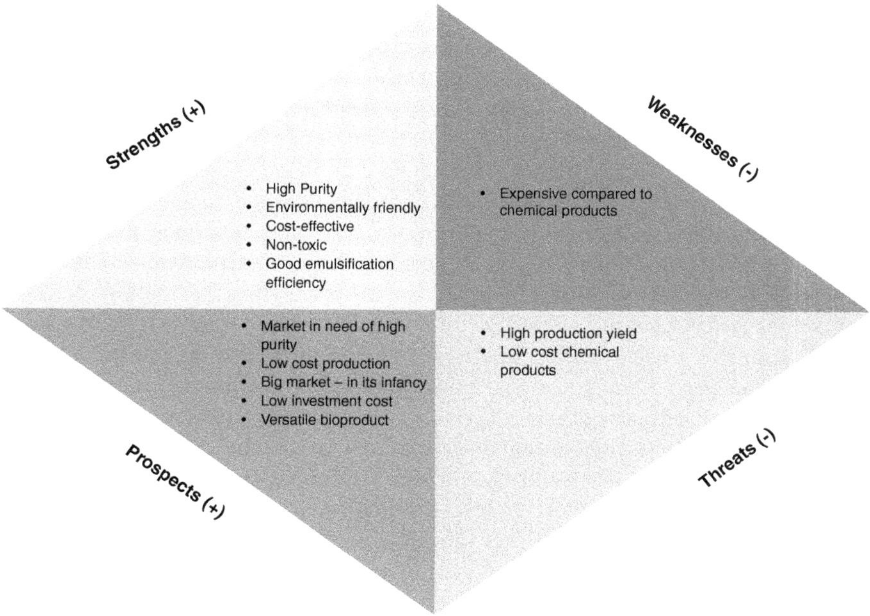

FIGURE 5.3 The strengths, weaknesses, prospects, and threats of biosurfactants (Jimoh & Lin, 2019a).

In microbially enhanced oil recovery (MEOR) investigations, BioSs from Serratia marcescens achieved 76% crude oil removal from a sand column under dynamic settings, in contrast to reduced removal rates in static conditions. Research indicated that Rhodococcus ruber BioSs were 1.4–2.3 times more efficacious in oil removal from the soil than synthetic surfactants. BioSs' amphiphilic properties decrease surface tension and improve hydrophobic substances' bioavailability, enhancing their efficacy in bioremediation. The extensive pollution of water sources by hydrocarbons from diesel, motor oil, and crude oil spills necessitates the employment of BioSs as an environmentally sustainable approach to enhance the mobility and solubility of hydrophobic contaminants, hence facilitating more effective cleanup. BioSs enhance the bioavailability of hydrocarbons in aquatic settings, facilitating emulsification, solubilization, and eventual degradation, contributing significantly to environmental sustainability.

5.2.3 Advanced Oxidation Processes (AOPs)

The chemical treatment of PW in the oil and gas sector uses chemical oxidants to convert organic pollutants into less toxic substances, including carbon dioxide (CO_2) and water (H_2O). Conventional oxidants include oxygen (O_2), ozone (O_3), hydrogen peroxide (H_2O_2), or their mixtures. Although these oxidants may dissolve several pollutants, PW often includes resistant chemicals such as benzene, toluene, ethylbenzene, xylene (BTEX), phenols, and acetic acid, which have limited reactivity with traditional oxidants, making them challenging to degrade. This constraint underscores the need for more sophisticated treatment methods, such as AOPs, which are efficient and cost-effective for the comprehensive oxidation of conventional and resistant organic pollutants in PW. AOPs are characterized by their capacity to produce highly reactive hydroxyl radicals (·OH), which may mineralize many organic contaminants into less hazardous intermediate compounds. The primary benefit of AOPs is their environmentally benign end products and the capability to get complete mineralization of intricate organic molecules. The technique may also harness solar energy to generate ·OH, reducing dependence on expensive and dangerous artificial light sources. AOPs are an attractive and economical technology for many applications (Jiménez et al., 2019).

O_3, a frequently used oxidant in AOPs, exhibits instability in aqueous environments and undergoes spontaneous disintegration via a complicated mechanism, producing ·OH that is very efficient in oxidizing diverse organic and inorganic compounds found in PW. O_3 degradation is expedited in alkaline environments, in the presence of catalysts, or when paired with H_2O_2. The breakdown of O_3 into free radicals is more effective at elevated pH levels, increasing its oxidative potential (Kumar & Gangawane, 2022b). This process may be optimized using catalytic ozonation, which employs catalysts to accelerate O_3 breakdown and augment the generation of ·OH. Catalytic ozonation can be a homogeneous and heterogeneous process. In homogeneous ozonation, transition metal ions such as Cr^{3+}, Zn^{2+}, and Fe^{2+} serve as catalysts to accelerate the breakdown of organic contaminants. Heterogeneous catalytic ozonation employs solid catalysts, including metals (e.g., Fe, Cu, Pd) and metal oxides (e.g., TiO_2, ZnO, MgO), to enhance the breakdown of O_3 on the surface of the catalyst. The efficacy of this procedure is contingent upon many elements, including the number of catalyst's sites, the medium's pH, the water's composition, and the characteristics of the organic pollutants (Pachhade et al., 2009).

Multiple studies have shown the efficacy of O_3-based treatment for PW in the oil and gas industry. A recent study investigated the effectiveness of a combination of O_3 and biological aerated filters in eliminating stubborn organic contaminants from wastewater. The findings demonstrated that this integrated process significantly decreased the toxicity of treated effluents, making it a viable method for secondary effluent refinement. Furthermore, experimental and modeling techniques were used to assess ozonation to eliminate PAHs, indicating augmenting O_3 dosages and greater temperatures enhanced the PAHs removal, primarily attributable to better O_3 transfer of mass, increased ·OH yields, and accelerated O_3 decomposition rates. Additionally, AOPs were used for treating PW, successfully eliminating virtually the organic constituents, including 78% of acetic acid and 74% of total organic carbon (TOC), via the integration of ozonation and H_2O_2 (peroxonation). The authors indicated that more optimization of O_3 and H_2O_2 dosages is necessary to scale the process properly for industrial purposes. These investigations emphasize the promise of O_3-based AOPs in PW treatment and stress the need for more studies to improve their efficacy and scalability (Jiménez et al., 2019).

The Fenton process, first articulated by Fenton, enhances the oxidative potential of H_2O_2 by combining ferrous ion (Fe^{2+}), forming the Fenton reagent. In this reaction, Fe^{2+} is a catalyst that activates H_2O_2, producing ·OH, which are powerful oxidizing agents. Introducing the Fenton reagent into wastewater generates ·OH that promotes the oxidation of aromatic and aliphatic compounds. The Fenton process is extensively used for the remediation of organic contaminants, yet the precise molecular mechanisms governing the oxidation of these molecules remain intricate and not entirely understood. The ·OH generated by the Fenton reagent is non-selective and strongly reactive, facilitating the destruction of many organic molecules. The extensive application and efficacy of the Fenton reaction make it an indispensable instrument in AOPs, especially for the remediation of persistent contaminants in wastewater. Additional study is required to clarify the specific molecular details of organic oxidation by Fenton reagent to enhance its applicability in environmental contexts (Babuponnusami & Muthukumar, 2014).

Numerous studies have evaluated the Fenton method's efficacy in treating oil and gas-PW. In a recent survey, AOPs were used for PW treatment, revealing that the standard Fenton process effectively eliminated 40% of TOC at 343.15 K without the presence of salts and 18% of TOC at 293.15 K. This technique demonstrated limited efficacy in removing more resistant chemicals, such as acetic acid. A comparative analysis of optimized in situ chemical oxidation was conducted, assessing the efficacy of the Fenton process against activated persulfate oxidation for treating groundwater contaminated with hydrocarbon. The investigation revealed that the Fenton process had superior reaction rates for benzene (1.65 L/h) and toluene (1.28 L/h) in contrast to activated persulfate oxidation, which recorded diminished reaction rates for benzene and toluene to be 130 and 100 mL/h, respectively. Notwithstanding these encouraging outcomes, the Fenton method has inherent limitations. Principal problems include the sluggish regeneration rate of Fe^{2+}, the want for elevated reagent doses, and the obligation to eliminate dissolved iron from the treated media. Furthermore, the process produces substantial quantities of iron sludge, resulting in additional expenses for removing and treating the secondary waste. A third drawback is that the Fenton reaction functions ideally in acidic circumstances, which may hamper its implementation in large-scale wastewater treatment systems. Rectifying these deficiencies is crucial for enhancing the method and making it more economically viable and ecologically sustainable for hydrocarbon pollution cleanup.

Complete mineralization of organic molecules in PW may be markedly improved by combining sunlight or UV radiation exposure with chemical-based AOPs. This integrated method generates ·OH photochemically using UV lamps with wavelengths ranging from 200 to 380 nm or by the application of sunlight. The photoperoxidation process offers a unique benefit, as the concurrent presence of UV radiation and H_2O_2 elevates the O_3 decomposition, thus enhancing the production rate of ·OH. The principal constraint obstructing the extensive implementation of this technique is the elevated expense related to O_3, H_2O_2, and ultraviolet light. As a result, this approach is often limited to severely polluted wastewater, where its cost-effectiveness is more defensible. Numerous research studies have been undertaken to investigate the efficiency of photolysis of the oxygen reactive species in treating PW. The use of various combinations of radiations was investigated for the quality improvement of biotreated oil refinery effluent prior to reverse osmosis treatment. They aimed to generate water that meets high-quality criteria for reuse. The results indicated that the UV/O_3 and UV/H_2O_2 processes achieved enhanced mineralization, with TOC removal efficiencies of 95% and 74%, respectively. This surpassed the performance of O_3 or H_2O_2 alone, which yielded TOC removal efficiencies below 20%, even at elevated O_3 concentrations (60 ppm) after 1 hour of reaction.

The UV/H_2O_2 method was examined for phenol degradation and the reduction of chemical oxygen demand (COD) in residual fluid catalytic cracking (RFCC) effluent. The results have shown that the UV/H_2O_2 system attained a 70.9% elimination efficiency for phenol and 4.39% for COD after a reaction duration of 3 hours. The efficacy of UV-185/H_2O_2 and UV-254/H_2O_2 procedures was assessed for the BTEX degradation in petrochemical effluent using a photoreactor. The results indicated that TOC removal efficiencies of 80% and 90% were achieved using UV-254/H_2O_2 and UV-185/H_2O_2, respectively, with a duration of 6 hours and 3 pH. The UV/O_3/H_2O_2 system was investigated for phenol degradation and the COD removal from RFCC refinery effluent. The research demonstrated that this integrated system was exceptionally effective, attaining phenol removal rates of 93.75% and COD removal rates of 12.2% after a reaction duration of 2 hours. These results underline the possibility of integrating photolysis with AOPs to increase PW pollutant degradation. However, cost and operational problems remain critical for large-scale applications (Azizah & Widiasa, 2018).

The simplicity, economic viability, and high efficacy of photo-Fenton systems have garnered considerable attention from studies treating generated water from hydrocarbon operations. The UV-Fenton system had the highest efficacy, with elimination efficiencies of 82%, 73%, 68%, and 95% for COD, TOC, biochemical oxygen demand (BOD_5), and dyes, respectively. The US-Fenton technique subsequently exhibited removal efficiencies of 79%, 70%, 66%, and 95% for the same contaminants. The traditional Fenton procedure performed marginally worse, with removal efficiencies of 70%, 58%, 51%, and 92%, respectively. The research demonstrated the effective removal of dyes and carbon from PW but advised further biological treatment or the implementation of a cost-effective built wetland for final refinement before recycling or environmental discharge of the treated water. In a similar study, the photo-Fenton method was used in a reactor subjected to sun and artificial UVA radiation to decompose organic molecules in phenolic effluents and PW. The research indicated sun irradiation surpassed artificial light, with a 76% effectiveness in eliminating organic pollutants. The authors suggested that solar energy may be used as a cost-efficient option to decrease the system's operating expenditures (Mota et al., 2018). A comparative analysis of photocatalysis using TiO_2 as a catalyst and the photo-Fenton method utilizing $FeSO_47H_2O$ as a catalyst was conducted for the treatment of both simulated PW, including phenol in saltwater, and real PW from oilfields in Rio Grande do Norte, Brazil. Their results demonstrated that the photo-Fenton technique had the best phenol removal effectiveness (99%) despite the elevated salt concentration in the saltwater (Silva et al., 2019).

The photo-Fenton technique was used with flotation for PW treatment from eight distinct production wells. The research sought to manage flowback PW for utilization sustainably. Their findings indicated substantial water conservation in hydraulic fracturing, water flooding, and EOR methods, with reductions of 85%, 70%, and 75% in freshwater demands, respectively. This indicates that just 15%, 30%, and 25% of freshwater would be required daily for ongoing fracturing, water flooding, and EOR procedures (Jin & Davarpanah, 2020). The electro-Fenton technique was used to eliminate COD from petrochemical effluent, optimizing variables using a nanoporous aluminum electrode. The investigation suggested an ideal COD removal of 65.03%, attained under the parameters of pH 2.96, a reaction duration of around 1.49 hours, a current density of 69.57 mA, a H_2O_2/Fe^{2+} molar ratio of 3.42, and a H_2O_2 volume fraction of 1.93 mL/L. The altered nanoporous aluminum oxide electrode improved the breakdown of H_2O_2 and the production of ·OH relative to standard aluminum electrodes. The continuous electro-Fenton process using a dimensionally stable Ti-RuO_2/IrO_2 anode and an activated carbon fiber felt cathode was investigated for the organic contaminants' degradation in PW. Their findings demonstrated a COD removal effectiveness of 73.33%, accompanied by an electrical energy usage of 3.2436 mJ/kg COD. The COD elimination efficiency enhanced to 81.1% and 86.0% with one and two 3W UVA lamps, respectively (Al-Khafaji & Mohammed, 2019).

The electro-Fenton method was used for the treatment of mother liquor from gas field wastewater, using Fe-Fe electrodes and incorporating H_2O_2 into the system. The research optimized many parameters, attaining a peak COD removal efficiency of 71.9% at a pH 3, a current density of 2.0 A, a H_2O_2 volume of 7 mL, and a reaction duration of 180 minutes. The least electrical energy consumption (17.028 mJ/kg COD) and specific electrical plate consumption (0.82 kg Fe/kg COD) were recorded at pH 3. The synergistic effects of ultrasound (US) and the Fenton process were examined on the remediation of total hydrocarbons in oil spill sludge. The research exhibited a peak reduction of petroleum hydrocarbons by 84.5% and 23.58% (nC_7-C_{10} & nC_{11}-C_{20}, respectively) under optimal conditions, which comprised a pH of 3, a sludge-to-water ratio of 1:1, 40% ultrasound amplitude, a H_2O_2/Fe^{2+} weight ratio of 10:1, and 10 minutes of ultrasonic treatment. The integrated US-Fenton approach significantly improved sludge degradation relative to ultrasonic or Fenton oxidation alone (Sivagami et al., 2019). Petrochemical effluent, including phenol and nitrophenol, was studied for its degradation and mineralization using the sono-photo-Fenton method, with starting COD and TOC values of 260 and 100 mg/L, respectively. The research used a comprehensive factorial experimental design to assess the impact of operational factors. The findings demonstrated removal efficiencies of 76.1% for COD and 68.7% for TOC under ideal circumstances, which included an Fe^{2+} concentration of 24 mg/L, a pH of 3, and a H_2O_2 concentration of 500 mg/L after 60 minutes of reaction (Shokri, 2018).

As shown by prior research, the photocatalytic oxidation method is acknowledged as a potential AOP for environmental cleanup. This approach is beneficial since it minimizes chemical use and decreases waste sludge production, a critical challenge in treating PW. Photocatalytic oxidation efficiently degrades resistant organic pollutants and converts metallic ions and metalloids in PW into less hazardous forms or species that may be readily eliminated in the following treatment phases. The fundamental photocatalytic oxidation process is titanium dioxide (TiO_2) excitation under UV or visible light exposure. This excitement results in the formation of electrons and holes in the conduction and valence bands, respectively, which subsequently facilitate reduction and oxidation processes. A primary advantage of this technology is its capacity to use sunshine as the irradiation source, lowering energy expenses. Furthermore, it provides regenerability, operates well at standard pressure and temperature, and can fully mineralize diverse organic molecules. Notwithstanding these benefits, the technique has drawbacks, notably the issue of achieving consistent irradiation throughout the entire catalytic surface on an industrial scale and retrieving suspended catalysts. These problems now obstruct the economic viability of extensive implementation (Lin et al., 2017).

Despite the limited use of photocatalytic oxidation in PW treatment, much research has proven its efficacy. The photocatalytic capabilities of nano TiO_2 powder affixed to glass slides (via thermal adhesion) against suspended TiO_2 were assessed in eliminating naphthalene from offshore PW. Their findings demonstrated that the immobilized catalyst showed enhanced efficacy in photo-oxidation and exhibited greater tolerance to the inhibitory effects of offshore PW compared to the suspended TiO_2 powder (Kumar & Gangawane, 2022a). The photocatalysis technique (TiO_2/UV) was used to eliminate dissolved organic materials, particularly phenol, from oil PW at a neutral pH of 7. Their research revealed the efficient elimination of phenol, substantiating the efficacy of photocatalytic oxidation in the remediation of oil-contaminated PW. The effectiveness of the heterogeneous photocatalysis technique (TiO_2/UV) was examined for treating oily industrial effluents. Under optimized conditions—modifications to reagent, temperature, oil content, catalyst amount, and pH—the research noted substantial improvements in water quality, especially regarding color, turbidity, and COD. The findings indicated an 80% removal efficiency of COD with TiO_2/UV and a significant 98% removal efficiency when used with the Fenton reagent (TiO_2/UV/Fe^{2+}/H_2O_2) (Mokhbi et al., 2019). The photocatalytic degradation of oil using titanium dioxide as a nano-photocatalyst was investigated under UV irradiation, applying batch and continuous methods in glass reactors. Their results indicated a total oil removal (100%) in the batch system, but the continuous system attained a removal effectiveness of 71%. These findings demonstrate the potential scalability of the photocatalytic oxidation method for oil removal in PW, albeit the efficiency in continuous systems may need additional improvement (Hassan & Al-Zobai, 2019).

Electrooxidation may transpire via direct or indirect anodic oxidation mechanisms. Oxygen evolution follows the adsorption of pollutants onto an electrode at a low voltage in direct anodic or electrochemical oxidation. This approach offers a more efficient alternative to traditional chemical coagulants used in wastewater treatment, which can produce substantial quantities of sludge with elevated bound water content. This sludge is challenging to dewater, resulting in elevated total dissolved solid (TDS) levels in the effluent, rendering it inappropriate for reuse or release. This problem highlights the want for innovative and resilient alternatives, such as electrocoagulation (EC), which presents advantages, such as little sludge generation, cost efficiency, safety, and adaptability compared to traditional coagulation techniques (Brillas & Martínez-Huitle, 2015).

EC operates by producing cations via the anodic oxidation of iron and aluminum, hence promoting the coagulation of pollutants in water. This method results in the creation of polymeric metal hydroxides with substantial positive charges. These species, in conjunction with negatively charged surfaces from hydroxides, oxyhydroxides, and oxides, facilitate the adsorption and elimination of pollutants. The EC approach incorporates many processes, such as oxidation, reduction, deposition, precipitation, flotation, and coagulation, to remove contaminants from wastewater. Recent research has concentrated on the efficacy of electrooxidation and electrochemical processes to treat generated water from oil and gas operations. A flow-by porous graphite electrode was used in an electrochemical cell to process organic compounds from produced water at a natural gas plant in one investigation. The researchers achieved removal efficiencies of 66.52% for COD, 68.2% for TDS, and total elimination of organic compounds up to C18, using a current density of 1.41 mA/cm^2 and an energy usage of 2.12 kWh/kg COD. It was noted that an increase in current density enhanced the elimination of TDS and COD, attributed to the production of ·OH during PW electrolysis, which acted as oxidizing agents (Abdel-Salam et al., 2018).

The use of electrochemical oxidation (EO) was investigated for the treatment of actual petrochemical effluent using laboratory and pre-pilot-size reactors with Ti/Pt and boron-doped diamond (BDD) anode materials. The findings indicated that over 97% of dissolved organic matter was eliminated via BDD in the pre-pilot reactor under ideal circumstances (15 mA/cm^2, pH 6.0, 25°C, 30 minutes of electrolysis). The research indicated an energy consumption of 3.37 kWh/m^3 and a corresponding cost of 0.57 USD/m^3, corroborating prior results about the economic effectiveness of BDD for rapid organic pollutant removal (de Oliveira Campos et al., 2018). The efficacy of the pulse electrocoagulation (PEC) method was assessed for its capability in treating PW from oilfield operations using alkali/surfactant/polymer flooding. The research showed substantial removal efficiency, achieving 98.3% for COD, 99.0% for oil and grease, 98.8% for turbidity, 98.1% for total suspended solids (TSS), and 94.3% for polyacrylamide using an Fe electrode. This indicates that PEC is an effective approach for enhancing water quality in oilfield wastewater treatment procedures. The findings were obtained under ideal working circumstances, including a pulse density cycle of 0.3, a current density of 35 mA/cm^2, a pulse frequency of 3.0 kHz, an electrode spacing of 1.0 cm, and a reaction time of 40 minutes, with a consumption of 3,000 g Fe/kg COD eliminated. The EC method was used for PW treatment, observing that the duration necessary to attain 90% oil removal efficiency decreased from 46 to 15 minutes with an increase in operating current density from 6.8 to 45.5 mA/cm^2. Their research emphasized the reactor type's efficiency, which used less energy than traditional electrochemical cells that relied on mechanical agitation (Ammar & Akbar, 2018).

An integrated electrocoagulation-chemical coagulation (EC-CC) method was examined for the treatment of water generated by steam-assisted gravity drainage. The analysis of variance (ANOVA) revealed that the treatment duration and concentration of the chemical coagulant were the main parameters influencing TOC removal, accounting for 40% and 26% of the efficiency, respectively. A maximum TOC removal rate of 39.8% was attained using a bipolar aluminum electrode configuration, with a current of 0.34 A at pH 8, a temperature of 60°C, a coagulant concentration of 200 mg/L, a mixing rate of 700 rpm, and a reaction duration of 90 minutes. This underscores the impact of process parameters on the efficacy of TOC reduction in water treatment derived from SAGD. EC was recently used in a batch method using a mild steel anode to eliminate oil and grease (O&G) from synthetic PW. The research used response surface methods to examine different operational parameters, achieving removal efficiencies of 99.7% for COD and 96.43% for O&G from an influent oil-in-water (O/W) emulsion concentration of 1,250 mg/L under ideal circumstances (Manilal et al., 2020).

Addressing hydrocarbon pollution is essential owing to the environmental hazards hydrocarbons provide, particularly to soil and aquatic ecosystems. Diverse techniques have been developed to tackle this problem, including BioSs, bioremediation, and AOPs. Each has unique benefits and drawbacks depending upon the degree of hydrocarbon pollution, as listed in Table 5.1.

The selection among BioSs, bioremediation, and AOPs is contingent upon many aspects, including the specific type of contamination, site conditions, cost, and the required speed of treatment. BioSs and bioremediation provide environmentally sustainable alternatives but with potentially extended timelines for efficacy. AOPs provide expedited and more intense treatment alternatives, but at increased expenses and the risk of secondary contamination. A combination of these technologies is often used in a phased or integrated approach to optimize efficiency and reduce environmental impact.

5.3 CHALLENGES AND FUTURE PROSPECTS

Despite substantial progress in using innovative and environmentally friendly technology for hydrocarbon pollution control, several obstacles remain. These problems stem from technological, economic, and environmental limitations, obstructing broad acceptance and efficacy:

Hydrocarbon pollutants, including crude oil, diesel, and motor oil, are chemically heterogeneous, including intricate combinations of aliphatic and aromatic chemicals, such as PAHs and volatile organic compounds (VOCs). These chemicals differ in toxicity, persistence, and breakdown rates. The challenge is establishing technology capable of precisely targeting this extensive array of chemicals. Certain hydrocarbons are stubborn and resistant to microbial breakdown, complicating bioremediation efforts. Moreover, the hydrophobic characteristics of hydrocarbons render them less susceptible to microbial degradation, necessitating novel approaches to enhance their bioavailability. The use of BioSs, bioemulsifiers, and microbial consortia for bioremediation has shown efficacy in laboratory conditions; however, scaling these methods to real-world situations may result in variable performance. Environmental conditions like temperature, pH, nutrient availability, and heavy metals or other co-contaminants influence the effectiveness of microbial decomposition. These factors often vary in natural ecosystems, complicating the maintenance of ideal conditions for bioremediation. Furthermore, indigenous microbial populations may lack the capability to effectively break down hydrocarbons, necessitating the introduction of other strains, which raises concerns about ecological consequences.

Notwithstanding the environmental advantages, the expense of producing and implementing green technologies, including BioSs and nanotechnologies, is a substantial obstacle. Numerous environmentally friendly alternatives, particularly microbiological and biotechnological solutions, remain in developing phases and lack economic competitiveness with conventional chemical surfactants. The industrial-scale manufacturing of BioSs and other biological agents may be costly owing to the intricacies of fermentation processes, purification, and scalability. This constrains their extensive use, especially in low- and middle-income nations, where cleanup costs must be reduced. At present, there are no universally recognized criteria for

TABLE 5.1
Advantages and Disadvantages of Biosurfactants, Bioremediation, and AOPs

Remediation Approach	Advantages	Disadvantages
BioSs	• BioSs are less harmful and biodegradable, they have less effect on the environment over the long run. • They increase the solubility of hydrophobic hydrocarbons, which makes them more accessible to the activity of microorganisms, which speeds up the breakdown process. • Ex situ (off-site) and in situ (on-site) circumstances are suitable for the use of BioSs for soil and water restoration. • BioSs are suitable for use in a wide variety of situations, such as oil spills or contaminated groundwater, since they are effective even when subjected to harsh circumstances (temperature, pH, and salinity). • The production of biosurfactants may sometimes be accomplished via the utilization of waste substrates, which has the potential to reduce the overall expenditure of hydrocarbon recovery.	• The efficacy of BioSs may fluctuate based on ambient parameters (e.g., temperature, pressure, and pH), influencing the reliability of the outcomes. • The industrial scaling of BioS synthesis, purification, and application remains suboptimal, limiting its extensive use in large-scale processes. • Not all microbes generate BioSs in substantial amounts, which may restrict their extensive applicability until genetic or biotechnological innovations enhance production.
Bioremediation	• Bioremediation is a sustainable and environmentally benign method that employs naturally existing microorganisms to remediate pollutants. • It is often more economical than physical or chemical hydrocarbon remediation approaches, particularly when conducted in situ. • Under ideal circumstances, bioremediation may achieve the total degradation of hydrocarbons into water, carbon dioxide, and biomass. • Microorganisms used in bioremediation exhibit adaptability to many contaminated environments and possess the capability to breakdown an extensive array of hydrocarbons. • In situ bioremediation may be implemented with minimum environmental disruption, since it operates immediately at the polluted location.	• Bioremediation may take weeks, months, or even years, depending upon the degree of pollution and environmental factors. • Microbial activity may diminish or cease under severe environmental circumstances (e.g., low temperatures, abnormal pH levels, or elevated quantities of hazardous chemicals). • To sustain microbial activity, the incorporation of nutrients (e.g., nitrogen, phosphorus) is often essential, which may elevate expenses and complicate the procedure. • Although bioremediation often transforms hydrocarbons into innocuous substances, the intermediate products generated during the degradation process may exhibit more toxicity than the original pollutants.

(Continued)

TABLE 5.1 (*Continued*)
Advantages and Disadvantages of Biosurfactants, Bioremediation, and AOPs

Remediation Approach	Advantages	Disadvantages
		• Certain polluted settings may be deficient in the requisite microbial population for hydrocarbon degradation, necessitating the introduction of exogenous microorganisms or bioaugmentation, which raises further logistical and environmental issues.
AOPs	• AOPs efficiently decompose hydrocarbons into smaller, less toxic compounds, making them appropriate for time-sensitive applications. • In some instances, AOPs may completely mineralize hydrocarbons into carbon dioxide and water, eradicating even little traces of pollution. • AOPs may address a diverse array of organic pollutants, including complex hydrocarbons that may withstand biological destruction. • AOPs are very efficacious in remediating polluted groundwater or industrial effluents containing dissolved or suspended hydrocarbons.	• AOPs can need substantial quantities of chemicals and energy, rendering them costly for extensive or prolonged cleanup initiatives. • Certain AOPs may produce harmful intermediates or by-products that need further treatment prior to the site being deemed totally remediated. • In soil contexts, AOPs may have difficulties in penetrating the subsurface, hence restricting their efficacy for extensive or deeply embedded hydrocarbon pollution. • The implementation of AOPs requires precise regulation of chemical dosage, reaction parameters, and system oversight, making it a technically challenging and intricate procedure. • The use of strong oxidizing chemicals may adversely affect non-target species or result in unforeseen environmental repercussions.

the use of environmentally friendly solutions in hydrocarbon pollution control. The absence of regulation impedes the extensive implementation of bioremediation methods since corporations may be reluctant to invest in technology without definitive governmental endorsement. Furthermore, the efficacy of various remediation methods is often assessed using disparate measures, complicating the comparison and optimization of solutions across diverse situations. Standardizing methodologies for assessing biodegradation rates, BioSs efficacy, and environmental effects is crucial for the area's progression. Although biological therapies are seen as more ecologically sustainable than chemical options, they are not devoid of dangers. Introducing non-native microbial strains for bioremediation may destabilize local ecosystems, resulting in unpredictable ecological repercussions. Concerns exist over the possible buildup of biodegradation by-products, which may have hazardous consequences. The long-term environmental effects of nanotechnology-based methods for hydrocarbon degradation are inadequately known, prompting worries over nanoparticle buildup in ecosystems.

To address these issues and attain sustainable management of hydrocarbon pollutants, other future avenues should be investigated.

Progress in biotechnology, including synthetic biology and genetic engineering, presents significant opportunities for the development of more efficient and economically viable BioSs and bioemulsifiers. Modifying microbial strains to augment BioS synthesis, boost resilience to adverse environmental conditions, and selectively target certain hydrocarbons can transform bioremediation methodologies. Moreover, metabolic engineering may facilitate the formation of microbial consortia, mainly designed to break down diverse hydrocarbons under varying environmental circumstances. Using multi-omics methodologies, including genomes, proteomics, metabolomics, and transcriptomics, may provide an enhanced understanding of the metabolic processes implicated in hydrocarbon breakdown (Ghosh et al., 2023). Comprehending the genetic and molecular foundations of hydrocarbon degradation in microbes may result in identifying innovative enzymes or metabolic pathways that exhibit enhanced efficiency in decomposing complex hydrocarbons. These findings might subsequently be used to develop bioengineered strains or enhance microbial communities for bioremediation.

Nanotechnology offers considerable promise to improve the breakdown of hydrocarbon pollutants. Nanoparticles may be engineered to enhance the bioavailability of hydrophobic hydrocarbons by enhancing their dispersion in the aqueous phase or enabling pseudo-solubilization. Furthermore, nanoparticles may be functionalized with enzymes or catalysts that decompose hydrocarbons more effectively than traditional approaches. Subsequent investigations should concentrate on creating biodegradable and environmentally benign nanomaterials that do not endanger human health or the ecosystem (Mohanty et al., 2024).

Integrating biological, chemical, and physical remediation strategies is a viable alternative to address the shortcomings of singular approaches. For instance, integrating BioSs with AOPs or photocatalytic degradation may expedite hydrocarbon decomposition. Integrating bioelectrochemical systems (BES) with microbial remediation methods may augment electron transport during hydrocarbon breakdown, enhancing efficiency (Kharisov et al., 2014).

Hybrid technologies that incorporate the advantages of many remediation methods may provide more resilient and versatile solutions. Developing real-time monitoring systems using biosensors, nanotechnology, and sophisticated analytics is essential for assessing hydrocarbon degradation progress. These systems may provide ongoing input on pollutant concentrations, microbial community activity, and environmental factors, allowing enhanced control and optimization of cleanup procedures. Artificial intelligence (AI) and machine learning algorithms may augment predictive modeling and process optimization, ensuring more effective and efficient remedial efforts. To facilitate the widespread use of eco-friendly technology, it is essential to establish explicit laws and regulations that promote sustainable practices in hydrocarbon pollution control. Governments must provide incentives, including tax reductions or subsidies, for enterprises investing in bioremediation and sustainable technology. Moreover, public awareness initiatives emphasizing these technologies' environmental and health advantages might stimulate demand for cleaner alternatives. International cooperation and knowledge-sharing platforms would enhance the interchange of best practices and expedite the establishment of global standards. The economic feasibility of advanced green technologies may be enhanced via process optimization and scalability. Improvements in fermentation technology, metabolic engineering, and process automation may lower the expenses related to BioSs manufacturing. Furthermore, the concept of a circular economy, whereby waste materials are used for BioS synthesis or microbial proliferation, may contribute to cost reduction and enhance the accessibility of these technologies. Collaborations among academics, businesses, and governments may facilitate the creation of pilot projects and case studies that illustrate the enduring economic and environmental advantages of green technology (Jimoh & Lin, 2019a).

5.4 CONCLUSIONS

In recent years, worldwide attention has turned to developing innovative and environmentally friendly solutions for managing hydrocarbon pollutants, driven by increasing environmental concerns and the need for sustainable industrial practices. Hydrocarbon contamination, mainly from the petroleum and natural gas sectors, threatens ecosystems and human health. Consequently, creative, efficient, eco-friendly solutions are crucial for tackling these difficulties. Innovative and environmentally sustainable methods, including AOPs, photocatalysis, EC, and bioremediation, have proven effective in diminishing the ecological impact of hydrocarbon pollutants. These technologies seek to decompose or eliminate dangerous organic pollutants, including petroleum hydrocarbons, PAHs, and other toxic compounds, while avoiding further environmental damage. In contrast to conventional techniques (e.g., chemical coagulation), these environmentally friendly solutions provide minimum waste, reduce sludge volumes, and diminish effluent toxicity, enhancing sustainability for prolonged environmental stewardship. Numerous sophisticated technologies, mainly those dependent on natural resources such as sunshine (solar photocatalysis) or those that improve energy utilization (PEC), exhibit greater energy efficiency than traditional techniques. Solar technologies specifically use renewable energy sources, decreasing operating expenses (OPEX) and energy usage. Specifically, technologies like EC and electrooxidation demonstrate reduced operating costs owing to their capacity to process substantial quantities of wastewater with deficient energy expenditure and limited chemical use. This has established them as scalable solutions for treating and reusing oilfield-PW, assisting enterprises in reducing freshwater consumption in operations like hydraulic fracturing and water flooding procedures.

The scalability of greener technology for industrial applications is critical in assessing their feasibility. Numerous nascent technologies, first evaluated in laboratory environments, are currently being modified for extensive applications, especially within the oil and gas industries. Scaling photocatalytic and electrochemical technologies to pre-pilot or pilot sizes has shown encouraging outcomes for practical use. Electrooxidation and photocatalysis have shown significant promise for large-scale applications in treating generated water and petrochemical wastes. Furthermore, several technologies have shown capable of adhering to rigorous environmental discharge criteria, which is crucial for enterprises functioning in ecologically sensitive areas. Moreover, bioremediation methods, which use microorganisms to decompose hydrocarbons, have gained popularity owing to their environmental sustainability and scalability. Biological treatments, particularly when integrated with chemical or physical treatment techniques, may provide efficient long-term solutions for remediating oil spills and controlling wastewater.

A principal benefit of modern technologies is their adaptability in addressing various hydrocarbon contaminants with differing complexity. Hydrocarbons in generated water, refinery effluent, and oil spills exhibit variability in molecular structure, toxicity, and solubility, complicating their management using traditional approaches alone. AOPs and EC have effectively degraded complex hydrocarbons in generated water, such as PAHs, phenols, and heavy metal pollutants. These methods may convert persistent organic pollutants into simpler, less harmful chemicals, improving the viability of later treatments or enabling direct environmental discharge where rules allow. Hybrid technologies integrating various treatment processes (e.g., EC with advanced oxidation or photocatalysis with Fenton processes) have enhanced removal efficiencies for multiple contaminants in intricate wastewater matrices, providing a holistic strategy for addressing hydrocarbon pollutants.

Notwithstanding the potential of these new technologies, several problems persist. The recovery and reuse of catalysts, especially in photocatalytic processes, are significant issues. The elevated expense of specific catalysts, like nano-scale titanium dioxide (TiO_2), constrains their extensive utilization. Moreover, attaining consistent irradiation across extensive reactor surfaces, particularly in photocatalysis, presents technical hurdles in industrial applications. Current research tackles these difficulties by creating more efficient and durable catalysts with new reactor designs that enhance energy efficiency and pollutant elimination. Incorporating renewable energy sources, such as solar energy, into these processes is expected to significantly diminish operating costs and carbon emissions, enhancing the appeal of these technologies for companies with substantial hydrocarbon waste streams. Subsequent studies must persist in refining operational parameters (e.g., pH, temperature, and reaction durations) to improve pollutant elimination while reducing prices and energy expenditure. Integrating sophisticated technology with biological treatment approaches has possibilities for holistic solutions that tackle hydrocarbon removal and the restoration of ecosystems impacted by hydrocarbon contamination. The transition to sustainable technology corresponds with the tenets of the circular economy, whereby waste materials are utilized or recycled instead of being disposed of. By purifying generated water to stringent requirements, oil and gas firms may recycle it, reducing freshwater use and easing the strain on natural water supplies. Technologies like EC, which efficiently eliminate pollutants, including oil, grease, and heavy metals from generated water, are essential for facilitating water reuse in industrial operations like hydraulic fracturing and chemical-enhanced oil recovery (EOR). These improvements enhance sustainable water management procedures, conserving water resources while preserving operating efficiency.

REFERENCES

Abdel-Salam, O. E., Abou Taleb, E. M., & Afify, A. A. (2018). Electrochemical treatment of chemical oxygen demand in produced water using flow-by porous graphite electrode. *Water and Environment Journal*, *32*(3), 404–411. https://doi.org/10.1111/wej.12336

Al-Khafaji, R. Q., & Mohammed, A. H. A. K. (2019). Optimization of continuous electro-fenton and photo electro-fenton processes to treat Iraqi oilfield produced water using surface response methodology. *IOP Conference Series: Materials Science and Engineering*, *518*(6). https://doi.org/10.1088/1757-899X/518/6/062007

Ammar, S. H., & Akbar, A. S. (2018). Oilfield produced water treatment in internal-loop air-lift reactor using electrocoagulation/flotation technique. *Chinese Journal of Chemical Engineering*, *26*(4), 879–885. https://doi.org/10.1016/j.cjche.2017.07.020

Azizah, A. N., & Widiasa, I. N. (2018). Advanced oxidation processes (AOPs) for refinery wastewater treatment contains high phenol concentration. *MATEC Web of Conferences*, *156*, 03012. https://doi.org/10.1051/matecconf/201815603012

Babuponnusami, A., & Muthukumar, K. (2014). A review on Fenton and improvements to the Fenton process for wastewater treatment. *Journal of Environmental Chemical Engineering*, *2*(1), 557–572. https://doi.org/10.1016/j.jece.2013.10.011

Bala, K., Dogra, A., Lata, P., & Das, A. P. (2024). Discussing sources and biological-chemical remediation approaches for microplastic pollution. In *Sustainable Microbial Technology for Synthetic and Cellulosic Microfiber Bioremediation*, 37–59. https://doi.org/10.1007/978-3-031-62600-5_3

Behera, I. D., Basak, G., Kumar, R. R., Sen, R., & Meikap, B. C. (2020). Treatment of petroleum refinery sludge by petroleum degrading bacterium Stenotrophomonas pavanii IRB19 as an efficient novel technology. *Journal of Environmental Science and Health, Part A*, *56*(2), 226–239. https://doi.org/10.1080/10934529.2020.1866924

Behera, I. D., Manoranjan, N., Subrata, B., Meikap, B. C., & Sen, R. (2021). Enhanced biodegradation of total petroleum hydrocarbons by implementing a novel two-step bioaugmentation strategy using indigenous bacterial consortium. *Journal of Environmental Management*, *292*, 112746. ISSN 0301–4797, https://doi.org/10.1016/j.jenvman.2021.112746

Behera, I. D., Nayak, M., Mishra, A., Meikap, B. C., & Sen, R. (2022). Strategic implementation of integrated bioaugmentation and biostimulation for efficient mitigation of petroleum hydrocarbon pollutants from terrestrial and aquatic environment. *Marine Pollution Bulletin*, *177*, 113492. ISSN 0025-326X, https://doi.org/10.1016/j.marpolbul.2022.113492

Brillas, E., & Martínez-Huitle, C. A. (2015). Decontamination of wastewaters containing synthetic organic dyes by electrochemical methods. An updated review. *Applied Catalysis B: Environmental*, *166–167*, 603–643. https://doi.org/10.1016/j.apcatb.2014.11.016

Chang, B.-V., Chang, I. T., & Yuan, S. Y. (2008). Anaerobic degradation of phenanthrene and pyrene in mangrove sediment. *Bulletin of Environmental Contamination and Toxicology*, *80*(2), 145–149. https://doi.org/10.1007/s00128-007-9333-1

Das, A., Dey, S., Bala, K., Gnanasekaran, R., & Das, A. P. (2024). Microbial bioremediation technology for sustainable treatment and management of synthetic microfiber waste. In *Sustainable Microbial Technology for Synthetic and Cellulosic Microfiber Bioremediation*, 221–238. https://doi.org/10.1007/978-3-031-62600-5_12

Das, A., Ray, S. K., Mohanty, M., Mohanty, J., Dey, S., & Das, A. P. (2024). Ecotoxicity of microplastic wastes and their sustainable management: A review. *Environmental Chemistry and Ecotoxicology*, *6*, 144–152. https://doi.org/10.1016/J.ENCECO.2024.05.003

Das, A. P., Sukla, L. B., Pradhan, N., & Nayak, S. (2011). Manganese biomining: A review. *Bioresource Technology*, *102*(16), 7381–7387. https://doi.org/10.1016/J.BIORTECH.2011.05.018

deOliveira Campos, V., do Amaral, L. B., Solano, A. M. S., de Araújo, D. M., Martínez-Huitle, C. A., & da Silva, D. R. (2018). Electrochemical treatment of produced water using Ti/Pt and BDD anode. *International Journal of Electrochemical Science*, *13*(8), 7894–7906. https://doi.org/10.20964/2018.08.44

Dey, S., Bissoyi, A., Koteswara Reddy, G., & Das, A. P. (2024). Synthetic microfibers: An urban emerging pollutant demanding for sustainable environmental solutions. In *Sustainable Microbial Technology for Synthetic and Cellulosic Microfiber Bioremediation*, 107–121. https://doi.org/10.1007/978-3-031-62600-5_7

Dey, S., Das, A., Mallick, K., Sahu, A., & Das, A. P. (2023). Environmental petroleum waste: Pollution, toxicity, sustainable remediation. 159–175. https://doi.org/10.1007/978-3-031-48220-5_7

Dey, S., Dash, N., Ray, S. K., Priyadarsini, S., Chhotaray, C., & Das, A. P. (2024). Occurrence and quantification of tire wear plastics pollutants from highway city dust of Bhubaneswar City of Odisha. In *Sustainable Microbial Technology for Synthetic and Cellulosic Microfiber Bioremediation*, 77–89. https://doi.org/10.1007/978-3-031-62600-5_5

Fakhru'l-Razi, A., Pendashteh, A., Abdullah, L. C., Biak, D. R. A., Madaeni, S. S., & Abidin, Z. Z. (2009). Review of technologies for oil and gas produced water treatment. *Journal of Hazardous Materials*, *170*(2–3), 530–551. https://doi.org/10.1016/j.jhazmat.2009.05.044

Ghosh, S., Tripathy, B., Dey, S., & Das, A. P. (2023). Comparative investigation of fungal and bacterial manganese biomining mechanisms. In *Manganese Mining Microorganisms*, 99–115. https://doi.org/10.1016/b978-0-12-822147-1.00016-9

Gudiña, E. J., Pereira, J. F. B., Costa, R., Evtuguin, D. V., Coutinho, J. A. P., Teixeira, J. A., & Rodrigues, L. R. (2015). Novel bioemulsifier produced by a Paenibacillus strain isolated from crude oil. *Microbial Cell Factories*, *14*(1), 14. https://doi.org/10.1186/s12934-015-0197-5

Hassan, A. A., & Al-Zobai, K. M. M. (2019). Chemical oxidation for oil separation from oilfield produced water under UV irradiation using titanium dioxide as a nano-photocatalyst by batch and continuous techniques. *International Journal of Chemical Engineering*, *2019*, 1–8. https://doi.org/10.1155/2019/9810728

Jha, S. S., Joshi, S. J., & S.J., G. (2016). Lipopeptide production by Bacillus subtilis R1 and its possible applications. *Brazilian Journal of Microbiology*, *47*(4), 955–964. https://doi.org/10.1016/j.bjm.2016.07.006

Jiménez, S., Andreozzi, M., Micó, M. M., Álvarez, M. G., & Contreras, S. (2019). Produced water treatment by advanced oxidation processes. *Science of the Total Environment*, *666*, 12–21. https://doi.org/10.1016/j.scitotenv.2019.02.128

Jimoh, A. A., & Lin, J. (2019a). Biosurfactant: A new frontier for greener technology and environmental sustainability. *Ecotoxicology and Environmental Safety*, *184*, 109607. https://doi.org/10.1016/j.ecoenv.2019.109607

Jin, Y., & Davarpanah, A. (2020). Using photo-fenton and floatation techniques for the sustainable management of flow-back produced water reuse in shale reservoirs exploration. *Water, Air, & Soil Pollution*, *231*(8), 441. https://doi.org/10.1007/s11270-020-04812-7

Kharisov, B. I., Dias, H. V. R., & Kharissova, O. V. (2014). Nanotechnology-based remediation of petroleum impurities from water. *Journal of Petroleum Science and Engineering*, *122*, 705–718. https://doi.org/10.1016/j.petrol.2014.09.013

Koolivand, A., Abtahi, H., Villaseñor, J., Saeedi, R., Godini, K., & Parhamfar, M. (2020). Effective scale-up of oily sludge bioremediation from a culture-based medium to a two-phase composting system using an isolated hydrocarbon-degrading bacterium: Effect of two-step bioaugmentation. *Journal of Material Cycles and Waste Management*, *22*(5), 1475–1483. https://doi.org/10.1007/s10163-020-01036-z

Kumar, A., & Gangawane, K. M. (2022a). Effect of precipitating agents on the magnetic and structural properties of the synthesized ferrimagnetic nanoparticles by co-precipitation method. *Powder Technology*, *401*, 117298. https://doi.org/10.1016/j.powtec.2022.117298

Kumar, A., & Gangawane, K. M. (2022b). Synthesis and effect on the surface morphology & magnetic properties of ferrimagnetic nanoparticles by different wet chemical synthesis methods. *Powder Technology*, *410*, 117867. https://doi.org/10.1016/j.powtec.2022.117867

Kumar, A., Gangawane, K. M., & Ramanjaneyulu, B. (2022). Ferrofluids for waste-water treatment. In *Advances in Chemical, Bio and Environmental Engineering*, 723–744. Springer. https://doi.org/10.1007/978-3-030-96554-9_48

Lahiri, D., Nag, M., Dey, A., Sarkar, T., Joshi, S., Pandit, S., Das, A. P., Pati, S., Pattanaik, S., Tilak, V. K., & Ray, R. R. (2021). Biofilm mediated degradation of petroleum products. *Geomicrobiology Journal*, 1–10. https://doi.org/10.1080/01490451.2021.1968979

Lászlová, K., Dudášová, H., Olejníková, P., Horváthová, G., Velická, Z., Horváthová, H., & Dercová, K. (2018). The application of biosurfactants in bioremediation of the aged sediment contaminated with polychlorinated biphenyls. *Water, Air, & Soil Pollution*, *229*(7), 219. https://doi.org/10.1007/s11270-018-3872-4

Li, J., & Yang, H. (2021). Polycyclic aromatic hydrocarbon improves the anaerobic biodegradation of benz [α] anthracene in sludge via boosting the microbial activity and bioavailability. *Pakistan Journal of Zoology*, *53*(6), 2445–2450. https://doi.org/10.17582/journal.pjz/20200106160152

Lin, L., Wang, H., Jiang, W., Mkaouar, A. R., & Xu, P. (2017). Comparison study on photocatalytic oxidation of pharmaceuticals by TiO_2-Fe and TiO_2-reduced graphene oxide nanocomposites immobilized on optical fibers. *Journal of Hazardous Materials*, *333*, 162–168. https://doi.org/10.1016/j.jhazmat.2017.02.044

Manilal, A. M., Soloman, P. A., & Basha, C. A. (2020). Removal of oil and grease from produced water using electrocoagulation. *Journal of Hazardous, Toxic, and Radioactive Waste*, *24*(1), 1–13. https://doi.org/10.1061/(ASCE)HZ.2153-5515.0000463

Martins, P. C., & Martins, V. G. (2018). Biosurfactant production from industrial wastes with potential remove of insoluble paint. *International Biodeterioration & Biodegradation*, *127*, 10–16. https://doi.org/10.1016/j.ibiod.2017.11.005

Mishra, S., Dash, D., & Das, A. P. (2024). Aquatic microbial diversity on plastisphere: Colonization and potential role in microplastic biodegradation. *Geomicrobiology Journal*, *41*(4), 312–323. https://doi.org/10.1080/01490451.2023.2209750

Mohanta, S., Pradhan, B., & Behera, I. D. (2023). Impact and remediation of petroleum hydrocarbon pollutants on agricultural land: A review. *Geomicrobiology Journal*, *41*(4), 345–359. https://doi.org/10.1080/01490451.2023.2243925

Mohanty, J., Dey, S., Bala, K., & Das, A. P. (2024). Recent advances in microplastic pollution for its sustainable management. In *Sustainable Management of Urban Plastic Waste through Circular Economic Approaches*, 27–48. https://doi.org/10.1201/9781003465447-3

Mokhbi, Y., Korichi, M., & Akchiche, Z. (2019). Combined photocatalytic and Fenton oxidation for oily wastewater treatment. *Applied Water Science*, *9*(2), 35. https://doi.org/10.1007/s13201-019-0916-x

Mota, A. L. N., Neto, L. G. L., Foletto, E. L., Chiavone-Filho, O., & do Nascimento, C. A. O. (2018). Analysis of solar and artificial UVA irradiations on the photo-Fenton treatment of phenolic effluent and oilfield produced water. *Chemical Engineering Communications*, *205*(11), 1594–1603. https://doi.org/10.1080/00986445.2018.1463524

Olajire, A. A. (2020). Recent advances on the treatment technology of oil and gas produced water for sustainable energy industry-mechanistic aspects and process chemistry perspectives. *Chemical Engineering Journal Advances*, *4*(October), 100049. https://doi.org/10.1016/j.ceja.2020.100049

Pachhade, K., Sandhya, S., & Swaminathan, K. (2009). Ozonation of reactive dye, Procion red MX-5B catalyzed by metal ions. *Journal of Hazardous Materials*, *167*(1–3), 313–318. https://doi.org/10.1016/j.jhazmat.2008.12.126

Pal, D., & Sen, S. (2023). Emerging petroleum pollutants and their adverse effects on the environment. In *Impact of Petroleum Waste on Environmental Pollution and its Sustainable Management through Circular Economy*, 103–137. https://doi.org/10.1007/978-3-031-48220-5_5

Parhamfar, M., Abtahi, H., Godini, K., Saeedi, R., Sartaj, M., Villaseñor, J., Coulon, F., Kumar, V., Soltanighias, T., Ghaznavi-Rad, E., & Koolivand, A. (2020). Biodegradation of heavy oily sludge by a two-step inoculation composting process using synergistic effect of indigenous isolated bacteria. *Process Biochemistry*, *91*(December 2019), 223–230. https://doi.org/10.1016/j.procbio.2019.12.014

Paul, A., Priyadarsini, S., Karakuş, S., Behera, J. K., Gnanasekaran, R., & Das, A. P. (2024). Fate and accumulation of synthetic microfibers in the environment. In *Sustainable Microbial Technology for Synthetic and Cellulosic Microfiber Bioremediation*, 61–76. https://doi.org/10.1007/978-3-031-62600-5_4

Sharma, D., & Saharan, B. S. (2016). Functional characterization of biomedical potential of biosurfactant produced by Lactobacillus helveticus. *Biotechnology Reports*, *11*, 27–35. https://doi.org/10.1016/j.btre.2016.05.001

Shokri, A. (2018). Application of Sono–photo-Fenton process for degradation of phenol derivatives in petrochemical wastewater using full factorial design of experiment. *International Journal of Industrial Chemistry*, *9*(4), 295–303. https://doi.org/10.1007/s40090-018-0159-y

Silva, P. C., Ferraz, N. P., Perpetuo, E. A., & Asencios, Y. J. O. (2019). Oil produced water treatment using advanced oxidative processes: Heterogeneous-photocatalysis and photo-Fenton. *Journal of Sedimentary Environments*, *4*(1), 99–107. https://doi.org/10.12957/jse.2019.40991

Sivagami, K., Anand, D., Divyapriya, G., & Nambi, I. (2019). Treatment of petroleum oil spill sludge using the combined ultrasound and Fenton oxidation process. *Ultrasonics Sonochemistry*, *51*(September 2018), 340–349. https://doi.org/10.1016/j.ultsonch.2018.09.007

6 Microbial Degradation of Polycyclic, Aromatic Hydrocarbon-Contaminated Sites

Aswetha Iyer, Alok Prasad Das, Melissa Chen, and Sevanan Murugan

6.1 INTRODUCTION

Polycyclic Aromatic Hydrocarbons (PAHs) represent a class of organic pollutants renowned for their profound toxicity, encompassing mutagenic, carcinogenic, teratogenic, and immunotoxic properties (Patel et al., 2020). Classified among persistent organic pollutants (POPs), PAHs are characterized by their different chemical properties that are in turn governed by their structural arrangement, comprising two or more fused benzene rings, with a minimum of two aromatic rings in their composition (Cao et al., 2023). According to the United States Environmental Protection Agency (US EPA), about 16 PAHs are priority categorized as emerging environmental pollutants based on their occurrence and toxicity (Ghosh et al., 2018, 2023). PHAs can be grouped into two types based on their molecular weights and properties- (1) *Low-molecular-weight* (LMW) PAHs, with two to three benzenoid rings, are generally deemed non-carcinogenic to humans, and (2) *High-molecular-weight* (HMW) PAHs, featuring four to six benzenoid rings, encompass seven known carcinogenic PAHs. The most popular PAH compounds include naphthalene, anthracene, and phenanthrene, and are omnipresent in the environment, arising from both natural and anthropogenic sources. Their hydrophobic nature predisposes them to accumulate in aquatic sediments, fostering bioaccumulation and escalating concentrations over time. Acute contact with PHAs and their residues can cause superficial damage to eyes and skin, and consistent exposure for a long duration may induce multiple organ failure and prove fatal (Behera et al., 2022).

6.1.1 Sources and Distribution of PAH

PAH pollutants are widespread, present in aquatic and terrestrial environments and in the air (Patel et al., 2020). These pollutants stem from two primary sources: anthropogenic emissions and natural emissions (Patel et al., 2020).

DOI: 10.1201/9781003508496-6

6.1.2 Natural Sources

While natural sources like biomass, coal (Jia et al., 2021), mud volcanoes, volcanic eruptions, natural forest fires, and lightning-induced moorland fires play a minor role, anthropogenic sources are the primary contributors to PAH pollution.

6.1.3 Anthropogenic Sources

These anthropogenic sources can be categorized into industrial, mobile, domestic, and agricultural emissions.

6.1.3.1 Industrial Emissions

Incomplete combustion during various industrial processes such as waste incineration, iron and steel production, aluminum production, cement manufacturing, coal-tar pitch production, dye manufacturing, asphalt industries, rubber tire manufacturing, fungicide and insecticide production, exhaust from refineries, and power production is the main source of PAH emissions. Other industrial sources include coal gasification, electric arc furnaces (EAFs), oxygen furnaces, diesel engines, and gasoline-powered engines in heavy machinery.

6.1.3.2 Mobile Emissions

Mobile sources encompass vehicle exhaust from aircraft, ships, trains, and both heavy and light off-road vehicles (Patel et al., 2020; Mojiri et al., 2019).

6.1.3.3 Domestic Emissions

Domestic emission sources encompass a range of household activities, including the combustion of garbage, coal, and wood as well as cooking on indoor stoves fueled by oil, gas, kerosene, or wood, and other forms of residential heating.

6.1.3.4 Agricultural Emissions

Agricultural emission sources primarily stem from open biomass burning, such as the disposal of garden waste and the combustion of crop residues (Fakinle et al., 2022). In rural areas, elevated levels of PAH pollution are primarily attributed to domestic and agricultural sources, whereas in urban settings, industrial, mobile, and domestic activities contribute significantly (Dey et al., 2023; Patel et al., 2020).

6.1.3.5 Anthropogenic Emissions

Anthropogenic sources overwhelmingly contribute to PAH pollution, stemming from the combined effects of combustion engines, residential heating, industrial operations, and biomass burning. The USEPA designates 16 priority PAHs for environmental analysis, with benzo(a)pyrene considered the benchmark due to its high carcinogenicity. Once in the atmosphere, PAHs distribute among gas, particle, and droplet phases, often traveling aerially and depositing across borders. Ultimately, they settle in soils and street dust before washing into aquatic environments via rain and stormwater. In water, PAHs preferentially accumulate in sediment, creating a complex environmental cycle and multiple exposure pathways (Paul et al., 2023).

Extensive global studies have uncovered high PAH levels in various environmental matrices, highlighting the widespread risk posed by these carcinogens (Hussain et al., 2018).

6.2 INDUSTRIES RELATED TO PAH

6.2.1 Steel and Iron Industries

The steel and iron industries are major contributors to PAH pollution due to the extensive use of carbonaceous materials and high-temperature processes involved in steelmaking. Throughout various stages of steel and iron production, including coke manufacturing, sintering, ironmaking, casting, mold pouring, and cooling, PAHs can be emitted into the environment (Sen et al., 2018). According to SAIL (Steel Authority of India Limited), coke manufacturing is particularly significant as it involves the heating of coal in the absence of air to produce coke, a fuel and reducing agent used in blast furnaces. This process releases PAHs as byproducts of coal pyrolysis and incomplete combustion (Zhang et al., 2023). Sintering, which involves heating iron ore fines with coke breeze or other carbonaceous materials to form a porous mass, can also lead to PAH emissions. During ironmaking in blast furnaces, coke and iron ore are melted and reduced to produce molten iron. The combustion of coke and other carbonaceous materials in the furnace generates PAHs, which can be released through stack emissions and fugitive sources. Similarly, casting, mold pouring, and cooling processes in steelmaking facilities can result in the formation and release of PAHs from the combustion of carbonaceous materials used as fuel or binders (Baraniecka et al., 2010). Efforts to mitigate PAH emissions in the steel and iron industries often focus on improving combustion efficiency, optimizing process parameters, and implementing pollution control technologies such as baghouses, scrubbers, and catalytic converters as used in waste power plants (Dey et al., 2024). However, the inherent reliance on carbonaceous materials in these processes poses ongoing challenges for reducing PAH pollution.

6.2.2 Electric Arc Furnace and Rolling Mill

EAFs play a crucial role in steelmaking, particularly in recycling scrap metal to produce new steel. EAFs use electric arcs to melt scrap steel and other raw materials, with carbon electrodes providing the necessary heat. This process can lead to significant PAH emissions, particularly at the electrode tips where intense heat promotes the formation of PAHs through the pyrolysis of organic compounds present in the scrap metal (Madias, 2016; Odabasi et al., 2009). In addition to EAFs, rolling mills are another source of PAH pollution in steelworks. Rolling mills shape and refine molten steel into various products through processes such as hot rolling (Panigrahi, 2001) and cold rolling. While the direct emissions of PAHs from rolling mills may be lower compared to EAFs, the combustion of fossil fuels for heating and powering machinery can still contribute to overall PAH emissions from the facility (Baraniecka et al., 2010). Efforts to reduce PAH emissions from EAFs and rolling mills often involve optimizing operating parameters, improving dust and fume collection systems, and adopting cleaner technologies such as Activated Charcoal Filters (Dat & Chang, 2017). Additionally, the use of alternative materials and fuels (Yilmaz et al., 2022) can help minimize PAH pollution in steelworks.

6.2.3 Petroleum Refining

Petroleum refining is a major industrial process that involves the conversion of crude oil into various petroleum products such as gasoline, diesel, and jet fuel (Robinson & Hsu, 2017). During refining, crude oil undergoes fractional distillation and other processes such as cracking, reforming, and hydroprocessing to separate and upgrade different hydrocarbon fractions. PAHs can be formed as byproducts of several refining processes, including cracking and reforming, where high temperatures and catalysts are used to break down large hydrocarbon molecules into smaller ones (Kuppusamy et al., 2020). Additionally, PAHs may be present in crude oil itself (Abdel-Shafy & Mansour, 2016; Zhao et al., 2014), and their concentration can vary depending on the source and composition of the crude.

6.2.4 Coal-Fired Power Plants

Coal-fired power plants are significant sources of PAH emissions due to the combustion of coal to generate electricity. PAHs are formed during the incomplete combustion of coal, particularly at high temperatures in the boiler and furnace. While modern coal-fired power plants are equipped with pollution control technologies such as electrostatic precipitators and flue gas desulfurization systems to reduce emissions of particulate matter and sulfur dioxide (Liang et al., 2020), PAH emissions remain a concern, especially in older and less efficient facilities.

6.2.5 Waste Management and Landfills

Waste management facilities, including landfills and waste incinerators, are significant sources of PAH pollution (Hsu et al., 2016). PAHs can be released during the decomposition of organic waste in landfills (Kalmykova et al., 2013), as well as during the incineration of municipal solid waste and hazardous waste (Li et al., 2019). In landfills, PAHs can leach from disposed-of materials into soil (Cai et al., 2019; Kuśmierz et al., 2016; Kalbe et al., 2008) and groundwater, posing risks to human health and the environment. In waste incinerators, PAHs are formed during the combustion of organic materials in waste streams, particularly those containing plastics, rubber, and other hydrocarbon-based materials (Hsu et al., 2021, 2016).

6.2.6 Chemical Manufacturing

Chemical manufacturing processes can generate PAHs as byproducts or intermediates in various reactions. PAHs may be formed during the production of certain chemicals, such as dyes, pesticides, pharmaceuticals, and plastics. Processes involving high-temperature reactions, combustion, or the use of aromatic compounds as raw materials are particularly prone to PAH formation (Hu et al., 2020).

6.2.7 Asphalt Production and Paving

Asphalt production plants and paving operations are sources of PAH emissions due to the heating and processing of bitumen, a dark, viscous material derived from crude oil. PAHs can be released during the heating and mixing of bitumen with aggregates to produce asphalt pavement materials. In asphalt production plants, PAH emissions may occur during the heating of bitumen in storage tanks, mixing tanks, and asphalt plants. Paving operations, including the laying and compaction of hot mix asphalt on roads and other surfaces, can also release PAHs into the air as the asphalt mixture is heated and compacted.

6.2.8 Brick and Ceramics Manufacturing

Brick and ceramics manufacturing processes involve the firing of clay and other raw materials in kilns at high temperatures to produce bricks, tiles, pottery, and other ceramic products (Ribeiro & Tulyaganov, 2021; Fernandes, 2019). PAHs can be released during the combustion of organic materials in the kiln, as well as from the use of coal or other carbonaceous fuels as energy sources.

6.3 ROUTEWAY OF PAH INTO THE ENVIRONMENT

6.3.1 Terrestrial Ecosystem

Sources of PAHs in terrestrial environments encompass both natural and anthropogenic origins, each contributing to the distribution and impact of these contaminants. Endogenous concentrations of PAHs in soil and vegetation, ranging from 1 to 20 μg/kg, arise from natural processes such as plant synthesis, forest fires, and volcanic activity. However, the predominant sources of PAHs are anthropogenic, with fossil fuel combustion being a primary contributor, estimated to release millions of kilograms of benzo(a)pyrene (BaP) annually (Edwards, 1983). Urbanization exacerbates PAH contamination in terrestrial ecosystems, with highly urbanized areas exhibiting elevated ecological risks associated with PAH exposure (Han et al., 2020). Furthermore, wildfires, though underexplored, emerge as significant contributors to PAHs in terrestrial environments, with immediate increases in PAH levels observed in soil and ash following fire events. The exploitation of fossil fuels and combustion processes releases pyrogenic PAHs, which may deposit in remote regions, including the Arctic, further exacerbating terrestrial contamination (Balmer et al., 2019).

Additionally, localized sources such as biomass burning and petroleum extraction activities contribute to PAH contamination in soils, highlighting a complex interplay between natural and anthropogenic sources. In terrestrial ecosystems, PAH releases from fossil fuel activities and related processes pose contamination risks, particularly with the expanding oil and gas industry. To address this, monitoring networks must include petrogenic PAHs near fuel transportation routes. Major PAH sources in Canada include forest fires, coal combustion for iron and steel production, and carbon electrode use in aluminum smelters. The hydrophobic and adsorptive properties of PAHs influence their behavior in soils and sediments, impacting their distribution, bioavailability, and degradation processes, thereby shaping the overall ecological risks associated with PAH contamination in terrestrial environments.

6.4 AQUATIC ECOSYSTEM

PAHs rank among the most prevalent contaminants encountered in aquatic ecosystems (Ozaki et al., 2019; Jaward et al., 2012). Typically co-occurring with other pollutants, they are detected in sediments, water sources, wastewater, and crustaceans. PAHs predominantly arise as byproducts of fuel combustion, although their discharge can also stem from agricultural fires, industrial waste, and cooking activities. PAHs vary in solubility in water according to molecular weight, with higher-molecular-weight PAHs exhibiting lower solubility. This characteristic often leads them to attach to particulate matter, serving as a primary transport pathway from land and air to aquatic environments and from surface waters to deeper layers (Moharana et al., 2024; Vagge et al., 2018). The entry of PAHs into water sources occurs through dry and wet deposition, road runoff, leaching from creosote-treated wood, industrial effluents, petroleum spills, and combustion of fossil fuels (Karyab et al., 2013). Also, forest fires act as drivers for the production and mobilization of PAHs, acting as a diffuse source of PAH contamination to the aquatic ecosystems (Campos & Abrantes, 2021).

Across diverse aquatic environments, PAH concentrations span a wide range. In drinking water, concentrations vary from 1.33 ng/L for benzo[a]pyrene in treated water in Tehran, Iran, to 139,000 ng/L for phenanthrene in untreated water in Lagos, Nigeria. River and lake concentrations range from 0.5 ng/L for benzo[a]pyrene in Northeastern China to 1,138,000 ng/L for pyrene in South Africa. Wastewater concentrations range from 14 ng/L for fluorene in domestic wastewater in Jordan to 8,310,000 ng/L for benzo[b]fluoranthene in domestic wastewater in South Africa. Seawater and groundwater concentrations range from 0.02 ng/L for chrysene in the Persian Gulf to 46,600 ng/L for indeno[1,2,3-cd] pyrene in the Timor Sea, Indonesia, and from 0.1 ng/L for benzo[a]pyrene in North China to 739.1 ng/L for naphthalene in Near Huai River, China. The highest PAH concentrations are typically observed at river mouths, decreasing toward both the river and sea with changing tides. Diagnostic ratios of PAHs indicate that coal and petroleum combustion are primary sources of PAHs in the Salt River. As POPs, PAHs, originating mainly from fossil fuel combustion and petroleum release, pose significant environmental threats due to their resistance to degradation and potential for bioaccumulation. Classified as priority pollutants by both the US Environmental Protection Agency and the European Union, PAHs represent ongoing challenges for ecosystem health and management.

6.5 ATMOSPHERE

The atmosphere serves as a reservoir for PAHs, derived from diverse natural and anthropogenic sources, with varying spatial and temporal distributions. Previous studies have highlighted the widespread presence of PAHs in coral reef regions. These compounds originate from incomplete combustion in traffic, industrial, agricultural, domestic, and natural processes, dispersing globally upon release into the atmosphere. Investigations in urban environments, such as Delhi-National Capital Region and Beijing, elucidate the sources and spatial distribution of PAHs, identifying pyrogenic and petrogenic emissions as major contributors (Hazarika et al., 2019). Additionally, in Zaragoza City, Spain, positive matrix factorization revealed coal combustion, gasoline, vehicular, and stationary emissions as significant sources of PAHs, with notable seasonal variations and directional impacts (Callén et al., 2013).

Combustion processes, including those involving coal, oil, wood, and petroleum products, stand as primary sources of atmospheric PAHs, generating these compounds through pyrosynthesis and pyrolysis mechanisms (Dat & Chang, 2017). While natural events like volcanic eruptions and forest fires contribute to atmospheric PAHs, anthropogenic activities dominate, particularly through incomplete combustion and pyrolysis (Dat & Chang, 2017). Thus, understanding the sources and dynamics of atmospheric PAHs is crucial for assessing their environmental fate and associated risks, highlighting the importance of comprehensive monitoring and mitigation strategies.

6.6 EFFECTS OF PAH ON THE ENVIRONMENT

Intensive human activities, particularly in coastal environments, have led to heightened environmental pollution. In Zhanjiang Harbor, China, a notable instance of this phenomenon, 16 priority PAHs were identified in sediments and marine species. Concentrations of these PAHs ranged from 151 to 453 ng/g dry weight (dw) in sediments and from 86.7 to 256 ng/g wet weight (ww) in organism tissues. Alarmingly, higher levels of PAHs were observed near the estuary. However, there has been a decline in PAH levels compared to pre-2012 levels. PAHs have emerged as significant concerns due to their adverse effects on various components of the ecosystem. These hydrocarbons, known for their carcinogenic potential, prompt researchers to employ toxic equivalency factors (TEFs) and risk quotients (RQs) to assess their impact on both organisms and ecosystems (Hussain et al., 2018). However, despite efforts to understand their risk, PAHs continue to pose challenges to environmental health. Human exposure to PAHs occurs through multiple routes, including inhalation, ingestion, and skin contact. Of these, diet emerges as a major contributor to PAH intake, with foods cooked over open flames and water being significant sources (Das et al., 2023; Sahoo et al., 2022; Sahu et al., 2024; Hussain et al., 2018). Moreover, occupational exposure to PAHs remains prevalent, particularly in industries such as coke plants and asphalt production, underscoring the widespread nature of this issue. Recent research has delved into the formation and transformation of PAH derivatives, which exhibit heightened toxicity compared to their parent compounds. These derivatives, containing oxygen, nitrogen, or sulfur, arise from both biological processes and combustion activities. Their presence underscores the intricate nature of PAH pollution and its implications for environmental health. In essence, the effects of PAHs on the environment are multifaceted and complex, spanning from coastal regions to broader ecosystems. Despite efforts to mitigate their impact, PAHs continue to pose significant challenges to environmental quality and human health, necessitating ongoing research and action to address this pressing issue.

6.7 BIOTOXICITY

PAHs pose diverse health risks, spanning mutagenicity, carcinogenicity, teratogenicity, and immunotoxicity (Abdel-Shafy & Mansour, 2016).

6.7.1 Immediate Risks

PAH exposure triggers acute health issues like irritation and gastrointestinal disturbances (Mallah et al., 2022; Sun et al., 2021; Abdel-Shafy & Mansour, 2016).

6.7.2 Long-Term Consequences

Chronic exposure leads to conditions such as organ damage and respiratory ailments (Abdel-Shafy & Mansour, 2016).

6.8 SOURCES AND EXPOSURE DYNAMICS

The mode, duration, and dosage of exposure significantly influence the severity of PAH toxicity. Different PAH sources entail varying health risks. Phototoxicity, or sunlight exposure, exacerbates PAH toxicity, causing DNA damage and other adverse effects (Von Koschembahr et al., 2020). In many instances, genotoxicity and carcinogenicity have been seen because of PAH metabolism by generation of reactive intermediates, inducing genotoxic effects and carcinogenesis (Gao et al., 2018). It has also been seen that PAH exposure yields embryotoxicity and reproductive system disruptions, affecting fertility and health outcomes. Additionally, PAH exposure also gives rise to immunotoxicity or compromised immune response by impairing immune system function, leading to cell suppression, altered cytokine production, and potential autoimmunity (Suzuki et al., 2020).

6.9 SUSTAINABLE APPROACHES TO TACKLE PAHS IN THE ENVIRONMENT

Conventional methods for PAHs degradation, such as physical, chemical, mechanical, volatilization, landfilling, and thermal degradation, are in use to date. However, these methods have shown adverse effects on the environment as well as on living organisms. Also, the relatively long timescale needed for the natural bioremediation of PAHs demands scientific intervention, especially when the rate of increase of contaminants and pollutants in the environment is alarming (Mishra et al., 2023; Mishra & Das, 2023; Paul et al., 2023; Pradhan et al., 2023; Ward et al., 2003). Sediments represent the main receptor of organic contaminants entering the aquatic ecosystem, including PAHs (Chen et al., 2020; Primost et al., 2018). The prime task of any bioremediation strategy is to decrease the bioavailability of the pollutants toward living entities (Ortega-Calvo et al., 2013). PAHs and other hydrocarbon compounds are generally decomposed and attenuated from the environment by microorganisms like bacteria and fungi (Ahmad, 2022). In recent years, bioremediation research has advanced in this direction.

6.9.1 Bioaugmentation, Biostimulation, and Phytoremediation

Approaches like augmentation, stimulation, and remediation paired with biological agents like microorganisms and plants are the most widely applied technologies for the removal of PAH-contaminated sediments. While biostimulation and bioaugmentation consist of the supplementation of organic matter or nutrients and enriched microorganisms to the sediment, respectively, to improve the PAH biodegradation under aerobic or anaerobic conditions, phytoremediation uses plants to remove PAHs from the contaminated sediment mainly via rhizodegradation (He & Chi, 2019).

However, it is well known that bioremediation efficiency is limited by PAH bioavailability, which is the fraction of PAHs accessible and degradable by microorganisms or plants.

6.9.2 Degradation of PAH Using Microorganisms

It has been documented that different microbial strains can degrade PAH, but all of these reports have one thing in common: the degradation process operates synergistically and is highly metabolism-driven . The limited number of key enzymes, particularly dioxygenase and dehydrogenase, the two vital enzymes during PAHs degradation, are the primary cause of the inefficiency in the degradation of PAHs by a single bacterium, as the complexity of the PAHs degradation process dictates the need for multiple enzymes to work together. Therefore, the cooperation of important enzymes in the bacterial mixture should be required to efficiently degrade PAHs. The most popular strains used for different types of PAHs are listed in Table 6.1.

Almost 99%–100% homology to some of the strains capable of degrading other complex molecules was found in the 16S rRNA typing of gene-nucleotide sequences of PHA-degrading strains, which were compared with sequences from the GenBank database. The community of degraders was revealed by evolutionary studies to be made up of four clusters that brought together members of the phyla Firmicutes, Bacteroidetes, Actinobacteria, and Proteobacteria. It was also reported that the microbial community prevailing in a particular habitat changed once the PAH degradation process initiated. The microorganisms recovered from the surface of the degraded polymers or co-polymers were in fact not primary degraders because of their inability to metabolize the larger polymer molecules. It was found that commensal organisms were present, which utilized oligomers, monomers, acetoacetate, and other products of degradation of high molecular-weight polymers, prevailing due to vital activities of primary true PHA degraders.

6.9.3 Conditions for Microbe-Driven PAH Degradation

Numerous studies have shown that degradation of PAHs occurred at different rates depending on the polymer composition, climate and weather conditions, temperature, pH, and microbial community composition (Volova et al., 2016).

TABLE 6.1
Polycyclic aromatic hydrocarbon (PAH)-degrading microbial strains, associated PAH compounds, mechanisms of degradation, and corresponding literature references.

Sr. No	Strain	PAH Type	Mode of Degradation	References
1.	a. *Mitsuaria* sp. b. *Chitinophaga* sp. c. *Acidovorax* sp.	P(3HB)	Biofilm and enzymatic degradation	Elbanna et al. (2004)
2.	a. *Roseateles depolymerans* b. *Cupriavidus* sp.	P(3HB/4HB)	Degradation by de-polymerization	Wang et al. (2021)
3.	a. *Roseomonas massiliae* b. *Delftia acidovorans* c. *Lysobacter gummosus* d. Lysobacter sp. e. *E. coli*	P(3HB/3HV)	Degradation by de-polymerization	Suyama et al. (1998)
4.	a. Pseudoxanthomonas sp., b. *Pseudomonas fluorescens*, c. *Ensifer adhaerens*, d. *Bacillus pumilus*, e. *Bacillus* sp	P(93HB/3HHx)	Degradation by oxygenation followed by phathlic acid pathway	Li et al. (2021)
5.	a. Roseateles depolymerans 4HB-4	Bioplastics like poly (tetramethylene succinate), poly(hexamethylene carbonate), and poly(εcaprolactone)	Carboxymethyl-cellulosic degradation	Kanungo et al. (2023)

6.9.3.1 Habitat

The PAHs present in freshwater lakes and tropical soils degraded over 50% of their original composition in three months, while brackish waters and soils in cold climates slowed down the degradation rate by taking almost a year to erode.

6.9.3.2 Polymer Composition

The duration of degradation is generally influenced by the physicochemical nature of PAHs, as each polymer has a unique degrader based on its surface. Because the amorphous regions of these natural polymers are first broken down, pure P(3HB) and P(3HB)/bacterial cellulose blends show an increase in crystallinity under aerobic degradation. During the degradation process, P(3HB/3HV) specimens showed an increase in degree of crystallinity, according to a study in 2015. In a similar vein, before this investigation, it was noted that P(3HB/3HHx) films degraded in anaerobic environments showed an increase in crystallinity. However, the rate of biodegradation slowed down in some settings, such as cold and marine environments due to the inert microbial activity of the common degraders. This is most likely caused by a decrease in the molecular weight, degree of crystallinity, polydispersity, and copolymer melting temperature (Madbouly et al., 2014).

6.9.3.3 Temperature

Poly(3-hydroxybutyrate), commonly called P(3HB), is one of the most popular polyester polymers derived biologically, along with co-polymers P(3HB/4HB), P(3HB/3HV), and P(3HB/3HHx). Identifying optimal environmental conditions for maximizing the breakdown of such biodegradable plastics is significant for effective waste management and environmental sustainability (Das et al., 2024; Mohanty et al., 2024; Singh Jadaun et al., 2022). The temperature varies with the regions and seasons in the environment.

Studies have shown that moderate temperature increases can enhance microbial activity and polymer breakdown. The temperature range of 16°C–30°C proved essential in the accelerated degradation of these co-polymers, especially in soil environments. However, when the temperature exceeded 35°C, the degradation rate slowed, potentially due to thermal stress impacting microbial function or enzyme activity, highlighting the need to optimize temperature conditions for effective biodegradation processes. Thus, 28°C of temperature was considered to be optimal for PHA degradation. The first seven days of the degradation process were crucial as 30% of the structure loss happened in all of the PHA films slowly, yet steadily. Among these, co-polymer P(3HB/4HB) films lost the most weight, followed by P(3HB) films that lost 10% of their initial mass. After that, the degradation rates of all PHAs increased. For 21 days, P(3HB/4HB) films had been degraded by 97%. However, exception cases were seen in this regard. A thermophilic *Streptomyces* species completely degraded P(3HB) at 50°C within a short span of less than ten days, highlighting the efficiency of thermophilic bacteria in polymer degradation. Other microorganisms that degraded the PHAs in elevated temperatures included *Leptothrix* sp. and *Paecilomyces lilacinus.*

6.9.3.4 pH

Normally, bacteria are pH-sensitive, most favored near neutral conditions (6.5–7.5). This also explains the fact that all the key enzymes involved are active at the neutral range that allows nutrient solubility and transportation. Since phthalic acid is an important metabolite after PAH degradation, it tends to lower the pH up to 4 or 3, leading to a decreased rate of PAH degradation. Although the bacterial activity was decreased with pH increase, the weakly alkaline medium could promote the degradation process. Therefore, the degradation rate decreased slightly in the weak base.

6.9.3.5 Microbial Community

The microbial community prevailing in the soil or sediments where PAH is also deposited is specific to the polymers. This ensures that the diversity is higher than in the non-PAH harboring areas in terms of percentage of the microbial species

(Zheng et al., 2018). Usually, about 80%–85% of degradation is achieved through a combination of bacteria and not one single strain alone. Also, the existence of related metabolism in the bacterial mixture, which is a cross-feeding relationship with metabolites during the bacterial mixture's degradation, implied better degradation of PAHs. The bacteria in synergism target the degradation process commonly under aerobic conditions through metabolism of oxygenase, including dioxygenase or monooxygenase enzymes (Ghosal et al., 2016). Another way to see why bacterial mixture works in favor of PAH degradation is that a network of key enzymes is created that is vital to the PAH degradation process where the single strain fails to do so.

6.10 CONCLUSION

The microbial biotransformation of PAHs presents a promising, environmentally sustainable approach for the remediation of sites contaminated with PAHs. Microorganisms, including various bacterial and fungal species, have exhibited the capability to metabolize and decompose these intricate and hazardous compounds into less toxic or inert forms, thereby serving as a crucial instrument for environmental restoration. The microbial degradation of PAHs can transpire under both aerobic and anaerobic conditions, though aerobic processes generally exhibit greater speed and efficiency. Notwithstanding the promise of microbial degradation, the process encounters significant obstacles, particularly due to the limited bioavailability of PAHs within contaminated matrices. The hydrophobic characteristics of PAHs result in their tendency to adsorb onto soil particles or to form complexes that are inaccessible to microbial organisms. This phenomenon frequently constrains the degradation rate, particularly in soils that are heavily polluted or compacted. Moreover, high-molecular-weight PAHs, which are characterized by their enhanced persistence and toxicity, are typically more recalcitrant to microbial decomposition compared to their lower-molecular-weight counterparts. The enhancement of microbial degradation can be accomplished through biostimulation—which involves the addition of nutrients to bolster microbial proliferation—and bioaugmentation, which entails the introduction of PAH-degrading microorganisms into contaminated environments. These methodologies have demonstrated potential in augmenting the efficacy of degradation processes across diverse environmental contexts. Furthermore, a comprehensive understanding of the specific metabolic pathways implicated in PAH degradation, along with the environmental determinants influencing microbial activity (such as temperature, pH, and oxygen concentration), is vital for the optimization of bioremediation strategies. In summary, microbial degradation of PAHs constitutes a feasible and sustainable approach for the remediation of contaminated sites; however, it necessitates a thorough comprehension of microbial ecology, the environmental parameters at play, and the physicochemical characteristics of the contaminants. Ongoing research into microbial consortia, the genetic engineering of more efficacious degraders, and the optimization of bioremediation methodologies will be imperative for enhancing the applicability and effectiveness of microbial mechanisms in PAH remediation.

REFERENCES

Abdel-Shafy, H. I., & Mansour, M. S. M. (2016). A review on polycyclic aromatic hydrocarbons: Source, environmental impact, effect on human health and remediation. *Egyptian Journal of Petroleum*, *25*(1), 107–123. https://doi.org/10.1016/j.ejpe.2015.03.011

Balmer, J. E., Hung, H., Yu, Y., Letcher, R. J., & Muir, D. C. G. (2019). Sources and environmental fate of pyrogenic polycyclic aromatic hydrocarbons (PAHs) in the Arctic. *Emerging Contaminants*, *5*, 128–142. https://doi.org/10.1016/j.emcon.2019.04.002

Baraniecka, J., Pyrzyńska, K., Szewczyńska, M., Pośniak, M., & Dobrzyńska, E. (2010). Emission of polycyclic aromatic hydrocarbons from selected processes in steelworks. *Journal of Hazardous Materials*, *183*(1–3), 111–115. https://doi.org/10.1016/j.jhazmat.2010.06.120

Cai, T., Ding, Y., Zhang, Z., Wang, X., Wang, T., Ren, Y., & Dong, Y. (2019). Effects of total organic carbon content and leaching water volume on migration behavior of polycyclic aromatic hydrocarbons in soils by column leaching tests. *Environmental Pollution*, *254*, 112981. https://doi.org/10.1016/j.envpol.2019.112981

Callén, M. S., López, J. M., Iturmendi, A., & Mastral, A. M. (2013). Nature and sources of particle associated polycyclic aromatic hydrocarbons (PAH) in the atmospheric environment of an urban area. *Environmental Pollution*, *183*, 166–174. https://doi.org/10.1016/j.envpol.2012.11.009

Cao, Y., Wang, J., Ming, X., Wang, B., & Lin, C. (2023). Spatial distribution and partition of polycyclic aromatic hydrocarbons (PAHs) in the water and sediment of the southern Bohai Sea: Yellow River and PAH property influences. *Water Research*, *248*, 120873. https://doi.org/10.1016/j.watres.2023.120873

Das, A., Ray, S. K., Mohanty, M., Mohanty, J., Dey, S., & Das, A. P. (2024). Ecotoxicity of microplastic wastes and their sustainable management: A review. *Environmental Chemistry and Ecotoxicology*, *6*, 144–152. https://doi.org/10.1016/J.ENCECO.2024.05.003

Das, A. P., Mohanty, A. K., van Hullebusch, E. D., Figueiredo, G., Das, A. P., Mohanty, A. K., van Hullebusch, E. D., & Figueiredo, G. (2023). Bioremediation as an emerging technology for the removal of synthetic microplastic pollutants from marine ecosystem. *MarPB*, *194*, 115297. https://doi.org/10.1016/J.MARPOLBUL.2023.115297

Dat, N.-D., & Chang, M. B. (2017). Review on characteristics of PAHs in atmosphere, anthropogenic sources and control technologies. *Science of the Total Environment*, *609*, 682–693. https://doi.org/10.1016/j.scitotenv.2017.07.204

Dey, S., Bissoyi, A., Koteswara Reddy, G., & Das, A. P. (2024). Synthetic microfibers: An urban emerging pollutant demanding for sustainable environmental solutions. In *Sustainable Microbial Technology for Synthetic and Cellulosic Microfiber Bioremediation*, 107–121. https://doi.org/10.1007/978-3-031-62600-5_7

Dey, S., Das, A., Mallick, K., Sahu, A., & Das, A. P. (2023). Environmental petroleum waste: Pollution, toxicity, sustainable remediation. In *Impact of Petroleum Waste on Environmental Pollution and Its Sustainable Management through Circular Economy*, 159–175. https://doi.org/10.1007/978-3-031-48220-5_7

Edwards, N. T. (1983). Polycyclic aromatic hydrocarbons (PAH's) in the terrestrial environment: A review. *Journal of Environmental Quality*, *12*(4), 427–441. https://doi.org/10.2134/jeq1983.00472425001200040001x

Elbanna, K., Lütke-Eversloh, T., Jendrossek, D., Luftmann, H., & Steinbüchel, A. (2004). Studies on the biodegradability of polythioester copolymers and homopolymers by polyhydroxyalkanoate (PHA)-degrading bacteria and PHA depolymerases. *Archives of Microbiology*, *182*, 212–225. https://doi.org/10.1007/s00203-004-0715-z

Fakinle, B. S., Odekanle, E. L., Ike-Ojukwu, C., Sonibare, O. O., Falowo, O. A., Olubiyo, F. W., Oke, D. O., & Aremu, C. O. (2022). Quantification and health impact assessment of polycyclic aromatic hydrocarbons (PAHs) emissions from crop residue combustion. *Heliyon*, *8*(3), e09113. https://doi.org/10.1016/j.heliyon.2022.e09113

Fernandes, F. M. (2019). Clay bricks. In *Long-Term Performance and Durability of Masonry Structures*, 3–19. Elsevier. https://doi.org/10.1016/B978-0-08-102110-1.00001-7

Gao, P., Da Silva, E., Hou, L., Denslow, N. D., Xiang, P., & Ma, L. Q. (2018). Human exposure to polycyclic aromatic hydrocarbons: Metabolomics perspective. *Environment International*, *119*, 466–477. https://doi.org/10.1016/j.envint.2018.07.017

Ghosh, S., Kumar, M. S., Bal, B., & Das, A. P. (2018). Application of bioengineering in revamping human health. In *Synthetic Biology: Omics Tools and Their Applications*, 21–37. https://doi.org/10.1007/978-981-10-8693-9_2

Ghosh, S., Tripathy, B., Dey, S., & Das, A. P. (2023). Comparative investigation of fungal and bacterial manganese biomining mechanisms. In *Manganese Mining Microorganisms*, 99–115. https://doi.org/10.1016/b978-0-12-822147-1.00016-9

Han, Y., Chen, Y., Feng, Y., Song, W., Cao, F., Zhang, Y., Li, Q., Yang, X., & Chen, J. (2020). Different formation mechanisms of PAH during wood and coal combustion under different temperatures. *Atmospheric Environment*, *222*, 117084.

Hazarika, N., Das, A., Kamal, V., Anwar, K., Srivastava, A., & Jain, V. K. (2019). Particle phase PAHs in the atmosphere of Delhi-NCR: With spatial distribution, source characterization and risk approximation. *Atmospheric Environment*, *200*, 329–342. https://doi.org/10.1016/j.atmosenv.2018.11.064

He, Y., & Chi, J. (2019). Pilot-scale demonstration of phytoremediation of PAH contaminated sediments by Hydrilla verticillata and Vallisneria spiralis. *Environmental Technology*. https://doi.org/10.1080/09593330.2017.1398783

Hsu, W. T., Liu, M. C., Hung, P. C., Chang, S. H., & Chang, M. B. (2016). PAH emissions from coal combustion and waste incineration. *Journal of Hazardous Materials*, *318*, 32–40. https://doi.org/10.1016/j.jhazmat.2016.06.038

Hsu, Y.-C., Chang, S.-H., & Chang, M. B. (2021). Emissions of PAHs, PCDD/Fs, dl-PCBs, chlorophenols and chlorobenzenes from municipal waste incinerator cofiring industrial waste. *Chemosphere*, *280*, 130645. https://doi.org/10.1016/j.chemosphere.2021.130645

Hu, Y., Xia, Y., Di Maio, F., Yu, F., & Yu, W. (2020). Investigation of polycyclic aromatic hydrocarbons (PAHs) formed in three-phase products from the pyrolysis of various wastewater sewage sludge. *Journal of Hazardous Materials*, *389*, 122045. https://doi.org/10.1016/j.jhazmat.2020.122045

Hussain, K., Hoque, R. R., Balachandran, S., Medhi, S., Idris, M. G., Rahman, M., & Hussain, F. L. (2018). Monitoring and risk analysis of PAHs in the environment. In C. M. Hussain (Ed.), *Handbook of Environmental Materials* Management, 1–35. Springer International Publishing. https://doi.org/10.1007/978-3-319-58538-3_29-2

Jaward, F. M., Alegria, H. A., Galindo Reyes, J. G., & Hoare, A. (2012). Levels of PAHs in the waters, sediments, and shrimps of Estero de Urias, an Estuary in Mexico, and their toxicological effects. *The Scientific World Journal*, *2012*, 1–9. https://doi.org/10.1100/2012/687034

Kalbe, U., Berger, W., Eckardt, J., & Simon, F.-G. (2008). Evaluation of leaching and extraction procedures for soil and waste. *Waste Management*, *28*(6), 1027–1038. https://doi.org/10.1016/j.wasman.2007.03.008

Kalmykova, Y., Björklund, K., Strömvall, A.-M., & Blom, L. (2013). Partitioning of polycyclic aromatic hydrocarbons, alkylphenols, bisphenol A and phthalates in landfill leachates and stormwater. *Water Research*, *47*(3), 1317–1328. https://doi.org/10.1016/j.watres.2012.11.054

Kanungo, J., Sahoo, T., Bal, M., & Behera, I. D. (2023). Performance of bioremediation strategy in waste lubricating oil pollutants: A review. *Geomicrobiology Journal*, *41*(4), 360–373. https://doi.org/10.1080/01490451.2023.2245395

Karyab, H., Yunesian, M., Nasseri, S., Mahvi, A. H., Ahmadkhaniha, R., Rastkari, N., & Nabizadeh, R. (2013). Polycyclic aromatic hydrocarbons in drinking water of Tehran, Iran. *Journal of Environmental Health Science and Engineering*, *11*(1), 25. https://doi.org/10.1186/2052-336X-11-25

Kuppusamy, S., Maddela, N. R., Megharaj, M., & Venkateswarlu, K. (2020). An overview of total petroleum hydrocarbons. In S. Kuppusamy, N. R. Maddela, M. Megharaj, & K. Venkateswarlu, *Total Petroleum Hydrocarbons*, 1–27. Springer International Publishing. https://doi.org/10.1007/978-3-030-24035-6_1

Kuśmierz, M., Oleszczuk, P., Kraska, P., Pałys, E., & Andruszczak, S. (2016). Persistence of polycyclic aromatic hydrocarbons (PAHs) in biochar-amended soil. *Chemosphere*, *146*, 272–279. https://doi.org/10.1016/j.chemosphere.2015.12.010

Li, H.-Y., Gao, P.-P., & Ni, H.-G. (2019). Emission characteristics of parent and halogenated PAHs in simulated municipal solid waste incineration. *Science of the Total Environment*, *665*, 11–17. https://doi.org/10.1016/j.scitotenv.2019.02.002

Li, N., Jiang, Q., Wang, F., Cui, P., Xie, J., Li, J., Wu, S., & Barbieri, D. M. (2021). Comparative assessment of asphalt volatile organic compounds emission from field to laboratory. *Journal of Cleaner Production*, *278*, 123479. https://doi.org/10.1016/j.jclepro.2020.123479

Liang, Y., Li, Q., Ding, X., Wu, D., Wang, F., Otsuki, T., Cheng, Y., Shen, T., Li, S., & Chen, J. (2020). Forward ultra-low emission for power plants via wet electrostatic precipitators and newly developed demisters: Filterable and condensable particulate matters. *Atmospheric Environment*, *225*, 117372. https://doi.org/10.1016/j.atmosenv.2020.117372

Madias, J. (2016). Electric arc furnace. In P. Cavaliere (Ed.), *Ironmaking and Steelmaking Processes*, 267–281. Springer International Publishing. https://doi.org/10.1007/978-3-319-39529-6_16

Mallah, M. A., Changxing, L., Mallah, M. A., Noreen, S., Liu, Y., Saeed, M., Xi, H., Ahmed, B., Feng, F., Mirjat, A. A., Wang, W., Jabar, A., Naveed, M., Li, J.-H., & Zhang, Q. (2022). Polycyclic aromatic hydrocarbon and its effects on human health: An overview. *Chemosphere*, *296*, 133948. https://doi.org/10.1016/j.chemosphere.2022.133948

Mohanty, J., Dey, S., Bala, K., & Das, A. P. (2024). Recent advances in microplastic pollution for its sustainable management. In *Sustainable Management of Urban Plastic Waste through Circular Economic Approaches*, 27–48. https://doi.org/10.1201/9781003465447-3

Moharana, S., Dey, S., Priyadarsini, S., Kumar, M. S., & Das, A. P. (2024). Identification and characterization of microplastic pollutants from the marine sediments of Paradeep Coast of Bay of Bengal, India for their sustainable management. In *Environmental Science and Engineering*, *Part F2333*, 57–73. https://doi.org/10.1007/978-3-031-51792-1_4

Odabasi, M., Bayram, A., Elbir, T., Seyfioglu, R., Dumanoglu, Y., Bozlaker, A., Demircioglu, H., Altiok, H., Yatkin, S., & Cetin, B. (2009). Electric arc furnaces for steel-making: Hot spots for persistent organic pollutants. *Environmental Science & Technology*, *43*(14), 5205–5211. https://doi.org/10.1021/es900863s

Ozaki, N., Yamauchi, T., Kindaichi, T., & Ohashi, A. (2019). Stormwater inflow loading of polycyclic aromatic hydrocarbons into urban domestic wastewater treatment plant for separate sewer system. *Water Science and Technology*, *79*(7), 1426–1436. https://doi.org/10.2166/wst.2019.143

Panigrahi, B. K. (2001). Processing of low carbon steel plate and hot strip: An overview. *Bulletin of Materials Science*, *24*(4), 361–371. https://doi.org/10.1007/BF02708632

Patel, A. B., Shaikh, S., Jain, K. R., Desai, C., & Madamwar, D. (2020). Polycyclic aromatic hydrocarbons: Sources, toxicity, and remediation approaches. *Frontiers in Microbiology*, *11*, 562813. https://doi.org/10.3389/fmicb.2020.562813

Paul, A., Dey, S., Ram, D. K., & Das, A. P. (2023). Hexavalent chromium pollution and its sustainable management through bioremediation. *Geomicrobiology Journal*, 1–11. https://doi.org/10.1080/01490451.2023.2218377

Pradhan, G., Tripathy, B., Ram, D. K., Digal, A. K., & Das, A. P. (2023). Bauxite mining waste pollution and its sustainable management through bioremediation. *Geomicrobiology Journal*, 1–10. https://doi.org/10.1080/01490451.2023.2235353

Ribeiro, M. J., & Tulyaganov, D. (2021). Traditional ceramics manufacturing. In F. Baino, M. Tomalino, & D. Tulyaganov (Eds.), *Ceramics, Glass and Glass-Ceramics*, 75–118. Springer International Publishing. https://doi.org/10.1007/978-3-030-85776-9_3

Robinson, P. R., & Hsu, C. S. (2017). Introduction to petroleum technology. In C. S. Hsu & P. R. Robinson (Eds.), *Springer Handbook of Petroleum Technology*, 1–83. Springer International Publishing. https://doi.org/10.1007/978-3-319-49347-3_1

Sahoo, P. P., Singh, S., Rout, P. K., Mishra, S., & Das, A. P. (2022). Microbial remediation of plastic pollutants generated from discarded and abandoned marine fishing nets. *Biotechnology and Genetic Engineering Reviews*. https://doi.org/10.1080/02648725.2022.2152629

Sahu, A., Dey, S., Mallick, K., Priyadarsini, S., Behera, I. D., & Das, A. P. (2024). A review on advanced technology for sustainable management of synthetic microplastic waste. In *Sustainable Management of Urban Plastic Waste Through Circular Economic Approaches*, 262–279. https://doi.org/10.1201/9781003465447-15

Sen, S., Narayana, J., Ravichandran, B., & Dhananjayan, V. (2018). Polyaromatic hydrocarbons depositions and their carcinogenic risk assessment in the foundry workers. *Aerosol Science and Engineering*, *2*(4), 173–181. https://doi.org/10.1007/s41810-018-0034-4

Singh Jadaun, J., Bansal, S., Sonthalia, A., Rai, A. K., & Singh, S. P. (2022). Biodegradation of plastics for sustainable environment. *Bioresource Technology*, *347*, 126697. https://doi.org/10.1016/J.BIORTECH.2022.126697

Sun, K., Song, Y., He, F., Jing, M., Tang, J., & Liu, R. (2021). A review of human and animals exposure to polycyclic aromatic hydrocarbons: Health risk and adverse effects, photo-induced toxicity and regulating effect of microplastics. *Science of the Total Environment*, *773*, 145403. https://doi.org/10.1016/j.scitotenv.2021.145403

Suyama T, Tokiwa Y, Ouichanpagdee P, Kanagawa T, & Kamagata Y (1998). Phylogenetic affiliation of soil bacteria that degrade aliphatic polyesters available commercially as biodegradable plastics. *Applied and Environmental Microbiology*, *64*, 5008–5011.

Suzuki, T., Hidaka, T., Kumagai, Y., & Yamamoto, M. (2020). Environmental pollutants and the immune response. *Nature Immunology*, *21*(12), 1486–1495. https://doi.org/10.1038/s41590-020-0802-6

Vagge, G., Cutroneo, L., Castellano, M., Canepa, G., Bertolotto, R. M., & Capello, M. (2018). The effects of dredging and environmental conditions on concentrations of polycyclic aromatic hydrocarbons in the water column. *Marine Pollution Bulletin*, *135*, 704–713. https://doi.org/10.1016/j.marpolbul.2018.08.006

Von Koschembahr, A., Youssef, A., Béal, D., Gudimard, L., Giot, J.-P., & Douki, T. (2020). Toxicity and DNA repair in normal human keratinocytes co-exposed to benzo[a]pyrene and sunlight. *Toxicology in Vitro*, *63*, 104744. https://doi.org/10.1016/j.tiv.2019.104744

Wang, M., Wang, C., Huang, S., & Yuan, H. (2021). Study on asphalt volatile organic compounds emission reduction: A state-of-the-art review. *Journal of Cleaner Production*, *318*, 128596. https://doi.org/10.1016/j.jclepro.2021.128596

Yilmaz, N., Vigil, F. M., & Donaldson, B. (2022). Fuel effects on PAH formation, toxicity and regulated pollutants: Detailed comparison of biodiesel blends with propanol, butanol and pentanol. *Science of the Total Environment*, *849*, 157839. https://doi.org/10.1016/j.scitotenv.2022.157839

Zhang, S., Li, H., He, R., Deng, W., Ma, S., Zhang, X., Li, G., & An, T. (2023). Spatial distribution, source identification, and human health risk assessment of PAHs and their derivatives in soils nearby the coke plants. *Science of the Total Environment*, *861*, 160588. https://doi.org/10.1016/j.scitotenv.2022.160588

Zhao, Y., Hong, B., Fan, Y., Wen, M., & Han, X. (2014). Accurate analysis of polycyclic aromatic hydrocarbons (PAHs) and alkylated PAHs homologs in crude oil for improving the gas chromatography/mass spectrometry performance. *Ecotoxicology and Environmental Safety*, *100*, 242–250. https://doi.org/10.1016/j.ecoenv.2013.10.018

Zheng, H., Xing, X., Hu, T., Zhang, Y., Zhang, J., Zhu, G., Li, Y., & Qi, S. (2018). Biomass burning contributed most to the human cancer risk exposed to the soil-bound PAHs from Chengdu Economic Region, western China. *Ecotoxicology and Environmental Safety*, *159*, 63–70.

7 Exploring the Effectiveness of Biological Remediation Techniques in the Cleanup of Crude Oil Spill Sites

Priyanka Singh, Samriddhi Singh, and Meenu Singh

7.1 INTRODUCTION

7.1.1 Overview of Crude Oil Spills and Their Environmental Impact

Crude oil is a naturally occurring, unrefined petroleum substance composed of hydrocarbon deposits and various other materials. It is extracted from underground reservoirs and redefined into various products such as gasoline, diesel, kerosene, and petrochemicals. This versatile energy source plays a crucial role in the global economy, powering industries, transportation, and heating. Crude oil spills, a catastrophic event that poses a significant threat to the environment, have garnered global attention due to their detrimental impact on ecosystems and human health. It refers to the accidental release of petroleum into the environment, typically occurring during transportation, extraction, or storage processes. Human error, equipment failure, natural disasters, or deliberate acts of sabotage can cause these spills. The frequency and distribution of crude oil spills vary globally, with regions like the Gulf of Mexico, the Niger Delta, and the Persian Gulf experiencing higher incidences due to their significant oil production and transportation activities. The environmental impact of crude oil spills is profound, with immediate effects on marine life and ecosystems. Oil spills coat marine organisms, such as fish, birds, and mammals, leading to suffocation, poisoning, and reduced reproductive success. Furthermore, the long-term consequences include biodiversity loss, habitat destruction, and disruptions in the food chain. Oil contamination can persist in the environment for years, affecting water quality and soil fertility and posing risks to human health through the consumption of contaminated seafood.

DOI: 10.1201/9781003508496-7

7.1.2 Importance of Effective Remediation Techniques

Effective remediation techniques for oil spills are critical for mitigating the extensive environmental, economic, and social impacts of such disasters. Swift and efficient responses help minimize damage to marine and coastal ecosystems, which can suffer from the toxic effects of crude oil, leading to long-term harm to wildlife, including fish, birds, and marine mammals. Effective remediation also aids in preserving the livelihood of communities reliant on fishing, tourism, and other coastal industries. Moreover, prompt cleanup can prevent oil from spreading to larger areas, reducing the overall environmental footprint of the spill. By employing advanced and effective remediation techniques, we can ensure that oil is removed from the environment as quickly and thoroughly as possible, thus aiding in the restoration of affected habitats and reducing the time needed for natural recovery processes. This helps not only in preserving biodiversity and maintaining ecosystem services but also in upholding public health and safety and safeguarding economic resources.

7.1.3 Introduction to Biological Remediation as a Sustainable Solution

Biological remediation, or bioremediation, offers a sustainable solution for environmental cleanups by leveraging natural processes to degrade, detoxify, or transform pollutants into harmless byproducts. This approach uses microorganisms and plants to break down pollutants in soil, water, and air. This can effectively restore ecosystems without the need for harmful chemicals or extensive physical intervention. By enhancing or accelerating these natural processes, bioremediation not only reduces pollution but also promotes biodiversity and ecosystem resilience. Moreover, it is cost-effective and minimizes environmental disruption, making it an attractive option for long-term environmental management and sustainability. Its application ranges from oil spill response to heavy metal detoxification, demonstrating its versatility and potential to address various forms of pollution while aligning with ecological conservation goals.

7.1.4 Purpose and Scope of the Chapter

The goal of authoring a book chapter on bioremediation strategies for oil spills is to unveil the ecological approach that stands as an antidote for one of the gravest forms of environmental contamination in a detailed and comprehensive way. This chapter intends to enlighten the readers on the mechanisms, efficiency, and field application that come along with using bioremediation while dealing with oil spill management. We will not only go through the scientific rationale underpinning bioremediation but also paint a picture through successful case studies, delving into the choice of biological agent ranging from bacteria and fungi, all the way to plants. The gamut covered involves looking at existing bioremediation technologies, teasing out challenges impeding their implementation (and likely solutions) and keeping an eye out for further innovative and improved prospects. These details are the focus of the chapter, with the aim of providing information on the potential and constraints of bioremediation for use as an ecological restoration and pollution control tool.

The intended audience is researchers, environmental professionals, policymakers, and students. The chapter strongly promotes (without any compromise) wider utilization of bioremediation in coordination with the integration into environmental managerial practices as well.

7.2 BACKGROUND AND LITERATURE REVIEW

7.2.1 Historical Context of Crude Oil Spills

Humans have utilized petroleum for thousands of years. Initially used in medicine and weapons, the material did not come into use until the Industrial Revolution, when it became valued as an illumination fuel plus lubricant—and then later as an energy source that replaced wood, coal, and animal power (Kanungo et al., 2023, Mohanta et al., 2023). Liquid petroleum had some significant advantages over other energy sources of the times: it was concentrated, and it could be transported easily from one place to another. As oil use boomed, its refined products came to be used to fuel automobiles, and a worldwide system of wells, ships, and storage terminals grew along with pipelines (Rafferty, 2023). There are occasions where oil is released into the environment due to extracting oil from the ground and moving it to refineries and beyond— caused by old equipment or bad luck. The biggest disaster has resulted in ten million of gallons of oil spilled, clogging fisheries, killing and injury animals, and decreasing tourism income. The following list below in Table 7.1 contains the top nine most damaging oil spills.

TABLE 7.1
Represents the Description of Oil Spill Incident

S.No.	Oil Spill Incident in History	Description
1	The *Amoco Cadiz* Oil Spill (1978)	The Amoco Cadiz, a very large crude carrier (VLCC), ran aground off the coast of Brittany, France, on March 16, 1978. This resulted in the spilling of its entire cargo of sixty-nine million gallons of light crude oil into the sea. About 321 km of French coast were polluted and caused extensive damage to marine life, local fishing industries, and coastal habitats.
2	The *Castillo de Bellver* Oil Spill (1983)	On August 6, 1983, the Spanish tanker Castillo de Bellver caught fire and broke into two off the coast of South Africa. Approximately 110,000 tons of oil were released into the Atlantic Ocean. The spill had significant environmental impact, though much of the oil dispersed in the open sea.
3	The Incidents at the Nowruz Oil Field (1983)	During the Iraq–Iran War, the Nowruz Oil Field in the Persian Gulf experienced multiple spills. On February 10, 1983, an oil platform was attacked, causing a major spill. Before it was eventually sealed in September 1983, the well spewed over 1,500 barrels (63,000 gallons) of oil into the Persian Gulf per day. Later that year, a tanker collision exacerbated the situation, releasing an estimated 733,000 barrels of crude oil into the gulf.

(*Continued*)

TABLE 7.1
Represents the Description of Oil Spill Incident

S.No.	Oil Spill Incident in History	Description
4	The Kolva River Spill (1994)	An oil pipeline break in the Russian Arctic due to corrosion was the source of the Kolva River oil spill. A dike kept the oil contained around the burst site for 8 months, but eventually broke way, causing some 84 million gallons of oil to leak into the Kolva River. Oil contaminated an area of 186 km^2, or around 72 $miles^2$, of marshes and tundra. Millions of gallons of oil were spilled into the environment via pipeline breaches in the area, some of which ended up in the Kolva River.
5	The Mingbulak (or Fergana Valley) Oil Spill (1992)	The largest inland oil spill occurred on March 2, 1922, in Uzbekistan's Fergana valley. It took 2 months for the oil to catch fire and burn until the well pressure decreased. Over 88 million gallons of the projected 88 million gallons released were kept out of the fire by berms and dikes and caused long-term environmental damage.
6	The *Atlantic Empress* Oil Spill (1979)	The Atlantic Express, another VLCC, collided with Aegean Captain off the coast of Trinidad and Tobago on July 19, 1979. The accident caused the release of about ninety million gallons of oil into the Atlantic Oceans, making it the largest spill in the history. The spill leads to serve marine pollution.
7	The *Ixtoc 1* Oil Spill (1979)	Between June 1979 and March 1980, the Ixtoc 1 catastrophe in Mexico spilled up to 140 million gallons of crude oil into the Bay of Campeche. The spill was caused by an explosion that occurred on the Ixtoc 1 platform, which was drilling exploratory wells in 164 feet (about 50 m) of water. A buildup of gas and oil in the pipe due to the drilling mud not moving properly was what triggered the explosion. The blowout preventer was bypassed, and a slurry of mud, oil, and natural gas surged up the pipe as workers tried to remove the drill so that material could flow back down the line and cover the hole. There was a leak of between 126 and 140 million gallons into the southern portion of Gulf of Mexico for about 9 months.
8	BP's *Deepwater Horizon* Oil Spill (2010)	On April 20, 2010, an explosion on the Deepwater Horizon oil rig, operated by BP in the Gulf of Mexico, caused the rig to sink and a subsea blowout to release around 134 million gallons of oil were released in American Gulf Coast, spanning from Texas to Florida, over 2,100 km (1,300 miles), was covered with oil.
9	The Persian Gulf War Oil Spill (1991)	During the Gulf War, Iraqi forces intentionally released oil from several tankers and oil terminals into the Persian Gulf. Estimates of the spill's size vary, but it is believed to be between 380 and 520 million gallons, making it the largest oil spill in history. The environmental damage was extensive, affecting coastal ecosystems and marine life.

* *Reference:* Rafferty (2023).

7.3 OVERVIEW OF TRADITIONAL REMEDIATION METHODS

An important risk to the ecosystem, marine life, and coastal towns is oil spill-related marine contamination. When oil leaks into the oceans, it forms an oil slick that upsets the surrounding environment by spreading quickly across the water surface. Due to the coating that this pollution causes on their bodies, marine species such as fish, birds, and marine mammals may suffer greatly. This impact includes impaired breathing, movement, and body temperature regulation. Planktons, larvae, and other fundamental components of the marine food web can be impacted by the poisonous components of oil, which can also infiltrate the water column. Oil pollution of beaches, mangroves, and marshlands causes long-term ecological harm as well as financial losses for coastal communities that depend on fishing and tourism. Although there have been oil spills for many years, sadly, cleaning methods have not improved much in that period. Techniques for responding to oil spills include chemical treatments like dispersion, thermal methods like in situ burning, mechanical means like booms, sorbents, and skimmers, and natural approaches like bioremediation (Pete et al., 2021). Booms and skimmers are still often utilized because they serve as physical barriers and collectors to stop oil from spreading and clearing the water's surface. To facilitate natural biodegradation, chemical dispersants are also used to break down oil into tiny droplets. Furthermore, biological agents that consume oil, such as bacteria, are being used to hasten the breakdown of oil. Two common biological processes are used to clean up oil spills: bioaugmentation, which includes the addition of certain types of bacteria strains that can break down oil components, and biostimulation, which involves adding nutrients or oxygen to encourage the growth of naturally occurring microorganisms that break down oil in the surrounding (Das et al., 2024; Das & Ghosh, 2023; Dey, Bissoyi, et al., 2024).

7.4 INTRODUCTION TO BIOLOGICAL REMEDIATION

One essential tactic for dealing with oil spills is biological remediation, also known as bioremediation, which uses the environment's natural processes to break down contaminants and repair the ecosystem. Physical and chemical procedures are not employed in any situation involving an oil spill because of the potential for environmental pollution, their excessive cost, and their low adsorption capability. Therefore, as a backup strategy, inexpensive biodegradable biomaterials are frequently employed to remove oil spills from the surface (Wolok et al., 2020). Because biological techniques utilize live organisms to carry out the rehabilitation of contaminated sites—particularly the cleaning of crude oil spills and the detoxification of these environments—they are favored over artificial methods that use many local microorganisms in water and soil (Prince Emeka Ndimele, 2018). These microbes change dangerous pollutants like carbon dioxide and water into less hazardous or non-toxic molecules by breaking down the components of the oil. An environmentally friendly method of cleaning up oil spills from the marine environment is to use bioremediation techniques. This technique, known as bioaugmentation or biostimulation, applies naturally occurring or genetically modified microorganisms to the contaminated location and/or enriches the polluted environment with nutrients (Mahsa Baniasadi, 2018). Although bioremediation takes longer than certain mechanical techniques, it is a vital tool in the control of marine oil spills since it may accomplish complete, long-term cleaning.

7.5 SUMMARY OF KEY RESEARCH STUDIES ON BIOLOGICAL REMEDIATION TECHNIQUES

Important studies on biological cleanup strategies for oil spills have shown how effective and promising these approaches may be in reducing environmental harm. Research has indicated that the introduction of certain microbial strains that break down oil, known as bioaugmentation, accelerates the rates at which hydrocarbons degrade in polluted settings. The significance of maximizing nutrition and oxygen levels to enhance the natural microbial activity and promote more effective oil breakdown has been brought to light by research on biostimulation (Arora, 2018; Das et al., 2015). Research on the use of genetically modified microbes has shown encouraging findings, demonstrating a quicker and more thorough breakdown of oil constituents. Furthermore, research on the synergistic benefits of mixing bioremediation with other techniques, such as mycoremediation—which uses fungi—and phytoremediation—which uses plants to absorb and break down contaminants—has shown increased cleanup results. All things considered, these investigations highlight the efficiency of biological cleanup methods and their indispensable position in all-encompassing plans for responding to oil spills.

7.6 TYPES OF BIOLOGICAL REMEDIATION TECHNIQUES

7.6.1 Bioremediation

Oil spills are often cleaned up using bioremediation, an eco-friendly method that makes use of microorganisms' innate capacity to break down and purify toxins. This process converts the poisonous hydrocarbons present in oil into less hazardous compounds like carbon dioxide and water by using bacteria, fungi, and algae. Natural bioremediation has been utilized for ages by civilizations to remediate wastewater; however, purposeful usage for the reduction of hazardous pollutants is a more recent development. Bioaugmentation—introducing certain bacteria that break down oil to the polluted area—and biostimulation—optimizing circumstances by supplying oxygen or nutrients to encourage the activity of native microorganisms—are two ways to improve the process (Das et al., 2024; Paul et al., 2023; Pradhan et al., 2023). Because bioremediation is affordable, environmentally friendly, and suitable for a variety of settings—including soils, shorelines, and aquatic bodies—it is beneficial. Bioremediation provides a comprehensive, long-term solution for oil spill cleaning, albeit one that often proceeds more slowly than some mechanical techniques. It also makes a substantial contribution to the restoration of impacted ecosystems.

7.6.2 Definition and Mechanism

The term "bioremediation" refers to the process of employing living things—microorganisms—as nutrition or energy sources to remove toxins from the environment (Baba et al., 2016). George Robinson is partially responsible for the application of bacteria in contemporary bioremediation. His work has been crucial in expanding the application of microbial processes to reduce pollution in the environment, especially when it comes to cleaning up oil spills (Behera et al., 2020, 2021, 2022). By his research and development work, Robinson contributed to the creation of methods that break down and detoxify oil. The most favored agent of hydrocarbon breakdown for bioremediation may have been bacteria due to their propensity for reproduction and widespread distribution (Macaulay, 2014). The principal approach is bioaugmentation, in which the polluted site is inoculated with certain strains of oil-degrading bacteria to expedite the breakdown process.

The goal of this strategy is to increase the pace or degree of degradation of complex pollutants by introducing microorganisms that break down contaminants (Madhurankhi Goswami, 2018). A further process is biostimulation, which is the addition of nutrients, such as oxygen, nitrogen, and phosphorus, to the natural microbial activity in order to provide the ideal circumstances for microbial growth and activity. These microbes eventually degrade the hydrocarbons by using them as a source of carbon and energy. Furthermore, the cooperative use of phytoremediation, in which plants aid in the uptake and degradation of pollutants in oceans, can be a promising solution. Five different methods of phytoremediation—phytostabilization, phytoextraction, phytodegradation, rhizofiltration, and phytovolatilization—can be used to remove contaminants. Together, these systems enable an extensive and long-lasting cleaning of oil spills, aiding in the restoration of contaminated areas.

7.6.3 Microorganisms Involved in Bioremediation

Microorganisms are found all across the biosphere due to their remarkable metabolic capacity and ease of growth in a variety of environmental settings (Abatenh et al., 2017). They may use a variety of organic and inorganic molecules as sources of carbon, energy, and nutrition because of their metabolic flexibility. Using living things to remove or neutralize pollutants from a contaminated region makes advantage of different microorganism's metabolic capacities, such as those of bacteria, fungi, and algae (Das et al., 2023; Dey, Dash, et al., 2024). This approach depends on encouraging the development of certain microflora or microbial consortia that are native to the polluted locations and capable of carrying out required tasks (Singh, 2014). It is possible to separate microorganisms from practically any type of environment. Microbes are adaptable and may develop in a variety of environments, including subzero temperatures, intense heat, desert climates, water, an abundance of oxygen, anaerobic circumstances, the presence of dangerous substances, and waste streams (Kensa, 2011).

Multiple crucial elements that govern the outcome of pollutant degradation impact the efficacy of microbial bioremediation. Temperature, pH, and moisture content are important environmental factors that affect microbial activity and metabolic rates. Certain parameters have ideal ranges for these factors. In nutrient-deficient settings, supplementation is frequently necessary due to the critical role that nutrients, especially nitrogen and phosphorus, play in microbial development and function. The most beneficial microbes for the bioremediation process will either be aerobic or anaerobic depending on the availability or absence of oxygen. Active members of microbial consortiums are defined as microorganisms that perform biodegradation

TABLE 7.2
Contains a Detailed Information about Microorganisms and Hydrocarbon Interaction

Microorganisms	Hydrocarbon	References
Penicillium chrysogenum	Monocyclic aromatic hydrocarbons, benzene, toluene, ethyl benzene and xylene, phenol compounds	Abatenh et al. (2017)
Gleophyllum striatum MTCC-1117	Polycyclic aromatic hydrocarbons pyrene, acenaphthene, anthracene, dibenothiophene, and 9-methyl anthracene	Yadav et al. (2011)
Coprinellus radians	PAHs, methylnaphthalenes, and dibenzofurans	Abatenh et al. (2017)
Rohodococcus sp. Strain Q15	Hydrocarbons	(Madhavi Tiwari (2022)
Mycobacter calcoaceticus	Naphthalene	Abu Barak Salleh (2003)
Mycobacterium sp.	Pyrene	Abu Barak Salleh (2003)
Phanerochaete chrysosporium	Biphenyl and triphenylmethane	Abatenh et al. (2017)

TABLE 7.3
Represents Microorganisms That Degrade Oil

Microorganisms	Compound	References
Fusarium sp.	Oil	Abatenh et al. (2017)
Aspergillus niger, Candida glabrata, Candida krusei, and Saccharomyces cerevisiae	Crude oil	Sahu and Das (2024)
Pseudomonas aeruginosa, P. putida, Arthrobacter sp., and Bacillus sp.	Diesel oil	Sukumar (2016)
Pseudomonas cepacia, Bacillus cereus, Bacillus coagulans, Citrobacter koseri, and Serratia fi caria	Diesel oil, crude oil	Abatenh et al. (2017)
Alcaligenes odorans, Bacillus subtilis, Corynebacterium propinquum, Pseudomonas aeruginosa	Oil	Etim (2012)
B. brevis, P. aeruginosa KH6, B. licheniformis, and B. sphaericus	Crude oil	Abatenh et al. (2017)

in a variety of settings (Singh, 2014). Table 7.2 lists the microorganisms that are involved in the breakdown of hydrocarbons.

A list of microorganisms that are important for oil remediation is mentioned in Table 7.3.

7.7 PHYTOREMEDIATION

7.7.1 Role of Plant in Oil Degradation

Using living plants to remove or break down pollutants from the environment is known as phytoremediation, and it is an environmentally friendly and sustainable technique (Angulu et al., 2018). This method makes use of plants' innate capacity to take in, retain, and cleanse pollutants via a variety of processes. Utilizing plants to remove pollutants can help cut down on the requirement for costly machinery, labor-intensive procedures, and energy-intensive methods (Lahiri et al., 2021). In addition to metabolizing organic pollutants and promoting their microbial breakdown in the root zone, plants can be employed to sequester inorganic and organic pollutants (Arthur et al., 2005). Numerous pollutants, such as metals, pesticides, explosives, and oil, can be cleaned up by plants. Additionally, the plants aid in retaining contaminants at the locations by preventing wind, rain, and groundwater from moving them away. Several approaches are included in phytoremediation, each of which is suited to a particular kind of pollution and its surroundings. By using plants' inherent abilities to absorb, collect, detoxify, or stabilize contaminants, these methods take advantage of such abilities. Table 7.4 gives the brief description of the different phytoremediation techniques. Through a process called phytoremediation, plants significantly contribute to the deterioration of oil in maritime environments. Due to its advantages over chemical techniques, phytoremediation has grown in popularity since the 1980s (Hussein et al., 2022).

TABLE 7.4
Show Different Phytoremediation Techniques

Phytoremediation	Description	References
Phytoextraction	This technology refers to plants that absorb metals from soil and translocate them to the harvestable shoots where they accumulate. The roots and shoots are subsequently harvested to remove the contaminants from the soil.	Moosavi and Seghatoleslami (2013)
Phytostabilization	Through precipitation in the rhizosphere or accumulation by roots, phytostabilization seeks to keep pollutants out of the vadose zone. By doing this, off-site pollution caused by their movement through soil dispersion, leaching, and erosion from wind and water is prevented.	El-Ramady et al. (2022)
Phytodegradation	Using microorganisms and plants to absorb, digest, and break down the organic pollutant is known as phytodegradation. This method involves decontaminating soil that has been polluted with organic substances by using plant roots in conjunction with microorganisms. The term "phytotransformation" also applies.	Moosavi and Seghatoleslami (2013)
Rhizofiltration	The process of using plant roots to absorb, concentrate, and precipitate harmful metals from contaminated ground water or polluted effluents is known as rhizofiltration. According to most studies, plants used in phytoremediation should only store metals in their roots. This is because, in all forms of phytoremediation, the plants' root systems interact with the pollutants or contaminated site to remove the pollutants and restore the region to a pollution-free state.	Rawat (2012)
Phytovolatilization	Through the process of phytovolatilization, pollutants are drawn up from the soil, changed into a volatile state, and then transpired into the atmosphere by means of plants. As expanding trees and other plants absorb water and the organic and inorganic pollutants, phytovolatilization takes place.	Singh (2005)

Reports regarding the in situ phytoremediation of submerged sediments polluted with PCBs and PAHs are unknown (Huesemann et al., 2009).

7.7.2 Types of Plant Used and Their Effectiveness

By taking in pollutants (mostly water-soluble) through their roots and moving them through different plant tissues where they can be digested, sequestered, or volatilized, plants function as solar-driven pumping and filtration systems (Moosavi and Seghatoleslami, 2013). There have been reports of over 400 plant species from at least 45 plant groups hyperaccumulating metals (Lasat, 2002). Brassica juncea is one of the species of plants that warrants special attention among the others since several tests have demonstrated its importance to the phytoextraction process of heavy metals from soil (Das et al., 2012; Mishra et al., 2022). Research has shown that B. juncea has a significant potential to collect Cd, primarily in the shoots, where a Cd level of 1,450 μg Cd/g dry weight was detected. This is three times more than the 555 μg/g dry weight observed in Brassica napus (Moosavi and Seghatoleslami, 2013). There are other various kinds of plant species that are used for the treatment of oil contaminants detail for the same is provided below in Table 7.5.

TABLE 7.5
Represents the Collective Data of Different Types of Plant Species with Their Effectiveness

Plant Species	Effectiveness
Seagrasses (e.g., Zostera spp. And Thalassia spp.)	Seagrasses are very good in stabilizing sediments and giving microorganisms that break down oil a place to live. Their vast root systems aid in microbial activity and soil oxygenation, which accelerates the decomposition of hydrocarbons. Seagrass meadows can help marine ecosystems recover by drastically lowering the amount of oil present in polluted regions.
Salt Marsh Plant (e.g., Spartina alterniflora and Juncus roemerianus)	Plants that grow in intertidal zones and have a high resistance to saltwater are known characteristics of salt marsh plants. They are essential in capturing oil and stopping its spread. These plants' oxygen- and organic-rich rhizosphere is home to a thriving bacterial population that breaks down oil. Because of this, salt marsh plants—especially those found near coasts—are very useful in bioremediation projects.
Mangroves (eg. Rhizophora spp. and Avicennia spp.)	Because of their intricate root systems, which offer a large surface area for microbial colonization, mangroves are very successful at degrading oil. Mangroves' pneumatophores, or aerial roots, allow gas exchange, which supports the maintenance of aerobic conditions that support the microbial breakdown of hydrocarbons. Additionally, mangroves serve as physical barriers, keeping oil contained and preventing it from getting to more vulnerable places. In tropical and subtropical areas, where they create thick forests along coasts, their role is particularly important.
Microalgae (eg. Chlollera spp. and Dunaliella spp.)	Hydrocarbons can be immediately absorbed by microalgae and then changed by metabolic processes. Additionally, during photosynthesis, they create oxygen, which helps bacteria degrade organic matter aerobically. Certain species of microalgae have the ability to break down oil components, which helps to reduce the amount of oil pollutants in the water column overall. Even though they could have a more limited effect than bigger plants, they are nevertheless an important part of the overall phytoremediation plan.

7.8 MYCOREMEDIATION

Mycoremediation is a type of bioremediation where pollutants from the environment, such as oil pollution in marine environments, are broken down or sequestered using fungus. The technique known as "mycoremediation" uses fungus to remove pollutants from the environment. These fungi grow long threads called hyphae that mate with rocks, soil, and tree roots to form a filamentous body (Dipanwita Biswas, 2023). The synthesis of extracellular enzymes like lignin peroxidases, manganese peroxidases, and laccases by fungi gives them a special capacity to degrade complex hydrocarbons. The difficult-to-break compounds present in oil may be broken down by these enzymes and converted into less dangerous materials. This process involves certain mechanisms for degradation such as biosorption, bioaccumulation, bioconversion, and biodegradation. Furthermore, the fungal mycelium has the ability to form a biofilm on oil surfaces, which increases the biodegradation process by giving enzymes a lot of surface area to work with. In addition to being efficient, the use of fungi in mycoremediation is also eco-friendly as it eliminates the need for chemical dispersants and other conventional cleanup techniques that might have a negative influence on the environment.

7.8.1 Fungi and Their Role in Breaking Hydrocarbon

Organic molecules that are made entirely of hydrogen and carbon atoms are known as hydrocarbons. They are important in the context of oil pollution since they are the main components of natural gas and petroleum. Land and water pollution caused by hydrocarbons are one of the significant problems in developing countries. Among the microorganisms previously identified from soil, water, and sediment that have been contaminated by hydrocarbons are fungi. Numerous fungi have been identified in areas polluted by petroleum hydrocarbons in recent years (Mahmud et al., 2022). Particularly, the crude oil has hundreds of distinct hydrocarbons, ranging from C1 (methane) to more than C30. Saturate, Aromatic, Resin, and Asphaltene (SARA) analysis reveals that, except for gases, the majority of aliphatic hydrocarbons in petroleum are alkanes. These include open-chain linear or branched alkanes and cycloalkanes, which are also referred to as naphthene (Daccò et al., 2020). Since fungi have enzyme systems, they are uniquely able to break down complicated hydrocarbon compounds. Extracellular enzymes that can dissolve the hard, intricate structures of hydrocarbons include lignin peroxidases, manganese peroxidases, and laccases, which are produced by fungi, especially white-rot fungus. White-rot fungi (WRF) are excellent candidates for use in bioremediation because they can break down a wide range of organic pollutants that are not easily broken down by other microorganisms. Fungi also possess vast mycelial networks that may infiltrate polluted surfaces and soils, expanding the surface area available for contact with contaminants and speeding up the degradation process. Fungi are important agents in environmental rehabilitation efforts because of their large natural attenuation and cleaning of hydrocarbon contaminants through these methods.

7.9 MECHANISMS OF ACTION

7.9.1 Microbial Metabolisms of Hydrocarbons

In ecosystems affected by oil spills and other hydrocarbon sources, the biodegradation of organic pollutants depends critically on the microbial metabolism of hydrocarbons. Hydrocarbons may be broken down by a variety of microorganisms, such as bacteria, fungi, and archaea, and used as sources of carbon and energy. Hydrocarbon-utilizing bacteria and fungi are widely distributed in both terrestrial and aquatic ecosystems, but their proportion to the overall heterotrophic community varies greatly. According to reports, their frequencies range from 6% to 82%, 0.13% to 50% for soil bacteria, and 0.003% to 100% for marine bacteria (Leahy and Colwell, 1990). Initial hydrocarbon activation is frequently followed by a sequence of oxidation and reduction processes that transform the hydrocarbon into intermediate molecules. These intermediate compounds are then further metabolized to produce biomass, CO_2, and water. The accelerated disappearance of petroleum that contaminates fields and streams, the degradation of some synthetic and natural rubber products, the spoiling of cooling oils, the deterioration of oiled or asphalt-surfaced highways, and the alteration of petroleum or its products stored in the presence of water can all be attributed to the microbial oxidation of hydrocarbons (Zobell, 1946). Especially when it comes to cleaning up contaminated areas and oil spills, this metabolic capacity is crucial for reducing environmental contamination.

7.9.2 Aerobic and Anaerobic Degradation Pathway

Both aerobic and anaerobic mechanisms, each involving unique microbial activities, are involved in the breakdown of hydrocarbons. However, hydrocarbon biodegradation slows down and becomes less extensive in anaerobic environments, and the range of substrates that are broken down usually narrows (Grishchenkov et al., 2000). The principal electron acceptor in aerobic circumstances is oxygen (Dey et al., 2023). A sequence of enzymatic processes is how microorganisms, including bacteria and fungi, decompose hydrocarbons. On the other hand, microbes use other electron acceptors, including nitrate, sulfate, or carbon dioxide, in anaerobic conditions, where oxygen is not present. Aerobic conditions provide the fastest and most thorough breakdown of most organic contaminants. The enzymatic activity of oxygenase and peroxidases catalyzes the first intracellular assault of organic contaminants, which involves the oxidative process of activation and incorporation of oxygen.

Organic pollutants are progressively transformed into central intermediary metabolic intermediates by peripheral degradation mechanisms, such as the tricarboxylic acid cycle. A few examples of the central precursor metabolites used in the biosynthesis of cell biomass include acetyl-CoA, succinate, and pyruvate. Gluconeogenesis synthesizes sugars that are needed for a variety of biosynthesis and growth (Olajire and Essien, 2014). Likewise, in anaerobic degradation, the process starts with the hydrocarbon being activated, usually by adding fumarate or other chemicals, and then proceeds through a series of reduction processes. This results in the creation of intermediate molecules, which are then refined into other simpler chemicals such as carbon dioxide and methane. To help with the bioremediation of polluted areas, both routes are essential for the natural attenuation of hydrocarbons in the environment.

7.9.3 Enzymatic Breakdown Processes

Enzymatic hydrocarbon degradation occurs in marine settings when certain bacteria use different enzymes to break down hydrocarbons. The hydrocarbons are transformed into alcohols, aldehydes, and carboxylic acids under aerobic circumstances by enzymes like monooxygenases and dioxygenases adding oxygen to the hydrocarbons. After further metabolism, carbon dioxide and water are produced from these intermediates. When there is no oxygen present, other enzymes help activate hydrocarbons by reacting with substitute electron acceptors such as nitrate or sulfate. This produces simpler molecules like carbon dioxide and methane. Anaerobic conditions are created when this process occurs. To bioremediate hydrocarbon contaminants in marine habitats, this enzymatic activity is necessary.

7.10 PLANT–MICROBE INTERACTIONS

7.10.1 Rhizoremediation and Root-Associated Microbial Activity

Rhizoremediation, or the employment of plants and the microbes they are linked to within the root zone, is one of the new techniques in bioremediation. It is particularly good at breaking down organic pollutants and has benefits for both ecology and engineering. Plant exudates can increase microbial activity in the rhizosphere because of the substances they produce through their roots. Under this process, plants interact with soil organisms to carry out the breakdown, detoxification, or immobilization of contaminants in polluted soils (Hassand et al., 2024). In this mutualistic interaction, microorganisms have a favorable environment to develop in while plants profit from improved nutrient absorption and protection from harmful chemicals. Plants can enhance the degradative capacity of rhizospheric microbes in several ways, such as through increased microbial population densities generally, specialized increases in microbial communities that break down petroleum hydrocarbons, increased expression of catabolic genes, increased catabolic gene transfer horizontally, and improved hydrophobic hydrocarbon bioavailability. Furthermore, some plants can improve microbial community conditions, stabilize the soil, and lessen erosion—all of which help to promote the breakdown of hydrocarbons. Oil pollution in polluted regions may be effectively and sustainably addressed by the synergistic technique known as phytoremediation.

7.10.2 Plant Uptake and Translocation of Oil Component

Hydrocarbons are absorbed by plants and moved throughout the system by means of translocation and uptake of oil components. Plants can absorb certain oil components through their roots when there is an oil spill. These compounds could include hydrocarbons that are lighter, more soluble, and able to enter root tissues. Not only do plants remove pollutants from the environment, but they also slow down groundwater flow and prevent pollution from spreading (Rafiee et al., 2017). Certain hydrocarbons may be transformed by plants into less harmful substances, which helps in detoxification. The kind of hydrocarbon, the species of plant, and the surrounding surroundings are some of the variables that affect this process' effectiveness. These factors affect the plant's ability to absorb and relocate hydrocarbons. Although this capacity is helpful for phytoremediation, it needs to be managed carefully to prevent phytotoxicity or food chain contamination. Plants aid in the natural attenuation of pollutants in polluted areas by absorbing and relocating oil components. This procedure aids in the breakdown of dangerous materials and lessens the effect they have on ecosystems. Nonetheless, maximizing the efficacy of phytoremediation initiatives requires careful consideration of site-specific factors and the selection of suitable plant species. The bioremediation of oil spills is mostly dependent on plant absorption and translocation, which helps the contaminated regions recover better.

7.11 FUNGAL DEGRADATION

The breakdown of hydrocarbons by different fungi, which release enzymes that decompose complex organic substances, is known as fungal degradation of oil spills. It has been found that fungi can break down oil in both marine and terrestrial environments (Simister et al., 2015). Enzymes like lignin peroxidases and manganese peroxidases, which may break down a variety of hydrocarbons, are produced by fungi, including black and white rot species. Fungal colonization and hydrocarbon use as a carbon source are two ways in which fungi contribute significantly to oil spill situations. They assist with environmental detoxification by dissolving these complex molecules into simpler, less hazardous ones. Improved nutrient cycling brought about by this degrading process also improves soil health and benefits the environment. The utilization of fungi in bioremediation techniques is essential because they can break down and access hydrocarbons that bacteria could find difficult to break down. Fungi are very useful in cleaning up oil-contaminated places because of their capacity to adapt and endure in a variety of challenging environments (Mohanty et al., 2016). Its capacity aids in the breakdown of contaminants and promotes the general rehabilitation of habitats that have been impacted. Utilizing fungal activity increases the efficiency of bioremediation operations, allowing oil pollutants to naturally attenuate and encouraging ecological regeneration in contaminated regions.

7.11.1 Mycelia Network and Enzymatic Activity

Fungal mycelium networks and their enzyme functions are primary components of biodegradation systems that take care of organic pollutants, among them hydrocarbons, which arise from oil spills. The hyphal network (mycelium) extends through the soil and contaminated substrates, facilitating transport of nutrients along with pollutants due to increased surface contact. It is through this vast network that fungi can probe great soil volumes and reach hydrocarbons that may be hidden or out of their reach. Fungi release an extensive range of external enzymes, which include but are not limited to lignin peroxidases, manganese peroxidases, and laccases—which act on the high molecular weight components by breaking them down into low molecular weight products. These simpler compounds are also less toxic; the enzyme systems act by generating free radicals, which tend to react with organic molecules in proximity (thus involving a chain reaction). This is how the biodetoxification process takes place: these oxidative reactions can occur under very mild conditions and degrade almost any kind of organic molecule. The enzymatic activity within the mycelial network not only aids in breaking down pollutants but also enhances nutrient cycling in the soil, facilitating the uptake of nutrients by plants and other microorganisms. This interaction fosters a healthier soil ecosystem, promoting the recovery of contaminated environments. Furthermore, the ability of fungal mycelia to withstand harsh environmental conditions, such as extreme pH levels and limited nutrient availability, makes them particularly effective in bioremediation. Their resilience and adaptability allow fungi to thrive in diverse environments, contributing to the natural attenuation of oil pollutants.

7.11.2 Degradation Pathway Specific to Fungi

Fungi employ a specialized degradation pathway to break down organic pollutants like hydrocarbons in environments such as oil spills. Central to this process are the extracellular enzymes secreted by fungi, including but not limited to lignin peroxidases, manganese peroxidases, and laccases. These enzymes play crucial roles in catalyzing oxidative reactions that initiate the breakdown of complex hydrocarbon molecules. Lignin peroxidases and manganese peroxidases, for instance, can oxidize a wide range of aromatic compounds present in hydrocarbons, breaking down their chemical structures. This initial oxidation step is followed by further enzymatic actions that cleave these compounds into smaller, more manageable fragments. These fragments can then be metabolized by the fungi as carbon and energy sources or assimilated into cellular structures. Additionally, fungi possess metabolic pathways that allow them to utilize various components of hydrocarbons for growth and reproduction. This metabolic versatility enables fungi to thrive in environments where hydrocarbons are abundant, contributing to their role in bioremediation efforts. The resilience of fungal species to diverse environmental conditions further enhances their effectiveness in degrading pollutants in contaminated sites. Overall, the degradation pathway specific to fungi involves the coordinated action of extracellular enzymes to degrade hydrocarbons into simpler compounds that can be utilized or further metabolized. This capability underscores fungi's significance in ecological processes and their potential as valuable tools in environmental biotechnology for mitigating the impacts of oil spills and other forms of pollution.

7.12 FACTORS INFLUENCING THE EFFECTIVENESS OF BIOLOGICAL REMEDIATION

Biological remediation, also known as bioremediation, is an eco-friendly approach that leverages living organisms such as bacteria, fungi, or plants to remove or neutralize pollutants from contaminated environments. The success of bioremediation processes is largely dependent on a variety of environmental conditions that influence the metabolism and growth of microorganisms involved. Key factors include temperature, pH, oxygen levels, and nutrient availability, all of which play crucial roles in determining the effectiveness of the remediation process.

7.13 TEMPERATURE

Temperature is a critical factor in bioremediation, as it directly affects microbial metabolism. Most bioremediative microorganisms have an optimal temperature range within which they perform most efficiently. For instance, bacteria such as *Pseudomonas* species, often used in bioremediation, thrive in moderate temperature ranges between 20°C and 30°C. At temperatures outside this range, microbial activity may decrease, leading to a slower remediation process. On the other hand, some extremophiles are capable of functioning in extreme temperatures, making temperature-tolerant species a valuable asset in diverse environmental contexts.

7.14 pH LEVELS

The pH of the environment significantly affects the survival and enzymatic activity of microbes involved in bioremediation. Most microorganisms prefer a neutral pH, ranging from 6 to 8, for optimal enzymatic functions. Extremes in pH, either highly acidic or basic conditions, can inhibit microbial growth and activity, rendering the remediation process ineffective. Specific organisms, such as acidophilic bacteria, can survive in low pH environments, making them useful for remediation in acidic conditions, such as mine drainage sites.

7.15 OXYGEN LEVELS

The availability of oxygen also plays a crucial role in the efficiency of bioremediation processes. Oxygen is essential for aerobic microorganisms, which are typically involved in the degradation of hydrocarbons and other organic pollutants. However, certain environments may lack oxygen, and in such cases, anaerobic bioremediation becomes necessary. Anaerobic bacteria, such as *Clostridium* species, can degrade pollutants in oxygen-deprived environments, such as subsurface soils or sediments, expanding the applicability of biological remediation to diverse settings.

7.16 NUTRIENT AVAILABILITY

The availability of nutrients such as nitrogen, phosphorus, and carbon is essential for the growth and metabolic activity of microorganisms used in bioremediation. Nutrient-limited environments can significantly reduce the effectiveness of the process, as microorganisms may not have enough resources to support their growth and pollutant degradation activities. Supplementing contaminated environments with appropriate nutrients, known as biostimulation, can enhance microbial activity and, subsequently, the bioremediation process.

7.17 ADVANCES AND INNOVATIONS IN BIOLOGICAL REMEDIATION

Biological remediation, or bioremediation, has emerged as a promising solution for mitigating environmental contamination caused by industrial, agricultural, and municipal waste. The process involves using microorganisms, plants, or enzymes to break down or neutralize harmful pollutants in soil, water, and air. As pollution becomes a more pervasive issue globally, advancements in bioremediation technologies are crucial for enhancing the efficiency and applicability of this approach. Recent innovations, such as genetic engineering, bioaugmentation, and phytoremediation, have significantly improved the effectiveness of bioremediation. This essay explores these innovations and their impact on the field of environmental restoration.

7.18 GENETIC ENGINEERING OF MICROORGANISMS

One of the most significant advances in biological remediation is the genetic engineering of microorganisms to enhance their ability to degrade pollutants. Traditionally, natural microbial communities were employed in remediation processes, but the metabolic capacity of these organisms was often limited. With the advent of genetic engineering, scientists have been able to modify microbial genomes to optimize their degradation pathways for specific pollutants (Mohanty et al., 2024; Ray et al., 2024). For example, recombinant bacteria such as genetically modified *Pseudomonas putida* have been engineered to break down toxic compounds like toluene, a common industrial solvent. These genetically modified organisms (GMOs) can be tailored to address a variety of contaminants, from hydrocarbons to heavy metals.

Moreover, gene editing technologies like CRISPR-Cas9 allow for precise manipulation of microbial DNA, enabling the development of microorganisms with enhanced resistance to environmental stressors such as extreme temperatures, pH, and salinity. These resilient microbes are particularly useful in remediation efforts in harsh or inhospitable environments, such as polar regions or contaminated mine tailings. Such innovations have the potential to revolutionize bioremediation by expanding its applicability to a wider range of contaminated sites.

7.18.1 Bioaugmentation and Biostimulation

Bioaugmentation involves the introduction of specific microbial strains or consortia to contaminated sites to accelerate the degradation of pollutants. This approach has gained attention due to its ability to enhance the natural microbial population's efficiency in breaking down complex or recalcitrant pollutants (Das & Chandran, 2011). Recent advancements in microbial isolation and culturing techniques have allowed scientists to identify and utilize highly specialized bacteria and fungi capable of degrading specific contaminants, such as polychlorinated biphenyls (PCBs) or petroleum hydrocarbons.

In tandem with bioaugmentation, biostimulation involves adding nutrients or electron donors to stimulate the indigenous microbial populations' metabolic activity.

Recent studies have demonstrated that the combination of bioaugmentation and biostimulation can yield synergistic effects, particularly in environments with limited nutrient availability. By optimizing the growth conditions for both introduced and native microbial communities, the remediation process can be significantly accelerated. This innovation has been particularly successful in treating oil spills and sites contaminated with organic solvents.

7.18.2 Phytoremediation

Phytoremediation is an innovative bioremediation technique that uses plants to absorb, sequester, or break down pollutants from soil and water. Certain plants, known as hyperaccumulators, have the unique ability to take up large quantities of heavy metals such as cadmium, arsenic, and lead (Joutey et al., 2013). One of the most well-known examples of a hyperaccumulator is *Thlaspi caerulescens*, a plant capable of extracting zinc and cadmium from contaminated soils.

Recent advances in biotechnology have allowed scientists to genetically engineer plants for enhanced phytoremediation capabilities. By modifying the genes responsible for pollutant uptake and metabolism, researchers have developed plants that can remove contaminants more efficiently and in greater quantities than their natural counterparts. Additionally, advances in agronomic practices, such as soil amendments and crop rotation, have improved the efficiency of phytoremediation, making it a viable option for large-scale environmental restoration projects.

7.18.3 Nanotechnology in Bioremediation

The integration of nanotechnology with bioremediation processes represents another exciting area of innovation. Nanoparticles, due to their small size and large surface area, have unique physicochemical properties that make them highly effective in catalyzing the degradation of pollutants. For example, iron nanoparticles have been used in conjunction with microbial communities to enhance the reduction of chlorinated organic compounds, such as trichloroethylene (TCE), a common groundwater contaminant .

Moreover, nanoparticles can serve as carriers for enzymes or bacteria, protecting them from environmental stressors and improving their delivery to contaminated sites. This "nano-bioremediation" approach has shown promise in the treatment of persistent organic pollutants (POPs) and heavy metal contamination in both soil and water environments (Joutey et al., 2013). As the field of nanotechnology continues to advance, it is likely that its integration with bioremediation will lead to even more effective and efficient pollution mitigation strategies.

7.18.4 Biosensors for Monitoring Remediation Progress

Recent innovations in biosensing technology have significantly improved the monitoring of bioremediation progress. Biosensors are analytical devices that use biological elements, such as enzymes or microorganisms, to detect specific pollutants or byproducts of microbial degradation. These devices provide real-time data on pollutant concentrations and microbial activity, allowing for more precise control of the bioremediation process (Das & Chandran, 2011).

The development of wireless and portable biosensors has enabled on-site monitoring, reducing the need for labor-intensive sampling and laboratory analysis (Shukla et al., 2010). These innovations have made it possible to track the success of remediation efforts more efficiently and accurately, leading to better decision-making in environmental management. As a result, biosensors have become an invaluable tool for optimizing bioremediation strategies in both urban and rural environments.

7.19 CHALLENGES AND FUTURE DIRECTIONS

Despite the numerous advancements in bioremediation, several challenges remain. One of the main obstacles is the unpredictability of microbial behavior in natural environments. Factors such as temperature fluctuations, moisture content, and competition with native species can affect the success of bioremediation efforts (Alavi and Nejad, 2020). To address these challenges, researchers are exploring the use of omics technologies, such as metagenomics and proteomics, to gain a deeper understanding of microbial communities and their interactions with pollutants. These insights will pave the way for more targeted and effective bioremediation approaches.

Furthermore, the development of multifunctional bioremediation systems that combine several techniques, such as phytoremediation and bioaugmentation, is a promising direction for future research. By integrating various biological, chemical, and physical processes, these systems have the potential to address complex contamination scenarios that would otherwise be difficult to remediate using a single approach.

7.20 FUTURE DIRECTIONS AND NEEDS FOR EMERGING TRENDS IN BIOLOGICAL REMEDIATION: INNOVATIONS IN TECHNOLOGY AND METHODOLOGY FOR LARGE-SCALE APPLICATIONS

Biological remediation, also known as bioremediation, has demonstrated considerable potential in addressing environmental contamination by leveraging microorganisms, plants, and biological processes to degrade or neutralize pollutants. While the field has made significant advances in recent years, particularly in the areas of genetic engineering, bioaugmentation, and phytoremediation, there remain many challenges and opportunities for improvement. As environmental threats from industrial pollution, agricultural runoff, and hazardous waste grow more acute, emerging trends in bioremediation technology and methodology will play a crucial role in large-scale applications. This essay explores the future directions of biological remediation, focusing on innovations in technology and methodology and their potential for broad-scale implementation.

7.21 EMERGING TRENDS IN BIOREMEDIATION TECHNOLOGIES

7.21.1 Genetic Engineering for Enhanced Microbial Degradation

Genetic engineering continues to be a cornerstone of emerging bioremediation technologies. With advancements in gene-editing tools like CRISPR-Cas9, the future of bioremediation will likely see the development of highly specialized microbial strains engineered for targeted degradation of complex pollutants. Microorganisms such as *Pseudomonas putida* have already been modified to degrade toxic compounds such as toluene, but future directions could include the development of microbial consortia designed to metabolize a wide array of POPs, including PCBs and polycyclic aromatic hydrocarbons (PAHs) (Ghosal et al., 2016).

Additionally, the integration of omics technologies—such as metagenomics, proteomics, and transcriptomics—will further enhance our understanding of microbial communities involved in remediation. This will enable scientists to design microorganisms with more robust metabolic pathways for pollutant degradation in diverse environmental conditions, from highly saline coastal areas to extreme temperature zones in the Arctic.

7.21.2 Nanotechnology in Bioremediation

Nanotechnology is another emerging trend that offers exciting possibilities for enhancing bioremediation processes. Nanoparticles, due to their small size and high surface area, can increase the bioavailability of pollutants, making them easier for microorganisms to degrade. For instance, iron nanoparticles have been used to reduce chlorinated compounds like TCE in groundwater. Moving forward, the development of multifunctional nanoparticles could enhance the catalytic activity of microorganisms or enzymes used in bioremediation (Joutey et al., 2013).

Nano-bioremediation systems, where nanoparticles act as carriers for enzymes or bacteria, can significantly improve the efficiency of pollutant breakdown, especially for persistent pollutants like heavy metals and hydrocarbons. The ability to apply these technologies on a larger scale will be pivotal to the future of remediation strategies in industrial and urban settings.

7.21.3 Biosensors and Real-Time Monitoring

The future of large-scale bioremediation will require efficient monitoring systems to track the progress of pollutant degradation in real time. Innovations in biosensing technologies have already shown promise in offering rapid, cost-effective solutions for on-site monitoring. Biosensors utilize biological elements such as enzymes, antibodies, or microbial cells to detect pollutants and their degradation byproducts in soil, water, and air. As wireless biosensors become more advanced, they will allow environmental engineers to monitor bioremediation efforts remotely and in real-time (Das & Chandran, 2011).

In the future, the integration of biosensors with artificial intelligence (AI) and machine learning could further optimize bioremediation processes. AI-powered models could analyze data from biosensors to predict the efficiency of remediation efforts and suggest adjustments in real-time, ensuring that remediation projects are not only effective but also efficient and sustainable.

7.22 METHODOLOGICAL INNOVATIONS

7.22.1 Bioaugmentation and Biostimulation for Enhanced Degradation

The combination of bioaugmentation and biostimulation has emerged as a powerful approach to accelerate the remediation process. While bioaugmentation involves introducing specialized microbial strains to a contaminated site, biostimulation enhances the growth and activity of indigenous microorganisms by providing them with essential nutrients or electron donors (Shukla et al., 2010). Recent innovations in microbial ecology and genomics have allowed scientists to fine-tune this process, making it more targeted and efficient.

Future research could focus on the development of microbial consortia designed for specific types of pollution. These consortia would consist of microorganisms with complementary metabolic pathways, allowing for the complete breakdown of complex pollutants like petroleum hydrocarbons or heavy metals. Additionally, the use of synthetic biology could lead to the creation of microbial communities that are both highly efficient and resilient to environmental stressors such as fluctuating temperatures, moisture levels, or oxygen availability.

7.22.2 Phytoremediation with Engineered Plants

Phytoremediation, the use of plants to absorb, sequester, or degrade pollutants from soil and water, has shown promise for large-scale environmental restoration projects. Hyperaccumulator plants, such as *Thlaspi caerulescens*, are capable of extracting heavy metals like cadmium and arsenic from contaminated soil. However, the future of phytoremediation lies in the genetic engineering of plants to enhance their capacity for pollutant uptake and degradation (Etim, 2012).

By introducing genes that improve a plant's ability to metabolize or sequester contaminants, researchers can develop plants specifically suited to different types of contamination. Furthermore, biotechnological advances in root architecture and symbiotic relationships between plants and soil microorganisms could further increase the efficacy of phytoremediation strategies, making them more scalable for larger, more complex contamination sites.

7.22.3 Scaling Bioremediation for Large-Scale Applications

The future of biological remediation lies in its ability to scale up for large-scale applications in diverse environments. Large-scale bioremediation projects face challenges such as inconsistent pollutant distribution, environmental variability, and high implementation costs (El-Ramady et al., 2022). To overcome these obstacles, future methodologies must focus on cost-effective solutions, such as the use of locally available microbial strains, modular treatment systems, and adaptive management practices.

Innovations in bioreactor design, for instance, could make large-scale bioremediation more feasible. Bioreactors provide controlled environments for microbial growth and pollutant degradation and could be deployed on contaminated sites to enhance the remediation process. Moreover, the development of mobile bioreactor systems could enable on-site treatment of hazardous waste in remote areas, further broadening the scope of bioremediation.

Furthermore, combining various bioremediation techniques, such as microbial degradation, phytoremediation, and nanotechnology, in a single treatment plan could lead to multifunctional systems capable of addressing complex contamination scenarios. These hybrid approaches will likely be necessary to manage large industrial sites, agricultural lands, and urban environments where multiple pollutants coexist.

7.23 CONCLUSION

As bioremediation continues to evolve, emerging trends in technology and methodology will play a vital role in making it a more efficient and scalable solution for environmental restoration. Innovations such as genetic engineering, nanotechnology, biosensing, and synthetic biology are transforming the field, allowing for more precise, targeted, and large-scale applications. As environmental challenges become increasingly complex, the ability to deploy these innovative technologies on a large scale will be essential for the future of pollution mitigation and sustainable development. Moving forward, interdisciplinary collaboration between environmental engineers, biotechnologists, and data scientists will be critical to overcoming the remaining challenges and maximizing the potential of bioremediation for global environmental health.

REFERENCES

Abatenh, E., Gizaw, B., Tsegaye, Z., & Wassie, M. (2017). The role of microorganisms in bioremediation: A review. *Open Journal of Environmental Biology*, *2*(1), 038–046.

Angulu, C. N., Atuluku S. A., Oladoja, E. O., Balogun, A. N., Attah-Olottah, R., & Adeniyi, K. A. (2018). Application of phytoremediation in the management of oil spillage: A review. *Global Journal of Earth and Environmental Science 3*(3), 16–22.

Arora, N. K. (2018). Bioremediation: A green approach for restoration of polluted ecosystems. *Environmental Sustainability*, *1*(4), 305–307. https://doi.org/10.1007/S42398-018-00036-Y

Arthur, E. L., Rice, P. J., Rice, P. J., & Anderson, T. A. (2005). Phytoremediation: An overview. *Critical Reviews in Plant Sciences, 24*(2), 109–122.

Baba, U., Mudasir, S., Sheikh, A. Q., & Nazir, R. (2016). Bioremediation: A management tool. *Journal of Bioremediation & Biodegradation*, *7*(2), 1000331.

Baniasadi, M., & Mousavi, S. M. (2018). A comprehensive review on the bioremediation of oil spills. *Microbial Action on Hydrocarbons*, 223–254.

Behera, I. D., Basak, G., Kumar, R. R., Sen, R., & Meikap, B. C. (2020). Treatment of petroleum refinery sludge by petroleum degrading bacterium *Stenotrophomonas pavanii* IRB19 as an efficient novel technology. *Journal of Environmental Science and Health, Part A*, *56*, 1–13. https://doi.org/10.1080/10934529.2020.1866924

Behera, I. D., Nayak, M., Biswas, S., Meikap, B. C., & Sen, R. (2021). Enhanced biodegradation of total petroleum hydrocarbons by implementing a novel two-step bioaugmentation strategy using indigenous bacterial consortium. *Journal of Environmental Management*, *292*, 112746. https://doi.org/10.1016/j.jenvman.2021.112746

Behera, I. D., Nayak, M., Mishra, A., Meikap, B. C., & Sen, R. (2022). Strategic implementation of integrated bioaugmentation and biostimulation for efficient mitigation of petroleum hydrocarbon pollutants from terrestrial and aquatic environment. *Marine Pollution Bulletin*, *177*, 113492. https://doi.org/10.1016/j.marpolbul.2022.113492

Daccò, C., Girometta, C., Asemoloye, M. D., Carpani, G., Picco, A. M., & Tosi, S. (2020). Key fungal degradation patterns, enzymes and their applications for the removal of aliphatic hydrocarbons in polluted soils: A review. *International Biodeterioration & Biodegradation*, *147*(10), 104866.

Das, A. P., & Ghosh, S. (2023). Microorganisms in manganese biomining processes. In *Manganese Mining Microorganisms*, 55–61. https://doi.org/10.1016/B978-0-12-822147-1.00013-3

Das, A. P., Ghosh, S., Mohanty, S., & Sukla, L. B. (2015). Advances in manganese pollution and its bioremediation. 313–328. https://doi.org/10.1007/978-3-319-19018-1_16

Das, A. P., Mohanty, A. K., van Hullebusch, E. D., Figueiredo, G., Das, A. P., Mohanty, A. K., van Hullebusch, E. D., & Figueiredo, G. (2023). Bioremediation as an emerging technology for the removal of synthetic microplastic pollutants from marine ecosystem. *Marine Pollution Bulletin*, *194*, 115297. https://doi.org/10.1016/J.MARPOLBUL.2023.115297

Das, A. P., Nayak, B., & Behera, I. D. (2024). Applications of microorganisms in advanced geomicrobiological and biotechnological exploration. *Geomicrobiology Journal*, *41*(4), 309–311. https://doi.org/10.1080/01490451.2024.2327059

Das, A. P., Sukla, L. B., & Pradhan, N. (2012). Microbial recovery of manganese using *Staphylococcus Epidermidis*. *International Journal of Nonferrous Metallurgy*, *1*(2), 9–12. https://doi.org/10.4236/IJNM.2012.12002

Dey, S., Bissoyi, A., Koteswara Reddy, G., & Das, A. P. (2024). Synthetic microfibers: An urban emerging pollutant demanding for sustainable environmental solutions. In *Sustainable Microbial Technology for Synthetic and Cellulosic Microfiber Bioremediation*, 107–121. https://doi.org/10.1007/978-3-031-62600-5_7

Dey, S., Das, A., Mallick, K., Sahu, A., & Das, A. P. (2023). Environmental petroleum waste: Pollution, toxicity, sustainable remediation. 159–175. https://doi.org/10.1007/978-3-031-48220-5_7

Dey, S., Dash, A., & Das, A. P. (2024). Bioremediation as a sustainable and circular solution for mining waste management and bioenergy generation. In *Sustainable Management of Mining Waste and Tailings: A Circular Economy Approach*, 107–119. https://www.taylorfrancis.com/chapters/edit/10.1201/9781003442455-6/bioremediation-sustainable-circular-solution-mining-waste-management-bioenergy-generation-sudeshna-dey-akanshya-dash-alok-prasad-das

Dipanwita Biswas, A. C. (2023). Mycoremediation is a potential strategy for environmental clean-up of heavy metal: A review. *Journal of Survey in Fisheries Sciences*, *10*(1S), 6316–6327.

El-Ramady, H., Abdalla, N., Fawzy, Z., Badgar, K., Llanaj, X., Törős, G., Hajdú, P., Eid, Y., & Prokisch, J. (2022). Biotechnology of oyster mushroom (Pleurotus ostreatus L.): A sustainable strategy for myco-remediation and bio-fermentation. *Sustainability*, *14*(6), 3667.

Etim, E. E. (2012). Phytoremediation and its mechanisms: A review. *International Journal of Environment and Bioenergy*, *2*(3), 120–136.

Ghosh, M., & Singh, S. P. (2005). A review on phytoremediation of heavy metals and utilization of it's by products. *Asian Journal on Energy and Environment*, *3*, 214–231.

Goswami, M., Chakraborty, P., Mukherjee, K., Mitra, G., Bhattacharyya, P., Dey, S., & Tribedi, P. (2018). Bioaugmentation and biostimulation: A potential strategies for environmental remediation. *Journal of Microbiology & Experimentation*, *6*, 223–231.

Grishchenkov, V. G., Townsend, R. T., McDonald, T. J., Autenrieth, R. L., Bonner, J. S., & Boronin, A. M. (2000). Degradation of petroleum hydrocarbons by facultative anaerobic bacteria under aerobic and anaerobic conditions. *Process Biochemistry, 35*(9), 889–896.

Hassand, M. H., Sarwari, A., Monib, A. W., Niazi, P., & Omirbekova, A. (2024). Microbial-plant interactions and their role in mitigating of oil pollution: A review. *European Journal of Theoretical and Applied Sciences, 2*(2), 11–22.

Huesemann, M. H., Hausmann, T. S., Fortman, T. J., Thom, R. M., & Cullinan, V. (2009). In situ phytoremediation of PAH- and PCB-contaminated marine sediments with eelgrass (Zostera marina). *Ecological Engineering, 35*(10), 1395–1404.

Hussein, Z. S., Hamido, N., Hegazy, A. K., El-Dessouky, M. A., Mohamed, N., & Safwat, G. (2022). Phytoremediation of crude petroleum oil pollution: A review. *Egyptian Journal of Botany*, *62*(3), 611–640.

Kanungo, J., Sahoo, S., Bal, M., & Behera, I. D. (2023). Performance of bioremediation strategy in waste lubricating oil pollutants: A review. *Geomicrobiology Journal*, *41*, 360–373.

Kensa, V. M. (2011). Bioremediation: An overview. *Journal of Industrial Pollution Control, 27*(2), 161–168. https://doi.org/10.1080/01490451.2023.2245395

Lahiri, D., Nag, M., Dey, A., Sarkar, T., Joshi, S., Pandit, S., Das, A. P., Pati, S., Pattanaik, S., Tilak, V. K., & Ray, R. R. (2021). Biofilm mediated degradation of petroleum products. *Geomicrobiology Journal*, 1–10. https://doi.org/10.1080/01490451.2021.1968979

Lasat, M. M. (2002). Phytoextraction of toxic metals: A review of biological mechanisms. *Journal of Environmental Quality*, *31*, 109–120.

Leahy, J. G., & Colwell, R. R. (1990). Microbial degradation of hydrocarbons in the environment. *Microbiological Reviews, 54*(3), 305–315.

Macaulay, B. (2014). Understanding the behaviour of oil-degrading micro-organism to enhance the microbial remediation of spilled petroleum. *Applied Ecology and Environmental Research*, *13*(1), 247–262.

Madhavi Tiwari, A. S. (2022). Mycoremediation of petroleum hydrocarbon contaminants: An overview. *Advances in Microbiology*, *1*, 106–119.

Mahmud, T., Sabo, I. A. A., Lambu, Z. N., Danlami, D., & Shehu, A. A. (2022). Hydrocarbon degradation potentials of fungi: A review. *Journal of Environmental Bioremediation and Toxicology*, *5*(1), 50–56.

Mishra, S., Dash, D., Al-Tawaha, A. R. M. S., & Das, A. P. (2022). A review on heavy metal ion adsorption on synthetic microfiber surface in aquatic environments. *Applied Biochemistry and Biotechnology*, *194*(10), 4639–4654. https://doi.org/10.1007/s12010-022-04029-w

Mohanta, S., Pradhan, B., & Behera, I. D. (2023). Impact and remediation of petroleum hydrocarbon pollutants on agricultural land: A review. *Geomicrobiology Journal*, *41*, 1–5. https://doi.org/10.1080/01490451.2023.2243925

Mohanty, J., Dey, S., Bala, K., & Das, A. P. (2024). Recent advances in microplastic pollution for its sustainable management. In *Sustainable Management of Urban Plastic Waste through Circular Economic Approaches*, 27–48. https://doi.org/10.1201/9781003465447-3

Mohanty, S., Ghosh, S., Nayak, S., & Das, A. P. (2016). Isolation, identification and screening of manganese solubilizing fungi from low-grade manganese ore deposits. *Geomicrobiological Journal*, *34*(4), 309–316. https://doi.org/10.1080/01490451.2016.1189016

Moosavi, S. G., & Seghatoleslami, M. J. (2013). Phytoremediation: A review. *Advance in Agriculture and Biology*, *1*(1), 5–11.

Olajire, A. A. & Essien, J. P. (2014). Aerobic degradation of petroleum components by microbial consortia. *Petroleum & Environmental Biotechnology, 5*, 1–22.

Paul, A., Dey, S., Ram, D. K., & Das, A. P. (2023). Hexavalent chromium pollution and its sustainable management through bioremediation. *Geomicrobiology Journal*, *41*, 1–11. https://doi.org/10.1080/01490451.2023.2218377

Pete, A. J., Bharti, B., & Benton, M. G. (2021). Nana-enhanced bioremediation for oil spill: Review. *ACS ES&T Engineering 1*(6), 928–946.
Pradhan, G., Tripathy, B., Ram, D. K., Digal, A. K., & Das, A. P. (2023). Bauxite mining waste pollution and its sustainable management through bioremediation. *Geomicrobiology Journal, 41*, 1–10. https://doi.org/10.1080/01490451.2023.2235353
Prince Emeka Ndimele, A. O. (Ed.). (2018). Chapter 24 - Remediation of crude oil spillage. In *The Political Ecology of Oil and Gas Activities in the Nigerian Aquatic Ecosystem*, 369–384. Cambridge: Academic Press.
Rafferty, J. P. (2023, September Friday). *9 of the Biggest Oil Spills in History*. Retrieved from https://www.britannica.com/story/9-of-the-biggest-oil-spills-in-history
Rafiee, M., Rad, M. J., & Afshari, A. (2017). Bioaccumulation and translocation factors of petroleum hydrocarbons in Aeluropus littoralis. *Environmental Health Engineering and Management Journal, 4*, 131–136.
Rawat, K. M. (2012). Rhizofiltration: A green technology for remediation of heavy metals. *International Journal of Innovations in Bio-Sciences, 2*(4), 193–199.
Ray, S. K., Dey, S., Das, A. P., & John, A. (2024). Microplastic waste management through greener technologies: A review. In *Sustainable Management of Urban Plastic Waste through Circular Economic Approaches*, 237–261. https://doi.org/10.1201/9781003465447-14
Sahu, S., & Das, A. P. (2024). Microplastics degradation mechanism by aquatic plastiomes: A review. 213–226. https://doi.org/10.1007/978-3-031-50840-0_12
Simister, R. L., Poutasse, C. M., Thurston, A. M., Reeve, J. L., Baker, M. C., & White, H. K. (2015). Degradation of oil by fungi isolated from Gulf of Mexico beaches. *Marine Pollution Bulletin, 100*, 327–333.
Singh, R., Singh, P., & Sharma, R. (2014). Microorganism as a tool of bioremediation technology for cleaning environment: A review. *Proceedings of the International Academy of Ecology and Environmental Sciences, 4*(1), 1–6.
Sukumar, S. A. (2016). Screening of diesel oil degrading bacteria from petroleum hydrocarbon contaminated soil. *International Journal of Advanced Research in Biological Sciences, 3*, 18–22.
Wolok, E., Barafi, J., Joshi, N., Girimonte, R., & Chakraborty, S. (2020). Study of bio-material for removal of the oil spill. *Arabian Journal of Geoscience, 13*, 1244.
Yadav, M., Singh, S. K., Sharma, J. K., & Yadav, K. D. S. (2011). Oxidation of polyaromatic hydrocarbons in systems containing water miscible organic solvents by the lignin peroxidase of Gleophyllum striatum MTCC-1117. *Environmental Technology, 32*, 1287–94.
Zobell, C. E. (1946). Action of microorganisms on hydrocarbons. *Bacteriological Review, 10*(1–2), 1–49.

8 Integrated Approaches to Minimize the Environmental Hydrocarbon Contaminants vis-à-vis Development

Somanath Behera and Maminee Panda

8.1 ENVIRONMENTAL HYDROCARBON CONTAMINANTS

Environmental hydrocarbon contamination has become a global concern, creating environmental pollution as a result of its widespread uses in industrial, agricultural, and other sectors through transportation, production, storage, accidental discharge, and workshop seepage (Imam et al., 2019).

Hydrocarbon contaminants are organic compounds consisting of hydrogen and carbon atoms, often found as pollutants in soil, water, and air (Bala et al., 2024; Dey, Bissoyi, et al., 2024; Dey, Dash, et al., 2024). They come in various types, depending on their molecular structure and origin, such as the following:

8.1.1 Aliphatic Hydrocarbons

They are of three types as per the bonds between the carbon atoms.

i. **Alkanes (Paraffins)**: Saturated hydrocarbons with single bonds between carbon atoms (e.g., methane, ethane, propane). They are typically less reactive and can contaminate groundwater and soil.
ii. **Alkenes (Olefins)**: They are the unsaturated group of hydrocarbons having at least one double bond between carbon atoms (e.g., ethylene, propylene). They are more reactive than alkanes and are often byproducts of industrial processes.
iii. **Alkynes**: Unsaturated hydrocarbons with at least one triple bond between carbon atoms (e.g., acetylene). They are less common as contaminants but can result from certain chemical reactions.

 DOI: 10.1201/9781003508496-8

8.1.2 Aromatic Hydrocarbons

These hydrocarbons have an aromatic ring structure. They are of two types:

i. **Monocyclic Aromatic Hydrocarbons (MAHs)**: Hydrocarbons containing a single aromatic ring, including benzene, toluene, ethylbenzene, and xylene. They are commonly found in gasoline and are toxic, posing significant health risks.
ii. **Polycyclic Aromatic Hydrocarbons (PAHs)**: These hydrocarbons have two or more aromatic rings, like naphthalene, anthracene, and benzo(a)pyrene. PAHs are generated from the incomplete combustion of organic matter and released from vehicle emissions, industrial processes, and wildfires. These are often persistent in the environment.

8.1.3 Chlorinated Hydrocarbons

Hydrocarbons that have been modified with one or more chlorine atoms are known as chlorinated hydrocarbons (e.g., dichloromethane, trichloroethylene, polychlorinated biphenyls (PCBs)). These are used in industrial solvents, pesticides, and refrigerants and are persistent environmental contaminants (Dey, Das, et al., 2023; Lahiri et al., 2021; Priyadarshini & Das, 2024).

8.1.4 Volatile Organic Compounds (VOCs)

Hydrocarbons that easily dissolve into the air at room temperature are known as volatile organic compounds (VOCs) such as benzene, toluene, and xylene. VOCs can be found in paints, cleaning agents, and fuels and contribute to air pollution.

8.1.5 Petroleum Hydrocarbons

Hydrocarbons derived from crude oil and petroleum products (e.g., gasoline, diesel, kerosene, lubricating oils). They are complex mixtures of various hydrocarbons of aliphatic and aromatic groups and are significant pollutants in soil and water due to spills and leaks.

8.1.6 Non-volatile Hydrocarbons

These are hydrocarbons with low vapor pressure that do not readily evaporate, such as heavy oils and tars. They can persist in the environment, especially in soil and sediments, and pose long-term contamination risks.

8.1.7 Natural Hydrocarbons

Naturally occurring hydrocarbons released into the environment through natural processes, such as the breakdown of organic matter, volcanic activity, or seeps of crude oil or gas.

8.1.8 Synthetic Hydrocarbons

Man-made hydrocarbons used in industrial processes, synthetic fuels, and lubricants. They can include various combinations of aliphatic, aromatic, and chlorinated hydrocarbons.

Hydrocarbon contaminants pose varying degrees of environmental and health risks depending on their type, concentration, and exposure pathways. Out of the above, the release of petroleum products, polycyclic aromatic hydrocarbons (PAHs), VOCs, and PCBs to the environment has become a global threat.

8.2 SOURCES OF HYDROCARBON CONTAMINANTS

Environmental hydrocarbon contaminants originate from both natural and anthropogenic (human-made) sources.

8.2.1 Natural Sources

i. **Natural Oil Seeps**: Crude oil naturally seeps from the ocean floor or other geological formations, releasing hydrocarbons into the environment.
ii. **Volcanic Activity**: Volcanic eruptions can release hydrocarbons and other organic compounds into the atmosphere.
iii. **Wildfires**: Burning of vegetation and forests releases hydrocarbons, including PAHs, into the air.
iv. **Biogenic Sources**: Decomposition of organic matter, such as plants and animals, can produce natural hydrocarbons like methane.

8.2.2 Anthropogenic Sources

i. **Oil and Gas Extraction**: Drilling, fracking, and production processes can release hydrocarbons, such as methane, benzene, and toluene, into the environment.
ii. **Transportation**: Vehicles, ships, and airplanes emit hydrocarbons from the combustion of fossil fuels like gasoline and diesel (Van Metre et al., 2000).
iii. **Industrial Processes**: Refineries, chemical manufacturing, and petrochemical plants release various hydrocarbons, including VOCs and PAHs.
iv. **Oil Spills**: Accidental or intentional discharge of crude oil or refined petroleum products into oceans, rivers, or lakes.
v. **Waste Disposal and Landfills**: Decaying of organic wastes in landfills yield methane, a strong greenhouse gas.
vi. **Agricultural Activities**: Use of fertilizers and pesticides can release hydrocarbons, such as methane and VOCs, into the atmosphere (Latimer et al., 1990).
vii. **Combustion of Biomass and Fossil Fuels**: Burning of wood, coal, and other biomass releases hydrocarbons and PAHs into the air.

8.2.3 Accidental Releases

i. **Pipeline Leaks**: Ruptures or leaks in oil and gas pipelines can release significant amounts of hydrocarbons into soil and water bodies.
ii. **Tank Accidents**: Accidental leaks or spills from oil storage tanks, tankers, and railcars.
iii. **Industrial Accidents**: Incidents like explosions, fires, or equipment failures in chemical plants or refineries can release hydrocarbons.

Each of these sources contributes to the existence of hydrocarbons in air, soil, and water, impacting ecosystems and human health.

8.3 EFFECTS OF HYDROCARBON CONTAMINANTS

Different hydrocarbons produced from petroleum, viz., crude oil, diesel, gasoline, heavy oil, and kerosene, are primarily utilized as conventional energy sources, despite acting as pollutants resulting in air, soil, and water pollution and being detrimental to living organisms (Dey, Tripathy, et al., 2023; Paul et al., 2024). Mostly these oil pollutants pollute the environment due to their extraction, production, transportation, utilization, and disposal. Chemical structure determines its biodegradation. The molecular formulation and the physical state of the compound may not be degraded by microbial enzymes, for example, compounds having an amine group, sulfonate group, and ether group. Hydrocarbons attached with aliphatic side chains are more prone to microbial attack. Complete biodegradation of crude oil is nearly impossible.

In addition to this, PAHs are occurring in the environment by processes such as petrogenic (oil drilling, spilling, industrial pollution, and traffic exhaust emissions) and pyrogenic processes (forest fires, volcanic eruptions, and incineration). PAHs like pyrene, benzopyrene, dibenzanthracene, dibenzopyrene, benzoperylene, and benzanthracene have long degradation periods; that's why these become deposited in soil, water, and air and eventually act as mutagenic as well as carcinogenic agents in cell/organisms (Lorenzi et al., 2011). In the environment, PAHs are getting transformed through different chemical and biological events. Chemical transformation and disintegration of PAHs caused by sunlight include processes like volatilization, nitration, and oxidation. Chemical degradation occurs in the atmosphere, also by halogenation catalyzed by the chloroperoxidase enzyme produced by some fungi. Fungi *Trametes versicolor* transforms PAHs into methoxylated PAHs (Zeng et al., 2013). In the atmosphere, hydrocarbons react with nitrogen oxides and nitric acid to form nitro-PAHs.

8.3.1 Effects of Hydrocarbon Contaminants on Air

Air becomes polluted by the evaporation of the hydrocarbons, which majorly affect the breathing activity, resulting in respiratory disorders in organisms. Air samples from Los Angeles contain PAHs ranging from 50 to –300 ng/m^3, as was reported by Zhu et al. (2008).

Furthermore, Oxy-PAHs are formed from the interaction of oxygen/ozone with PAHs generated either by UV radiation or from oxygen radicals during the combustion of diesel. These processes help in the formation of PCBs, polychlorinated dibenzo-p-dioxins, and polychlorinated didenzofurans. These genotoxic substances may act as mutagens (Hajizadeh et al., 2011).

Though petroleum hydrocarbon is a great source of energy, it causes the environment to be contaminated with its hazardous impacts on organisms.

8.3.2 Effects of Hydrocarbon Contaminants on Soil

Contamination of petroleum products in soil destroys soil physicochemical and biological properties, ceasing growth and development of living organisms, including both plants and animals (Kanungo et al., 2023). Spilling of petroleum hydrocarbon on soil is the most extensive source of soil contamination, by which physicochemical and biological properties of soil are declined. Pesticides (herbicides, fungicides, and insecticides) are another source of hydrocarbon contamination in soil, as these are directly applied to the soil. Deficiency of many macro- and micro-molecules like C, N, Mg, CO, Fe, Ne, Cl, P, K, H_2O, and O_2 in soil due to the contaminants ceasing the growth and development of plants and animals. Soil enzyme activities of microbes are associated with soil biochemical reactions and are adversely affected by the toxicity of the hydrocarbons. Maximum, that is, about 93% degradation of crude oil is possible by the White rot fungus *Polyporus spp.* (Hadibarata & Tachibana, 2010).

8.3.3 Effects of Hydrocarbon Contaminants on Water

Hydrocarbon pollution happens due to accidental release of hydrocarbons into the sea. Oil spills not only contaminate water by floating on it but also the air as volatile hydrocarbons evaporate. It makes a waterproof film on the water surface that acts as a barrier and hinders gaseous exchange between the environment and water, which is dangerous to plants, animals, and human beings. Kvenvolden and Isin Cooper reported that about 600,000 metric tons of crude oil seepage per year occurs, with a range of 200,000 to 600,000 metric tons/year. Oil spills release petroleum hydrocarbons into the sea, causing pollution to coastal environments (Behera et al., 2022). Clean-up efforts for oil spills are undertaken, but complete isolation is not done. Due to their low solubility in water, some petroleum hydrocarbons float in water, making films, facilitating assortment of inorganic and organic matters, and preventing oxygen transportation.

8.3.4 Effects of Hydrocarbon Contaminants on Living Organisms

Oil spills prevent oxygen penetration in water of marine ecosystem. Researchers also reported that diesel fuel toxicity resulted in reductions in species richness, species evenness, and phylogenetic diversity.

8.3.4.1 Effects of Hydrocarbon Contaminants on Plants

Labud et al. (2007) observed that the presence of hydrocarbon contaminants reduced microbial biomass. The adverse impact of crude oil on plant growth is determined by measuring plant height, weight, and leaf area, as well as coating of roots, which is directly proportional to the degree of contamination. This results in the ultimate death of plants. Also, diesel oil has a significant adverse effect on soil fertility. Nitrifying and denitrifying bacteria are more crumbled by diesel than petrol, thus having chronic or acute effects on plants. Both oxy- and nitro-PAHs are converted to diol-epoxide species of PAHs in the presence of the cytochrome P_{450} system. They have the properties to develop mutations leading to cancer (Vepachedu et al., 2001). They adduct with guanosine residues and hinder the DNA proofreading of polymerase enzymes (Huberman et al., 1976). Chlorinated PAHs are again converted to oxy-PAHs through a chlorine-oxygenation mechanism catalyzed by the P_{450} cytochrome system (Isin and Guengerich, 2007).

8.3.4.2 Effects of Hydrocarbon Contaminants on Animals

The outer protective layer of the feathers in seabirds gets damaged by the oil present in water. The contaminated water is absorbed through the gills of the fish and accumulates in different organs, which are then consumed by humans, causing bio-magnification.

8.3.4.3 Effects of Hydrocarbon Contaminants on Human Health

Human health is also adversely affected by its chronic and acute exposure through inhalation, ingestion, and dermal contact. Occupational exposure to diesel oil can cause health impacts. Activities such as manual filling and discharge, marine diesel bunkering, tank dipping, filter cleaning in refineries, cleaning and repairing pipelines and tanks, manufacturing and servicing diesel engines, and laboratory handling of diesel oils can lead to direct contact of humans with hydrocarbons. Skin exposures and pulmonary exposure may occur during activities such as refueling vehicles and manual siphoning. The effects of hydrocarbons on humans are both acute and chronic. Chronic skin exposure to diesel may result in dermatitis, erythema, blistering, superficial burns, hyperkeratosis, cardiac arrhythmias, and even cancer. Inhalation of diesel can cause acute problems such as headache, dizziness, drowsiness, vomiting, and euphoria (Behera et al., 2020, 2021). Ingestion can lead to irritation in the gastrointestinal tract. Inhalation into the lungs can cause pneumonitis, choking, breathlessness, cyanosis, coughing, and fever, among other symptoms. Ocular exposure to diesel results in pain and hyperemia. Nephrotoxicity can lead to acute kidney failure, renal tubular necrosis, oliguria, anuria, nausea, and diarrhea (Das et al., 2024). Neurotoxicity symptom may include drowsiness, neurasthenia (chronic fatigue syndrome), decreased sensorimotor nerve conduction, and paralysis.

Various harmful effects result from the use of gasoline on soil and water ecosystems. Due to inhalation, it may cause headache, dizziness, drowsiness, respiratory abnormalities, skin disorders, eye defects, diarroea, and ultimately death of the concerned human. Persistent and prolonged exposure to it also adversely affects the nervous system (Lal & Khanna, 1996). However, kerosene is not particularly poisonous, but it can lead to short-term lung damage. Chronic exposure can also cause skin damage.

Also, nitro-PAHs are produced by the combustion process and form dinitro species like dinitropyrene and mono-nitro-PAHs, causing asthma, allergies, and lung cancer (Tokiwa et al., 1984). Nitro-PAHs exhausted from diesel induce frame-shift mutations, which account for 30%–40% of the mutagenicity. Guanosine adducts react with DNA to form 1-N-(deoxyguanosin-8-yl) amino-6-nitropyrene (Smith et al., 1995), which causes mutations leading to G:C > A:T transitions, G:C > T:A transversions, and base deletion at C:G sites (Hashimoto et al., 2006).

In addition to these, polycyclic hydrocarbons have various severe ill effects on humans. The immune system deteriorated because of this. Cancer risk and hormonal disruption damage the developmental and reproductive processes. Contaminated water with hydrocarbons enters into animal body through absorption, breathing, and ingestion.

Moreover, apart from these, hydrocarbons also bring about transformation at cellular, subcellular, and genetic levels, ultimately causing many more chronic diseases.

8.4 ANALYTICAL ADVANCES FOR DETECTING HYDROCARBON CONTAMINANTS

Analytical advances in the study of environmental hydrocarbon contaminants have significantly increased the capability to identify, quantify, and recognize the behavior of these contaminants in various ecosystems. The analytical advances are as follows:

8.4.1 Advanced Detection Techniques

Gas Chromatography-Mass Spectrometry (GC-MS), High Resolution Mass Spectrometry (HRMS), and Fluorescence Spectroscopy can be used to detect the diversity of hydrocarbon contaminants in the environment (Bojkovic et al., 2021).

8.4.1.1 High-Resolution Mass Spectrometry (HRMS)

HRMS, including Time-of-Flight (TOF) and Orbitrap mass spectrometers, allows for highly accurate and sensitive detection of hydrocarbon contaminants, including PAHs and aliphatic hydrocarbons. These instruments can detect contaminants at trace levels, providing detailed information on their molecular composition.

8.4.1.2 Gas Chromatography-Mass Spectrometry (GC-MS) Improvements

Newer GC-MS techniques, such as comprehensive two-dimensional gas chromatography (GC×GC-MS), offer enhanced partition and identification of complex hydrocarbon mixtures (Barman et al., 2000). This instrument helps differentiate between closely related compounds and identify unknown contaminants.

8.4.1.3 Fluorescence Spectroscopy

Advanced fluorescence techniques are employed for rapid screening and detection of PAHs in water and soil samples. This method is particularly useful for real-time monitoring and remote sensing applications.

8.4.2 In Situ Sensing and Remote Monitoring

8.4.2.1 Portable Sensors

Developments in portable gas sensors and biosensors allow for on-site detection of volatile and semi-volatile hydrocarbons. These sensors are often based on nanomaterials, such as carbon nanotubes or metal-organic frameworks, providing high sensitivity and selectivity.

8.4.2.2 Remote Sensing Technologies

Satellite and drone-based remote sensing technologies, utilizing infrared spectroscopy and laser-induced fluorescence, are being used to detect hydrocarbon spills and monitor large-scale contamination events in oceans, rivers, and terrestrial environments (Tripathy et al., 2022, 2024).

8.4.3 Machine Learning with Data Analysis

8.4.3.1 Machine Learning Algorithms

The machine learning and application of artificial intelligence in data analysis enable more efficient processing and interpretation of complex datasets related to hydrocarbon contaminants (Adeyeri, 2024). These algorithms can recognize patterns, predict contaminant behavior, and assess risks based on large volumes of data.

8.4.3.2 Data Integration from Multiple Sources

Combining data from different analytical methods, such as spectroscopic, chromatographic, and remote sensing, enhances the overall understanding of contaminant distribution, sources, and fate in the environment.

8.4.4 Microbial and Biological Assays

8.4.4.1 Bioassays and Bio-Reporters

The use of genetically engineered microorganisms and biological assays helps to detect specific hydrocarbons or their metabolites. These tools can indicate the presence of contaminants through fluorescence or colorimetric changes.

8.4.4.2 Molecular Techniques

Techniques like quantitative polymerase chain reaction (qPCR) and metagenomics are used to study various microbes involved in the biological degradation of hydrocarbons, providing insights into natural attenuation processes and potential bioremediation strategies.

8.4.5 High-Throughput Screening and Automation

8.4.5.1 Automation of Sample Preparation

Robotic systems and automated platforms for sample extraction, clean-up, and analysis reduce human error, increase throughput, and improve reproducibility. This automation is particularly important for large-scale environmental monitoring programs.

8.4.5.2 High-Throughput Screening Techniques

Methods like liquid-liquid extraction and solid-phase microextraction (SPME), coupled with advanced chromatographic techniques, allow for the rapid screening of large numbers of samples from the contaminated sites.

8.4.6 Isotope Ratio Analysis

Compound-Specific Isotope Analysis (CSIA) is used to know the sources and transformation pathways of hydrocarbons in the contaminated site. By measuring the ratios of stable isotopes (e.g., carbon, hydrogen), this method provides insight into the degradation processes and helps distinguish between biogenic and anthropogenic sources of contaminants (Birch et al., 2021).

8.4.7 Environmental Forensics

Fingerprinting and Source Apportionment in environmental forensics, such as hydrocarbon fingerprinting, help in identifying the source of contamination (e.g., natural seeps, industrial discharges, and oil spills). This is critical for legal and regulatory purposes, as well as for designing effective remediation strategies.

8.4.8 Emerging Analytical Techniques

8.4.8.1 Electrochemical and Optical Sensors

Emerging technologies in electrochemical and optical sensing are being developed for detecting hydrocarbons at lower concentrations and in diverse matrices, including water, air, and sediments.

8.4.8.2 Nano-enhanced Techniques

Nanotechnology-based methods, such as nano-sorbents and nano-filtration, offer enhanced sensitivity and specificity for detecting and removing hydrocarbon contaminants.

These analytical advancements provide an overall understanding of the presence, distribution, and impact of hydrocarbon contaminants in the environment. They enable faster, more accurate monitoring, enhance our ability to mitigate pollution, and contribute to developing new strategies for environmental protection and remediation.

8.5 REMOVAL OF ENVIRONMENTAL HYDROCARBON CONTAMINANTS

To remove environmental hydrocarbon contaminants, various remediation methods are employed based on the type of contamination, location, and environmental factors. The main methods include physical methods, chemical methods, thermal methods, and bio-remediation.

8.5.1 Physical Methods

It mostly focuses on the direct, mechanical, or material-based separation of hydrocarbons from the contaminated environment by any of the following methods:

8.5.1.1 Skimming and Booming

This method is used for oil spills, where barriers and skimmers remove oil from water's surfaces.

8.5.1.2 Washing of Soil

It involves washing contaminated soil with a liquid solution to separate hydrocarbons from soil particles (Li et al., 2016).

8.5.1.3 Sorption

It uses materials like activated carbon, peat, or synthetic sorbents to absorb hydrocarbons (Adeniji et al., 2017).

Though these physical methods of remediation of hydrocarbon contaminants provide immediate results, they are the temporary solutions that are highly costly and labor-intensive processes.

8.5.2 Chemical Methods

These are the methods using chemical reactions to transform or break down hydrocarbons into less harmful substances. It employs the following methods:

8.5.2.1 Oxidation

It uses oxidants (e.g., hydrogen peroxide, ozone) to break down hydrocarbons into less harmful substances.

8.5.2.2 Surfactant-Enhanced Remediation

It uses surfactants to increase hydrocarbon solubility, facilitating removal.

The chemical methods for removing hydrocarbon contaminants are designed for a wide range of contaminants, but the surfactants used can be potentially toxic to living beings. There is a chance of incomplete degradation, which may require high-cost specialized equipment.

8.5.3 Thermal Methods

It can be done in the following two methods:

8.5.3.1 Thermal Desorption

In this method, contaminated materials are heated to vaporize hydrocarbons, which are then captured and treated (Zhao et al., 2019).

8.5.3.2 Incineration

In this method, contaminated materials are burned at high temperatures to destroy hydrocarbons, often used for severe contamination.

These methods are highly costly and not meant for site specific applicability.

8.5.4 Bioremediation

Bioremediation is an environmentally friendly approach by using living organisms (such as bacteria, fungi, yeast, or plants) to degrade the harmful contaminants in contaminated sites (Mishra et al., 2021). In this process, the focus is on breaking down hydrocarbon compounds made up of hydrogen and carbon—found in pollutants like oil, gasoline, and other petroleum-based substances. This method is highly employed due to its cost-effectiveness and sustainability.

8.6 MECHANISM OF BIOREMEDIATION

According to the type of living organisms used, the mechanism of bioremediation can be classified into two types:

8.6.1 Microbial Bioremediation

It uses microbes (bacteria, fungi, or algae) to degrade hydrocarbons into less harmful substances. They use contaminants as a source of energy and transform them into less harmful substances like CO_2, H_2O, and biomass. It can be enhanced with bio-augmentation (adding specific microbes) or bio-stimulation (adding nutrients).

It involves two types of degradation processes.

8.6.2 Microbial Degradation

Microorganisms metabolize hydrocarbons as a source of carbon and energy. Different species have varying capacities to degrade different hydrocarbons. During metabolism, it can undergo either aerobic or anaerobic metabolism.

8.6.2.1 Aerobic Bioremediation

It happens in the presence of oxygen, where aerobic microorganisms use oxygen to break down hydrocarbons. Being an electron acceptor, oxygen is essential for the oxidation of hydrocarbons. The microorganisms produce enzymes such as oxygenases and dehydrogenases that catalyze the breakdown of hydrocarbons into simpler molecules like carbon dioxide (CO_2), water (H_2O), and biomass.

8.6.2.2 Anaerobic Bioremediation

In environments lacking oxygen, anaerobic microorganisms consume other electron acceptors (such as nitrate, sulfate, iron, or CO_2) to degrade hydrocarbons (Abbasian et al., 2016). Anaerobic degradation is generally slower than aerobic degradation but is important in oxygen-poor environments.

8.6.3 Enzymatic Degradation

Microorganisms produce specific enzymes that break down hydrocarbons. The degradation pathways involve Hydroxylation and Oxidation. Enzymes like monooxygenases and dioxygenases add oxygen atoms to the hydrocarbon molecules, initiating the breakdown process. This often leads to the formation of alcohols, aldehydes, and carboxylic acids. Intermediate products are further degraded by other enzymes, eventually leading to mineralization of the hydrocarbons to CO_2 and H_2O (in the case of aerobic degradation) or simpler compounds like CH_4 (in the case of anaerobic degradation).

8.6.4 Phytoremediation

It involves using plants to absorb, accumulate, and break down hydrocarbons from contaminated soil or water (Meagher, 2000). Some plants release root exudates that stimulate microbes present in the rhizosphere to enhance the breakdown of hydrocarbon contaminants.

8.7 METHODS OF BIOREMEDIATION

The methods of bioremediation for hydrocarbon contaminants are categorized mainly into *in situ* (treatment at the contamination site) and *ex situ* (treatment after removal from the site) techniques. These methods aim to enhance the natural degradation process by providing optimal conditions for microorganisms to degrade hydrocarbons effectively.

8.7.1 *In Situ* Bioremediation Methods

8.7.1.1 Bioventing

It involves the controlled supply of air or oxygen into contaminated soil to enhance the aerobic degradation of hydrocarbons by indigenous microorganisms. Primarily, it is used for unsaturated soils contaminated with lighter hydrocarbons like gasoline. The main advantage of using this method is cost-effectiveness, minimal site disturbance, and a reduced risk of contaminant exposure.

8.7.1.2 Biosparging

In this method, air or oxygen is inserted directly into the saturated zone (groundwater) to increase oxygen concentration, thereby enhancing microbial degradation. This method is effective for treating groundwater and soils contaminated with dissolved hydrocarbons. It is relatively low-cost and can treat contaminants in both soil and groundwater.

8.7.1.3 Bio-Augmentation

It involves adding specific strains or consortia of microorganisms that have a high capacity to degrade hydrocarbons to the contaminated site (Sharma, 2012). In certain cases, genetically engineered microbes are added to the contaminated water to get better results (Gentry et al., 2004). For instance, *Pseudomonas spp.* LB400, a genetically modified bacterium, enhances the treatment of sites contaminated with PCBs by modifying substrate specificity (Erickson & Mondello, 1993). This method is used when indigenous microbes are not effective or abundant enough to degrade certain contaminants. It can target specific contaminants and increase the degradation rate.

8.7.1.4 Biostimulation

By adding nutrients such as N, P, and other trace elements to the contaminated site, it stimulates the growth and activity of indigenous microorganisms that degrade hydrocarbons. This method is suitable for soils or water bodies where microbial populations are present but lack sufficient nutrients to function effectively. It increases microbial activity and enhances the biodegradation rate (Zahed et al., 2010).

8.7.1.5 Natural Attenuation

These are the natural processes of biodegradation to reduce contaminant concentrations over time, such as dispersion, dilution, sorption, volatilization, and chemical or biological stabilization. It is used in sites where contamination levels are relatively low and without an immediate threat to human health or the environment. This method involves low-cost, minimal disturbance to the site.

8.7.2 *Ex Situ* Bioremediation Methods

8.7.2.1 Bio-Piles

In this method, soil with hydrocarbon contaminants is excavated and piled in a controlled area. Air and nutrients are supplied to stimulate microbial activity. It is suitable for treating soils contaminated with petroleum hydrocarbons and other organic contaminants. In controlled conditions, this method enhances biodegradation faster than some in situ methods.

8.7.2.2 Window Composting

Contaminated soil is formed into long piles (windrows) that are regularly turned to aerate and distribute moisture and nutrients. It is often used for soils contaminated with heavy hydrocarbons, like those from oil spills. It is a simple technology that is effective for a vast range of hydrocarbon contaminants and can treat large volumes of soil.

8.7.2.3 Land Farming

Soil with hydrocarbon contaminants is placed on a prepared bed and periodically raked to aerate the soil, with the addition of nutrients and moisture as needed. It is used for treating surface soil contamination, especially hydrocarbons and pesticides. This method is relatively low-cost, simple to implement, and can handle large volumes of soil.

8.7.2.4 Bioreactors

Contaminated soil, water, or sludge is placed in a controlled reactor vessel where environmental factors (such as the concentration of oxygen, temperature, pH, and nutrients) are optimized for microbial activity. This method is suitable for treating both solid and liquid waste contaminated with hydrocarbons. The highly controlled environment allows for maximum degradation efficiency and faster treatment times.

8.7.2.5 Slurry-Phase Bioremediation

Contaminated soil, also known as sludge, is mixed with water to form a slurry, which is then sent to a bioreactor for treatment under controlled conditions. This method is effective for treating highly contaminated soils or sludges that are not suitable for in situ methods. It is preferred due to its rapid treatment, high degradation efficiency, and good control over process conditions.

8.8 MICROBES INVOLVED IN BIOREMEDIATION OF HYDROCARBON CONTAMINANTS

Bioremediation involves various microbes such as bacteria, fungi, and algae for the degradation of hydrocarbon contaminants.

8.8.1 Bacteria

Bacteria are the most common microorganisms for the biodegradation of hydrocarbons due to their versatility and adaptability. Tremblay et al. (2017) suggested that more than 79 genera of bacteria can degrade hydrocarbon contaminants. Under sufficient nutrient availability and suitable environments, microorganisms can mineralize petroleum hydrocarbons completely within one day (under culture conditions) to several days (under natural conditions) (Zheng et al., 2018). Some of the notable bacteria include:

8.8.1.1 Pseudomonas spp.

These are among the most effective hydrocarbon-degrading bacteria. They have versatile metabolic pathways that allow them to break down various hydrocarbons, including alkanes, aromatic hydrocarbons, and PAHs. Common species are *Pseudomonas aeruginosa and Pseudomonas putida.* They produce bio-surfactants that enhance the availability of hydrocarbons for degradation (Wilson & Bradley, 1996).

8.8.1.2 Alcanivorax spp.

These bacteria are particularly effective in degrading alkanes, a major component of crude oil, for example, *Alcanivorax borkumensis*. They thrive in marine environments and are often dominant in oil-contaminated waters.

8.8.1.3 Rhodococcus spp.

Rhodococcus bacteria have a wide substrate range, including the ability to degrade complex hydrocarbons like PAHs and aromatic hydrocarbons. They are known for their high tolerance to toxic compounds. Common species are *Rhodococcus erythropolis and Rhodococcus opacus.*

8.8.1.4 Mycobacterium spp.

Certain species can degrade complex hydrocarbons, including PAHs (Zeng et al., 2013). These bacteria are often found in soils contaminated with petroleum and other hydrocarbons, for example, *Mycobacterium vanbaalenii.*

8.8.1.5 Acinetobacter spp.

Acinetobacter bacteria are capable of degrading a wide variety of hydrocarbons, including aromatic and aliphatic hydrocarbons (Xu et al., 2017). They are found in contaminated soil and water, for example, *Acinetobacter calcoaceticus.*

8.8.1.6 Sphingomonas spp.

Sphingomonas species (e.g., *Sphingomonas paucimobilis*) are known for their ability to degrade aromatic hydrocarbons, including PAHs. They are commonly found in soil and aquatic environments.

8.8.1.7 Marinobacter spp.

These bacteria are halo-tolerant and thrive in marine environments (Yakimov et al., 2007). They are capable of degrading both aliphatic and aromatic hydrocarbons, for example, *Marinobacter hydrocarbonoclasticus.*

8.8.1.8 Burkholderia spp.

Burkholderia bacteria are versatile in degrading various hydrocarbons and are often used in the bioremediation of contaminated soil and water. For instance, *Burkholderia cepacia* causes alkyl-aromatic hydrocarbon degradation (Adebusoye et al., 2007).

8.8.1.9 Nocardia spp.

These are actinobacteria capable of degrading aromatic hydrocarbons and are often found in soil environments contaminated with petroleum products (Azadi & Shojaei, 2020). Common species are *Nocardia asteroids and Nocardia cyriacigeorgica.*

8.8.1.10 Bacillus spp.

These bacteria are known for their robust nature and ability to degrade a variety of hydrocarbons in both aerobic and anaerobic conditions. Common species include *Bacillus subtilis* and *Bacillus cereus.*

These bacteria, individually or in consortia, transform hydrocarbon pollutants into less harmful substances in natural and engineered bioremediation processes, contributing to environmental clean-up efforts.

8.8.2.11 Fungi

Fungi also play a significant role in the degradation of hydrocarbons, particularly in hydrocarbon-contaminated soil.

8.8.2.12 White-Rot Fungi

These fungi (e.g., *Phanerochaete chrysosporium*) produce enzymes such as lignin peroxidases and manganese peroxidases, which can degrade complex hydrocarbons, including PAHs.

8.8.2.13 Aspergillus spp.

It degrades various hydrocarbons, including n-alkanes and aromatic hydrocarbons. They are effective in both soil and water environments (Hassaine & Bordjiba, 2019).

8.8.2.14 Penicillium spp.

It is known for its ability to degrade hydrocarbons and produce extracellular enzymes that break down complex organic pollutants (Pinto & Moore, 2000).

8.8.2.15 Algae

Though algae are not typically primary agents in hydrocarbon degradation, certain algae, like *Chlorella* and *Scenedesmus*, can enhance bioremediation by producing oxygen and creating a favorable environment for aerobic bacteria (El-Sheekh et al., 2013).

These microbes play a crucial role in mitigating the environmental impact of hydrocarbon pollution, and their use is a sustainable approach to restoring contaminated ecosystems.

8.9 FACTORS AFFECTING PETROLEUM HYDROCARBON DEGRADATION

The main objective of remediating these hydrocarbon-contaminated sites is to minimize the adverse effects of contaminants on human health and the environment by various effective techniques (Perelo, 2010). Actually, bioremediation technology is very challenging as it is controlled by various environmental and biological factors. It is influenced by the composition and concentration of contaminants, type of soil, and the type of microorganisms present. Environmental parameters such as temperature, pH value, oxygen availability, and nutrient content of the soil also determine the bioremediation process, either by promoting the growth of microbes or reducing the accessibility of pollutants to microbial degradation (Priyadarshini et al., 2024).

Temperature influences the rate of hydrocarbon degradation and the physicochemical structure of oil increasing the bioavailability of hydrocarbons and the metabolic activity of the microbes. Atlas (1975) investigated that the viscosity of oil increased with decreasing temperature, while the volatility decreased, leading to a reduction in the biodegradation process. The biodegradation rate typically decreases with lower temperatures.

Nutrients like N, P, and Fe are required for successful biodegradation of hydrocarbons. However, excessive nutrients may hinder the biodegradation process. The use of organic fertilizer, such as poultry manure, in contaminated soil can enhance the biodegradation rate (Okolo et al., 2005). Maki et al. (2001) suggested that photooxidation increases the biodegradation of petroleum hydrocarbons by enhancing their bioavailability to microbes.

Further, the biodegradation also depends upon the composition of the hydrocarbon contaminants. According to Okoh and Ezeronye (2002), heavier and bulkier crude oils are more difficult to biodegrade in comparison to lighter ones. When the molecular weight of hydrocarbon contaminants is high, the solubility declines and the bioavailability of hydrocarbons for microbial activity becomes less.

8.10 CONCLUSION

The hydrocarbon contaminants are setting an alarming call to the society, which is disparaging to the health of the environment. The remediation methods used to remove those contaminants must be cost-effective with sustainability. Furthermore, bioremediation is a prominent technology for the treatment of contaminated environments with hydrocarbons, offering a sustainable and potentially cost-effective solution. However, its success depends on a careful assessment of site conditions, contaminant characteristics, and the appropriate selection of bioremediation strategies. The choice of method is selected based on the type and concentration of hydrocarbons, condition of the site (soil type, permeability, and depth of contamination), the presence of suitable microorganisms, environmental regulations, and cost considerations. These bioremediation methods provide a range of options to address hydrocarbon contamination, each with its own strengths and limitations, allowing for tailored approaches to specific environmental and contamination conditions. Development is needed but not at the cost of the environment. Strict policies and constant monitoring prevent contamination at the source level. Greener technology can be promoted to minimize dependency on hydrocarbon products and work toward a sustainable future.

REFERENCES

Abbasian, F., Lockington, R., Megharaj, M., & Naidu, R. (2016). A review on the genetics of aliphatic and aromatic hydrocarbon degradation. *Applied Biochemistry and Biotechnology*, *178*, 224–250.

Adebusoye, S. A., Ilori, M. O., Amund, O. O., Teniola, O. D., & Olatope, S. O. (2007). Microbial degradation of petroleum hydrocarbons in a polluted tropical stream. *World Journal of Microbiology and Biotechnology*, *23*, 1149–1159.

Adeniji, A. O., Okoh, O. O., & Okoh, A. I. (2017). Analytical methods for the determination of the distribution of total petroleum hydrocarbons in the water and sediment of aquatic systems: A review. *Journal of Chemistry*, *2017*(1), 5178937.

Adeyeri, T. B. (2024). Enhancing financial analysis through artificial intelligence: A comprehensive review. *Journal of Science & Technology*, *5*(2), 102–120.

Atlas, R. M. (1975). Effect of the temperature and crude oil composition on petroleum biodegradation. *Applied Microbiology*, *30*, 396–403.

Azadi, D., & Shojaei, H. (2020). Biodegradation of polycyclic aromatic hydrocarbons, phenol and sodium sulfate by Nocardia species isolated and characterized from Iranian ecosystems. *Scientific Reports*, *10*(1), 21860.

Bala, K., Dogra, A., Lata, P., & Das, A. P. (2024). Discussing sources and biological-chemical remediation approaches for microplastic pollution. In *Sustainable Microbial Technology for Synthetic and Cellulosic Microfiber Bioremediation*, 37–59. https://doi.org/10.1007/978-3-031-62600-5_3

Barman, B. N., Cebolla, V. L., & Membrado, L. (2000). Chromatographic techniques for petroleum and related products. *Critical Reviews in Analytical Chemistry*, *30*(2–3), 75–120.

Behera, I. D., Basak, G., Kumar, R. R., Sen, R., & Meikap, B. C. (2020). Treatment of petroleum refinery sludge by petroleum degrading bacterium *Stenotrophomonas pavanii* IRB19 as an efficient novel technology. *Journal of Environmental Science and Health, Part A*, 1–13. https://doi.org/10.1080/10934529.2020.1866924

Behera, I. D., Nayak, M., Biswas, S., Meikap, B. C., & Sen, R. (2021). Enhanced biodegradation of total petroleum hydrocarbons by implementing a novel two-step bioaugmentation strategy using indigenous bacterial consortium. *Journal of Environmental Management*, *292*, 112746. https://doi.org/10.1016/j.jenvman.2021.112746

Behera, I. D., Nayak, M., Mishra, A., Meikap, B. C., & Sen, R. (2022). Strategic implementation of integrated bioaugmentation and biostimulation for efficient mitigation of petroleum hydrocarbon pollutants from terrestrial and aquatic environment. *Marine Pollution Bulletin*, *177*, 113492. https://doi.org/10.1016/j.marpolbul.2022.113492

Birch, Q. T., Potter, P. M., Pinto, P. X., Dionysiou, D. D., & Al-Abed, S. R. (2021). Isotope ratio mass spectrometry and spectroscopic techniques for microplastics characterization. *Talanta*, *224*, 121743.

Bojkovic, A., Vermeire, F. H., Kuzmanovic, M., Dao Thi, H., & Van Geem, K. M. (2021). Analytics driving kinetics: Advanced mass spectrometric characterization of petroleum products. *Energy & Fuels*, *36*(1), 6–59.

Das, A., Ray, S. K., Mohanty, M., Mohanty, J., Dey, S., & Das, A. P. (2024). Ecotoxicity of microplastic wastes and their sustainable management: A review. *Environmental Chemistry and Ecotoxicology*, *6*, 144–152. https://doi.org/10.1016/J.ENCECO.2024.05.003

Dey, S., Bissoyi, A., Koteswara Reddy, G., & Das, A. P. (2024). Synthetic microfibers: An urban emerging pollutant demanding for sustainable environmental solutions. In *Sustainable Microbial Technology for Synthetic and Cellulosic Microfiber Bioremediation*, 107–121. https://doi.org/10.1007/978-3-031-62600-5_7

Dey, S., Das, A., Mallick, K., Sahu, A., & Das, A. P. (2023). Environmental petroleum waste: Pollution, toxicity, sustainable remediation. 159–175. https://doi.org/10.1007/978-3-031-48220-5_7

Dey, S., Dash, N., Ray, S. K., Priyadarsini, S., Chhotaray, C., & Das, A. P. (2024). Occurrence and quantification of tire wear plastics pollutants from highway city dust of Bhubaneswar City of Odisha. In *Sustainable Microbial Technology for Synthetic and Cellulosic Microfiber Bioremediation*, 77–89. https://doi.org/10.1007/978-3-031-62600-5_5

Dey, S., Tripathy, B., Kumar, M. S., & Das, A. P. (2023). Ecotoxicological consequences of manganese mining pollutants and their biological remediation. In *Environmental Chemistry and Ecotoxicology*. https://doi.org/10.1016/J.ENCECO.2023.01.001

El-Sheekh, M. M., Hamouda, R. A., & Nizam, A. A. (2013). Biodegradation of crude oil by Scenedesmus obliquus and Chlorella vulgaris growing under heterotrophic conditions. *International Biodeterioration & Biodegradation*, *82*, 67–72.

Erickson, B. D., & Mondello, F. J. (1993). Enhanced biodegradation of polychlorinated biphenyls after site-directed mutagenesis of a biphenyl dioxygenase gene. *Applied and Environmental Microbiology*, *59*(11), 3858–3862.

Gentry, T., Rensing, C., & Pepper, I. A. N. (2004). New approaches for bioaugmentation as a remediation technology. *Critical Reviews in Environmental Science and Technology*, *34*(5), 447–494.

Hadibarata, T., & Tachibana, S. (2010). Characterization of phenanthrene degradation by strain Polyporus sp. S133. *Journal of Environmental Sciences*, *22*(1), 142–149.

Hajizadeh, Y., Onwudili, J. A., & Williams, P. T. (2011). PCDD/F formation from oxy-PAH precursors in waste incinerator fly ash. *Chemosphere*, *85*(11), 1672–1681.

Hashimoto, A. H., Amanuma, K., Hiyoshi, K., Takano, H., Masumura, K. I., Nohmi, T., & Aoki, Y. (2006). In vivo mutagenesis in the lungs of gpt-delta transgenic mice treated intratracheally with 1, 6-dinitropyrene. *Environmental and molecular mutagenesis*, *47*(4), 277–283.

Hassaine, A., & Bordjiba, O. (2019). Removal of hydrocarbons from liquid media by Aspergillus Niger van Tieghem. *Acta Ecologica Sinica*, *39*(4), 300–305.

Huberman, E., Sachs, L., Yang, S. K., & Gelboin, V. (1976). Identification of mutagenic metabolites of benzo (a) pyrene in mammalian cells. *Proceedings of the National Academy of Sciences*, *73*(2), 607–611.

Imam, A., Suman, S. K., Ghosh, D., & Kanaujia, P. K. (2019). Analytical approaches used in monitoring the bioremediation of hydrocarbons in petroleum-contaminated soil and sludge. *TrAC Trends in Analytical Chemistry*, *118*, 50–64.

Isin, E. M., & Guengerich, F. P. (2007). Complex reactions catalyzed by cytochrome P450 enzymes. *Biochimica et Biophysica Acta (BBA)-General Subjects*, *1770*(3), 314–329.

Kanungo, J., Sahoo, S., Bal, M., & Behera, I. D. (2023). Performance of bioremediation strategy in waste lubricating oil pollutants: A review. *Geomicrobiology Journal*. https://doi.org/10.1080/01490451.2023.2245395

Labud, V., Garcia, C., & Hernandez, T. (2007). Effect of hydrocarbon pollution on the microbial properties of a sandy and a clay soil. *Chemosphere*, *66*(10), 1863–1871.
Lahiri, D., Nag, M., Dey, A., Sarkar, T., Joshi, S., Pandit, S., Das, A. P., Pati, S., Pattanaik, S., Tilak, V. K., & Ray, R. R. (2021). Biofilm mediated degradation of petroleum products. *Geomicrobiology Journal*, 1–10. https://doi.org/10.1080/01490451.2021.1968979
Lal, B., & Khanna, S. (1996). Degradation of crude oil by *Acinetobacter calcoaceticus* and *Alcaligenes odorans*. *Journal of Applied Microbiology*, *81*(4), 355–362.
Latimer, J. S., Hoffman, E. J., Hoffman, G., Fasching, J. L., & Quinn, J. G. (1990). Sources of petroleum hydrocarbons in urban runoff. *Water, Air, and Soil Pollution*, *52*, 1–21.
Li, P., Cai, Q., Lin, W., Chen, B., & Zhang, B. (2016). Offshore oil spill response practices and emerging challenges. *Marine Pollution Bulletin*, *110*(1), 6–27.
Lorenzi, D., Entwistle, J. A., Cave, M., & Dean, J. R. (2011). Determination of polycyclic aromatic hydrocarbons in urban street dust: implications for human health. *Chemosphere*, *83*(7), 970–977.
Maki, H., Sasaki, T., & Harayama, S. (2001). Photo-oxidation of biodegraded crude oil and toxicity of the photo-oxidized products. *Chemosphere*, *44*(5), 1145–1151.
Meagher, R. B. (2000). Phytoremediation of toxic elemental and organic pollutants. *Current Opinion in Plant Biology*, *3*(2), 153–162.
Mishra, S., Swain, S., Sahoo, M., Mishra, S., & Das, A. P. (2021). Microbial colonization and degradation of microplastics in aquatic ecosystem: A review. *Geomicrobiology Journal*. https://doi.org/10.1080/01490451.2021.1983670
Okoh, A. I., & Ezeronye, O. U. (2002). Assessment of the potentials of some bacterial isolates for application in the bioremediation of petroleum hydrocarbon polluted soil. *African Journal of Biotechnology*, *1*(4), 215–226.
Okolo, J. C., Amadi, E. N., & Odu, C. T. I. (2005). Effects of soil treatments containing poultry manure on crude oil degradation in a sandy loam soil. *Applied Ecology and Environmental Research*, *3*(1), 47–53.
Paul, A., Priyadarsini, S., Karakuş, S., Behera, J. K., Gnanasekaran, R., & Das, A. P. (2024). Fate and accumulation of synthetic microfibers in the environment. In *Sustainable Microbial Technology for Synthetic and Cellulosic Microfiber Bioremediation*, 61–76. https://doi.org/10.1007/978-3-031-62600-5_4
Perelo, L. W. (2010). In situ and bioremediation of organic pollutants in aquatic sediments. *Journal of Hazardous Materials*, *177*(1–3), 81–89.
Pinto, L. J., & Moore, M. M. (2000). Release of polycyclic aromatic hydrocarbons from contaminated soils by surfactant and remediation of this effluent by Penicillium spp. *Environmental Toxicology and Chemistry: An International Journal*, *19*(7), 1741–1748.
Priyadarshini, B., & Das, A. P. (2024). The adverse impact of synthetic microplastics released from disposable face masks pollutants on aquatic microbiota, aquatic species and humans: A review. In *Environmental Science and Engineering*, *Part F2827*, 67–81. https://doi.org/10.1007/978-3-031-50840-0_5
Priyadarshini, B., Santhosh Kumar, M., & Das, A. P. (2024). Microbial degradation of synthetic plastics released from COVID-19 generated single use disposable face masks. In *Environmental Science and Engineering*, *Part F2827*, 1–16. https://doi.org/10.1007/978-3-031-50840-0_1
Sharma, S. (2012). Bioremediation: Features, strategies and applications. *Asian Journal of Pharmacy and Life Science*, *2231*, 4423.
Smith, B. A., Fullerton, N. F., Heflich, R. H., & Beland, F. A. (1995). DNA adduct formation and T-lymphocyte mutation induction in F344 rats implanted with tumorigenic doses of 1, 6-dinitropyrene. *Cancer Research*, *55*(11), 2316–2324.
Tokiwa, H., Otofuji, T., Horikawa, K., Kitamori, S., Otsuka, H., Manabe, Y., Kinouchi, T., & Ohnishi, Y. (1984). 1,6-Dinitropyrene: Mutagenicity in Salmonella and carcinogenicity in BALB/c mice. *Journal of the National Cancer Institute*, *73*(6), 1359–1363.

Tremblay, J., Yergeau, E., Fortin, N., Cobanli, S., Elias, M., King, T. L., Lee, K., & Greer, C. W. (2017). Chemical dispersants enhance the activity of oil-and gas condensate-degrading marine bacteria. *The ISME Journal, 11*(12), 2793–2808.

Tripathy, B., Dash, A., & Das, A. P. (2022). Detection of environmental microfiber pollutants through vibrational spectroscopic techniques: Recent advances of environmental monitoring and future prospects. *Critical Reviews of Analytical Chemistry, 54*, 1–11. https://doi.org/10.1080/10408347.2022.2144994

Tripathy, B., Sahoo, P. P., Sundaray, H., & Das, A. P. (2024). In-vitro biodegradation of discarded marine microplastics across the eastern coast of the Bay of Bengal, India using Exiguobacterium sp. *Environmental Chemistry and Ecotoxicology, 6*, 236–247. https://doi.org/10.1016/J.ENCECO.2024.07.003

VanMetre, P. C., Mahler, B. J., & Furlong, E. T. (2000). Urban sprawl leaves its PAH signature. *Environmental Science & Technology, 34*(19), 4064–4070.

Vepachedu, S., Ya, N., Yagi, H., Sayer, J. M., & Jerina, D. M. (2001). Marked differences in base selectivity between DNA and the free nucleotides upon adduct formation from Bay-and Fjord-Region Diol Epoxides Volume 13, Number 9, September 2000, pp 883−890. *Chemical Research in Toxicology, 14*(1), 148–148.

Wilson, N. G., & Bradley, G. (1996). The effect of immobilization on rhamnolipid production by Pseudomonas fluorescens. *Journal of Applied Microbiology, 81*(5), 525–530.

Xu, X., Zhai, Z., Li, H., Wang, Q., Han, X., & Yu, H. (2017). Synergetic effect of bio-photocatalytic hybrid system: g-C_3N_4 and Acinetobacter sp. JLS1 for enhanced degradation of C16 alkane. *Chemical Engineering Journal, 323*, 520–529.

Yakimov, M. M., Timmis, K. N., & Golyshin, P. N. (2007). Obligate oil-degrading marine bacteria. *Current Opinion in Biotechnology, 18*(3), 257–266.

Zahed, M. A., Aziz, H. A., Isa, M. H., & Mohajeri, L. (2010). Effect of initial oil concentration and dispersant on crude oil biodegradation in contaminated seawater. *Bulletin of Environmental Contamination and Toxicology, 84*, 438–442.

Zeng, J., Lin, X., Zhang, J., Zhu, H., Chen, H., & Wong, M. H. (2013). Successive transformation of benzo [a] pyrene by laccase of Trametes versicolor and pyrene-degrading Mycobacterium strains. *Applied Microbiology and Biotechnology, 97*, 3183–3194.

Zhao, C., Dong, Y., Feng, Y., Li, Y., & Dong, Y. (2019). Thermal desorption for remediation of contaminated soil: A review. *Chemosphere, 221*, 841–855.

Zheng, J., Feng, J. Q., Zhou, L., Mbadinga, S. M., Gu, J. D., & Mu, B. Z. (2018). Characterization of bacterial composition and diversity in a long-term petroleum contaminated soil and isolation of high-efficiency alkane-degrading strains using an improved medium. *World Journal of Microbiology and Biotechnology, 34*, 1–11.

Zhu, Y., Fung, D. C., Kennedy, N., Hinds, W. C., & Eiguren-Fernandez, A. (2008). Measurements of ultrafine particles and other vehicular pollutants inside a mobile exposure system on Los Angeles freeways. *Journal of the Air & Waste Management Association, 58*(3), 424–434.

9 Phytoremediation of Hydrocarbon Pollutants for Sustainable Management

Rajalaxmi Badra, Sailaja Priyadarsini, Amel Gacem, Ipsita Dipamitra Behera, Maha Awjan Alreshidi, Noushi Zaidi, Ramakrishnan Gnansekaran, and Alok Prasad Das

9.1 INTRODUCTION

Hydrocarbons are organic compounds consisting entirely of hydrogen and carbon atoms, being less polar and less soluble than molecules like methanol and have distinct solubility (Behera et al., 2020). These are the toxic pollutants like plastics (Mishra, Dash, & Das, 2022), heavy metals (HMs) (Das et al., 2011, 2015), and electronic waste that contribute to increase in environmental pollution. Hydrocarbon degradation involves more than solubilization; microbes such as Bacillus and Pseudomonas species excrete emulsifiers, increasing substrate surface area (Behera et al., 2020; Dey, Das, et al., 2023). Microbial cell surface modifications also aid in absorbing hydrophobic substances. Additionally, altering hydrocarbons' physicochemical properties can change the behavior of individual hydrocarbons and mixtures. Low-carbon remediation techniques can help nations working to clean up contaminated soil and support international environmental objectives. In this sense, it was discovered that a comprehensive and integrated strategy, incorporating bioremediation along with techniques like chemical oxidation, EKR, and phytoremediation, was a more viable option for cleaning up oil-contaminated areas in the Niger Delta (Azuazu et al., 2023). Warming is a major adverse effect of climate change, alongside issues like dwindling water supplies, reduced agricultural productivity, food insecurity, melting glaciers, and biodiversity loss (Dey, Bissoyi, et al., 2024; Dey, Dash, et al., 2024). Historically, declining agricultural output has been linked to worsening water quality and climate anomalies (Das & Mishra, 2010; Mishra et al., 2024). Crop productivity is heavily influenced by long-term trends in precipitation and temperature, annual climate variations, growth stage shocks, and extreme weather events (Srivastav et al., 2021). The framework under which organizations function and the expectations placed on management to help resolve sustainability concerns outside of organizational boundaries are both impacted by how society views sustainable development. Both macro-level

DOI: 10.1201/9781003508496-9

events, like new environmental constraints, and meso-level circumstances, like social industry standards and rules, have an impact on businesses. Additionally, businesses must address significant sustainability concerns at the macro level, such as the greenhouse effect, and contribute to the meso-level sustainability transformation of markets. The demands of managers and management accounting for sustainable information are growing and evolving as a result of these advancements. The pace at which sludge is produced in the petroleum industry has been rising due to rising energy consumption. It has been demonstrated that the sludge contains potentially harmful ingredients that could harm the environment and public health (Behera et al., 2020). As a result, there must be tight controls on their release into the environment. There has been investigation into various treatment and disposal techniques aimed at reducing sludge volume and toxicity and extracting petroleum products. However, very few systems can meet stringent environmental laws while using a significant amount of energy, water, and chemicals because of the stubborn nature of the sludge. Currently, there are no accessible, waste-free, cost-effective technologies for treating PS. Waste products used include fish, wood chips, distillery effluent, nut shells, green mango peels, wheat straw, rice husks, and animal dung. Plants such as Onobrychis viciifolia, Zea mays, Secale cereale, Vetiver zizanioide, and others show higher effectiveness compared to unplanted groups. It is best to use an integrated bioremediation strategy that includes biostimulation and bioaugmentation to remove hazardous petroleum hydrocarbons and their derivatives at a reasonable cost. The challenges and future directions of the novel, creative remediation solutions for the sustainable control of pollutants of heavy metals (PHs) in the environment (Prabhakar et al., 2019). Although there is a lot of promise in using nanotechnology to break down pollutants, its practicality depends on economic efficiency. The use of microbial electrochemical techniques to bioremediate and biodegrade soil contaminated with pH by utilizing reactive bacteria. A strategy like this promotes the creation of sustainable restoration plans and is in line with environmental sustainability (Kakkar & Kumar, 2024). Petroleum wastes include oxides, nitrogen, aliphatic and aromatic chemicals, and amines. However, due to variations in the local geology and biological processes, the composition may vary from location to location. The need for petroleum products is growing, which eventually raises concerns about their disposal in the environment. This affects people, soil, aquatic life, plants, and other elements of the environment because there is no appropriate way to dispose of the waste. It causes enzymatic malfunction that inhibits seed germination in plants and modifies the concentration of chlorophyll, which in turn impacts photosynthesis. It affects the respiratory system in humans, causing lung damage or pneumonia, irritated skin and eyes, headaches, dizziness, and other symptoms. In a similar vein, it impacts aquatic life through altering the reproductive process, reducing egg production, and lowering oxygen levels, all of which have an impact on aquatic creatures, flora, and fauna. The rehabilitation of petroleum hydrocarbon wastes involves the application of several physicochemical and biological techniques (Kanungo et al., 2024). Petroleum hydrocarbons affect agricultural land by altering soil porosity, permeability, and water-holding capacity. They can cause soil compaction, reduced infiltration, and increased runoff, impacting crop yield. Additionally, hydrocarbons reduce soil aeration, leading to anaerobic conditions harmful to the microbiome and plant growth, and alter soil pH, nutrient availability, and chemical properties. Petroleum hydrocarbons can alter the biological properties of the soil by modifying microbial activity, variety, and composition. This could lead to decreased soil fertility and productivity. Petroleum hydrocarbon pollution's effects on agricultural productivity depend on a number of elements, including the kind of hydrocarbon, the characteristics of the soil, and the type of crop. The production of certain crops may be impacted by the fact that they are substantially more vulnerable to hydrocarbon contamination than others (Mohanta et al., 2024). The effectiveness of biochar in soil and water environments for oil remediation and highlight its various mechanisms and roles. In particular, biochar improved the biodegradation of petroleum hydrocarbons by altering their bioavailability, thus cleaning up the oil-contaminated soil (Wei et al., 2024).

9.2 HYDROCARBON POLLUTANTS

Many nations across the world have taken action to decarbonize human activities as part of the fight against climate change. Therefore, it is anticipated that cutting back on the burning of fossil fuels will significantly cut the amount of greenhouse gas emissions produced and help to slow down global warming. This countermeasure, while helpful in lowering carbon dioxide (CO_2) production to combat climate change, regrettably does not mean that hydrocarbon pollution is gone because there are still several sources of hydrocarbon pollution. Every year, industrial and petroleum activities unintentionally discharge between 1.7 and 8.8 million tons of oil into the environment, leaving behind a multiplicity of hydrocarbon-contaminated areas that need to be restored. Decarbonization is expected to have a favorable impact on reducing the industrial release of hydrocarbons into the environment by reducing the frequency and consequences of oil spills (Campos et al., 2019). In addition to the constant burning of fossil fuels, which releases hydrocarbons directly into the atmosphere, human activity also releases hydrocarbons indirectly into the environment through wildfires that release polycyclic aromatic hydrocarbons, or polycyclic aromatic hydrocarbons (PAHs). Climate change is making "mega fires," in particular, more common. These fires burn enormous sections of forest. Naturally, hydrocarbons are constantly released into the environment via biogenic sources as well as natural sources like volcanic eruptions and oil seeds from the sea (Campos et al., 2019). Unavoidably, the operations of the petroleum industry produce a significant amount of waste known as petroleum industry sludge (PIS). The rising energy demand has led to an increase in the generation rate. It can produce energy. It has been demonstrated that PIS contains potentially harmful ingredients that could harm the environment and the general public's health. This waste needs to be treated and disposed of globally. It has been shown that a wide range of treatment techniques can recover petroleum components while lowering sludge volume and toxicity. Treatment methods can lessen the harmful elements in sludge and the harm they do to the environment and human health. The application of PIS inefficiently produces rubber tires, carbon-clay composites, and metal catalysts for use in sensors, electronic devices, and microorganism-based biofertilizers (Jagaba et al., 2022). Although the use of bacteria to clean up oil spills is quickly developing, environmental issues and regulatory obstacles still need to be addressed.

Develop a novel approach to utilize native soil microbial communities for bioremediation by introducing the community with catabolic genes that can break down petroleum hydrocarbons. In Escherichia coli, overexpressing three enzymes resulted in the breakdown of 60%–99% of target hydrocarbon substrates (French et al., 2020). The investigation of the cultivable fungal diversity in hydrocarbon-contaminated coastal sediments showed that while fungal diversity is hidden in coastal sediment, its metabolic potential, particularly concerning its ability to remove PAHs, has not yet been fully explored. Six fungal genera contained a significant percentage of the isolated strains (48%), which demonstrated PAH tolerance with a removal capacity (degradation and/or sorption) exceeding 60% (Álvarez-Barragán et al., 2021). Phytoremediation is a low-cost, environmentally friendly alternative to chemical and physical methods for eliminating toxins from soil. The potential of the plant species Alhagi camelorum for HM and total petroleum hydrocarbon (TPH) phytoremediation of oil-contaminated soil, specifically for chromium (Cr), cadmium (Cd), nickel (Ni), and lead (Pb), has been studied (Nemati et al., 2024). Although crude oil is one of the most important energy sources in the world, the processes involved in its large-scale production, transportation, consumption, and disposal greatly contribute to environmental degradation. The most significant soil contaminants in nations with continuous exploration, facilities, refineries, and abundant oil resources are believed to be the seepage, infiltration, and leakage of petroleum pollutants and derivatives. In regions where oil is extracted, petroleum hydrocarbon-related soil pollution is a serious hazard. The consequences of hydrocarbon contamination on the environment have received a lot of attention in the last few decades. Petroleum hydrocarbons can have a harmful impact on the biological activities that soil microorganisms carry out, which can result in toxicity. The quality of the soil may suffer as a result of this problem. Moreover, these substances may have adverse effects on the chemical composition of the soil. Petroleum hydrocarbons are blocking the pore spaces in the soil, which raises the temperature of the soil and reduces the amount of microorganisms that are beneficial. Furthermore, these pollutants prevent plants from absorbing nutrients, eventually leading to plant death. Innovative techniques for the bioremediation of soil contaminants have been studied and implemented recently, with promising study outcomes (Okparanma et al., 2022). Using inexpensive biological techniques, plants are used in phytoremediation to remove and decompose pollutants from polluted soil, such as HMs and other hazardous organic materials. This method feeds, transfers, deactivates, and immobilizes soil contaminants by combining plant activity with the microbial community. Five different methods of phytoremediation, phytostabilization, rhizofiltration, phytoextraction, phytodegradation, and phytovolatilization can be used to remove contaminants. This technique is not only well-received by the public but also new and sustainable, making it economical and appropriate, especially for developing nations.

9.2.1 Sources of Hydrocarbon Pollutants

Crude oil contains organic molecules called petroleum hydrocarbons, which are made of carbon and hydrogen. They are ubiquitous environmental pollutants caused by human activity such as oil spills, leaking storage tanks, and transportation mishaps.

Farmland is not exempt from the detrimental effects of this contamination on the ecosystem, which might include water, air, and soil degradation. The type of petroleum hydrocarbon and the current climate can all have an impact on how agricultural land is affected by petroleum hydrocarbons. Petroleum hydrocarbons have an impact on the porosity, permeability, and water-holding capacity of soil. Hydrocarbons can reduce soil aeration, leading to anaerobic conditions that harm the soil microbiome and stunt plant growth, while also altering the soil's pH, nutrient availability, and chemical properties (Mohanta et al., 2024). Polycyclic aromatic hydrocarbons, or PAHs, are petrogenic sources of coal and coal gangue that have a negative effect on the environment, such as slack gangue, lump gangue, slime, cleaned coal, and raw coal. Compared to coals, coal gangues demonstrated a larger percentage and a lower degree of 4–6-ring PAH alkylation (Li et al., 2024). The most notable soil pollutants are HMs, which are recognized for their detrimental impacts on human health as well as their ability to reduce agricultural output and product quality. Although these metals are considered necessary nutrients for plants in tiny quantities and are found in soil in trace amounts naturally, high concentrations of these metals can be hazardous, endangering both human health and plant life (Ahmadov et al., 2020). Because HMs have a lengthy half-life and are not biodegradable, they can have a negative long-term impact on the biological processes in the soil. Furthermore, they have the capacity to cause a range of illnesses in people. Because HMs are not biodegradable, they are among the most dangerous categories of contaminants in the environment. These mineral salts can enter aquatic environments through air deposition, rainfall runoff, sewage discharge, and wastewater discharge. HMs have a strong tendency to gather inside the bodies of aquatic species, which causes their buildup (Ahmadov et al., 2020). Due to their consumption, pollution passes through trophic levels and eventually reaches humans, who are at the top of the food chain and hence at risk for health problems. When HMs are present in environmental pollution, there can be a number of health consequences that result, such as heart attacks and strokes, genetic disorders, cancer, increased kidney problems, mental and psychological disorders, birth defects, low IQs, and difficulty focusing (Rasouli-Sadaghiani et al., 2019; Yu et al., 2021). Because hydrocarbons have an extensive impact on all life forms, their contamination is a major concern. Because crude oil is widely utilized and has associated transportation and disposal issues, pollution from increased usage of the fuel is common. A diverse blend of aliphatic, aromatic, and heterocyclic chemicals can be found in crude oil. In addition to being naturally occurring, HM concentrations in soil are rising as a result of human activity like pesticide use and oil refinement. HM and metalloid concentrations in some places are so high that they are adversely affecting the local natural environment (Das & Mishra, 2010). The explanation is that the activity of microbes is restricted by HMs and metalloids, which makes them unsuitable for hydrocarbon breakdown and thus less effective (Das, Bal, et al., 2015; Das & Singh, 2011; Mishra et al., 2009). Therefore, environmental remediation, which entails removing pollutants from soil, air, and water, is essential. Hydrocarbon contamination is primarily caused by mishaps on oil platforms and ships that transport hydrocarbons, but it can also result from water spills into the ocean from tanker vessel wash bays. Petroleum products and crude oil create a waterproof layer over water that prevents the exchange of oxygen between the environment and water, causing harm to people, animals, and plants. PAHs are considered pervasive pollutants. There are one hundred different polycyclic aromatic hydrocarbon compounds in total. While few PAHs are utilized in industry, those that are mostly employed in the manufacture of insecticides, dyes, plastics, and medications. When organic materials are partially burned, polycyclic aromatic hydrocarbons are created. Because of their toxicity, sluggish rate of biodegradation in the environment, and carcinogenicity, polychlorinated biphenyls (PCBs) are widely considered to be the worst pollutants (Srivastava et al., 2019). Finding plant species that can clean up contaminated soils is essential in oil-rich locations. Native plant species that have evolved to thrive in contaminated areas, especially those that can withstand salinity, drought, and contamination, may have useful phytoextraction potential. Based on prior studies and field observations, the *A. camelorum* plant was chosen in this study to perform phytoremediation of TPHs and HMs. This plant, which is a member of the Fabaceae family, is well-known for its deep roots and resistance to drought and cold. It grows best in the southern parts of Iran (Zangoie et al., 2023). Iran's vegetation cover is made up of shrubby plants in addition to annual and perennial herbaceous species. The soils in this location are vulnerable to severe oil pollution due to the daily extraction of crude oil, as well as refining and oil transportation activities. In addition, there may be health concerns for people living close by, as well as for surrounding farms and crop harvesting (Zahermand et al., 2021).

9.2.2 Effects of Hydrocarbon Pollutants on Terrestrial Ecosystem

After air and water, soil is regarded as the third most significant component of the ecosystem worldwide. Given that soil provides 95% of the food consumed by humans, healthy soil is essential to life as we know it. Planning for the preservation of healthy soil is crucial to ensuring human survival. When pollutants are added to soil, the soil's quality varies and environmental issues arise. Pollutant emissions and human activities are the main causes of environmental contamination, which is the fundamental cause of many environmental issues. Of these, one that stands out as a major problem is the contamination of soil with petroleum compounds in areas that produce oil and in the vicinity of oil refineries. Unfortunately, HMs, which are persistent pollutants because they cannot be removed or broken down, are among the many contaminants that soil acts as a major sink for (Das, Ghosh, et al., 2015; Das, Sukla, et al., 2012). These elements are naturally found in the crust of the Earth in a variety of solid forms found in soils and sediments, as well as dissolved in water. Certain HMs such as copper (Cu), iron (Fe), nickel (Ni), manganese (Mn), zinc (Zn), and molybdenum (Mo) are poisonous to humans, plants, and microbes at high concentrations, while low amounts are required for metabolic function (Das et al., 2012;). Large quantities of mine waste, contaminated bare soil devoid of plant cover, or severely degraded plant development conditions are all found in mining regions, which are also subject to erosion and leaching episodes and represent a major environmental risk (Dey, Tripathy, et al., 2023; Mohanty et al., 2017, 2018; Cornelissen et al., 2019). To increase soil structure, nutritional status, water-holding capacity, and pH buffering, inorganic materials like fly ash, phosphate-based compounds, and lime are used along with organic amendments like compost, biochar, biosolids, and biosolids. Modifying metal-rich tailings with livestock manure or bio-waste improves soil for plant growth and facilitates metal recovery from waste. Plants contribute to the preservation of the landscape's aesthetic qualities in addition to increasing soil fertility, stabilizing soil in open pits and slopes, lowering HM bioavailability and toxicity, stopping the spread of HMs through wind erosion and groundwater leaching, and promoting the ecosystem's ecological development (Zhang et al., 2019). Implementing HM phytoremediation requires selecting plant species with traits suited to local soil and climate, including tolerance to high HM concentrations, salinity, or pH, along with rapid growth, high biomass production, and resilience, often favoring metalliferous autochthonous species (Yan et al., 2020). Soil degradation in agricultural areas, linked to the long-term use of impure pesticides, fertilizers, and organic amendments high in HMs like Cu, Cd, and Pb, threatens crop output and exacerbates poverty and malnutrition in vulnerable regions. In arid or semi-arid regions, it is common practice to irrigate crops with wastewater that is rich in C, N, or other nutrients. This reduces the need for chemical fertilizers, but it also often leads to the buildup of HMs in the soil unless the crops have been treated beforehand to remove the metal load. Furthermore, it is important to consider how these amendments promote root development, which gives plants access to bigger pools of accessible HMs. In addition to their detrimental effects on plants, HMs also modify important aspects of soil function, reducing biological activity and nutrient availability, decreasing crop yield and soil fertility, and changing the composition of the microbial community (Zhang et al., 2019). For a plant to be classified as a hyper-accumulator, its dry weight foliar tissue must have a minimum concentration of a trace element of the following: As, Pb or rare earth elements; greater than 3,000 mg/kg for Zn, 100 mg/kg for Cd, thallium (Tl), or Se, greater than 300 mg/kg for Co, Cu, or Cr, greater than 1,000 mg/kg for Ni, greater than 10,000 mg/kg for Mn (Reeves et al., 2018). Despite substantial soil HM pollution, tomatoes, potatoes, and cabbage displayed low accumulation values and hence pose fewer health hazards. Pre-treating wastewater and choosing the right crops are essential to reducing the dangers to human health when using contaminated soil for crop production (Cao et al., 2022). Phytoremediation is a promising method for remediating contaminated soil or water and recovering valuable materials, especially using fast-growing plants like Salix, Populus, and species like Thlaspi, Odontarrhena, and Pteris vittata, which accumulate specific trace elements. Li also enters the food chain and is absorbed by all plants (Bolan et al., 2021). For vaporization, rotary dryers or thermal screw dryers are used. Rotary dryers indirectly heat the soil by rotating a cylinder, while thermal screw dryers directly circulate hot oil or steam through the soil as it moves through an auger. Thermal desorption can be performed at either low or high temperatures. Low temperature is commonly employed for the extraction of organic pollutants from soil. The pollutants that are released into the atmosphere are gathered in vapor condensation systems, where they undergo partial degradation into non-toxic byproducts. HTTD is used to remove HMs; air pollution scrubber units are used to collect the off-gas, which needs additional treatment to reduce any toxicity. In economically developing nations, thermal desorption is subject to many constraints. The price of the thermal desorption unit and the additional expense of heating it represent major drawbacks. Another drawback is that off-gas typically needs to be handled as a hazardous item and requires additional cleanup.

9.2.3 Effects of Hydrocarbon Pollutants on Aquatic Ecosystem

The presence of dangerous chemicals, mainly biological or chemical, in surface or groundwater that are either naturally occurring or caused by human activity is commonly referred to as water pollution. Similar to soil pollution, surface and groundwater pollution is an old issue that was made worse by the expansion of human settlements. Water pollution, in contrast to soil contamination, can spread quickly and worldwide via the water cycle. The severity of water contamination increased significantly with global industrialization. Plastics and artificial pesticides are dangerous human-caused water contaminants. Water pollution is classified into the following categories by many organizations that regulate water quality: chemical, effluent, industrial-specific, microbiological, and radiochemical. In general, there are two types of chemical pollution: organic and inorganic. Chemical pollutants can change salinity and toxicity, alter pH, and increase chemical oxygen demand. The majority of the time, wastewater contamination is linked to municipal operations and frequently consists of a volatile mixture of contaminants. Thermal pollution and sediment pollution are examples of industrial pollution. Conventional groundwater remediation techniques can be carried out either in situ or ex situ. Using a well to physically pump contaminated water out of the ground and collect it in containers for cleanup or disposal is the simplest and most popular form of ex situ remediation. Pumping systems are great for emerging economies since they are easy to operate and reasonably priced. Sadly, there isn't a universal technique for extracting water from soil. The features of the site, such as the kind of soil and the location of the water in the soil profile, have a significant impact on the design of wells and pumping systems. Pumping allows for flexibility by enabling either on-site or off-site treatment of the contaminated water. Standard onsite or off-site water treatment is used in ex situ treatments of groundwater to store or neutralize liquid hazardous wastes. A remediation method called in situ air sparging was created for situations involving saturated soils and shallow groundwater pollution. Currently, the only target contaminant that is effectively treated with air sparging is organic pollution. The process of air sparging involves physically pumping air into the groundwater. The pollutants are volatilized by the air bubbles, allowing vapor phase technology to extract them. Methods of chemical degradation, such as oxidation, could improve the process. Site-specificity, which depends on the composition of the soil and the level of water saturation, is the process's main drawback. Air injection wells need to be planned specifically for the location. Its application has great promise for developing nations (Lwin et al., 2024). Purification is the most basic conventional method for rehabilitating lotic aquatic systems, such as tidal zones and rivers. It involves pumping clean water into the system to reduce the concentration of pollutants, dispersing them downstream or into tidal outflows. Although this method does not remove pollutants, it dilutes them to levels below toxicity thresholds, thereby promoting natural biological, chemical, and physical degradation processes. The effectiveness of this approach can be enhanced by applying optimal control theory, which can significantly improve water quality. However, this technology remains underutilized in many rapidly growing emerging countries. Additionally, in situ methods can further enhance the purification process. Polluted lotic water can be pumped into containers or diverted to retention ponds as part of another conventional remediation process. These methods can be used for on-site or off-site treatment, utilizing a range of wastewater purification techniques as well as hazardous material neutralization or disposal techniques. The fact that dilution and diversion techniques only reduce pollution in the water and leave contaminants in the soils of riverbanks and benthic areas is one of their main drawbacks. Since lentic systems contain pollutants, cleaning them up is a relatively easier operation that nonetheless makes use of many conventional techniques for cleaning up flowing water. Furthermore, huge lakes have demonstrated the effectiveness of in situ flocculation when utilized both independently and in combination with other conventional cleanup techniques. Chemical analysis alone is insufficient for assessing and predicting the impact of chemicals on aquatic organisms and ecosystems due to the growing number of chemical stressors and their interactions. Ecological indices have a limited ability to identify the causes of environmental stress. To directly address biological effects and link them to chemical exposure, complementary approaches like (eco)toxicological testing are needed. These assays expose biological components such as individuals, populations, cells, and communities to chemical mixtures to observe biological effects. The detailed figure explaining the major sources, effects, and phytoremediation of hydrocarbons is shown in Figure 9.1.

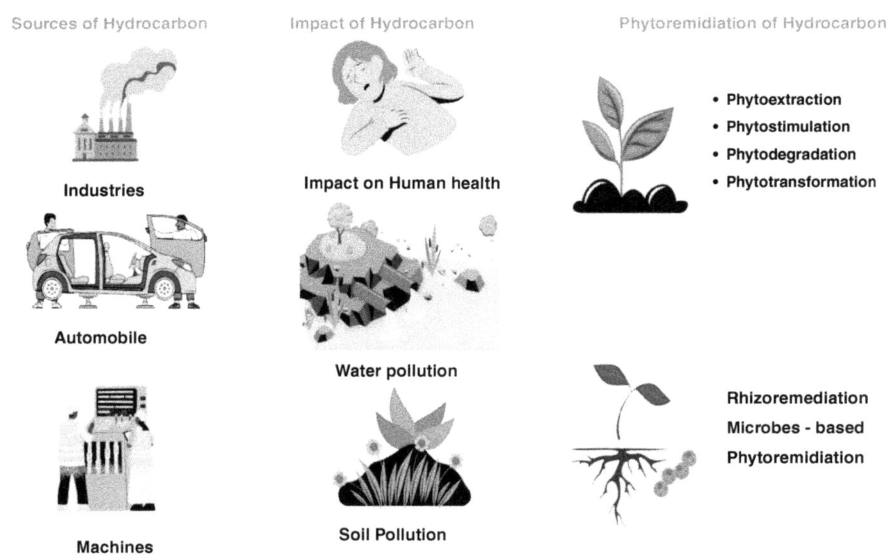

FIGURE 9.1 The comprehensive figure illustrating the primary sources, impacts, and phytoremediation strategies for hydrocarbons.

9.3 PHYTOREMEDIATION

One of the suggested sustainable pollution remediation best practices under the sustainable development objectives is phytoremediation, especially for nations with growing economies. Samples are frequently sent to chemical testing facilities and maintained. Portable testing laboratories can also be used for on-site testing. Standard assays are typically used in soil pollution screening tests to identify pollutants and determine their levels (Behera et al., 2021). The complexity of this process varies, especially when the area has multiple sources of contamination (Shmaefsky, 2020). To eliminate soil pollutants, the newly developed technique of electroremediation can be applied alone or in conjunction with other remediation techniques. Using a low-voltage direct current charge, this method is most practical when used in situ to remove HMs from soil. The contaminated site is surrounded by slotted PVC-lined wells into which the electrodes are placed. The HMs are drawn to the anodes and cathodes, where they are electrokinetically migrated and immobilized in wells. By creating an acidic pH in the polluted soil, the electrodes also aid in the migration of the HMs. The effectiveness of this procedure is limited by the difficulty in creating a sufficient electrokinetic field and maintaining a consistent pH in many different types of soils. Furthermore, the process is impractical for large remediation sites, which are likely located in developing nations (Reddy et al., 2019). Groundwater remediation and surface water remediation are two categories into which traditional methods of treating water pollution are frequently separated. Many of the techniques used in soil remediation can also be applied to remove pollutants from groundwater. Although better access to the pollution makes water techniques easier to implement, it is challenging to limit the contamination events that occur (Bell, 2019).

9.3.1 Phytoremediation of Petroleum Hydrocarbons for Sustainable Management

A novel technique called phytoremediation employs carefully chosen metal-accumulating plants to purge radioactively polluted soil of HMs. Compared to the soil removal and burial techniques now in use, phytoremediation presents an alluring and affordable option. Plants can achieve high biomass and metal accumulation rates from soils by combining properly chosen metal-accumulating agricultural plants with creative soil additives (Horne, 2020). Remediation of polluted soil is crucial to eliminate HMs from the ecosystem. That being said, the process of metal remediation is extremely difficult and intricate. The application of conventional techniques, such as immobilization or extraction by physico-chemical methods like soil erosion, ion exchange, chemical oxidation or reduction, precipitation, electro-kinetic treatment, pollutant stabilization, excavation, and incineration, may have a negative effect on the structure of the soil and possibly result in the destruction of natural habitats. With the use of plants' capacity for HM tolerance and the rhizospheric microflora they support, phytoremediation is a green engineering technique that breaks down or sequesters metal contaminants found in soil and water. The most effective plants with the potential for phytoremediation are those that can generate useful biomass even when the soil contains HMs (Kanungo et al., 2024; Priyadarshini et al., 2024). Phytoextraction and phytostabilization are the most suitable methods for phytoremediation of HM-contaminated areas. Through the process of bioremediation, dangerous substances can be broken down or eliminated from the environment using plants, microorganisms, microbial enzymes, or enzymes produced by plants. Toxin-containing substances are eliminated from surface and groundwater, as well as from contaminated sediment, soil, and the atmosphere by employing higher-living species in the ecosystem. Due to their potential for transport and propensity to accumulate toxins, green plants can be used in phytoremediation to absorb or render harmless organic compounds, radioactive substances, and HMs from soil and water. Herbs such as Thlaspi caerulescens, plants like Salix spp., or a combination of the two, are frequently used in phytoremediation (Allamin, 2021). Phytoremediation is becoming increasingly popular in both the public and business sectors as a more economical and ecologically friendly alternative to traditional remediation processes. In very vast areas where the application of traditional technologies is not feasible or cost-effective, phytoremediation can be used effectively. Because of its exceptional resistance to extreme environmental factors including fluctuating pH 1.6 levels and high concentrations of HMs, vetiver grass (C. zizanioides) has been the subject of much research into its usage as a hyperaccumulator. Vetiver has been shown to be efficient at absorbing impurities and nutrients, especially phosphate (P) and nitrogen (N). Hemp has been shown to be a phytoremediator of HMs, radionuclides and PAHs. It has also been shown to be a useful feedstock for bioenergy processes such as biodiesel, bioethanol, biogas, and CHP. Hemp is a good contender for making a profit from phytoremediation operations due to its significant aboveground biomass output, deep taproot depth, and commercial potential for bioenergy products. Selecting the best hemp genetics for the particular pollutant at hand requires a case-by-case analysis of hemp's viability for this dual use. The majority of the plants showed the capacity to grow at high concentrations of TPHs, suggesting that they can be applied to soil contaminated with petroleum hydrocarbons through phytoremediation. In the polluted site, *Aspilia africana, Bryophyllum pinnatum, Paspalum vaginatum, Paspalum scrobiculatum, Spondias mombin, Cyperus rotundus, Uvaria chamae,* and *Eragrostis atrovirens* all showed signs of phytoremediation. Microbes can flourish on ant roots thanks to their rich specialization, which allows them to eliminate pollutants through the action of root exudates. Rhizoremediation is an inexpensive method of removing pollutants when vast polluted regions can be safely removed slowly and the level of contamination is not particularly high. Atropha spp. have the capacity to absorb HMs into their roots and shoots, which suggests that they could function as a phytoremediator (Awotedu & Ogunbamowo, 2019). There are novel approaches to environmental restoration that involve using plants to change and break down xenobiotic organic contaminants. The level of cadmium (Cd) pollution in river sediments is increasing. Phytoremediation has been used to treat Cd-contaminated sediments, but remediation efficiency still needs to be improved. Phyto means plant in Greek, and remedium means able to cure or restore in Latin. These two terms together form the term phytoremediation. Plant species are chosen for phytoremediation based on a variety of factors, including soil composition, root depth, polluted soil, and local climate (Haq et al., 2020). More metal buildup in the soil is harmful to plants. Their roots have the capacity to take up large concentrations of metal ions from the soil, which they can then transfer to their aerial biomass and impact critical metabolisms. Plants that are resistant to growing in soil contaminated with metals have evolved physiological defenses against metal entrance into root cells and to deal with HM concentrations inside the plant. One of the primary causes of soil pollution is thought to be HMs, including Cd, Cu, Ni, Cr, Zn, and Pb. The behavior of microorganisms in soil is influenced by HMs (Das & Mishra, 2010). They fluctuate in terms of diversity, population size, and soil-based activity. Applications for phytoremediation include rhizofiltration, phytoextraction, phytostabilization, phytodegradation, and phytovolatilization (Zaghloul, M., 2020). Endophytes have shown promise in improving the physiological state of plants under metallic stress, promoting root and shoot growth, and increasing metal uptake in the biomass of the shoots of a broad range of hyperaccumulating plants. Endophyte-assisted phytoremediation is a potentially effective technique for cleaning up naturally metal-rich or contaminated soil. The details of the plant species involved in the phytoremediation are described in Table 9.1. One workable solution for eliminating pollution from polluted soil and water is phytoremediation, a type of bioremediation. Originating from the need for a low-cost, sustainable, and ecologically friendly substitute for conventional methods for cleaning up chemical and physical pollutants. In the beginning, bioremediation involved extracting or breaking down organic and inorganic pollutants from soil and water in situ using bacteria and other microorganisms (Priyadarsini & Das, 2024, 2025). It then progressed to additional applications in conjunction with conventional chemical and physical cleanup techniques. Using aquatic plants (species like *Eichhornia crassipes* and cultivars like *Lemna minor* and *Spirodela polyrhiza)* for phytoremediation of aquatic settings is an efficient, effective, and low-cost method of cleaning pollution and water of nitrogen and other chemical residues.

TABLE 9.1
Describe the Details of the Plant Species That Are Involved in the Phytoremediation

Sl. No.	Common Name	Scientific Name	Plant Type	References
1	Italian ryegrass	Lolium multiflorum L.	Plants with fibrous root	
2	Sorghum	Sorghum vulgae Pers.	Plants with fibrous root	Banks et al. (2003)
3	Maize	Zea mays L.	Plants with fibrous root	Chaineau et al. (2000)
4	Alfa	Medicago sativa L.	Plants belonging to leguminous	Pradhan et al. (1999)
5	Bermuda grass	Cynodon dactylon L.	Plants with fibrous root	
6	Southern crabgrass	Digitaria ciliaris	Plants with fibrous root	Olson and Fletcher (2000)
7	Red clover	Trifolium pratense L.	Plants belonging to leguminous	Olson and Fletcher (2000)
8	Spinach	Spinacia oleracea L.	Plants belong to tap root	Kaimi et al. (2007)
9	Sun flower	Helianthus annu	Plants belong to tap root	Kaimi et al. (2007)
10	Rice	Oriza sativa L.	Plants with Fibrous root	Kaimi et al. (2007)

In order to make ecosystem repair economically feasible, biotechnological integration can be a vital first step. Similar to single-plant phytoremediation, mixed-plant phytoremediation involves a specific relationship between contaminants, plants, and the environment. It is critical that plants are conditioned to thrive in a contaminated environment, particularly when it comes to working together to eradicate different types of pollution (Wang et al., 2019). Controlling the water content in contaminated soil or adjusting the anaerobic/aerobic status of the environment is typically required. The advantages of mixed-plant phytoremediation are greatly influenced by the types of pollutants found in the environmental media and the plant species selected. Soil is stressed as a crucial non-renewable natural resource in most international policy frameworks. Some soil is naturally rich in HMs. However, some anthropogenic activities are known to cause the redistribution of HMs, which may have negative effects on the ecosystem and/or human health (Sánchez et al., 2023). The limits for lead in agricultural soil on the Chinese mainland are ≤250 mg/kg (pH 7.5). Different nations have different regulatory limits for harmful metal concentrations in agricultural soil. For example, Pb has a limit of less than 300 mg/kg in Australia and less than 70 mg/kg in Canada (Verma et al., 2022). Plant aerial tissues may potentially become contaminated if HMs are dropped directly on biomass above ground. Since plants form the base of the terrestrial food chain, it is important to monitor and reduce the amounts of HMs in plants as soon as possible (Alengebawy et al., 2021; Verma et al., 2022). HMs or metalloids have been found in more than half of the over 10 million sites with recorded soil contamination worldwide. There are an estimated 2.8 million locations in Europe where HM contamination may exist; in China, hazardous pollutants are present in 19% of agricultural soil; and in India, anthropogenic sources are responsible for 80% of polluted soil in the states of Maharashtra, Gujarat, and Telangana (Hou et al., 2020). Excavation of contaminated soil, transportation to the treatment region, and dumping of the treated soil at approved sites are necessary for ex situ methods. Although the process is generally more controlled and can produce better results faster than other methods, it is costly and disturbs the soil. At the polluted site, in situ technologies are used, soil disturbance is reduced, and workers and residents are exposed to the toxins compared to ex situ methods (Dhaliwal et al., 2020).

9.4 CONCLUSION

Phytoremediation offers a promising approach for the sustainable management of hydrocarbon pollutants. This method effectively mitigates the environmental impact of hydrocarbon pollutants by utilizing plants to absorb, degrade, or contain contaminants. The various types of phytoremediation, such as phytoextraction, phytodegradation, and phytostabilization, can be tailored to address different hydrocarbon pollutants and their specific effects on soil and water. This eco-friendly strategy not only cleans contaminated sites but also promotes long-term sustainability by minimizing ecological damage and supporting natural recovery processes.

REFERENCES

Ahmadov, M., Humbatov, F., Mammadzada, S., Balayev, V., Ibadov, N., & Ibrahimov, Q. (2020). Assessment of heavy metal pollution in coastal sediments of the western Caspian Sea. *Environmental Monitoring and Assessment*, *192*, 1–18. https://doi.org/10.1007/s10661-020-08401-3

Alengebawy, A., Abdelkhalek, S. T., Qureshi, S. R., & Wang, M. Q. (2021). Heavy metals and pesticides toxicity in agricultural soil and plants: Ecological risks and human health implications. *Toxics*, *9*(3), 42. https://doi.org/10.3390/toxics9030042

Allamin, I. A. (2021). Phytoremediation of heavy metals in contaminated soils: A review. *Journal of Biochemistry, Microbiology and Biotechnology*, *9*(2), 7–14. https://doi.org/10.54987/jobimb.v9i2.610

varez-Barragán, J., Cravo-Laureau, C., Wick, L. Y., & Duran, R. (2021). Fungi in PAH-contaminated marine sediments: Cultivable diversity and tolerance capacity towards PAH. *Marine Pollution Bulletin*, *164*, 112082. https://doi.org/10.1016/j.marpolbul.2021.112082

Awotedu, O. L., & Ogunbamowo, P. O. (2019). Comparative heavy metal uptake and phytoremediation potential of three jatropha species. *Environment & Ecosystem Science*, *3*(2), 26–30. https://doi.org/10.4102/safp.v45i10.1874

Azuazu, I. N., Sam, K., Campo, P., & Coulon, F. (2023). Challenges and opportunities for low-carbon remediation in the Niger Delta: Towards sustainable environmental management. *Science of the Total Environment*, 165739. https://doi.org/10.1016/j.scitotenv.2023.165739

Banks, M. K., Kulakow, P., Schwab, A. P., Chen, Z., & Rathbone, K. (2003). Degradation of crude oil in the rhizosphere of Sorghum bicolor. *International Journal of Phytoremediation*, *5*(3), 225–234. https://doi.org/10.1201/9781003075745

Behera, I. D., Basak, G., Kumar, R. R., Sen, R., & Meikap, B. C. (2020). Treatment of petroleum refinery sludge by petroleum degrading bacterium Stenotrophomonas pavanii IRB19 as an efficient novel technology. *Journal of Environmental Science and Health - Part A Toxic/Hazardous Substances and Environmental Engineering*, *56*(2). https://doi.org/10.1080/10934529.2020.1866924

Behera, I. D., Nayak, M., Biswas, S., Meikap, B. C., & Sen, R. (2021). Enhanced biodegradation of total petroleum hydrocarbons by implementing a novel two-step bioaugmentation strategy using indigenous bacterial consortium. *Journal of Environmental Management*, *292*, 112746. https://doi.org/10.1016/j.jenvman.2021.112746

Bolan, N., Hoang, S. A., Tanveer, M., Wang, L., Bolan, S., Sooriyakumar, P., ... & Rinklebe, J. (2021). From mine to mind and mobiles: Lithium contamination and its risk management. *Environmental Pollution*, *290*, 118067. https://doi.org/10.1016/j.envpol.2021.118067

Campos, I., Abrantes, N., Pereira, P., Micaelo, A. C., Vale, C., & Keizer, J. J. (2019). Forest fires as potential triggers for production and mobilization of polycyclic aromatic hydrocarbons to the terrestrial ecosystem. *Land Degradation & Development*, *30*(18), 2360–2370. https://doi.org/10.1002/ldr.3427

Cao, Z., Wei, B., Yang, L., Yu, J., Meng, M., Chen, Q., & Li, F. (2022). Different crop rotation patterns vary heavy metals behavior in soils under plastic sheds: Evidence from field research. *Process Safety and Environmental Protection*, *162*, 543–552. https://doi.org/10.3390/ijerph17155359

Chaineau, C. H., Morel, J. L., & Oudot, J. (2000). Biodegradation of fuel oil hydrocarbons in the rhizosphere of maize. *Journal of Environmental Quality*, *29*(2), 569–578. American Society of Agronomy, Crop Science Society of America, and Soil Science Society of America.

Cornelissen, H., Watson, I., Adam, E., & Malefetse, T. (2019). Challenges and strategies of abandoned mine rehabilitation in South Africa: The case of asbestos mine rehabilitation. *Journal of Geochemical Exploration, 205*, 106354. https://doi.org/10.1016/j.gexplo.2019.106354

Das, A. P., & Ghosh, S. (2023d). Microorganisms in manganese biomining processes. *Manganese Mining Microorganisms*, 55–61. https://doi.org/10.1016/B978-0-12-822147-1.00013-3

Das, A. P., & Singh, S. (2011). Occupational health assessment of chromite toxicity among Indian miners. *Indian Journal of Occupational and Environmental Medicine, 15*(1), 6. https://doi.org/10.4103/0019-5278.82998

Das, A. P., Bal, B., & Mahapatra, P. S. (2015). Chromogenic biosensors for pathogen detection. In P. Prokopovich (Ed.), *Biological and Pharmaceutical Applications of Nanomaterials* (1st ed.). CRC Press, Taylor & Francis. https://doi.org/10.1007/bf01774917

Das, A. P., Ghosh, S., Mohanty, S., & Sukla, L. B. (2015). Advances in manganese pollution and its bioremediation. In L. B. Sukla, N. Pradhan, S. Panda, & B. K. Mishra (Eds.), *Environmental Microbe Biotechnology* (1st ed., pp. 313–328). Springer, Cham. https://doi.org/10.1007/978-3-319-19018-1_16

Das, A. P., Sukla, L. B., & Pradhan, N. (2012). Microbial Recovery of Manganese using *Staphylococcus Epidermidis*. *International Journal of Nonferrous Metallurgy, 2012*(02), 9–12. https://doi.org/10.4236/IJNM.2012.12002

Das, A. P., Swain, S., Panda, S., Pradhan, N., & Sukla, L. B. (2012). Reductive acid leaching of low grade manganese ores. *Geomaterials, 2012*(04), 70–72. https://doi.org/10.4236/GM.2012.24011

Das, A., & Mishra, S. (2010). Biodegradation of the metallic carcinogen hexavalent chromium Cr(VI) by an indigenously isolated bacterial strain. *Journal of Carcinogenesis, 9*.

Dey, S., Bissoyi, A., Koteswara Reddy, G., & Das, A. P. (2024). Synthetic microfibers: An urban emerging pollutant demanding for sustainable environmental solutions. In *Sustainable Microbial Technology for Synthetic and Cellulosic Microfiber Bioremediation* (pp. 107–121). https://doi.org/10.1007/978-3-031-62600-5_7

Dey, S., Das, A., Mallick, K., Sahu, A., & Das, A. P. (2023). Environmental petroleum waste: Pollution, toxicity, sustainable remediation. (pp. 159–175). https://doi.org/10.1007/978-3-031-48220-5_7

Dey, S., Dash, A., & Das, A. P. (2024). Bioremediation as a sustainable and circular solution for mining waste management and bioenergy generation. In *Sustainable Management of Mining Waste and Tailings: A Circular Economy Approach* (pp. 107–119). https://doi.org/10.1201/9781003442455-6

Dey, S., Tripathy, B., Kumar, M. S., & Das, A. P. (2023). Ecotoxicological consequences of manganese mining pollutants and their biological remediation. *Environmental Chemistry and Ecotoxicology*. https://doi.org/10.1016/J.ENCECO.2023.01.001

Dhaliwal, S. S., Singh, J., Taneja, P. K., & Mandal, A. (2020). Remediation techniques for removal of heavy metals from the soil contaminated through different sources: A review. *Environmental Science and Pollution Research, 27*(2), 1319–1333. https://doi.org/10.1007/s11356-019-06967-1

French, K. E., Zhou, Z., & Terry, N. (2020). Horizontal 'gene drives' harness indigenous bacteria for bioremediation. *Scientific Reports, 10*(1), 15091. https://doi.org/10.1093/fs/liii.1.54-a

Haq, S., Bhatti, A. A., Dar, Z. A., & Bhat, S. A. (2020). Phytoremediation of heavy metals: An eco-friendly and sustainable approach. In *Bioremediation and Biotechnology: Sustainable Approaches to Pollution Degradation* (pp. 215–231). https://doi.org/10.4018/978-1-6684-3714-8.ch009

Horne, A. J. (2020). Phytoremediation by constructed wetlands. In *Phytoremediation of Contaminated Soil and Water* (pp. 13–39). CRC Press. https://doi.org/10.1201/9780367803148-2

Hou, D., O'Connor, D., Igalavithana, A. D., Alessi, D. S., Luo, J., Tsang, D. C., … & Ok, Y. S. (2020). Metal contamination and bioremediation of agricultural soils for food safety and sustainability. *Nature Reviews Earth & Environment*, *1*(7), 366–381. https://doi.org/10.1038/s43017-020-0061-y

Kaimi, E., Mukaidani, T., & Tamaki, M. (2007). Screening of twelve plant species for phytoremediation of petroleum hydrocarbon-contaminated soil. *Plant Production Science*, *10*(2), 211–218. https://doi.org/10.1626/pps.10.211

Kakkar, A., & Kumar, S. (2024). Remediation technologies for petroleum hydrocarbons from the environment. In *Impact of Petroleum Waste on Environmental Pollution and Its Sustainable Management through Circular Economy* (pp. 205–233). Springer Nature Switzerland, Cham. https://doi.org/10.1007/978-3-031-48220-5_10

Kanungo, J., Sahoo, T., Swain, L. P., & Behera, I. D. (2024). Toxicity of persistent hydrocarbon pollutants, sources and sustainable remediation process. In *Impact of Petroleum Waste on Environmental Pollution and Its Sustainable Management through Circular Economy* (pp. 39–65). Springer Nature Switzerland, Cham. https://doi.org/10.1007/978-3-031-48220-5_2

Li, S., Gao, X., Zhu, S., & Liang, H. (2024). Polycyclic aromatic hydrocarbons (PAHs) in coal preparation plant products: A contributor to environmental pollution. *Science of the Total Environment*, *906*, 167887. https://doi.org/10.1109/iccce55785.2022.10036238

Lwin, C. S., Kim, Y. N., Lee, M., Jung, H. I., & Kim, K. R. (2024). In situ immobilization of potentially toxic elements in arable soil by adding soil amendments and the best ways to maximize their use efficiency. *Journal of Soil Science and Plant Nutrition*, *24*(1), 115–134. https://doi.org/10.1007/s42729-024-01630-2

Mishra, S., Das, A. P., & Seragadam, P. (2009). Microbial remediation of hexavalent chromium from chromite contaminated mines of Sukinda Valley, Orissa (India). *Journal of Environmental Research and Development*, 1122–1127.

Mishra, S., Dash, D., & Das, A. P. (2022). Detection, characterization and possible biofragmentation of synthetic microfibers released from domestic laundering wastewater as an emerging source of marine pollution. *Marine Pollution Bulletin*, *185*, 114254. https://doi.org/10.1016/J.MARPOLBUL.2022.114254

Mishra, S., Dash, D., & Das, A. P. (2024). Isolation and molecular characterization of polyester-degrading bacterial strains from municipal wastewater and their application in microplastic degradation in contaminated wastewater. *Geomicrobiology Journal*. https://doi.org/10.1080/01490451.2024.2376101

Mohanta, S., Pradhan, B., & Behera, I. D. (2024). Impact and Remediation of Petroleum Hydrocarbon Pollutants on Agricultural Land: A Review. *Geomicrobiology Journal*, *41*(4), 345–359. https://doi.org/10.1080/01490451.2023.2243925

Mohanty, S., Ghosh, S., Bal, B., & Das, A. P. (2018). A review of biotechnology processes applied for manganese recovery from wastes. *Reviews in Environmental Science and Bio/Technology*, *17*(4), 791–811. https://doi.org/10.1007/S11157-018-9482-1

Mohanty, S., Ghosh, S., Nayak, S., & Das, A. P. (2017). Bioleaching of manganese by Aspergillus sp. isolated from mining deposits. *Chemosphere*, *172*, 302–309. https://doi.org/10.1016/J.CHEMOSPHERE.2016.12.136

Nemati, B., Baneshi, M. M., Akbari, H., Dehghani, R., & Mostafaii, G. (2024). Phytoremediation of pollutants in oil-contaminated soils by Alhagi camelorum: Evaluation and modeling. *Scientific Reports*, *14*(1), 5502. https://doi.org/10.1038/s41598-024-56214-y

Okparanma, R. N., Ukoha-Onuoha, E., & Ayotamuno, J. M. (2022). Predicting selected kinetic parameters of hydrocarbon interactions with alum-based water treatment residuals. *Results in Engineering*, *16*, 100726. https://doi.org/10.1016/j.rineng.2022.100726

Prabhakar, A., Mishra, S., & Das, A. P. (2019). Isolation and identification of lead (Pb) solubilizing bacteria from automobile waste and its potential for recovery of lead from end of life waste batteries. *Geomicrobiological Journal*, *36*(10), 894–903. https://doi.org/10.1080/01490451.2019.1654044

Pradhan, J., Das, S. N., & Thakur, R. S. (1999). Adsorption of hexavalent chromium from aqueous solution by using activated red mud. *Journal of Colloid and Interface Science*, *217*(1), 137–141. https://www.sciencedirect.com/science/article/abs/pii/S0021979799962884

Pradhan, S. P., Conrad, J. R., Paterek, J. R., & Srivastava, V. J. (1998). Potential of phytoremediation for treatment of PAHs in soil at MGP sites. *Journal of Soil Contamination*, *7*(4), 467–480. https://doi.org/10.1080/10588339891334401

Priyadarsini, S., & Das, A. P. (2024). Analytical and structural characterization of waste lithium-ion batteries for their effective recycling strategy. *Environmental Chemistry and Ecotoxicology*. https://doi.org/10.1016/j.enceco.2024.12.004

Priyadarsini, S., & Das, A. P. (2025). Lithium bioleaching: A review on microbial-assisted sustainable technology for lithium bio-circularity. *Journal of Water Process Engineering*, *69*, 106744. https://doi.org/10.1016/j.jwpe.2024.106744

Rasouli-Sadaghiani, M. H., Karimi, H., Ashrafi Saeidlou, S., & Khodaverdiloo, H. (2019). The effect of humic acid on the phytoremediation efficiency of Pb in the contaminated soils by wormwood plant (Artemicia absantium). *JWSS-Isfahan University of Technology*, *22*(4), 261–278. https://doi.org/10.29252/jstnar.22.4.261

Reddy, K. R., Cameselle, C., & Adams, J. A. (2019). *Sustainable Engineering: Drivers, Metrics, Tools, and Applications*. John Wiley & Sons.

Reeves, R. D., Baker, A. J., Jaffré, T., Erskine, P. D., Echevarria, G., & van Der Ent, A. (2018). A global database for plants that hyperaccumulate metal and metalloid trace elements. *New Phytologist*, *218*(2), 407–411. https://doi.org/10.1111/nph.14907

Shmaefsky, B. R. (2020). Principles of phytoremediation. In *Phytoremediation: In-situ Applications*, 1–26. https://doi.org/10.1007/978-3-030-00099-8_1

Srivastav, A. L., Dhyani, R., Ranjan, M., Madhav, S., & Sillanpää, M. (2021). Climate-resilient strategies for sustainable management of water resources and agriculture. *Environmental Science and Pollution Research*, *28*(31), 41576–41595. https://doi.org/10.1007/s11356-021-14332-4

Verma, A., Gaharwar, U. S., Priyadarshini, E., & Rajamani, P. (2022). Metal accumulation and health risk assessment in wastewater used for irrigation around the Agra Canal in Faridabad, India. *Environmental Science and Pollution Research*, *29*(6), 8623–8637. https://doi.org/10.1007/s11356-021-16088-3

Wang, C., Gu, L., Ge, S., Liu, X., Zhang, X., & Chen, X. (2019). Remediation potential of immobilized bacterial consortium with biochar as carrier in pyrene-Cr (VI) co-contaminated soil. *Environmental Technology*, *40*(18), 2345–2353. https://doi.org/10.1080/09593330.2018.1441328

Wei, Z., Wei, Y., Liu, Y., Niu, S., Xu, Y., Park, J. H., & Wang, J. J. (2024). Biochar-based materials as remediation strategy in petroleum hydrocarbon-contaminated soil and water: Performances, mechanisms, and environmental impact. *Journal of Environmental Sciences*, *138*, 350–372. https://doi.org/10.1016/j.jes.2023.04.008

Yan, A., Wang, Y., Tan, S. N., Mohd Yusof, M. L., Ghosh, S., & Chen, Z. (2020). Phytoremediation: A promising approach for revegetation of heavy metal-polluted land. *Frontiers in Plant Science*, *11*, 359. https://doi.org/10.3389/fpls.2020.00359

Yu, Z., Liu, E., Lin, Q., Zhang, E., Yang, F., Wei, C., & Shen, J. (2021). Comprehensive assessment of heavy metal pollution and ecological risk in lake sediment by combining total concentration and chemical partitioning. *Environmental Pollution*, *269*, 116212. https://doi.org/10.1016/j.envpol.2020.116212

Zaghloul, M. (2020). Phytoremediation of heavy metals principles, mechanisms, enhancements with several efficiency enhancer methods and perspectives: A review. *Middle East Journal of Agriculture Research*, *9*(1), 186–214. https://doi.org/10.36632/mejar/2020.9.1.17

Zahermand, S., Vafaeian, M., & Bazyar, M. H. (2021). Ability of Alfalfa plant to refine soil contaminated with nickel and lead. https://doi.org/10.15446/esrj.v24n2.76217

Zangoie, M., Parsa, S., Jami Al-Ahmadi, M., & Izanloo, A. (2023). Effect of irrigation water salinity and plant density on forage quality of two camelthorn (Alhagi camelorum Fisch.) ecotypes. *Environmental Stresses in Crop Sciences*, *16*(4), 987–1004. https://doi.org/10.22077/escs.2023.5251.2127
Zhang, C., Wang, W., Duan, A., Zeng, G., Huang, D., Lai, C., … & Yang, Y. (2019). Adsorption behavior of engineered carbons and carbon nanomaterials for metal endocrine disruptors: Experiments and theoretical calculation. *Chemosphere*, *222*, 184–194. https://doi.org/10.1016/j.chemosphere.2019.01.128

10 Fungal Bioremediation to Treat Hydrocarbon-Contaminated Sites

Current Focus of Technology and Its Assessment

Kumari Aditi, Kiran Bala, Arushi Dogra, Sarfaraz Ali, Alok Prasad Das, and Raj Kumar

10.1 INTRODUCTION

The human society is witnessing an era of rapid industrialization. As a result, toxic pollutants are being emitted on a daily basis in the environment, causing severe damage to various life forms (Gaur et al., 2014). Levels of major environmental contaminants such as petroleum, heavy metals, polymers, organic wastes, biodiesels, *etc.* have increased significantly due to their excessive industrial production and usage, endangering public health (Das et al., 2011). Various remediation techniques are currently being deployed to eliminate hazardous environmental pollutants. Among them, bioremediation remains the most preferred, environmentally friendly, and cost-effective method to efficiently curb elevated pollutant levels. In order to naturally break down contaminants, bioremediation techniques predominantly employ biological agents of either indigenous or extraneous origin, such as higher plants, microorganisms (particularly bacteria), their byproducts, such as enzymes, algae, and fungi (Gaur et al., 2014). Bioremediation can be performed either *in-situ* in the affected areas with contaminated soil, water bodies, sludges, *etc.* or *ex-situ* in bioreactors, through biopile-mediated technique, land farming, and using the windrow method to degrade and mineralize pollutants. These techniques transform the contaminants into simpler substances like water (H_2O), carbon dioxide (CO_2), nitrogen dioxide (NO_2), biomass, *etc.*, thus, polluted areas are naturally detoxified and significant reduction occurs in the solubility and properties of hazardous pollutants (Chakraborty et al., 2023; Zhang et al., 2017; Azubuike et al., 2016).

Hydrocarbons are very common, toxic and hydrophobic pollutants released into the environment by a variety of sources like coal, crude oil, refined petroleum products, including fuel oil, petrol, diesel, gasoline, kerosene, lubricating oils, tar, grease, *etc.* Anthropogenic activities such as extensive usage of harmful organic compounds, reckless dumping and accidental oil spillage mishaps during transportation or storage,

 DOI: 10.1201/9781003508496-10

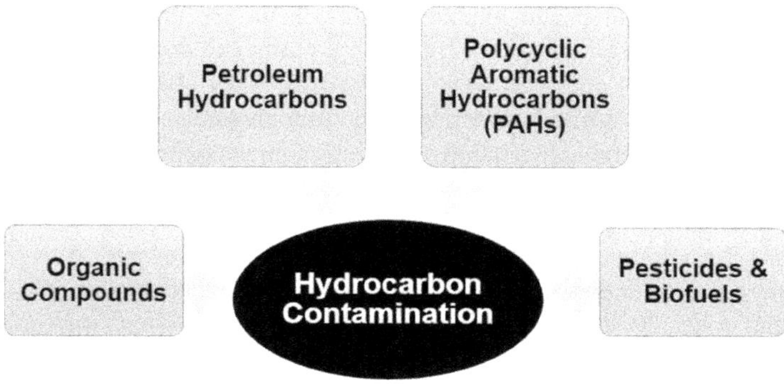

FIGURE 10.1 Different sources of hydrocarbon contamination in the environment.

and unhindered application of pesticides have increased hydrocarbon contamination in natural habitats (Naseri et al., 2014; Mishra et al., 2019). Hydrocarbons pose severe environmental threats due to their high persistence. The majority of hydrocarbon pollution is caused by petroleum hydrocarbons (PHs) and polycyclic aromatic hydrocarbons (PAHs); however, emergences of new fuel alternatives such as biofuels or biodiesel, biogas, bioethanol, natural gas, *etc.*, further contribute to the hydrocarbon pool and augment their bioremediation challenge (Figure 10.1). Several techniques are currently being used to remediate areas contaminated with PHs and PAHs (Xu et al., 2013). These include physical methods such as thermal desorption, soil excavation, soil replacement, *etc.*, chemical remediation methods such as soil flushing, electrokinetic remediation, incineration, *etc.*, and biological remediation such as land farming, soil vapor extraction, photolysis, phytoremediation, microbial remediation and oil-spill cleanups (Varjani, 2017). The physicochemical methods applied to reclaim the decontaminated soil usually lead to major site disturbances, release hazardous byproducts, and are expensive (Bala et al., 2024; Das, Dey, et al., 2024; Dey, Bissoyi, et al., 2024; Dey, Dash, et al., 2024). Hence, biological approaches prove to be ecologically more acceptable and sustainable to clean up the hydrocarbon-contaminated areas (Gaur et al., 2014; Varjani, 2017). Decontamination methods are chosen depending on their specificity toward hydrocarbons and their efficacy in addressing the environmental and health concerns associated with hydrocarbon pollution (Mohanta et al., 2023).

Fungal species have been used for a long time to restore wastelands following any natural catastrophes such as wildfires or accidental oil spillage. Certain fungi possess the ability to co-metabolize hydrocarbons into CO_2 and oxidized intermediates which are less toxic and bioavailable; hence, fungi are considered potent bioremediation agents for the degradation of hydrocarbons (Dey, Das, et al., 2023; Lahiri et al., 2021). Fungi are known to be resistant to environmental changes and have an advantage over bacteria, as they do not require constant water-filled channels for their translocation within soil. Fungi can modify their hyphae by exposing cysteine-rich proteins called hydrophobins which provide hydrophobic coating all over the hyphal surface. Because of such coating, fungi are capable of breaching air–water interface in soil and able to expand their hyphal network inside the fine pores in aggregated hydrocarbon-polluted soil (Wosten et al., 1999). Fungi can directly reach soil-bound hydrophobic pollutants and start their metabolism. They also facilitate the dispersion of useful soil-degrading bacteria by providing channels, liquid films and water droplets surrounding their continuous and interconnected hyphal surfaces. These vast fungal networks enhance the soil bioremediation process by increasing hydrocarbon bioavailability to bacterial species found in soil (Wick et al., 2007; Federici et al., 2007). Basidiomycetes *Pleurotus ostreatus* and Ascomycetes *Botryosphaeria rhodina* are a few examples of hydrocarbon-degrading fungal species (Covino et al., 2016; Yadav et al., 2022). Bioremediation technologies based on fungi are increasingly being developed to mitigate hydrocarbon contamination. Fungi account for the third popular choice for degradation of hydrocarbons after bacteria and other processes involving enzymatic breakdown (Quintella et al., 2019). Various patents have been granted approval in the USA and China in the last decade, and several techniques are being developed based on fungal platforms which offer high critical mass and show huge future growth potential. Various emerging trends are focusing on the amalgamation of microbial and fungal agents in order to remediate hydrocarbon contamination (Mishra et al., 2021; Raj & Das, 2023; Singh et al., 2020). Such combinations show robustness toward fluctuating environmental conditions and have higher effectiveness. In this chapter, we will shed light on the assessment of various fungi-based bioremediation technologies developed to neutralize hydrocarbon contamination in affected areas.

10.2 PETROLEUM HYDROCARBONS AND POLYCYCLIC AROMATIC HYDROCARBONS

Petroleum is a complex mixture of low and high molecular weight hydrocarbons such as saturated and branched-chain alkanes and alkenes, homocyclic and heterocyclic compounds, aromatic rings complexes with heteroatoms like nitrogen, sulfur, oxygen and heavy metals, large aromatic rings and hydrocarbon molecules carrying different functional groups (Srivastava et al., 2019). These are highly lipophilic compounds. The low molecular weight hydrocarbons are readily absorbed through biological membranes and mucosal layers. PAHs or polyarenes are another group of ubiquitous PHs which pose severe negative impacts on the ecosystem, animals and human health (Das, Ray, et al., 2024). Specific PAHs have been recognized as high priority pollutants by the US Environmental Protection Agency (USEPA), due to their high toxicity, prevalence and persistent nature (Table 10.1) (Safdari et al., 2018; Hussar et al., 2012). PAHs are released from oil refineries, incomplete combustion of organic compounds in vehicle engines, coal, steel and aluminum plants, coke furnaces, burning of municipal and rural solid wastes, forest fires, volcanoes, *etc.* (Kadri et al., 2017; Mastral & Callen, 2000). PAHs exhibit carcinogenic, cytotoxic, teratogenic, genotoxic and ecotoxic effects as shown in several *in-vitro* and *in-vivo* studies performed in plants, animals and humans (Anyakora, 2007).

TABLE 10.1
List of PAHs Designated as High Priority Pollutants by the US Environmental Protection Agency (USEPA)

High Priority Pollutants (Polycyclic Aromatic Hydrocarbons)	
Naphthalene (NAP)	Benzo[*a*]Anthracene (B[*a*]A)
Acenaphthylene (ACY)	Chrysene (CHRY)
Acenaphthene (ACE)	Benzo[*b*]Fluoranthene (B[*b*]F)
Fluorene (FLU)	Benzo[*k*]Fluoranthene (B[*k*]F)
Phenanthrene (PHEN)	Benzo[*a*]Pyrene (B[*a*]P)
Anthracene (ANTH)	Benzo[*g,h,i*]Perylene (B[*ghi*]P)
Fluoranthene (FLTH)	Indeno[1,2,3-*c,d*]Pyrene (IND)
Pyrene (PYR)	Dibenz[*a,h*]Anthracene (D[*ah*]A)

PHs and PAHs bind strongly and irreversibly to the organic matter and soil particles to form complexes, causing agglomeration of soil and obstruction of soil pores (Rengarajan et al., 2015). Anaerobic environment builds up within polluted soil which hinders the natural process of hydrocarbon biodegradation (Ramadass et al., 2015; Chang et al., 2013). Fine and clumped soil particles exhibit lower permeability and reduced oxygen diffusion than coarse soil particles, thus optimal gaseous exchange serves as a critical factor in microorganism-mediated degradation of hydrocarbons (Olivella & Gens, 2000). Soil pH, hydrophobicity, particle structure and ionic strength are altered due to elevated hydrocarbon contamination, and, disrupted soil texture along with reduced gaseous exchange negatively impact the native soil microbiota, microbial-metabolism, therefore, deteriorating the soil quality and hampering the overall process of environmental preservation (Dacco et al., 2020).

10.3 MYCOREMEDIATION OF HYDROCARBON CONTAMINANTS

Mycoremediation refers to fungi-assisted remediation of environmental contaminants. Mycoremediation is an eco-friendly and cost-effective strategy employing fungi or their derivatives to break down contaminants such as PHs, PAHs, pesticides, heavy metals, organic compounds and even radiation, *etc*. Fungi have tremendous growth capacity, robust hyphal networks and a large surface area to volume ratio due to which they play a key role in the restructuring of ecosystems (Behera et al., 2020, 2021, 2022). They maintain innumerable life forms that are replenishing for the environment. Certain fungi release potent extracellular lignolytic enzymes with low substrate specificity that effectively degrade a wide variety of hydrocarbon contaminants into simpler forms (Chakraborty et al., 2023). Fungi show high adaptability toward fluctuating environmental temperature, pH, metal-binding proteins and resistance to heavy metal toxicity (Khan et al., 2019). Mycoremediation techniques use mycelium, the vegetative component of fungi, to restore contaminated sites. Mycelium is the underground root network of fungi which consists of a mass

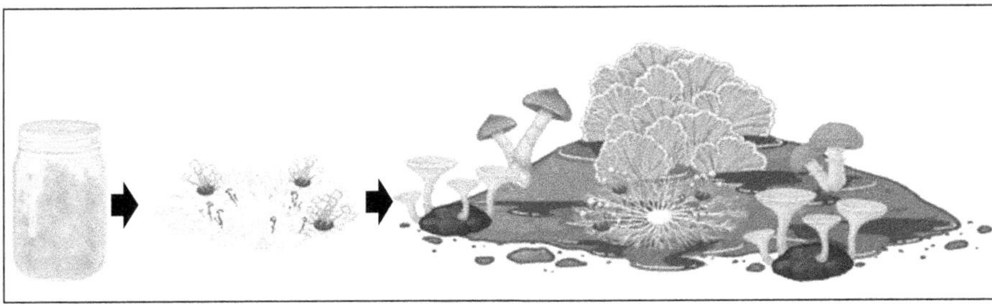

FIGURE 10.2 Mycoremediation of a hydrocarbon-contaminated site.

of whitish filamentous branched structures called hyphae. Hyphae remain interconnected in decomposing environments such as soil or rotting tea trunks, absorb the nutrients and supply them to the mycelium. Mycelium is considered one of the most resilient structures in nature as it can withstand both fire and waterlogging. Fungi degrade the majority of soil contaminants prior to the formation of spore-producing fruiting bodies and absorb the nutrient molecules within a few weeks of growth (Figure 10.2).

Fungi show a distinct ability to degrade and reduce chemical contaminants such as PHs and PAHs in the environment. Fungi are chemoorganotrophic organisms; however, they cannot use large PHs and PAHs as their direct carbon source, and instead they utilize them in their co-metabolic pathways (Gadd, 2017). Fungal mycelia grow vast interconnected networks, facilitating the spread of bioremediating bacteria over long distances in soil (Banitz et al., 2013). Fungal mycelium also increases the bioavailability of hydrocarbons in soil (Harms et al., 2011).

10.4 ENZYMATIC DEGRADATION BY FUNGAL SPECIES

Two broad groups of fungi called ligninolytic and non-ligninolytic are shown to be involved in the breakdown and restoration of PHs and PAHs contaminated sites. Ligninolytic fungi mainly include basidiomycetes that produce enzymes necessary to degrade lignin present in woods. Basidiomycetes are also known as mushroom-forming fungi and they are potent natural decomposers due to their rapid growth and large biomass production (Chakraborty et al., 2023). Few examples of PHs and PAHs degrading basidiomycetes are *Agaricus bisporus* or button mushrooms, *P. ostreatus* or oyster mushrooms, *Pleurotus pulmonarius* or Indian/Italian/Lung oyster or Phoenix mushroom, *Trametes versicolor* or *Coriolus versicolor* also known as Turkey tail or white-rot fungus. White-rot fungi are a heterogeneous group of ligninolytic basidiomycetes that are involved in degrading soil contaminated with hydrocarbons (Isikhuemhen et al., 2012). These include species such as *Phanerochaete, Pleurotus* and *Trametes* (Mancera-Lopez et al., 2008). They produce cytochrome P450 which can oxidize both low and high molecular weight PAHs (Masaphy et al., 1996). Three types of cytochrome P450 genes CYP63A1,

A2 and A3 are found which code for cytochrome P450. Expression of these genes depends upon availability of nutrients, culture conditions and presence of inducer molecules as substrate. The ligninolytic white-rot fungi also produce three key enzymes known as lignin peroxidases (LiP), manganese peroxidases (MnP) and laccases which can oxidize PAHs into PAH diphenols which can be oxidized to quinones spontaneously. LiP oxidize PAHs in the presence of H_2O_2, MnP catalyze Mn-dependent peroxidation of unsaturated lipids and oxidizes PAHs, and laccase, a copper-containing enzyme, oxidizes PAHs in the presence of mediators like phenol, cysteine, reduced glutathione, *etc.* (Johannes & Majcherczyk, 2000; Bogan et al., 1996; Hammel et al., 1986). Fungi, such as *Phanerochaete chrysosporium*, *Lentinus squarrosulus* and *Pleurotus tuber-regium*, are also used in the mycoremediation of environmental PAHs (Chakraborty et al., 2023). Non-ligninolytic fungi include zygomycetes, ascomycetes, hypomycetes, deuteromycetes, *etc.* They cannot produce enzymes but are well-known PAH degraders. Non-ligninolytic fungi *Cunninghamella elegans* produce cytochrome P450 and are used to degrade PAHs with two to five aromatic carbon rings. Some fungi may cleave the PAHs rings, produce CO_2 and some fungi are able to use PAHs as a carbon source. Other fungal species such as two species of *Aspergillus* called *Aspergillus niger* and *Aspergillus flavus*, *Varicosporium elodeae* and *Cochliobolus lunatus* have been shown to effectively degrade crude oil–contaminated soil (Gaur et al., 2022). Hence, mycoremediation essentially utilizes the inherent decomposition ability of fungi to rebuild and replenish the ecosystem (Figure 10.3).

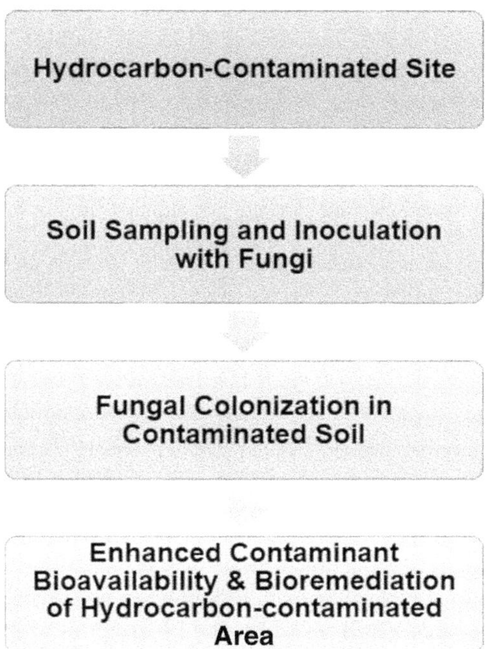

FIGURE 10.3 Steps involved in fungal bioremediation of hydrocarbon-contaminated sites.

10.5 TECHNOLOGICAL POTENTIAL OF FUNGI AS A BIOREMEDIATION AGENT

Bioremediation of hydrocarbon-contaminated soil using fungi shows the highest potential for technological developments. We assess the technological readiness in the context of fungal bioremediation of hydrocarbon-contaminated sites to see whether fungi actually exhibit enough potential to rise from laboratory research to wide-scale field applications. Recently, techniques based on fungal reclamation of hydrocarbon-contaminated sites are emerging in countries like China, the USA, Japan, India and the United Kingdom. China has registered the highest number of patents on fungal bioremediation followed by the USA. Active research as well as patent filings on fungi-based bioremediation techniques have increased equally in percentage, and the number of such patents is increasing annually. In the last decade, around 36% of patents filed on fungal-remediation platforms were granted approval, and 28% of them awaited approval (Quintella et al., 2019). Many of these granted patent technologies are available in the public domain for worldwide use. Research focusing on fungal bioremediation of hydrocarbon-contaminated sites has gained momentum, and interestingly, they have become the second most preferred choice for soil bioremediation after bacteria. Fungal research has helped in gaining useful insights into developing new technologies pertaining to hydrocarbon decontamination. Ever-increasing exploitation of hydrocarbon sources and high incidences of accidental mishaps have caused considerable soil and ecological deterioration. Therefore, green technologies are urgently needed to effectively remediate hydrocarbon-polluted soil. Fungi are ubiquitous and resilient keystone organisms with immense potential to neutralize hydrocarbon moieties and produce huge amounts of biomass; therefore, advanced technologies aimed toward utilizing the inherent capability of fungi and degrade such persistent contaminants must be developed to reclaim the polluted environment in a sustainable manner (Steffen & Tuomela, 2011).

A number of previously accepted patented technologies have shown the use of white-rot fungal spores to effectively remediate soil contaminated with PAHs (Cheng et al., 2015). The most common types of fungi used in patented technologies include species such as *Acremonium, Penicillium, Phanerochaete, Aspergillus, Chrysosporium, etc.* (Quintella et al., 2019). Among these, *Penicillium* and *Aspergillus* are the most commonly available and widely known fungal species which are extensively used in the bioremediation of crude oil–contaminated sites. A multipurpose fungal delivery system has been developed by Cisco Technology which ensures fungal spore delivery for various purposes such as ecological rehabilitation, mycoremediation, agricultural yield improvement, *etc.* In another Chinese invention, a solid fungal preparation was made containing a mix of white-rot fungal species and other common fungi to degrade PHs (Li et al., 2001).

Fungi can be used in both *in-situ* and *ex-situ* bioremediation processes. In *in-situ* strategy, fungi are introduced in contaminated sites where they remediate hydrocarbons. Various inventions aimed at *in-situ* bioremediation of petroleum-contaminated soil were developed which utilize a complex of bacterial and fungal consortia to promote biomass conversion. *Phanerochaete chrysosporium* was used as the predominant fungus in combination with bacterial species such as *Bacillus subtilis,*

Pseudomonas, etc. (Liu et al., 2007a). *Ex-situ* mycoremediation is carried out in an artificial environment such as bioreactors where physicochemical conditions are tightly regulated and microorganism growth is facilitated. In *ex-situ* methods too, a combination of bacterial and fungal species is used to achieve bioremediation (Jia et al., 2013; Liu et al., 2007b). Composite inoculants have also been used which comprise fungal and bacterial species such as *Aspergillus, Trichoderma, Penicillium, Bacillus acidophaeus* in optimized proportions, and such composites successfully remediate deep-oil–contaminated soil (Zhang, 2016; Kanungo et al., 2023).

10.6 BIOAUGMENTATION AND BIOSTIMULATION

Two enhancement techniques called bioaugmentation and biostimulation are frequently used in the *in-situ* mycoremediation process to accelerate hydrocarbon degradation. In bioaugmentation, inoculation of preadapted or genetically modified single bacterial or fungal strain or consortia of microorganisms equipped to perform specific degradation processes is introduced in hydrocarbon-contaminated soil, in order to enhance the biodegradation process. Bioaugmentation is usually performed when the native soil community despite having a large population is unable to break down high molecular weight hydrocarbons and is unable to emulsify such insoluble contaminants (Mancera-Lopez et al., 2008; D'Annibale et al., 2006). However, competition from indigenous autochthonous species and varying physicochemical soil properties pose major challenges to the efficacy of bioaugmentation. Such concerns are addressed by administering an inoculum of autochthonous fungi or microbial strain or their consortia which are site-specific to enhance the efficiency of bioaugmentation (Dai et al., 2023). Several bioaugmentation studies have been performed using filamentous fungi which can easily degrade low molecular weight hydrocarbons. On the contrary, degradation of complex hydrocarbon structures is not feasible by a few strains, so they require consortia of both bacterial and fungal species to show high success rates (Boonchan et al., 2000). Reinoculation of soil contaminated with high levels of crude oil with pure fungal culture containing *Acremonium* species and fungal–bacterial consortia containing *Acremonium–B. subtilis* has been shown to be effective in remediating heavily polluted soil (Ma et al., 2015).

In another enhancement technique called biostimulation, the soil is enriched with nutrients and growth-promoting factors, which leads to the activation of native microorganism population in soil, and thus stimulates hydrocarbon remediation. Studies have been performed using filamentous fungi strains such as *Aspergillus sydowii*, *Penicillium funiculosum* and *Rhizopus* isolated from aged hydrocarbon-contaminated soil. They have shown that these fungi are able to degrade a much higher percentage of soil hydrocarbons using bioaugmentation and biostimulation strategies (Mancera-Lopez et al., 2008).

10.7 CONCLUSION

Numerous exciting evidences suggest that fungi are incredibly capable of cleaning up soil which are densely contaminated with either crude oil, refined petroleum products, PAHs or any other type of hydrocarbons. Such immense potential of fungi is highly encouraging for the field of soil remediation. We ought to think about employing fungal remediation in cases where mycelium is easily accessible and inexpensive, such as waste from mushroom-rearing facilities, and consider restoration of local contaminated sites. In order to justify the wider adoption of mycoremediation technologies, further research is necessary as fungal remediation is still in its nascent stages. Several nonvolatile contaminants can serve as putative candidates for mycoremediation. Fungal technologies provide realistic opportunities to save soil from further degradation. The natural ability of fungi to break down a wide range of toxic compounds provides us with an excellent tool to replenish areas that are majorly destroyed in lieu of expanding industrialization and urbanization. We should step forward and aim toward developing safe and eco-friendly bioremediation alternatives like mycoremediation technologies to repair the damage that has already been afflicted to the environment.

REFERENCES

Anyakora, C. (Ed.) (2007). *Environmental Impact of Polynuclear Aromatic Hydrocarbons*. Kerala: Research Signpost.

Azubuike, C. C., Chikere, C. B., & Okpokwasili, G. C. (2016). Bioremediation techniques-classification based on site of application: principles, advantages, limitations and prospects. *World Journal of Microbiology and Biotechnology*, 32(11). https://doi.org/10.1007/s11274-016-2137-x

Bala, K., Dogra, A., Lata, P., & Das, A. P. (2024). Discussing sources and biological-chemical remediation approaches for microplastic pollution. In *Sustainable Microbial Technology for Synthetic and Cellulosic Microfiber Bioremediation*, 37–59. https://doi.org/10.1007/978-3-031-62600-5_3

Banitz, T., Johst, K., Wick, L. Y., Schamfub, S., Harms, H., & Frank, K. (2013). Highways versus pipelines: contributions of two fungal transport mechanisms to efficient bioremediation. *Environmental Microbiology Reports*, 5(2). https://doi.org/10.1111/1758-2229.12002

Behera, I. D., Basak, G., Kumar, R. R., Sen, R., & Meikap, B. C. (2020). Treatment of petroleum refinery sludge by petroleum degrading bacterium Stenotrophomonas pavanii IRB19 as an efficient novel technology. *Journal of Environmental Science and Health, Part A*, 1–13. https://doi.org/10.1080/10934529.2020.1866924

Behera, I. D., Nayak, M., Biswas, S., Meikap, B. C., & Sen, R. (2021). Enhanced biodegradation of total petroleum hydrocarbons by implementing a novel two-step bioaugmentation strategy using indigenous bacterial consortium. *Journal of Environmental Management*, 292, 112746. https://doi.org/10.1016/j.jenvman.2021.112746

Behera, I. D., Nayak, M., Mishra, A., Meikap, B. C., & Sen, R. (2022). Strategic implementation of integrated bioaugmentation and biostimulation for efficient mitigation of petroleum hydrocarbon pollutants from terrestrial and aquatic environment. *Marine Pollution Bulletin*, 177, 113492. https://doi.org/10.1016/j.marpolbul.2022.113492.

Bogan, B. W., Lamar, R. T., & Hammel, K. E. (1996). Fluorene oxidation in vivo by Phanerochaete chrysosporium and in vitro during manganese peroxidase dependent lipid peroxidation. *Applied and Environmental Microbiology*, 62(5). https://doi.org/10.1128/aem.62.5.1788-1792.1996

Boonchan, S., Britz, M. L., & Stanley, G. A. (2000). Degradation and mineralization of high molecular-weight polycyclic aromatic hydrocarbons by defined fungal–bacterial cocultures. *Applied and Environmental Microbiology*, 66(3). https://doi.org/10.1128/AEM.66.3.1007-1019.2000

Chakraborty, A., Mandal, S., Mandal, K., Das, D., Bose, S. K., Laha, A., & Ghosh, S. (2023). Potential approach of mushrooms in bioremediation: A short review. *Journal of Advanced Zoology*, 44(S5). https://doi.org/10.53555/jaz.v44iS5.3214

Chang, W., Akbari, A., Snelgrove, J., Frigon, D., & Ghoshal, S. 2013. Biodegradation of petroleum hydrocarbons in contaminated clayey soils from a sub-arctic site: the role of aggregate size and microstructure. *Chemosphere*, 91(11). https://doi.org/10.1016/j.chemosphere.2012.12.058

Cheng, M., Lai, C., Ceng, G., Huang, D., Liu, Y., Xu, P., Zhang, C., Zhu, Y., & Huang, C. Inventors. Hunan University. Assignee. (2015). Biological Remediation Method for Polycyclic Aromatic Hydrocarbon and Heavy Metal Compound Contaminated Soil. China Patent CN105013815.

Covino, S., Stella, T., D'Annibale, A., Llado, S., Baldrian, P., Cvancarova, M., Cajthaml, T., & Petruccioli, M. (2016). Comparative assessment of fungal augmentation treatments of a fine-textured and historically oil contaminated soil. *Science of the Total Environment*, 566–567. https://doi.org/10.1016/j.scitotenv.2016.05.018

D'Annibale, A., Rosetto, F., Leonardi, V., Federici, F., & Petruccioli, M. (2006). Role of autochthonous filamentous fungi in bioremediation of a soil historically contaminated with aromatic hydrocarbons. *Applied and Environmental Microbiology*, 72(1). https://doi.org/10.1128/AEM.72.1.28-36.2006

Dacco, C., Girometta, C., Asemoloye, M. D., Carpani, G., Picco, A. M., & Tosi, S. (2020). Key fungal degradation patterns, enzymes and their applications for the removal of aliphatic hydrocarbons in polluted soils: a review. *International Biodeterioration & Biodegradation*, 147. https://doi.org/10.1016/j.ibiod.2019.104866

Dai, Y., Li, J., Yang, X., Wang, S., Zhao, X., Wang, Y., Zhang, D., Luo, C., & Zhang, G. (2023). New insight into the mechanisms of autochthonous fungal bioaugmentation of phenanthrene in petroleum contaminated soil by stable isotope probing. *Journal of Hazardous Materials*, 452. https://doi.org/10.1016/j.jhazmat.2023.131271

Das, A., Dey, S., Bala, K., Gnanasekaran, R., & Das, A. P. (2024). Microbial bioremediation technology for sustainable treatment and management of synthetic microfiber waste. In *Sustainable Microbial Technology for Synthetic and Cellulosic Microfiber Bioremediation*, 221–238. https://doi.org/10.1007/978-3-031-62600-5_12

Das, A., Ray, S. K., Mohanty, M., Mohanty, J., Dey, S., & Das, A. P. (2024). Ecotoxicity of microplastic wastes and their sustainable management: a review. *Environmental Chemistry and Ecotoxicology*, 6, 144–152. https://doi.org/10.1016/J.ENCECO.2024.05.003

Das, A. P., Sukla, L. B., Pradhan, N., & Nayak, S. (2011). Manganese biomining: a review. *Bioresource Technology*, 102(16), 7381–7387. https://doi.org/10.1016/J.BIORTECH.2011.05.018

Dey, S., Bissoyi, A., Koteswara Reddy, G., & Das, A. P. (2024). Synthetic microfibers: an urban emerging pollutant demanding for sustainable environmental solutions. In *Sustainable Microbial Technology for Synthetic and Cellulosic Microfiber Bioremediation*, 107–121. https://doi.org/10.1007/978-3-031-62600-5_7

Dey, S., Das, A., Mallick, K., Sahu, A., & Das, A. P. (2023). Environmental petroleum waste: pollution, toxicity, sustainable remediation. 159–175. https://doi.org/10.1007/978-3-031-48220-5_7

Dey, S., Dash, N., Ray, S. K., Priyadarsini, S., Chhotaray, C., & Das, A. P. (2024). Occurrence and quantification of tire wear plastics pollutants from highway city dust of Bhubaneswar City of Odisha. In *Sustainable Microbial Technology for Synthetic and Cellulosic Microfiber Bioremediation*, 77–89. https://doi.org/10.1007/978-3-031-62600-5_5

Federici, E., Leonardi, V., Giubilei, M. A., Quaratino, D., Spaccapelo, R., D'Annibale, A., & Petruccioli, M. (2007). Addition of allochthonous fungi to a historically contaminated soil affects both remediation efficiency and bacterial diversity. *Applied Microbiology and Biotechnology*, 77(1). https://doi.org/10.1007/s00253-007-1143-1

Gadd, G. M. (2017). Geomicrobiology of the built environment. *Nature Microbiology*, 2(4). https://doi.org/10.1038/nmicrobiol.2016.275

Gaur, N., Flora, G., Yadav, M., & Tiwari, A. (2014). A review with recent advancements on bioremediation-based abolition of heavy metals. *Environmental Science: Processes & Impacts*, 16(2). https://doi.org/10.1039/c3em00491k

Gaur, V. K., Tripathi, V., & Manickam, N. (2022). Chapter 18 - Bacterial- and fungal-mediated biodegradation of petroleum hydrocarbons in soil. In *Development in Wastewater Treatment Research and Processes*. Elsevier. https://doi.org/10.1016/B978-0-323-85839-7.00008-6

Hammel, K. E., Kalyanaraman, B., & Kirk, T. K. (1986). Oxidation of polycyclic aromatic hydrocarbons and dibenzo[p]dioxins by Phanerochaete chrysosporium ligninase. 261(36). https://doi.org/10.1016/S0021-9258(19)75982-1

Harms, H., Schlosser, D., & Wick, L. Y. (2011). Untapped potential: exploiting fungi in bioremediation of hazardous chemicals. *The Journal of Bilological Chemistry*, 9. https://doi.org/10.1038/nrmicro2519

Hussar, E., Richards, S., Lin, Z. Q., Dixon, R. P., & Johnson, K. A. (2012). Human health risk assessment of 16 priority polycyclic aromatic hydrocarbons in soils of Chattanooga, Tennessee, USA. *Water, Air and Soil Pollution,* 223(9). https://doi.org/10.1007/s11270-012-1265-7

Isikhuemhen, O. S., Mikiashvili, N. A., Adenipekun, C. O., Ohimain, E. I., & Shahbazi, G. (2012). The tropical white rot fungus, Lentinus squarrosulus Mont.: lignocellulolytic enzymes activities and sugar release from cornstalks under solid state fermentation. *World Journal of Microbiology and Biotechnology*, 28(5). https://doi.org/10.1007/s11274-011-0998-6

Jia, H., Song, X., Yan, S., Qu, L., & Wang, Q. Inventors. Northeast Forestry University. Assignee. (2013). Electrically-assistant Microbial Remediation Method of Oil Polluted Soil. China patent CN104056857B.

Johannes, C., & Majcherczyk, A. (2000). Natural mediators in the oxidation of polycyclic aromatic hydrocarbons by laccase mediator systems. *Applied and Environmental Microbiology*, 66(2). https://doi.org/10.1128/AEM.66.2.524-528.2000

Kadri, T., Rouissi, T., Brar, S. K., Cledon, M., Sarma, S., & Verma, M. (2017). Biodegradation of polycyclic aromatic hydrocarbons (PAHs) by fungal enzymes: a review. *Journal of Environmental Sciences*, 51. https://doi.org/10.1016/j.jes.2016.08.023

Kanungo, J., Sahoo, S., Bal, M., & Behera, I. D. (2023). Performance of bioremediation strategy in waste lubricating oil pollutants: A review. *Geomicrobiology Journal*, 41(4). https://doi.org/10.1080/01490451.2023.2245395

Khan, I., Aftab, M., Shakir, S., Ali, M., Qayyum, S., Rehman, M. U., Haleem, K. S., & Touseef, I. (2019). Mycoremediation of heavy metal (Cd and Cr)–polluted soil through indigenous metallotolerant fungal isolates. *Environmental Monitoring and Assessment*, 191(9). https://doi.org/10.1007/s10661-019-7769-5

Lahiri, D., Nag, M., Dey, A., Sarkar, T., Joshi, S., Pandit, S., Das, A. P., Pati, S., Pattanaik, S., Tilak, V. K., & Ray, R. R. (2021). Biofilm mediated degradation of petroleum products. *Geomicrobiology Journal*, 1–10. https://doi.org/10.1080/01490451.2021.1968979

Li, P., Zhang, H., & Yao, D. Inventors. Institute of Applied Ecology, Chinese Academy of Sciences. Assignees. (2001). Process for Preparing Solid Fungus to Degrade Petroleum. China patent CN1382758.

Liu, Z., Zhang, K., Xu, Y., Hua, X., & Zhang, M. Inventors. China Petroleum & Chemical and Tsinghua University. Assignees. (2007a). In-situ Biological Repairing Method for Biomass Intensified Petroleum Contaminative Soil. China patent CN101104177B.

Liu, Z., Zhang, K., Xu, Y., Hua, X., & Zhang, M. Inventors. China Petroleum & Chemical. Assignee. (2007b). Biology in Situ Renovation Method for Biomass Strengthen Mix Contaminated Soil of Stone Oil-Salt. China patent CN101172286.

Ma, X. K., Ding, N., & Peterson, E. C. (2015). Bioaugmentation of soil contaminated with high-level crude oil through inoculation with mixed cultures including Acremonium sp. *Biodegradation*, 26. https://doi.org/10.1007/s10532-015-9732-7

Mancera-Lopez, M. E., Esparza-Garcia, F., Chavez-Gomez, B., Rodriguez-Vazquez, R., Saucedo-Castaneda, G., & Barrera-Cortes, J. (2008). Bioremediation of an aged hydrocarbon-contaminated soil by a combined system of biostimulation–bioaugmentation with filamentous fungi. *International Biodeterioration & Biodegradation*, 61(2). https://doi.org/10.1016/j.ibiod.2007.05.012

Masaphy, S., Levanon, D., Henis, Y., Venkateswarlu, K., & Kelly, S. L. (1996). Evidence for cytochrome P-450 and P450-mediated benzo[a]pyrene hydroxylation in the white rot fungus Phanerochaete chrysosporium. *FEMS Microbiology Letters*, 135(1). https://doi.org/10.1111/j.1574-6968.1996.tb07965.x

Mastral, A. M., & Callen, M. S. (2000). A review on polycyclic aromatic hydrocarbon (PAH) emissions from energy generation. *Environmental Science & Technology*, 34. https://doi.org/10.1021/es001028d

Mishra, S., Rath, C. C., & Das, A. P. (2019). Marine microfiber pollution: a review on present status and future challenges. *Marine Pollution Bulletin*, 140, 188–197. https://doi.org/10.1016/J.MARPOLBUL.2019.01.039

Mishra, S., Rout, P. K., & Das, A. P. (2021). Emerging microfiber pollution and its remediation. In R. Prasad (Ed.), *Environmental and Microbial Biotechnology*, 247–266. Springer, Singapore. https://doi.org/10.1007/978-981-15-5499-5_9

Mohanta, S., Pradhan, B., & Behera, I. D. (2023). Impact and remediation of petroleum hydrocarbon pollutants on agricultural land: a review. *Geomicrobiology Journal*. 2023 Aug 10:1–5. https://doi.org/10.1080/01490451.2023.2243925

Naseri, M., Barabadi, A., & Barabady, J. (2014). Bioremediation treatment of hydrocarbon-contaminated Arctic soils: influencing parameters. *Environmental Science and Pollution Research*, 21(19). https://doi.org/10.1007/s11356-014-3122-2

Olivella, S., & Gens, A. 2000. Vapour transport in low permeability unsaturated soils with capillary effects. *Transport in Porous Media*, 40. https://doi.org/10.1023/A:1006749505937

Quintella, M. C., Mata, M. T., & Lima, L. C. P. (2019). Overview of bioremediation with technology assessment and emphasis on fungal bioremediation of oil contaminated soils. *Journal of Environmental Management*, 241. https://doi.org/10.1016/j.jenvman.2019.04.019

Raj, K., & Das, A. P. (2023). Lead pollution: Impact on environment and human health and approach for a sustainable solution. *Environmental Chemistry and Ecotoxicology*, *5*. https://doi.org/10.1016/J.ENCECO.2023.02.001

Ramadass, K., Smith, E., Palanisami, T., Mathieson, G., Srivastava, P., Megharaj, M., & Naidu, R. (2015). Evaluation of constraints in bioremediation of weathered hydrocarbon-contaminated arid soils through microcosm biopile study. *International Journal of Environmental Science and Technology*, 12. https://doi.org/10.1007/s13762-015-0793-2

Rengarajan, T., Rajendran, P., Nandakumar, N., Lokeshkumar, B., Rajendran, P., & Nishigaki, I. (2015). Exposure to polycyclic aromatic hydrocarbons with special focus on cancer. *Asian Pacific Journal of Tropical Biomedicine*, 5(3). https://doi.org/10.1016/S2221-1691(15)30003-4

Safdari, M. S., Kariminia, H. R., Rahmati, M., Fazlollahi, F., Polasko, A., Mahendra, S., Wilding, W. V., & Fletcher, T. H. (2018). Development of bioreactors for comparative study of natural attenuation, biostimulation, and bioaugmentation of petroleum-hydrocarbon contaminated soil. *Journal of Hazardous Materials*, 342. https://doi.org/10.1016/j.jhazmat.2017.08.044

Singh, R. K., Tripathi, R., Ranjan, A., & Srivastava, A. K. (2020). Fungi as potential candidates for bioremediation. In *Abatement of Environmental Pollutants Trends and Strategies*, 177-191. https://doi.org/10.1016/B978-0-12-818095-2.00009-6

Srivastava, M., Srivastava, A., Yadav, A., & Rawat, V. (2019). Source and control of hydrocarbon pollution. IntechOpen. https://doi.org/10.5772/intechopen.86487

Steffen, K., & Tuomela, M. (2011). Fungal soil bioremediation: developments towards largescale applications. Springer. https://doi.org/10.1007/978-3-642-11458-8_22

Varjani, S. J. (2017). Microbial degradation of petroleum hydrocarbons. *Bioresource Technology*, 223. https://doi.org/10.1016/j.biortech.2016.10.037

Wick, L. Y., Remer, R., Wurz, B., Reichenbach, J., Braun, S., Schafer, F., & Harms, H. (2007). Effect of fungal hyphae on the access of bacteria to phenanthrene in soil. *Environmental Science & Technology*, 41(2). https://doi.org/10.1021/es061407s

Wosten, H. A., Wetter, M. A. V., Lugones, L. G., van der Mei, H. C., Busscher, H. J., & Wessels, J. G. H. (1999). How a fungus escapes the water to grow into the air. *Brief Communications*, 9(2). https://doi.org/10.1016/S0960-9822(99)80019-0

Xu, N., Bao, M., Sun, P., & Li, Y. (2013). Study on bioadsorption and biodegradation of petroleum hydrocarbons by a microbial consortium. *Bioresource Technology*, 149. https://doi.org/10.1016/j.biortech.2013.09.024

Yadav, R., Singh, S., Kumar, A., & Singh, A. N. (2022). Chapter 15 - Phytoremediation: a wonderful cost-effective tool. In *Advances in Environmental Pollution Research, Cost Effective Technologies for Solid Waste and Wastewater Treatment*. Elsevier. https://doi.org/10.1016/B978-0-12-822933-0.00008-5

Zhang, F., Inventor. XiAn Huanuo Environment Protection. Assignee. (2016). Composite Inoculant for Deep Oil Contaminated Soil Bioremediation Process. China patent CN106244479.

Zhang, S., Gedalanga, P. B., & Mahendra, S. (2017). Advances in bioremediation of 1, 4-dioxane-contaminated waters. *Journal of Environmental Management*, 204(Pt 2). https://doi.org/10.1016/j.jenvman.2017.05.033

11 Advances in Microbial Bioremediation Strategies for Petroleum Hydrocarbon-Contaminated Environments

Abhinay Thakur and Ashish Kumar

11.1 INTRODUCTION

Petroleum hydrocarbons refer to a broad class of nonionic organic chemicals that are obtained from crude oil and as we know crude oil is formed from the remains of marine plants and animals over millions of years. These hydrocarbons differ in the arrangement of carbon and the number of carbon atoms: the two general types of hydrocarbon are aliphatic – the carbon atoms are arranged in linear or branched chain fashion and aromatic hydrocarbon the carbon atoms are arranged in cyclic or ring structures (Ghorbanian et al., 2014; Ivshina et al., 2016; Salgado et al., 2019). The chemical composition of petroleum hydrocarbon defines the physical characteristics of the substance, for instance, volatility, solubility and stable resistance in any environment. For instance, methane and ethane that have lighter molecular weights are gaseous at standard conditions and very mobile while the heavy compounds, specifically bitumen or tar, are mobile and exist in the soils and sediments (Dragoi et al., 2021; Xia et al., 2020). Paraffinic and olefinic compounds on average are less active and more biodegradable than average aromatic hydrocarbons. Benzene, toluene, ethylbenzene and xylene (BTEX) are more chemically stable and sometimes more toxic than aliphatic HAPs. Polycyclic aromatic hydrocarbons (PAHs) which include two or even more fused-aromatic rings are most recalcitrant and are often problematic in bioremediation because they are hard to break down by microbes and because they bioconcentrate in tissues (Behera et al., 2021; Das, Nayak, et al., 2024; Mishra et al., 2021). Some of the petroleum hydrocarbons are persistent and toxic hence making significant impacts with regard to their ability to pollute the environment and their effects on ecosystems and the health of man. Petroleum hydrocarbons have wide applications in the commercial and industrial sectors: as fuel – gasoline, diesel, kerosene; as a feedstock for producing plastics, chemicals, lubricants and many others (Alsharyani

DOI: 10.1201/9781003508496-11

& Muruganandam, 2024; Bhuyan et al., 2023; Lee, Cao, et al., 2019; Ugochukwu et al., 2018). Consequently, there is a likelihood for pollution throughout the petroleum life span including extraction, processing, distribution, stockpiling and dumping (Kanungo, Sahoo, Bal, et al., 2023; Kanungo, Sahoo, Swain, et al., 2023). This implies that remediation of any area depends on the possible petroleum hydrocarbons present which is a diversified list, and each class of hydrocarbons behaves differently in the environment as well as requires a different approach when it comes to remediation (Paul et al., 2024; Sahu, Dey, et al., 2024). Petroleum hydrocarbon contamination comes from a number of sources, planned and unplanned. The principal causes or incidents include oil slick, pipeline leak, tank burst and unintentional release of petroleum products in handling or transportation. Accidents from offshore oil drilling and tanker shipping have attracted huge concern because they cause major pollution in the marine environment. These events spew large volumes of hydrocarbons into the marine environment, shorelines, sea life and marine/benthic habitats, sometimes within the span of years to decades (Dey, Dash, Ray, et al., 2024; Mallick et al., 2023; Mohanty et al., 2024; Sahoo et al., 2022). Pipelines and underground storage tanks are other major sources of petroleum hydrocarbon pollution from leaking pipes. Since pipelines are used for carrying crude oil and different types of products over great lengths, there are instances that pipelines have defects and the spillage of crude and processed products into the ground and water sources are unavoidable. Underground storage tanks such as those used to hold fuel in service stations are usually coupled with two major problems: corrosion and leakage that makes hydrocarbons percolate into the subsurface and pollute drinking water. Other causes of hydrocarbon pollution include: the discharge of hydrocarbons by industrial facilities and refineries and by urban runoff (Qv et al., 2024; Wang et al., 2010). Petroleum-related industries, for instance, refineries, are sites for potential hydrocarbon emission since they involve the processing of crude oil with different products which inconceivably produce waste products that contain hydrocarbons and oily sludge among others. Auto emissions also become a problem with city runoff and streets; they leak oil in areas of traffic, parking lots and garages that enter water structures such as rivers, lakes and seashores. In addition, the absorption of used motor oils, lubricants and industrial solvents releases hydrocarbons into the environment (Dey, Bissoyi, et al., 2024; John & Das, 2024). The socially irresponsible management of used oils, a product mixed with hydrocarbons and additives, often results in improper disposal through draining, landfilling or dumping on open lands; hence polluting the soil and water (Koolivand et al., 2020; Mirjani et al., 2021). Oily wastes are deliberately dumped into the environment by industries or individuals interested in quick wealth through the sale of the waste, since they are cheap and can be produced in large quantities without the stringent legal backing on environmental waste management as there in developed countries. Hydrocarbon pollution is attributed to upstream oil extraction activities, especially in offshore facilities (Behera et al., 2020; Dey et al., 2023; Lahiri et al., 2021). Disposed in the daily operational activities, produced water that is used in the process of extraction of oils on the surface contains dissolved and dispersed hydrocarbons which are relentlessly discharged and spread out into the marine regions. Further, during drilling, there are cases of blowouts or release of oil which causes a great pollutive impact on ecology as well as on the economy. Figure 11.1 shows the several types of hydrocarbon sources in the environment.

FIGURE 11.1 Sources of hydrocarbons.

Petroleum hydrocarbons form a diverse family of chemicals that are discharged into various environments and undergo a series of changes in their physical, chemical and biological characteristics that define their environmental existence (Koolivand et al., 2019; Li et al., 2024). Several processes affect the behavior of hydrocarbons in the environment including the chemical structure of the hydrocarbons, the molecular weight of the hydrocarbon and other environmental factors such as temperature, hydrogen ion concentration and availability of oxygen or other electron acceptors (Das et al., 2015; Jagatee et al., 2024; Nag et al., 2023; Sahu, Mallick, et al., 2024). Knowledge of these processes is important for modeling the severity of contamination, evaluating threats and designing potential cleanup methods. Walker hydrocarbons in the atmosphere include methane and propane in that the compounds dissolve within the atmosphere when dispersed they cause air pollution and the formation of ground ozone levels. Nevertheless, they are severely consumed in other photochemical reactions and their major consumption has been attributed to the hydroxyl radical. The fate of the heavier hydrocarbons, however, is governed by factors such as solubility, adsorption to particles and microbial activity in soil and water (Lamy et al., 2013; Mangse et al., 2020). Depending on their character, the hydrocarbons can dissolve in water, float on the water's surface, or be adsorbed by suspended particles in the water. BTEX is soluble and can dissolve in the water column, thereby becoming available for uptake by aquatic organisms or to be transported to other locations by river currents or tidal waters. Higher molecular weight hydrocarbons like PAH can be adsorbed into sediments in which they remain stable for long periods affecting bottom-dwelling organisms. Slicks on the water surface can rapidly spread creating a thin film that hinders gas exchange and light transmission to the water making photosynthesizing organisms like algae and water plants.

Petroleum hydrocarbons possess a variable mobility in soil media based on their physical and chemical characteristics. Some members of BTEX may evaporate from the upper layer of soil while others may deposit themselves on the particles within the substrate and stay there. The organic matter in the soil and the content of clay and soil moisture affect the extent to which hydrocarbons are immobilized or mobilized in the soil profile. In coarse-grained soils where water content is relatively high, hydrocarbons may be washed down to the water table, contaminating drinking water. The microbial degradation process is one of the significant factors determining the PAH behavior in soil and aquatic systems (Das, Dey, et al., 2024; Das, Ray, et al., 2024; Das et al., 2023; Raj & Das, 2023). Some bacteria, archaea and fungi are indigenous microorganisms capable of carrying out enzymatic pathways in the process of metabolizing hydrocarbons derived from carbon sources (Fetanat & Tayebi, 2022; Kim et al., 2023). However, the rate of degradation depends upon environmental factors to a large extent. Aerobic degradation processes that require oxygen are usually faster and more efficient than anaerobic or oxygen-starved ones. Under low oxygen conditions such as in waterlogged environments including saturated soils or deep water sediments, rates of hydrocarbon degradation decrease considerably, and advanced microbial communities with novel mechanisms are utilized that can accept other oxidizing agents such as nitrate, sulfate, or iron.

Petroleum hydrocarbons have a significant impact on ecology since soil and water environments are polluted. In the soil environment, the hydrocarbon can change the physical and chemical properties of the soil by reducing the fertility and microbial processes (Das & Ghosh, 2022; Das, Nayak, et al., 2024; Das & Ghosh, 2023). It was also found that hydrocarbons hinder plant growth inasmuch as they limit the nutrients available for plant growth and make the root environment toxic. Hydrocarbons also adversely affect the soil-dwelling organisms – these include earthworms and various insects, the presence of which influences their behavior or, in some cases, reduces their number and thus produces an undesirable impact on the equilibrium of the ecosystem (Barathi et al., 2023; Wei et al., 2022). Thus, the impact of the hydrocarbon on aquatic life is the direct impact which is likely to hit the base of the food chain components, including plankton, algae and benthic invertebrates. Oil tends to suffocate such plants because essential conditions such as oxygen and light to support photosynthesis may be scarce, and when fish and other marine creatures' gills are covered by the oil, they cannot breathe. Long-term effects on health and potential biomagnification of hydrocarbons in the food chain are evident because these substances can also bioaccumulate in the tissues of various aquatic organisms. This can lead to the pollution of commercially profitable fish species, hence a threat to human end consumers. Petroleum hydrocarbons have long-term impacts on the water bottom and the organisms dwelling in it, since hydrocarbons trapped in sediments may take a year or more to diffuse into the water column or be consumed by the bottom dwellers. This will cause long-term contamination for species residing in sediments such as shellfish and crustaceans whose development will be negatively affected by hydrocarbon poisoning. Hydrocarbon contamination is often reflected in the loss of biological species and distortions of ecosystem functions, including nutrient cycling and water purification.

Petroleum hydrocarbons have various effects on the environment and human health below are the effects. Petroleum hydrocarbons also have direct impacts on human health in that they can pollute drinking water supplies or be emitted into the ambient atmosphere as VOCs. Petroleum hydrocarbons can be ingested, inhaled, or absorbed through the skin, and the health impact caused by exposure depends on the type of petroleum hydrocarbon, concentration and duration of exposure (Safdari et al., 2018). Short-term effects of exposure to volatile hydrocarbons like the BTEX are irritation of the respiratory system, dizziness, headache, and in severe cases, effect on the central nervous system. Some examples of the cancer-causing substances found in exposure to these hydrocarbons include benzene, which, if people are exposed to it for long, will increase their risk of leukemia. Microorganisms in contaminated water or food cause gastrointestinal illness, hepatic and renal disorders, as well as other systemic illnesses. Health risks, since the accumulation of petroleum hydrocarbon pollution has potential social and economic consequences, especially to social groups that are dependent on natural resources to support their livelihoods. The adverse effects of oil spillage and other contaminations lead to the loss of fishing and tourism services and agriculture, which have resulted in economic impacts and displacement of people. Hydrocarbon pollution can also impair the availability of clean water, fertile soil and aggressive impacts on environmental quality can pose threats that affect food security and health.

11.2 BIOLOGICAL TREATMENT TECHNOLOGIES: AN OVERVIEW

11.2.1 What Is Bioremediation?

Bioremediation is defined as the process of reintroduction of polluted sites by using biological organisms to degrade or eliminate hazardous compounds from the environment. Bioremediation is the process that originated from the metabolic potential of these organisms, whereby contaminants are used as carbon sources or energy sources. By enzymatic actions, contaminants like hydrocarbons are converted to less toxic or nontoxic substances including water, carbon dioxide and biomass (Guarino et al., 2017; Wang et al., 2021). Bioremediation may be either indigenous or exogenous, based on the magnitude and the extent of the pollution, and the types of existing conditions. It utilizes microbial endurance and growth across different habitats for the elimination of hazardous wastes. In petroleum hydrocarbon-contaminated sites, for example, some bacteria will have the ability to metabolize hydrocarbons, but in the aerobic process, they will use oxygen, while in anaerobic conditions, nitrate or sulfate will be used by the bacteria. Aerobic and anaerobic bioremediation are other advantages of microbes because various forms of bioremediation can be adopted depending on the environment. Secondly, different kinds of fungi and plants also participate in bioremediation, which can involve alteration of the contaminants or adsorption of the pollutants in mycorestoration and phytorestoration, respectively (Okparanma et al., 2022). Bioremediation can happen in any environment and can be classified into four major types depending on environmental circumstances and pollutant type. The popular categories are natural attenuation, bioaugmentation and biostimulation. Natural attenuation uses naturally existing microorganisms in the environment to break chemical pollutants without any addition or intercession. However, when indigenous microorganisms are either scarce or incapable of degrading particular pollutants at the site, bioaugmentation can be used by adding specific, appropriate microorganisms from other sources. On the other hand, bios stimulation entails the introduction of nutrients, oxygen, or any other composites that would stimulate the indigenous microbes (Priyadarsini & Das, 2025). These approaches can be in situ where the contaminated site is treated without the need to move soil and or water or ex situ where contaminated material is transported for treatment. The proposed mechanism for the microbial breakdown of alkanes and aromatic hydrocarbons involves several biochemical pathways that enable microorganisms to utilize these compounds as a source of energy and carbon (Figure 11.2). The process typically starts with the activation of these hydrocarbons through oxygenation or hydroxylation reactions, catalyzed by specific enzymes like monooxygenases and dioxygenases. For alkanes, this usually results in the formation of alcohols, which are further oxidized to aldehydes and eventually to fatty acids. These fatty acids then enter the β-oxidation cycle, a metabolic pathway where they are progressively broken down into acetyl-CoA units, which can be utilized in the Krebs cycle for energy production. In the case of aromatic hydrocarbons, their microbial degradation is more complex due to the stability of the aromatic ring structure. Microorganisms employ specialized enzymes to activate these compounds, often by incorporating of oxygen atoms to destabilize the ring. This leads to the formation of dihydrodiols, which are further converted into catechols or other intermediates. These intermediates undergo ring cleavage through either ortho- or meta-pathways, resulting in simpler molecules that can enter the central metabolism of the microbial cell. This mechanism highlights the versatility and adaptability of microorganisms in degrading various types of hydrocarbons. The ability to metabolize both alkanes and aromatics allows microbes to play a crucial role in bioremediation, particularly in environments contaminated with petroleum-based pollutants.

Bioremediation depends on factors such as the type of pollutants, CD, environmental conditions, including temperature, pH level, DO level and suitable microorganism population. For instance, we have hydrocarbons which are organic compounds that can be decomposed by different types of microorganisms and can therefore be bioremediated. Thus, while all hydrocarbons are taken into account, not all degrade at the same rate; they are relatively slow by virtue of being referred to as PAHs. In such a situation, controlled bioremediation approaches, including, for instance, the application of genetically manipulated microorganisms, may be used to facilitate degradation (Krucoń et al., 2023; Wu et al., 2019). Bioremediation is suitable for most polluted sites, and, therefore, it is categorized as an important method of cleaning contamination in soil, water, sediment and even air. Specifically, the bioremediation technique has been applied in a variety of situations, including the treatment of marine oil spills, and industrial and agricultural soils. Another advantage of bioremediation is that it is environmentally friendly and more economical compared to the traditional methods of remediation, such as incineration, chemical treatment, or excavations.

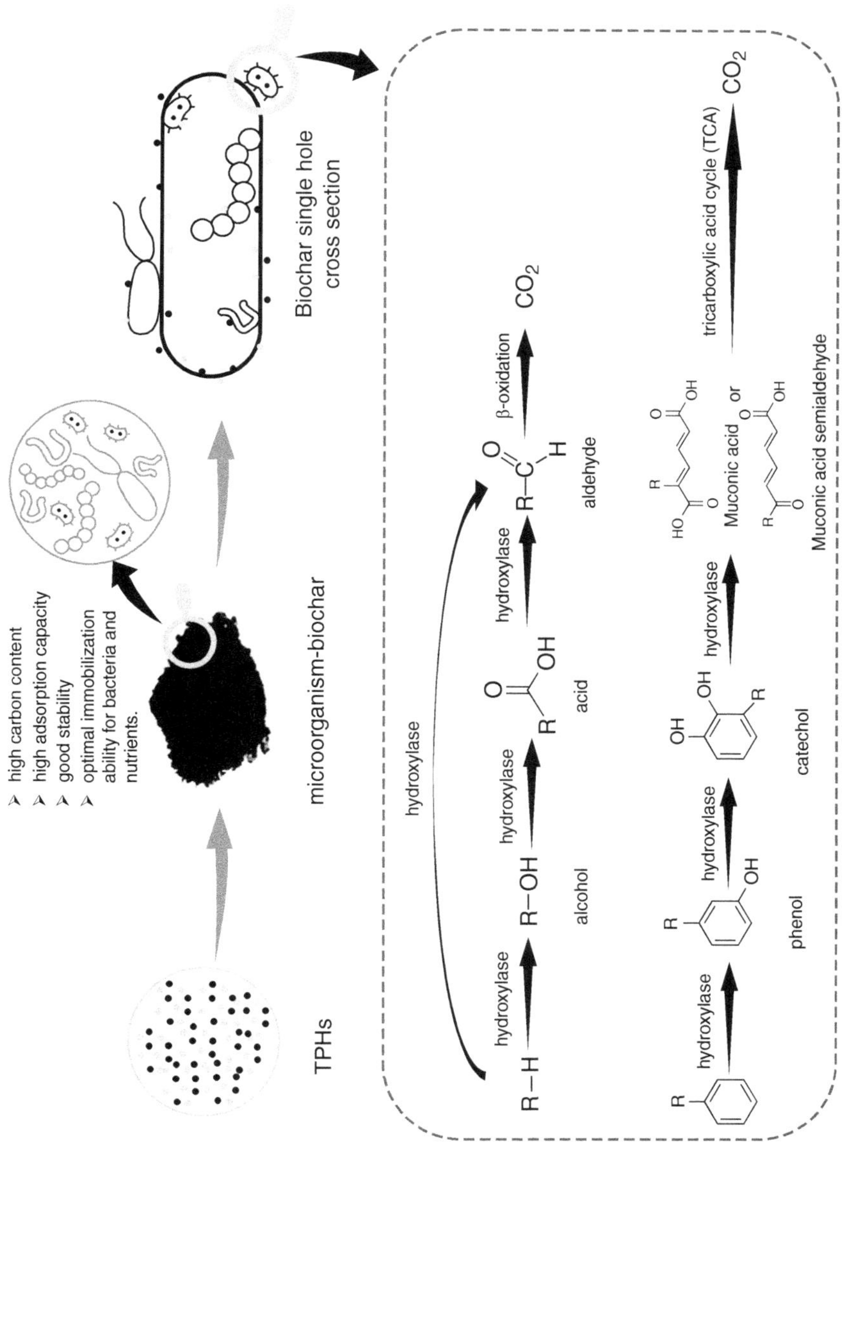

FIGURE 11.2 Shows the microbial degradation mechanism of alkanes and aromatic hydrocarbons.

11.2.2 Advantages of Biological Treatments

It is therefore much dated that using biological treatment technologies, particularly bioremediation, provides several advantages over the other normal physical and chemical removal methods. The first advantage resulting from these methods is environmental conservation. Biological approaches are those that depend on inherent factors that are unlikely to require imports, such as chemicals or energy. This is in contrast to the more drastic treatments like, more so, incineration or chemical oxidation, as they lead to secondary pollution, energy intensity, or toxic formation (Saini et al., 2021; Wang et al., 2024). Bioremediation, on the other hand, is a more effective technique as it transforms pollutants into harmless end products, such as carbon dioxide, water and biomass with negligible effect on the environment. Yet another advantage of biological treatments is their low cost. Some of the prevailing remediation technologies demand a lot of capital on equipment, labor and energy, among other things. In contrast, bioremediation utilizes the inherent biochemical mechanisms of microorganisms, plants and fungi that may be activated or accelerated with modest amounts of relatively inexpensive additions such as nutrients or oxygen (Poi et al., 2018). Bioremediation is generally favorable for large or distant polluted zones because of its low power and facility demands, which may be beyond money-related or physically accomplishable for orthodox arrangements. Biological treatments can also treat materials in situ many times, so there are little or no transportation costs for the contaminated material involved. Biological treatments also offer advantages in flexibility and control as compared to many of the traditional interventions. The beauty of bioremediation technologies is that they are based on living organisms, hence they can be dosed to suit the pollutant and the environment where the contamination will take place. Daily fluctuations in contaminants, temperature, pH, or other conditions may impose a shift in microbial populations, leading to an enhancement of treatment efficiency. This flexibility is particularly useful at contaminated sites that contain a variety of more than one type of contaminant with different characteristics. Conventional techniques may call for the individual removal of each pollutant, whereas biochemical treatments can deal with multiple pollutants in most cases.

Moreover, biological remediation is preferable to most other physicochemical techniques because it is less destructive. Operational activities such as excavation or chemical treatment of contaminated soil, for example, have impacts like disturbing the local ecosystem, displacing people within the area, and producing noise and dust (Al Souki et al., 2021; Zhang et al., 2024). In this sense, compared with ex situ remediation, in situ bioremediation can be completed without disturbing the natural environment; thus, various ecosystems can be restored during the process of pollutants' degradation (Dey, Dash, & Das, 2024; Mishra et al., 2024; Ray et al., 2024). For this reason, biological treatment is ideal for environments that require the protection of the ecosystem, for instance, wetlands, forests, or agricultural lands. Also, the application of the biologische Behandlung method is in line with circular economy principles, especially the recovery of natural systems. For example, bioremediation includes composting and vermiculture, which not only remove pollutants but also provide useful goodies like bioorganic fertilizers which improve soil quality and fertility. This regenerative quality of biological treatments is in stark contrast with the common remediation techniques which in many cases generate other waste by-products that still need to be handled further. Still, it could be noted that like most methodologies, biological treatments are not entirely without their drawbacks. It returns the natural environment to its usage before pollution at a rate that may be slower than in chemical or physical methods especially in cold areas or in an anaerobic phase where microbial action is limited. Additionally, some contaminants, especially heavy metals or certain synthetic compounds, are biodegradable with difficulty and may need to undergo some other treatment as well. However, traditional bioremediation methods have some limitations that prevent their wide application, but constant improvement of new techniques, including genetic engineering and bioreactor systems, helps to solve these problems and to widen the spectrum of pollutants that can be treated by biological methods.

11.2.3 Overview of Conventional and Emerging Technologies

Biological treatment technologies refer to a vast category of conventional and advanced technologies for treating wastewater, of which each with its advantages and disadvantages. Traditional bioremediation methods include natural attenuation, biostimulation and bioaugmentation, and they have been applied for a long time and still give impressive results in most cases. Natural attenuation employs the contaminant biodegradation ability of native microorganisms without adding any other probed treatments (Khudur et al., 2019; Yousaf et al., 2022). This approach is most efficient where the levels of contaminant concentration are fairly low, and inherent microbial activity is high. Nevertheless, natural attenuation is sometimes too slow to help site owners meet the required time frames for site remediation, especially in areas with a high degree of contamination. Biostimulation is a more active type of bioremediation in which selected nutrients, oxygen or some other promoting substances are introduced into the contaminated site to stimulate the growth and level of metabolic activities of indigenous microorganisms. For instance, at petroleum hydrocarbon-contaminated places the use of nitrogen and phosphorus fertilizers boosts the activity of hydrocarbon-utilizing bacteria thus enhancing the degradation of oil products. Biostimulation can also be relatively easy to facilitate and potentially inexpensive; however, proper management and monitoring are needed to avoid capability enhancement of unwanted nontarget microbes or buildup of undesirable levels of biochemical oxygen demand in aquatic systems. Bioaugmentation, another traditional method is the addition of specific microbes, whose ability to degrade the contaminants, to the site. This approach is normally applied where indigenous microbial populations are scarce or deficient in the required metabolic capabilities to assimilate the forecast contaminants. For instance, in sites with chlorinated solvents or Polychlorinated Biphenyls (PCB), bioaugmentation by particular bacteria or fungi will advance the breakdown process (Gielnik et al., 2021; Nawaz et al., 2023). However, bioaugmentation can be problematic in practice, as the inoculated microorganisms rarely become integrated into the local microbial community, which may outcompete them.

Novel technologies of bioremediation revealed in this work are concerned with the extension of bioremediation options and overcoming the existing method drawbacks.

Another developed arm of research largely explored is the use of genetically engineered microorganisms (GEMs) with tailored metabolic efficiency to break down resistance pollutants. GEMs have been designed to depollute difficult hydrocarbon molecules, heavy metals and synthetic chemicals that do not undergo biodegradation or any other natural process (Liu et al., 2019; Yang et al., 2023). As GEMs may pose risks to the environment and may trigger various regulatory concerns, the improvements in synthetic biology and engineering of microbes are making it even possible to apply these organisms for specific bioremediation objectives. Another new innovative invention is related to the use of bioelectrochemical systems (BES) for bioremediation. These systems take advantage of the biochemical pathways of bacterial metabolism to produce electricity as well as reduce contaminants. As in a BES, electrogenic bacteria chemosmotically transfer electrons obtained from the degradation of organic contaminants to an electrode to develop an electric current. It is not only a new way of dealing with contaminated sites but contains the potential for energy recovery as a positive by-product. At this time, BESs are still considered a fledgling technology; however, research into their potential has identified applications in the treatment of wastewater, oil-contaminated land and industrial effluents. Apart from microbial-based technologies, another novel area of bioremediation utilization includes treatment using plants and fungi. Phytoremediation or the process of using plants to remove, degrade and/or immobilize contaminants has received considerable interest due to its prospects of managing diverse types of pollutants including heavy metals, petroleum hydrocarbons and pesticides. Some plants called hyperaccumulators have the tolerance to absorb chemicals from the soil while other plants release chemicals from the roots that stimulate the growth of pollutant-degrading microbes in the root zone. Phytoremediation is cost-effective, nontoxic and enhances the sites' appearance, hence can be used to remediate almost all the urban and agricultural gray-fields. However, it is usually a long-drawn process and may take several years before a significant reduction of contaminant levels can be realized.

Another of the relevant technologies that is still in the process of development is mycoremediation, using fungi to break down pollutants. White-rot fungi form and release strong extracellular enzymes which are able to degrade complicated organic materials, such as lignin, which is chemically related to many hazardous environmental contaminants (Ling et al., 2023; Widera et al., 2024). This enzyme can also break down several impurities, including hydrocarbons, pesticides and dyes. Small-scale field application of mycoremediation has been established to be effective through laboratory experiments and the method still needs to be fine-tuned for efficiency in contaminated sites. Improvement in the bioreactor has also opened up more areas for possibilities in the biological way of tackling pollutants. Ecosystems are favorable for microbial or fungal degradation of pollutants since they offer controlled conditions in bioreactors for degradation compared to in situ methods. The bioreactors can be classified as either aerobic, anaerobic or a hybrid of both and are used in the treatment of industrial wastewaters, contaminated soils and Harvested wood products (HWP). As mentioned earlier, modern designs like the membrane bioreactor and honeycomb-packed bed bioreactor are more efficient and scalable than the traditional designs of the past allowing for large-scale use. Finally, synergistic methods that combine biological with physical or chemical methods are being enhanced as a method of managing the weaknesses of single systems. For instance, integrating bioremediation with chemical oxidation speeds up the degradation of refractory pollutants, integrating phytoremediation with soil vapor extraction for the enhancement of the elimination of volatile organic compounds from contaminated soil. These integrated models provide the prospects for the development of more extensive and efficient strategies for the elimination of contamination, especially in cases where the extent of pollution is rather complicated.

11.3 MICROBIAL DEGRADATION OF PETROLEUM HYDROCARBONS

11.3.1 Microbial Metabolism and Hydrocarbon Degradation Pathways

Bacterial degradation of petroleum compounds consists of a sequence of metabolic reactions in which the organisms degrade complex organic substrates to metabolites that could be used for anabolic synthesis and energy production. The metabolic activities in microorganisms depend on the type of hydrocarbon in question whether it is aliphatic, aromatic or cyclic as well as the conditions of degradation. The degradation of hydrocarbons typically follows two primary metabolic routes: aerobic and anaerobic ways. This way, in aerobic degradation, oxygen acts as the terminal electron acceptor and the widest range of hydrocarbons can be oxidized to CO_2, H_2O and new individual biomass. This one starts with the activation of hydrocarbons through oxygenases; these enzymes fix molecular oxygen in the hydrocarbon molecule. For instance, in the oxidation of alkanes, alkane monooxygenases place an oxygen atom on the carbon–hydrogen bond changing the alkane into an alcohol (Chiu et al., 2017; Lee, Tsang, et al., 2019; Martínez-Rabelo et al., 2023). The alcohol is then oxidized to an aldehyde, which is then converted to a fatty acid which is then metabolized to acetyl-CoA, which forms part of the β oxidation. Further, acetyl-CoA enters the tricarboxylic acid cycle and is completely oxidized to liberate energy present in the ATP molecule. BTEX are also typical of the aerobic degradation process but in these compounds, distinct enzymes are used to begin the attack on the aromatic ring. Monooxygenases and dioxygenases set a hydroxyl group into the ring, which makes catechols or other similar compounds. These intermediate products are then cleaved by such enzymes as catechol 1,2-dioxygenase or catechol 2,3-dioxygenase which split the aromatic compounds into small organic acids that can then be metabolized through the nine-carbon core metabolism and become completely mineralized. Hydrocarbon degradation through anaerobic processes takes place in environments characterized by low oxygen availability, including deep-seabed sediments, anoxic soils or aquifers. In anaerobic degradation, by use of other electron acceptors such as nitrate, sulfate, Fe^{3+} or CO_2, the hydrocarbon is broken down. For instance, nitrate-reducing bacteria may metabolize alkanes and aromatics through denitrification, the process where nitrate is converted into nitrogen gas while the hydrocarbon is oxidized. Sulfate reducing bacteria (SRBs) are not insignificant in aquatic conditions as they have the ability to transform sulfate into an electron acceptor to dissolve hydrocarbons. The degradation process in an anaerobic environment is slower than in an aerobic environment because anaerobic metabolic reactions have lower energy (Liu et al., 2023).

One of the most significant problems in the anaerobic degradation of hydrocarbons is to activate the chemically rather unreactive hydrocarbon molecules without help from oxygen. To counter this, some microorganisms have synthesized unique enzymes and are called specialized enzymes which include benzylsuccinate synthase that binds fumarate with the hydrocarbons and begins the process of degradation. This fumarate addition is a unique step in degrading toluene and other aromatic hydrocarbons under anaerobic conditions. The subsequent reactions lead to the formation of organic acids such as succinate which can then be metabolized through the fumarate reduction pathway or other anaerobic pathways. The pathways of biodegradation of PAHs differ from those of naphthalene because PAHs are more inert and prove more difficult to degrade but contain the same kind of reaction mechanisms as a result of having multiple fused-aromatic rings. Biodegradation of PAHs occurs by estradiol oxidation and subsequent unraveling of the aromatic moieties. Hormone receptor activators, such as the dioxygenases, have a role in the initial step of breaking down the PAHs to diols, which are metabolized to carboxylated intermediates, and assimilated into the central metabolic pathway. As mentioned earlier, the complete biotransformation of PAHs into the mineral phase frequently involves joint efforts of several microbial varieties that become involved in the process at different phases.

11.3.2 Key Microorganisms Involved in Bioremediation

Petroleum hydrocarbon is degraded by a variety of microorganisms which constitute bacteria and archaea, fungi and even algae. In these, bacteria and fungi are the most explored, and their capacities for the degradation of hydrocarbons are well-established. The main strains employed in the process of bioremediation are usually adapted to degrade certain types of hydrocarbon and bioremediation efficiency depends on the environmental medium and the presence of other microorganisms.

- **Bacteria**: The study has shown that bacteria are the chief agents of hydrocarbon biodegradation under aerobic and anaerobic conditions. Some bacterial species contain intrinsic capabilities to take up and degrade aliphatic and aromatic hydrocarbons, thus they are well-defined bioremediation microbes. The most famous types of hydrocarbon-degrading bacteria include Pseudomonas, Alcanivorax Mycobacterium, Rhodococcus and Acinetobacter. These bacteria have a number of different metabolisms that enable them to degrade hydrocarbons in different constraints of the environment (Siddiqui et al., 2023; Wang et al., 2020). For instance, Pseudomonas species are highly adaptable and can metabolize virtually all the hydrocarbon alkanes, soluble BTEX and PAHs. *Pseudomonas aeruginosa* and *Pseudomonas putida* are among the most common bacteria that are found in oil-contaminated sites and these bacteria are known for their ability to degrade hydrocarbons. These bacteria synthesize several oxygenases and other enzymes that facilitate the cleavage of different hydrocarbons into smaller molecules that can be metabolized. Alcanivorax species are especially associated with marine environments to which they are adept in associating with oil-contaminated water. Most of these bacteria are obligate hydrocarbon degraders, thus they obtain almost all of their carbon and energy from hydrocarbons. Alcanivorax borkumensis is one of the best-studied species in this genus and is known to be involved in the biodegradation of oil in the sea. The fact that it can metabolize alkanes and other hydrocarbons, in addition to the high growth rate it indicates in oil-contaminated areas, makes it a major player in ocean bioremediation strategies. SRB plays an essential role in anaerobic hydrocarbon degradation throughout marine and subsurface environments where sulfate concentrations are high. SRB, including the Desulfovibrio and Desulfobacter genera, incorporate hydrocarbon compound oxidations to sulfate and reduce the sulfate to hydrogen sulfide. These bacteria are of immense value in the degradation of hydrocarbons under anoxic conditions such as in crude oil deposits and suboceanic deposits.

- **Fungi**: Similarly, the ability of fungi, filamentous fungi and yeast, to degrade petroleum hydrocarbons has been evidenced in soil samples. The most frequently researched hydrocarbon-degrading fungi comprise Aspergillus, Penicillium, Fusarium, as well as Candida. These fungi secrete extracellular enzymes/active/with an ability to oxidatively degrade complex hydrocarbons like PAHs, including laccases and peroxidases. Fungi are capable of degrading a broad class of substrates especially the high molecular weight HMW HC that are less susceptible to bacterial attack (Li et al., 2019; Zou et al., 2024). One example of white-rot fungi includes *Phanerochaete chrysosporium* because of its capability to degrade an organic polymer, known as lignin, using lignin peroxidase and manganese peroxidase. These enzymes are also able to degrade virtually all types of organo-halogen compounds such as PAHs and other non-amenable hydrocarbon compounds. Soil and sediment-borne white-rot fungi have received considerable attention in the recent past due to their depuration properties, especially in the removal of PAHs.
- **Archaea**: However, the role of archaea in hydrocarbon degradation is also recognized although they have been studied less than bacteria and fungi; archaea were found at hot springs, deep-sea hydrothermal vents and oil reservoirs. For instance, methanogenic archaea reduced hydrocarbons in oil deposits and played a role in the generation of methane during methanogenesis. These microorganisms are, however, more useful in circumstances where oxygen and other electron acceptors are scarce. For efficient degradation of complex hydrocarbon mixtures, the bacterial, archeal and fungal interactions are vital in the microbial life groups. These microorganisms are often involved in mutualistic associations in which products of metabolism by one species become the precursors for the next species in the chain, thus promoting more effective degradation of the hydrocarbons.

11.4 FACTORS INFLUENCING MICROBIAL DEGRADATION

Several environmental and physicochemical factors affect the efficiency of microbial degradation of petroleum hydrocarbons. Knowledge of these factors is crucial in the management of bioremediation and the achievement of the best solution for the bioremediation of polluted sites.

- **Oxygen Availability**: Among all the elements, oxygen ranks among the most important determinants in the rate at which hydrocarbons decompose, especially in aerobic conditions. Aerobic microbes require molecular oxygen as the final electron acceptor in their metabolic sequences depending on which oxygen is the limiting factor that governs the activities and efficiency of hydrocarbon biodegradation (Sun et al., 2023; Wang et al., 2023). Soil and water, which are located in regions with plenty of oxygen, are accredited to aerobic degradation processes and hence enhanced and efficient degradation of hydrocarbons. Nevertheless, in anoxic conditions, for example, in the lower depths of the soil, waterlogged sediments and groundwaters, the anaerobic processes of degradation come into operation, which are generally slower and less efficient than the aerobic processes. In some techniques of bioremediation, namely biostimulation, oxygen is added to the system so that aerobic microorganisms are boosted. This can be done in ways such as soil oxygenation, oxygen-releasing chemicals or air injection into the polluted water-bearing formation. Such interventions can very much advance the degradation process by supplying aerobic microorganisms with the oxygen they need.
- **Nutrient Availability**: Accessibility of nutrients that include nitrogen, phosphorus, as well as potassium, at times are other factors that affect microbial degradation. These nutrients are essential for both the growth and the metabolic processes of hydrocarbon-degrading microorganisms. In many contaminated environments, water in the vicinity of the oil spill or highly polluted water contains more or less hydrocarbons as opposed to nutrients, hence nutrient limitation. This kind of imbalance helps slow the degradation rate as nutrients are needed to break down hydrocarbons with accompanying microbial action. Most of the biostimulation techniques take nutrient amendment with fertilizers to areas of nutrient constraint as the primary approach. Hence, microbial communities can boost their organic mass and the degrading performance of hydrocarbons in response to the provided nitrogen and phosphorus supplies. However, nutrient addition should be controlled to prevent fixations of the target species and nontarget microorganisms or the eutrophication of water leading to toxic algal blooms and hypoxic conditions.
- **Temperature**: Temperature, therefore, forms an important factor in microbial activity and rates of degradation of the hydrocarbons. All microorganisms have temperatures at which their parameters for metabolism work at their best optimal conditions. In most hydrocarbon-degrading bacteria and fungi, the optimum temperature falls within the range of 15°C–30°C. In the colder climate, hydrocarbon degradation is slow as most microbial activity is lower in the Arctic and Antarctic regions, while the oils break down poorly because they become viscous making it difficult to bioavailability. Nevertheless, it has been pointed out that modern psychrophilic microorganisms, which are acclimatized to cold climates, are capable of breaking down hydrocarbons, although not at as rapid a rate as mesophilic microorganisms. Conversely, very much higher temperatures, such as those

in thermal springs or hydrothermal ecosystems allow for the delivery of hydrocarbons by thermophiles. These are microorganisms that prefer to live in environments with temperatures more than 60°C and can metabolize hydrocarbons in locations that are unfavorable for all other living organisms (Sui et al., 2021).

- **pH and Salinity**: The level of pH also affects microbial degradation, mainly as the large majority of hydrocarbon-degrading microbes thrive best in neutral to slightly alkaline environments. Very high and very low pH factors can slow down the rates of microorganisms and enzymes which in turn slow down the rate of hydrocarbon breakdown (Chachina et al., 2016; Mohanakrishna et al., 2018). In other instances, such as in acidic soils, liming or other measures to alter the pH of the substrate may be called for in order to create an environment suitable for the growth of beneficial microorganisms. Another component that is significant for aquatic habitats is salinity for which variation in marine settings might be significant. Microorganisms like Halomonas and Marinobacter genera are isolated from high salinity environments and can easily degrade hydrocarbons in marine areas. In any case, the rates of microbial degradation are significantly affected by salinity; while in hypersaline environments like salt marshes or evaporating ponds, salinity concentrations may reduce microbial growth, leading to slow degradation rates.

11.5 MYCOREMEDIATION: ROLE OF FUNGI IN HYDROCARBON DEGRADATION

11.5.1 Algal Bioremediation and Hydrocarbon Assimilation

Besides fungi, microalgae are reported to have the potential to be used in the bioremediation of petroleum hydrocarbons. Pro handshake with algal bioremediation since algae can absorb and metabolize the hydrocarbon as its carbon and energy source, thus a perfect species in addressing the issue of oil pollution in water habitats. Compared to many other organisms, algae have metabolic capabilities that enable them to synthesize enzymes that convert various hydrocarbons, such as aliphatic and aromatic compounds, among other structures and can grow in many conditions. Micro and macro algae were also identified as capable of efficiently photographing and metabolizing hydrocarbons (Kogbara et al., 2016; Poi et al., 2017; Villeneuve et al., 2004). Chlorella, Scenedesmus and Spirulina are species of microalgae especially suited for the uptake of nutrients and organic pollutants due to their fast growth and ability to respond to stress factors. As for their source of carbon, they use hydrocarbons that are metabolized into biomass through photosynthesis and cellular processes. Besides minimizing hydrocarbon concentrations in the water, this process makes it possible to obtain biomass that has found various uses in biofuel production for vehicles, animal consumption and bioplastics. Macroalgae, known as seaweeds, also have a major function in algal bioremediation. Such larger algal species harbor hydrocarbons on the surface of their fronds, as well as in their tissues, in varying concentrations. In this chapter, Sargassum and Fucus species prove to have the potential to accumulate and store hydrocarbons; therefore, they may be used in future coastal bioremediation processes. Macros are used in the bioremediation of the oil and support the restoration of such habitats as well as the conservation of diversity. One of the significant advantages of algal bioremediation is its dual functionality: As they have the ability to absorb pollutants from the environment, algae release oxygen and capture carbon dioxide through photosynthesis. This characteristic makes algal bioremediation favorable for the challenges of pollution and climate change. Further, the biomass generated during the bioremediation process is also useful for the efficient use of resources in bioenergy generation for sustainability. For instance, Cai et al. (2021) explored the biodegradation capabilities of an Acinetobacter strain, specifically Acinetobacter sp. SCYY-5, which was isolated from oil sludge, a common environmental pollutant. The research aims to tackle the persistent issue of total petroleum hydrocarbon (TPH) contamination, which poses significant risks to ecosystems. The isolated strain was characterized using 16S rDNA sequencing, confirming its identity. Notably, Acinetobacter sp. SCYY-5 exhibited strong resistance to heavy metals and could metabolize a range of carbon and nitrogen sources, highlighting its adaptability and making it a strong candidate for bioremediation efforts. To maximize the strain's efficiency in breaking down TPHs, the researchers employed response surface methodology (RSM), a robust statistical tool designed to optimize various factors influencing biodegradation. This method allowed the fine-tuning of both microbial and environmental conditions, leading to a notable enhancement in the strain's degradation performance. By introducing additional nutrients into the system, they observed a significant increase in TPH removal, with the efficiency reaching 79.94%. This improvement is attributed to the enhanced microbial activity and the co-metabolic interactions required for the effective breakdown of hydrocarbons. To visually represent these optimizations, the researchers utilized Design-Expert 8.0 software to generate three-dimensional (3D) response surface plots and two-dimensional (2D) contour plots (Figure 11.3). These graphical representations illustrated the interactions between key factors, such as temperature, pH and NaCl concentration, on TPH degradation efficiency. For instance, the response surface plots showed that when one factor, like pH, is held constant, TPH removal initially increases with increasing temperature before eventually decreasing. A similar pattern was observed for pH variation at constant temperature. These insights, derived from 17 experimental groups designed by RSM, provided an effective framework for adjusting these variables to achieve optimal bioremediation performance. The findings from this study underscore the potential of Acinetobacter sp. SCYY-5 as part of a well-optimized bioremediation strategy for cleaning up petroleum-contaminated environments, particularly oil sludge. The use of advanced tools like RSM not only improved the efficiency of biodegradation but also made the process more viable for real-world applications.

Xue et al. (2023) introduced an innovative technique called electromagnetic induction low-temperature thermal desorption (EMI LTTD) for the remediation of soils contaminated with petroleum hydrocarbons. The primary objective of the study was to investigate how various factors, such as temperature, initial contaminant concentration, soil particle size and soil thickness, influence the removal efficiency of TPHs. In addition, the study examined the changes in soil structure and morphology

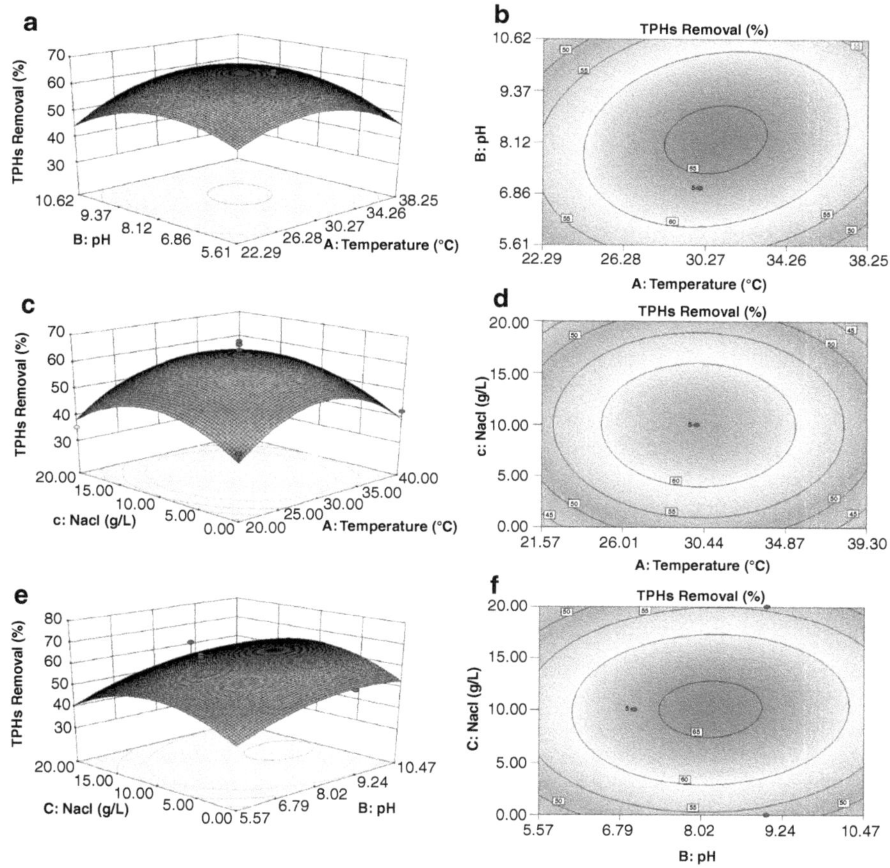

FIGURE 11.3 RSM response surface and contour plots showing TPH removal as a function of (a,b) temperature (A) and pH (B), (c,d) temperature (A) and NaCl (C), and (e,f) pH (B) and NaCl (C). (Adapted from Cai et al., 2021 under CCBY 4.0.)

during the desorption process, as well as the underlying mechanisms driving hydrocarbon removal. The results demonstrated that increasing the heating temperature significantly improved the removal rate of hydrocarbons. During the first 20 minutes of the process, between 93.44% and 96.91% of TPH was removed when the temperature ranged from 200°C to 300°C, showing the effectiveness of thermal desorption in eliminating most hydrocarbons within a short period. However, factors such as the initial concentration of contaminants, soil particle size and the thickness of the soil layer were found to have a minimal impact on the removal efficiency, suggesting that temperature is the most critical variable in the process. A desorption kinetic study revealed that the behavior of hydrocarbon removal followed a first-order model, meaning the rate of desorption was proportional to the concentration of hydrocarbons remaining in the soil. This provided a clear understanding of how hydrocarbons volatilize and desorb over time during the thermal treatment. Furthermore, through

RSM, the researchers identified optimal conditions for the EMI LTTD process, which included a temperature of 216°C, a residence time of 21 minutes and a moisture content of 18%. These conditions were determined to maximize the efficiency of the desorption process and could be applied in practical field operations. Additionally, the analysis of hydrocarbon composition after EMI LTTD treatment showed that lighter hydrocarbons, specifically in the C10–C16 and C17–C22 carbon number fractions, remained in the soil, while heavier hydrocarbons in the C23–C28 range were completely removed. This suggests that the method is particularly effective in eliminating heavier, more persistent hydrocarbons, which are typically more challenging to remediate. The proposed mechanism for TPH removal through EMI LTTD involves the evaporation of hydrocarbons at any temperature, while heavier hydrocarbons undergo partial cracking near the induction medium. This cracking leads to the re-adsorption of some lighter hydrocarbons onto the soil particles, indicating that the process not only removes contaminants but may also alter their chemical structure. To assess the potential for reusing the treated soil, a buckwheat germination and growth test was conducted. The results indicated that the soil treated with EMI LTTD is suitable for agricultural use, demonstrating the viability of this method for both environmental remediation and land reclamation purposes. The study highlights the potential of EMI LTTD as an effective and practical solution for the treatment of petroleum-contaminated soils, offering a balance between high removal efficiency and the preservation of soil quality for future reuse. Similarly, Ambaye et al. (2024) examined and compared the environmental and economic performances of two treatment technologies – electro-Fenton and bio-slurry – used for remediating petroleum hydrocarbon-contaminated soil. Traditionally, remediation approaches for such contaminated soils have been assessed based on their removal efficiency, with less emphasis on their broader environmental and economic impacts. In this research, a life cycle assessment (LCA) was employed to provide a comprehensive analysis of the environmental impacts, while a life cycle cost (LCC) analysis was used to evaluate the economic performance of each process. The chapter utilized both primary and secondary data to assess 18 potential environmental impacts using the ReCiPe 2016 characterization method in the SimaPro 9.5 software. Results indicated that the electro-Fenton process emerged as the more environmentally sustainable option. However, it was found that the primary contributors to its environmental footprint were the chemicals and energy required during the treatment process. In particular, energy consumption was identified as the most significant contributor, accounting for 90% of the total environmental impact. In contrast, the bio-slurry process was associated with a higher overall environmental impact, primarily due to its heavy reliance on biosurfactants and electricity. These two factors contributed to approximately 60% of the total environmental burden in the bio-slurry process. Despite its higher environmental impact, the bio-slurry process was identified as the more cost-effective option, with an LCC of $7.13/kg of soil treated. On the other hand, the electro-Fenton process, while environmentally favorable, was significantly more expensive, with an LCC of $661.25/kg. This high cost was largely attributed to the use of boron-doped diamond (BDD) as the electrode material, which alone cost $323/kg of soil treated. The research highlights the trade-offs between environmental sustainability and economic feasibility in the remediation of petroleum hydrocarbon-contaminated soils. While the electro-Fenton process is more environmentally friendly, it comes at a much higher financial cost. In contrast, the bio-slurry process, though less sustainable in terms of environmental impact, presents a more economically viable solution for large-scale soil remediation projects. This study underscores the importance of considering both environmental and economic factors when selecting remediation technologies, especially in large-scale applications where cost-effectiveness and environmental sustainability must be balanced.

However, there are also the drawbacks of algal bioremediation. It is shown that the efficiency of algae in hydrocarbons depends on abiotic factors such as nitrogen, phosphorus, light factor and temperature. Sometimes the concentration levels of hydrocarbons can be toxic to the algal species as opposed to causing mere stress to the species. Also, the nature of toxic compounds present in given hydrocarbons negatively affects algal growth and its metabolism which makes the bioremediation process more challenging. The practical consideration of deploying algal bioremediation is best done with an understanding of algal physiology and ecology. Findings have indicated that in a situation where two or more algal species are cultured simultaneously then the process of hydrocarbon degradation can be accelerated since the species can foster different metabolic routes that make the process very efficient. Moreover, the possibility of engineering new algal strains with higher competence for hydrocarbon degradation appears to provide new opportunities to enhance bioremediation effectiveness.

11.6 POTENTIAL AND LIMITATIONS

There is what one can describe as a high likelihood of mycoremediation and algal bioremediation in removing or reducing hydrocarbons in water and soil. The two organisms: fungi and algae have exhibited great potential as bioremediation of hydrocarbons for cost-efficient, environmental and effective means to 'clean up' oil. Mycoremediation can be a highly effective and affordable strategy in cleaning polluted soils and sediments, and algal bioremediation can be a major player in the cleanup of oil slicks and other waterborne pollutants (Cavazzoli et al., 2022; Reunamo et al., 2013). A primary benefit of both mycoremediation and algal bioremediation is that they are in situ processes, minimizing the required excavation and transportation of contaminated material. This characteristic not only makes remediation cheaper but also reduces its impact on the rest of the ecosystem. Moreover, both strategies can result in the production of less toxic by-products that can assist the improvement of the general ecosystem health, and thereby enhancing the recovery of ecosystem homogeneity. Another strength of bioremediation is the production of biomass. The present status of research indicates that fungal and algal biomass can be harvested and used for a variety of purposes, including biofuel, bioplastic and animal feed. Pestec's coupling of bioremediation and biomass corresponds to the circular economy approach, in which waste can be turned into useful inputs. However, there are some difficulties that cannot be overcome in the application of mycoremediation and algal bioremediation processes on a large scale. The system is not standardized since the die-off of different fungal and algal species is different depending on the environmental conditions of the polluted site. In addition, the possible biotoxicity in connection with some fungi and algae poses a critical necessity to conduct tests on the chosen species and check whether they will bring more harm to the environment. There are also difficulties in the scale of bioremediation actions. In general, much of the research in mycoremediation and algal bioremediation has been limited to laboratory and pilot-scale studies, the results of which should be scaled up appropriately. Site characteristics, the concentrations of the pollutant and the overall environment within which the bioremediation is to take place are other considerations.

11.7 SUSTAINABILITY AND CIRCULAR ECONOMY IN BIOREMEDIATION

The difficulties arising from environmental pollution and more so industrial pollution and urbanization necessitate the development of efficient methods of addressing polluted sites. Bioremediation has become one of the strategies of interest for the elimination of pollution, especially petroliferous hydrocarbons and heavy metals by the usage of microorganisms, fungi and plants. This approach is fully consistent with the decision-making paradigms of sustainable development and the circular economy, focusing on the rationality of integrated technologies capable of addressing the problem of contaminated territories and the cyclical use of resources (Kumar et al., 2022; Wei et al., 2024). This chapter seeks to explicate the interaction between bioremediation, sustainability and circular economies, paying attention to the ecological, social and economic aspects. Sustainability refers to a situation in which the current population, a population of former generations at least, can satisfy its needs but not at the cost of future generations' ability to do the same. It acknowledges the fact that the environment, economics and equity are mutually dependent. Bioremediation is sustainable in the following way because it presents a more natural way of remediating contaminated sites than using chemicals and digging (Ajona & Vasanthi, 2021; Martínez Álvarez et al., 2015; Ou et al., 2024). These traditional methodologies are likely to cause huge interruptions to the environment by, for instance, destroying natural habitats and producing hazardous waste. Second, alternatively, unlike chemical treatment, bioremediation harnesses natural biological processes in the degradation of pollutants, thus disrupting ecosystems in the course of the process. This is as per sustainable standards as it supports ecological reconciliation and the kit is known for supporting biodiversity. Besides, bioremediation is effective in improving the quality of soils and water in the environment and, therefore, can improve ecological systems. As living and abiotic resources are interconnected, ecology also demonstrates an evident utility as ecosystems offer important services such as purification of air and water, availability of nutrients and storage of carbon for human use. By making use of bioremediation to restore damaged environments, we enhance the quality of the natural resources and bolster the toughness of ecosystems dependent on environmental conditions, including climate change and habitat destruction. Such an approach is the key issue of the so-called environmental management and is being focused on as a critical part of the sustainability strategy. The link between bioremediation and the sustainable development goals (SDGs) shows that the use of bioremediation can enhance the new concept of sustainable development globally. The global goals of sustainable development also known as SDGs were adopted by the United Nations member states in 2015 and make up a canvass for approaching the world's biggest challenges from poverty to inequality, climate change and even environmental issues. Bioremediation impacts most beneficially on six of the UN SDGs including SDG 6 on clean and sustainable water, SDG 12 on responsible production and consumption, as well as SDG 15, on viable terrestrial ecosystems (Tran et al., 2021; Zhen et al., 2019).

The sixth goal in relation to clean water and sanitation for all calls for the important role of water quality for human health and ecosystems. Impure water sources represent a major hazard to the health of the people and cause diseases, and affect people's ability to earn a livelihood. Phytoremediation and microbial remediation, when implemented efficiently, reduce the level of contaminants in water bodies, thus the quality of water in those areas is maintained so that people can freely access water for use. Using bioremediation to remediate sustained pollution in water sources averts the objectives set for sixth SDGs, ensuring sustainable water supplies for future generations. In this context, goal 12 is concerned with the consumption and production patterns that must be sustainable and hazardous waste minimization. Bioremediation fits into this concept well because it not only provides a technology for eliminating or reducing local pollutant load but also encourages the recycling of nutrients and organic matter in ecosystems. For instance, bioremediation processes may cause a change of the contaminants to forms that are nonhazardous or perhaps less dangerous to health, which may then be used by plants and microorganisms to make the soil more fertile and productive (Meneceur et al., 2023; Paul Choudhury et al., 2022). These two make up this cyclical process, thus encouraging the use of natural methods in farming rather than the chemically fertilized ones that a lot of producers currently employ, hence food cycles for our survival. Also, Goal 15 focuses on protecting, restoring and promoting the sustainable use of terrestrial ecosystems; preventing desertification; and, by 2020, taking legal and appropriate proportionate measures to achieve adequate protection and sustainable use of the world's forests and halt and reverse land degradation and the termination of biodiversity loss. The technique of bioremediation is useful when dealing with polluted soil and other ecosystems with the aim of restoring indigenous vegetation and wildlife. Hence, bioremediation fosters improved health and diversification of soil and, therefore, the ecosystems' functionality to support the population. Bioremediation of ecosystems to restore them is also in line with targets set under SDG 15, showing the relation between the environment and human welfare (Bensalah et al., 2024; Sharma et al., 2025; Thakur et al., 2024; Thakur & Kumar, 2024). Another very important factor in the sustainability of bioremediation is that it has added to the circular economy. The principles of a circular economy are all about reducing waste and loss and recycling used materials to create a new, more sustainable economy. Bioremediation is compatible with these principles in that avails a technique for regaining valuable resources from waste products and contaminated areas. For instance, the biomass formed in bioremediation, including that produced by microbial or algal remediation can be produced and utilized in biofuel, animal feed and bioplastic. The physical metamorphosis of waste into other valuable assets proves the idea in the circular economy of reducing waste while increasing resource utilization. Furthermore, bioremediation may prove more useful in efforts to make further benefits with the materials of contaminated places. For instance, some bioremediation processes can remove metals and minerals from contaminated soils and sediments to be used again in different industrial processes. This recovery of materials also helps to decrease the need for virgin resources, as well as the negative effects characteristic of the extraction and processing of materials (Desalegn et al., 2018; Guo et al., 2024). With the concept of bioremediation applied to the recovery of resources, we can assist in the creation of a closed-loop economy. Despite the high dependence on microbial degradation of pollutants, bioremediation is equally efficient in providing several environmental values. Thus, bioremediation improves the quality of the physical environment and those functions that make ecosystems valuable to life such as sequestration of carbon, cycling of nutrients and provision of habitat. Sustainable ecosystems have an ameliorative effect on climate change by absorbing carbon dioxide, which is a greenhouse gas from the atmosphere. Moreover, ecosystems, where the restoration is carried out, contribute to the population of certain types of species increasing the resistance to possible changes in the environment. They further demonstrate the importance of bioremediation in pursuing ecological improvement as a core environmental goal.

It is also noteworthy that bioremediation has great economic benefits that cannot be reduced to the scales. Other conventional remediation processes may be expensive due to requisites in terms of labor, machinery and costs of disposal (Savaş Kaya Hassane Lgaz & Ahmed, 2023; Sharma, Thakur, Sharma, Jakhar, et al., 2023; Sharma, Thakur, Sharma, Kumar, et al., 2023). Thus, bioremediation mostly could be cheaper and much more suitable for massive pollution episodes. This approach operates under natural practice by using natural material and organisms in processing, hence, overcoming the costs of using chemicals. In the same way, the bioremediation process might create new revenue streams, including bioremediation technologies, biomass for biofuels and rehabilitation of devitalized territories for agriculture or tourism purposes. Bioremediation technologies and practices mean considerable long-term investment which would cut down the expenses incurred over environmental hazards and improve property value. Cleaned-up sites can be repurposed for sustainable development, leading to economic revitalization and job creation in local communities. The transition to sustainable practices in remediation can also align with broader economic goals, including green job creation and the promotion of environmentally friendly industries.

11.8 CHALLENGES AND FUTURE PERSPECTIVES

Bioremediation has been accepted in the recent past as a sound ecological remedy for containing pollution from sources such as petroleum hydrocarbons and other pollutants. All the same, bioremediation has several challenges that need to be overcome to improve its effectiveness and versatility. These challenges could be classified into technical challenges and regulatory constraints, and the ability to continuously develop the intensified biological treatment technologies (Al-Mutairi et al., 2008; Upadhyay et al., 2024). It is necessary to comprehend these issues as well as to analyze the prospects of bioremediation to enhance the process of its development and usage for solving environmental problems. Limitations of bioremediation technicalities mainly stem from the nature and condition of the contaminated soils, pollutants may include diverse chemical nature and varying concentrations. A major technological challenge is the lack of knowledge concerning the metabolism of microbes and the pathways of chemical compound degradation. Despite the emerging knowledge of the essential microorganisms in the bioremediation process, the mechanisms, or pathways through which complex hydrocarbons are metabolized remain unknown. Such a lack of information hampers the use of bioremediation technologies because it prevents the proper choice and optimization of microbial strains suitable for the given contaminants. Similarly, bioremediation has generally been known to depend on parameters including temperature, pH, DO and nutrient status in the contaminated environment (Wu et al., 2020; Xue et al., 2022). For example, many microorganisms capable of hydrocarbon degradation have a specific dependence on environmental factors, and these factors greatly affect the biodegradation process if they aren't ideal. Most of the time, polluted environments host irregular surfaces and thus cannot be treated uniformly. This can at times give rise to different outcomes and reduce the potential of employing bioremediation techniques in a range of applications, such as industrial polluted lands, agricultural fields, as well as many ecosystems.

The other technical drawback is the formation of toxic media during the degradation process. The fact is that among the representatives of microbial populations that have a high efficiency of petroleum hydrocarbon degradation, some metabolites formed during biodegradation may be even more toxic than the initial contaminants. This issue epitomizes the importance of conducting extensive studies of bioremediation routes to reduce the likelihood of accumulation of toxic methylation products that are a potential health hazard to human beings or the environment. Solving these technical difficulties calls for the use of microbiology, environmental science and engineering for a better approach to bioremediation strategy. The considerations of up-and-coming trends in biological treatment technologies suggest the dynamism of bioremediation. There is a growing popularity of a process called bioaugmentation, where special microbial species are introduced to increase the degradation process. This approach can be highly helpful where indigenous microbial populations are likely to be inadequate for the degradation of pollutants. One main advantage of bioaugmentation is that it can be applied selectively against individual contaminants, thus making the bioremediation process more effective. However, the long-term outcomes from bioaugmentation are still under investigation since the introduced microorganisms have to outcompete indigenous populations and effectively become a part of the associated polluted matrix (Al-Mutairi et al., 2008; Desalegn et al., 2018; Guo et al., 2024; Meneceur et al., 2023). Another new direction is the biosurfactants – surface-active agents of microbial origin that can increase the mobility of water-insoluble pollutants. The reduction of the interfacial tension between water and oil makes it possible to enhance the availability of petroleum hydrocarbon- bon to microbes through dispersion by biosurfactants. The incorporation of biosurfactants enhances the ability of bioremediation in difficult circumstances including oil slicks and extremely polluted locations. However, the commercialization of biosurfactants presents challenges related to production costs and regulatory approvals, necessitating further research and development in this area. In addition to these trends, the integration of biotechnology and molecular techniques into bioremediation research is expanding the understanding of microbial communities and their interactions with contaminants.

Advanced techniques such as metagenomics, transcriptomics and proteomics enable researchers to characterize microbial populations and their functional potentials in contaminated environments. These tools can identify specific genes and metabolic pathways involved in biodegradation, providing insights into how to optimize bioremediation strategies. This knowledge is crucial for enhancing the effectiveness of bioremediation and addressing the technical challenges associated with complex pollutants.

The future of bioremediation also hinges on addressing the opportunities for research and innovation. Collaborative efforts between academia, industry and regulatory agencies are essential for driving advancements in bioremediation technologies (Wu et al., 2020; Xue et al., 2022). Research initiatives that focus on the development of GEMs, biostimulants and novel delivery systems for bioremediation agents can significantly enhance the efficacy of these approaches. For example, synthetic biology offers the potential to design microorganisms with tailored metabolic pathways for the efficient degradation of specific pollutants, increasing the overall success rate of bioremediation efforts. Furthermore, interdisciplinary research that combines insights from environmental science, microbiology, chemistry and engineering can lead to innovative solutions for complex contamination scenarios. For instance, integrating phytoremediation techniques with microbial bioremediation can enhance pollutant degradation by utilizing the synergistic effects of plants and microorganisms. This holistic approach can address the challenges posed by heterogeneous contaminated environments and improve the overall efficiency of remediation efforts. Another area ripe for innovation is the development of real-time monitoring technologies for bioremediation processes (Agnello et al., 2016; Kaur et al., 2023). The ability to assess the progress of biodegradation in situ can provide valuable information for optimizing treatment strategies and ensuring compliance with regulatory standards. Advances in sensor technologies, remote sensing and data analytics can facilitate the continuous monitoring of bioremediation efforts, enabling timely adjustments to treatment protocols as needed. In addition to technical and research-related advancements, there is a growing recognition of the importance of public awareness and stakeholder engagement in the successful implementation of bioremediation technologies. Building trust and understanding among local communities and stakeholders can enhance the acceptance and support of bioremediation projects. Education and outreach efforts that communicate the benefits of bioremediation, as well as the science behind these technologies, can foster a more informed public and facilitate collaboration among various stakeholders.

11.9 CONCLUSIONS

As we face increasing environmental pollution from industrial activities and urbanization, bioremediation emerges as a sustainable and effective solution for mitigating the impacts of contaminants, particularly petroleum hydrocarbons, on ecosystems and human health. This chapter has explored the technical and regulatory challenges that bioremediation technologies encounter, highlighting the complexities of contaminated environments and the need for tailored approaches to address diverse pollutant types and concentrations. The lack of standardized regulatory frameworks can hinder the adoption of innovative bioremediation practices, emphasizing the necessity for greater awareness and understanding among policymakers and stakeholders. Emerging trends in biological treatment technologies signify the ongoing evolution of bioremediation practices. Bioaugmentation and the use of biosurfactants represent promising advancements that can enhance the efficiency of bioremediation processes, particularly in challenging scenarios such as oil spills and heavily contaminated sites. Furthermore, the integration of molecular techniques and biotechnology into bioremediation research is expanding our understanding of microbial communities and their interactions with pollutants. This knowledge is crucial for optimizing bioremediation strategies and improving the overall success rate of remediation efforts. However, it is essential to recognize that the commercialization of these technologies must address production costs and regulatory requirements, which remain significant barriers to their widespread application. Opportunities for research and innovation abound within the realm of bioremediation, calling for collaborative efforts between academia, industry and regulatory agencies. Interdisciplinary approaches that combine insights from various scientific fields can lead to the development of innovative solutions to complex contamination challenges. For example, integrating phytoremediation with microbial bioremediation may yield synergistic effects that enhance pollutant degradation and ecosystem restoration. Moreover, advancements in real-time monitoring technologies can facilitate continuous assessment of bioremediation processes, enabling timely adjustments to treatment protocols and ensuring compliance with regulatory standards.

Public awareness and stakeholder engagement are paramount to the successful implementation of bioremediation technologies. Building trust and understanding among local communities and stakeholders can enhance the acceptance and support of bioremediation projects. Education and outreach initiatives that effectively communicate the benefits of bioremediation and the underlying science can foster informed public discourse and encourage collaboration among various stakeholders. As we look to the future, the potential of bioremediation to restore polluted ecosystems and promote sustainability is immense. By addressing the challenges and leveraging the opportunities within bioremediation practices, we can develop more effective and sustainable solutions for environmental contamination. The alignment of bioremediation with principles of sustainability and the circular economy offers a pathway to not only mitigate the impacts of pollution but also to recover valuable resources and enhance ecosystem services. Ultimately, the integration of innovative bioremediation strategies into environmental management frameworks will contribute to a cleaner and healthier planet, ensuring that future generations can thrive in a sustainable environment. The ongoing commitment to research, innovation and collaboration in bioremediation practices will be essential in overcoming existing barriers and achieving our collective environmental goals.

REFERENCES

Agnello, A. C., Bagard, M., van Hullebusch, E. D., Esposito, G., & Huguenot, D. (2016). Comparative bioremediation of heavy metals and petroleum hydrocarbons co-contaminated soil by natural attenuation, phytoremediation, bioaugmentation and bioaugmentation-assisted phytoremediation. *Science of the Total Environment*, *563–564*, 693–703. https://doi.org/10.1016/j.scitotenv.2015.10.061

Ajona, M., & Vasanthi, P. (2021). Bioremediation of petroleum contaminated soils: A review. *Materials Today: Proceedings*, *45*, 7117–7122. https://doi.org/10.1016/j.matpr.2021.01.949

Al Souki, K. S., Burdová, H., Mamirova, A., Kuráň, P., Kříženecká, S., Oravová, L., Tolaszová, J., Nebeská, D., Popelka, J., Ust'ak, S., Honzík, R., & Trögl, J. (2021). Evaluation of the Miscanthus × giganteus short term impacts on enhancing the quality of agricultural soils affected by single and/or multiple contaminants. *Environmental Technology & Innovation*, *24*, 101890. https://doi.org/10.1016/j.eti.2021.101890

Al-Mutairi, N., Bufarsan, A., & Al-Rukaibi, F. (2008). Ecorisk evaluation and treatability potential of soils contaminated with petroleum hydrocarbon-based fuels. *Chemosphere*, *74*(1), 142–148. https://doi.org/10.1016/j.chemosphere.2008.08.020

Alsharyani, A. K., & Muruganandam, L. (2024). Fabrication of zinc oxide nanorods for photocatalytic degradation of docosane, a petroleum pollutant, under solar light simulator. *RSC Advances*, *14*(13), 9038–9049. https://doi.org/10.1039/d4ra00672k

Ambaye, T. G., Formicola, F., Sbaffoni, S., Lima, A. T. M., Franzetti, A., & Vaccari, M. (2024). Environmental and economic performance of chemical and biological processes for treating petroleum hydrocarbon-contaminated soil: An experimental study. *Journal of Environmental Chemical Engineering*, *12*(5), 113672. https://doi.org/10.1016/j.jece.2024.113672

Barathi, S., J, G., Rathinasamy, G., Sabapathi, N., Aruljothi, K. N., Lee, J., & Kandasamy, S. (2023). Recent trends in polycyclic aromatic hydrocarbons pollution distribution and counteracting bio-remediation strategies. *Chemosphere*, *337*, 139396. https://doi.org/10.1016/j.chemosphere.2023.139396

Behera, I. D., Basak, G., Kumar, R. R., Sen, R., & Meikap, B. C. (2020). Treatment of petroleum refinery sludge by petroleum degrading bacterium Stenotrophomonas pavanii IRB19 as an efficient novel technology. *Journal of Environmental Science and Health - Part A Toxic/Hazardous Substances and Environmental Engineering*, *56*(2). https://doi.org/10.1080/10934529.2020.1866924

Behera, I. D., Nayak, M., Biswas, S., Meikap, B. C., & Sen, R. (2021). Enhanced biodegradation of total petroleum hydrocarbons by implementing a novel two-step bioaugmentation strategy using indigenous bacterial consortium. *Journal of Environmental Management*, *292*, 112746. https://doi.org/10.1016/J.JENVMAN.2021.112746

Bensalah, J., Thakur, A., & Kumar, A. (2024). Investigating the adsorption processes of polymer resins for the removal of micropollutants: A comprehensive review in the field of environmental remediation. *Environmental Research*, *254*, 119128. https://doi.org/10.1016/j.envres.2024.119128

Bhuyan, R., Namdeo, A., Verma, J. S., & Geed, S. R. (2023). Isolation of indole biodegrading bacterial species from petroleum contaminated site of northeast India: Growth kinetics and crude oil biodegradation assessment. *Journal of Water Process Engineering*, *52*, 103544. https://doi.org/10.1016/j.jwpe.2023.103544

Cai, Y., Wang, R., Rao, P., Wu, B., Yan, L., Hu, L., Park, S., Ryu, M., & Zhou, X. (2021). Bioremediation of petroleum hydrocarbons using acinetobacter sp. SCYY-5 isolated from contaminated oil sludge: Strategy and effectiveness study. *International Journal of Environmental Research and Public Health*, *18*(2), 1–14. https://doi.org/10.3390/ijerph18020819

Cavazzoli, S., Selonen, V., Rantalainen, A.-L., Sinkkonen, A., Romantschuk, M., & Squartini, A. (2022). Natural additives contribute to hydrocarbon and heavy metal co-contaminated soil remediation. *Environmental Pollution*, *307*, 119569. https://doi.org/10.1016/j.envpol.2022.119569

Chachina, S. B., Voronkova, N. A., & Baklanova, O. N. (2016). Biological remediation of the petroleum and diesel contaminated soil with earthworms Eisenia fetida. *Procedia Engineering*, *152*, 122–133. https://doi.org/10.1016/j.proeng.2016.07.642

Chiu, H. Y., Verpoort, F., Liu, J. K., Chang, Y. M., & Kao, C. M. (2017). Using intrinsic bioremediation for petroleum–hydrocarbon contaminated groundwater cleanup and migration containment: Effectiveness and mechanism evaluation. *Journal of the Taiwan Institute of Chemical Engineers*, *72*, 53–61. https://doi.org/10.1016/j.jtice.2017.01.002

Das, A., Dey, S., & Das, A. P. (2024). Ecotoxicological impacts of synthetic microfiber pollutants and development of sustainable mitigation strategies. In *Environmental Chemistry and Ecotoxicology*. https://doi.org/10.1016/J.ENCECO.2024.12.003

Das, A., & Ghosh, S. (2022). Role of microorganisms in extenuation of mining and industrial wastes. *Geomicrobiology Journal*, *39*, 173–175. https://doi.org/10.1080/01490451.2022.2038953

Das, A. P., & Ghosh, S. (2023). Microorganisms in manganese biomining processes. In *Manganese Mining Microorganisms*, 55–61. https://doi.org/10.1016/B978-0-12-822147-1.00013-3

Das, A. P., Ghosh, S., Mohanty, S., & Sukla, L. B. (2015). Advances in manganese pollution and its bioremediation. In L. B. Sukla, N. Pradhan, S. Panda, & B. K. Mishra (Eds.), *Environmental Microbe Biotechnology* (1st ed., pp. 313–328). Springer, Cham. https://doi.org/10.1007/978-3-319-19018-1_16

Das, A. P., Mallavarapu, M., & Ghosh, S. (2023). Ecotoxicity of mining pollutants on the environment and their remediation. *Environmental Chemistry and Ecotoxicology*, *5*, 165–167. https://doi.org/10.1016/J.ENCECO.2023.08.002

Das, A. P., Nayak, B., & Behera, I. D. (2024). Applications of microorganisms in advanced geomicrobiological and biotechnological exploration. *Geomicrobiology Journal*, *41*(4), 309–311. https://doi.org/10.1080/01490451.2024.2327059

Das, A., Ray, S. K., Mohanty, M., Mohanty, J., Dey, S., & Das, A. P. (2024). Ecotoxicity of microplastic wastes and their sustainable management: A review. *Environmental Chemistry and Ecotoxicology*, *6*, 144–152. https://doi.org/10.1016/J.ENCECO.2024.05.003

Desalegn, B., Megharaj, M., Chen, Z., & Naidu, R. (2018). Green mango peel-nanozerovalent iron activated persulfate oxidation of petroleum hydrocarbons in oil sludge contaminated soil. *Environmental Technology & Innovation*, *11*, 142–152. https://doi.org/10.1016/j.eti.2018.05.007

Dey, S., Bissoyi, A., Koteswara Reddy, G., & Das, A. P. (2024). Synthetic microfibers: An urban emerging pollutant demanding for sustainable environmental solutions. In *Sustainable Microbial Technology for Synthetic and Cellulosic Microfiber Bioremediation* (pp. 107–121). https://doi.org/10.1007/978-3-031-62600-5_7

Dey, S., Dash, A., & Das, A. P. (2024). Bioremediation as a sustainable and circular solution for mining waste management and bioenergy generation. In *Sustainable Management of Mining Waste and Tailings: A Circular Economy Approach* (pp. 107–119). https://doi.org/10.1201/9781003442455-6

Dey, S., Das, A., Mallick, K., Sahu, A., & Das, A. P. (2023). Environmental petroleum waste: Pollution, toxicity, sustainable remediation. (pp. 159–175). https://doi.org/10.1007/978-3-031-48220-5_7

Dey, S., Dash, N., Ray, S. K., Priyadarsini, S., Chhotaray, C., & Das, A. P. (2024). Occurrence and quantification of tire wear plastics pollutants from highway city dust of Bhubaneswar City of Odisha. In *Sustainable Microbial Technology for Synthetic and Cellulosic Microfiber Bioremediation* (pp. 77–89). https://doi.org/10.1007/978-3-031-62600-5_5

Dragoi, E.-N., Godini, K., & Koolivand, A. (2021). Modeling of oily sludge composting process by using artificial neural networks and differential evolution: Prediction of removal of petroleum hydrocarbons and organic carbon. *Environmental Technology & Innovation*, *21*, 101338. https://doi.org/10.1016/j.eti.2020.101338

Fetanat, A., & Tayebi, M. (2022). A picture fuzzy set-based decision support system for treatment technologies prioritization of petroleum refinery effluents: A circular water economy transition towards oil & gas industry. *Separation and Purification Technology*, *303*, 122220. https://doi.org/10.1016/j.seppur.2022.122220

Ghorbanian, M., Moussavi, G., & Farzadkia, M. (2014). Investigating the performance of an up-flow anoxic fixed-bed bioreactor and a sequencing anoxic batch reactor for the biodegradation of hydrocarbons in petroleum-contaminated saline water. *International Biodeterioration & Biodegradation*, *90*, 106–114. https://doi.org/10.1016/j.ibiod.2014.02.009

Gielnik, A., Pechaud, Y., Huguenot, D., Cébron, A., Esposito, G., & van Hullebusch, E. D. (2021). Functional potential of sewage sludge digestate microbes to degrade aliphatic hydrocarbons during bioremediation of a petroleum hydrocarbons contaminated soil. *Journal of Environmental Management*, *280*, 111648. https://doi.org/10.1016/j.jenvman.2020.111648

Guarino, C., Spada, V., & Sciarrillo, R. (2017). Assessment of three approaches of bioremediation (natural attenuation, landfarming and bioagumentation: Assistited landfarming) for a petroleum hydrocarbons contaminated soil. *Chemosphere*, *170*, 10–16. https://doi.org/10.1016/j.chemosphere.2016.11.165

Guo, H., Qin, Q., Hu, M., Chang, J.-S., & Lee, D.-J. (2024). Treatment of refinery wastewater: Current status and prospects. *Journal of Environmental Chemical Engineering*, *12*(2), 112508. https://doi.org/10.1016/j.jece.2024.112508

Ivshina, I., Kostina, L., Krivoruchko, A., Kuyukina, M., Peshkur, T., Anderson, P., & Cunningham, C. (2016). Removal of polycyclic aromatic hydrocarbons in soil spiked with model mixtures of petroleum hydrocarbons and heterocycles using biosurfactants from Rhodococcus ruber IEGM 231. *Journal of Hazardous Materials*, *312*, 8–17. https://doi.org/10.1016/j.jhazmat.2016.03.007

Jagatee, S., Priyadarshini, S., Rath, C. C., & Das, A. P. (2024). Synthetic microfiber: An enduring environmental problem linked to sustainable development. In *Environmental Science and Engineering, Part F2333* (pp. 93–112). https://doi.org/10.1007/978-3-031-51792-1_6

John, A., & Das, A. P. (2024). Government strategies for achieving circular economy for plastic waste management in Southeast Asia. In *Sustainable Management of Urban Plastic Waste through Circular Economic Approaches* (pp. 72–82). https://doi.org/10.1201/9781003465447-6

Kanungo, J., Sahoo, T., Bal, M., & Behera, I. D. (2023). Performance of bioremediation strategy in waste lubricating oil pollutants: A review. *Geomicrobiology Journal*. https://doi.org/10.1080/01490451.2023.2245395

Kanungo, J., Sahoo, T., Swain, L. P., & Behera, I. D. (2023). Toxicity of persistent hydrocarbon pollutants, sources and sustainable remediation process. (pp. 39–65). https://doi.org/10.1007/978-3-031-48220-5_2

Kaur, R., Gupta, S., Tripathi, V., Chauhan, A., Parashar, D., Shankar, P., & Kashyap, V. (2023). Microbiome based approaches for the degradation of polycyclic aromatic hydrocarbons (PAHs): A current perception. *Chemosphere*, *341*, 139951. https://doi.org/10.1016/j.chemosphere.2023.139951

Kaya, S., Lgaz, H., Thakkur, A., Kumar, A., Işın, D. Ö., Karakuş, N., & Ahmed, S. B. (2023). Molecular insights into the corrosion inhibition mechanism of omeprazole and tinidazole: A theoretical investigation. *Molecular Simulation*, *49*(17), 1632–1646. https://doi.org/10.1080/08927022.2023.2256888

Khudur, L. S., Shahsavari, E., Webster, G. T., Nugegoda, D., & Ball, A. S. (2019). The impact of lead co-contamination on ecotoxicity and the bacterial community during the bioremediation of total petroleum hydrocarbon-contaminated soils. *Environmental Pollution*, *253*, 939–948. https://doi.org/10.1016/j.envpol.2019.07.107

Kim, K., Kwon, H., Park, J., Lee, H., & Choi, Y. (2023). Thermal treatment of petroleum-contaminated marine sediment according to oxygen availability and temperature: Product quality as a potential plant-growth medium. *Chemosphere*, *324*, 138347. https://doi.org/10.1016/j.chemosphere.2023.138347

Kogbara, R. B., Ayotamuno, J. M., Onuomah, I., Ehio, V., & Damka, T. D. (2016). Stabilisation/solidification and bioaugmentation treatment of petroleum drill cuttings. *Applied Geochemistry*, *71*, 1–8. https://doi.org/10.1016/j.apgeochem.2016.05.010

Koolivand, A., Abtahi, H., Parhamfar, M., Didehdar, M., Saeedi, R., & Fahimirad, S. (2019). Biodegradation of high concentrations of petroleum compounds by using indigenous bacteria isolated from petroleum hydrocarbons-rich sludge: Effective scale-up from liquid medium to composting process. *Journal of Environmental Management*, *248*, 109228. https://doi.org/10.1016/j.jenvman.2019.06.129

Koolivand, A., Saeedi, R., Coulon, F., Kumar, V., Villaseñor, J., Asghari, F., & Hesampoor, F. (2020). Bioremediation of petroleum hydrocarbons by vermicomposting process bioaugmentated with indigenous bacterial consortium isolated from petroleum oily sludge. *Ecotoxicology and Environmental Safety*, *198*, 110645. https://doi.org/10.1016/j.ecoenv.2020.110645

Krucoń, T., Ruszkowska, Z., Pilecka, W., Szych, A., & Drewniak, Ł. (2023). Bioprospecting of the Antarctic Bacillus subtilis strain for potential application in leaching hydrocarbons and trace elements from contaminated environments based on functional and genomic analysis. *Environmental Research*, *227*, 115785. https://doi.org/10.1016/j.envres.2023.115785

Kumar, V., Agrawal, S., Bhat, S. A., Américo-Pinheiro, J. H. P., Shahi, S. K., & Kumar, S. (2022). Environmental impact, health hazards, and plant-microbes synergism in remediation of emerging contaminants. *Cleaner Chemical Engineering*, *2*, 100030. https://doi.org/10.1016/j.clce.2022.100030

Lahiri, D., Nag, M., Dey, A., Sarkar, T., Joshi, S., Pandit, S., Das, A. P., Pati, S., Pattanaik, S., Tilak, V. K., & Ray, R. R. (2021). Biofilm mediated degradation of petroleum products. *Geomicrobiology Journal*, 1–10. https://doi.org/10.1080/01490451.2021.1968979

Lamy, E., Tran, T. C., Mottelet, S., Pauss, A., & Schoefs, O. (2013). Relationships of respiratory quotient to microbial biomass and hydrocarbon contaminant degradation during soil bioremediation. *International Biodeterioration & Biodegradation*, *83*, 85–91. https://doi.org/10.1016/j.ibiod.2013.04.015

Lee, T. H., Cao, W. Z., Tsang, D. C. W., Sheu, Y. T., Shia, K. F., & Kao, C. M. (2019). Emulsified polycolloid substrate biobarrier for benzene and petroleum-hydrocarbon plume containment and migration control: A field-scale study. *Science of the Total Environment*, *666*, 839–848. https://doi.org/10.1016/j.scitotenv.2019.02.160

Lee, T. H., Tsang, D. C. W., Chen, W. H., Verpoort, F., Sheu, Y. T., & Kao, C. M. (2019). Application of an emulsified polycolloid substrate biobarrier to remediate petroleum-hydrocarbon contaminated groundwater. *Chemosphere*, *219*, 444–455. https://doi.org/10.1016/j.chemosphere.2018.12.028

Li, X., Li, Y., Zhang, X., Zhao, X., Sun, Y., Weng, L., & Li, Y. (2019). Long-term effect of biochar amendment on the biodegradation of petroleum hydrocarbons in soil microbial fuel cells. *Science of the Total Environment*, *651*, 796–806. https://doi.org/10.1016/j.scitotenv.2018.09.098

Li, X., Wang, Y., Li, N., & Su, Y. (2024). Enhancing the phytoremediation of petroleum-contaminated arid saline-alkali soil through irrigation-intercropping system combined with surfactants. *Ecological Engineering*, *203*, 107245. https://doi.org/10.1016/j.ecoleng.2024.107245

Ling, H., Hou, J., Du, M., Zhang, Y., Liu, W., Christie, P., & Luo, Y. (2023). Surfactant-enhanced bioremediation of petroleum-contaminated soil and microbial community response: A field study. *Chemosphere*, *322*, 138225. https://doi.org/10.1016/j.chemosphere.2023.138225

Liu, F., Tu, T., Li, S., Cai, M., Huang, X., & Zheng, F. (2019). Relationship between plankton-based β-carotene and biodegradable adaptablity to petroleum-derived hydrocarbon. *Chemosphere*, *237*, 124430. https://doi.org/10.1016/j.chemosphere.2019.124430

Liu, Z., Li, Z., Chen, S., & Zhou, W. (2023). Enhanced phytoremediation of petroleum-contaminated soil by biochar and urea. *Journal of Hazardous Materials*, *453*, 131404. https://doi.org/10.1016/j.jhazmat.2023.131404

Mallick, K., Sahu, A., Dubey, N. K., & Das, A. P. (2023). Harvesting marine plastic pollutants-derived renewable energy: A comprehensive review on applied energy and sustainable approach. *Journal of Environmental Management*, *348*, 119371. https://doi.org/10.1016/J.JENVMAN.2023.119371

Mangse, G., Werner, D., Meynet, P., & Ogbaga, C. C. (2020). Microbial community responses to different volatile petroleum hydrocarbon class mixtures in an aerobic sandy soil. *Environmental Pollution*, *264*, 114738. https://doi.org/10.1016/j.envpol.2020.114738

Martínez Álvarez, L. M., Lo Balbo, A., Mac Cormack, W. P., & Ruberto, L. A. M. (2015). Bioremediation of a petroleum hydrocarbon-contaminated Antarctic soil: Optimization of a biostimulation strategy using response-surface methodology (RSM). *Cold Regions Science and Technology*, *119*, 61–67. https://doi.org/10.1016/j.coldregions.2015.07.005

Martínez-Rabelo, F., Gómez-Guzmán, L. A., García-Segura, D. R., Villegas-García, E., Rodriguez-Campos, J., Velázquez-Fernández, J. B., Hernández-Castellanos, B., Barois, I., & Contreras-Ramos, S. M. (2023). Hydrocarbon bioremediation in a pilot-scale: A combination of bioaugmentation, phytoremediation, and vermiremediation. *Environmental Technology & Innovation*, *31*, 103210. https://doi.org/10.1016/j.eti.2023.103210

Meneceur, S., Bouafia, A., Laouini, S. E., Mohammed, H. A., Daoudi, H., Chami, S., Hasan, G. G., Abdullah, J. A. A., & Salmi, C. (2023). Removal efficiency of heavy metals, oily in water, total suspended solids, and chemical oxygen demand from industrial petroleum wastewater by modern green nanocomposite methods. *Journal of Environmental Chemical Engineering*, *11*(6), 111209. https://doi.org/10.1016/j.jece.2023.111209

Mirjani, M., Soleimani, M., & Salari, V. (2021). Toxicity assessment of total petroleum hydrocarbons in aquatic environments using the bioluminescent bacterium Aliivibrio fischeri. *Ecotoxicology and Environmental Safety*, *207*, 111554. https://doi.org/10.1016/j.ecoenv.2020.111554

Mishra, A., Siddiqi, H., Kumari, U., Behera, I. D., Mukherjee, S., & Meikap, B. C. (2021). Pyrolysis of waste lubricating oil/waste motor oil to generate high-grade fuel oil: A comprehensive review. *Renewable and Sustainable Energy Reviews*, *150*, 111446. https://doi.org/10.1016/J.RSER.2021.111446

Mishra, S., Dash, D., & Das, A. P. (2024). Aquatic microbial diversity on plastisphere: Colonization and potential role in microplastic biodegradation. *Geomicrobiology Journal*, *41*(4), 312–323. https://doi.org/10.1080/01490451.2023.2209750

Mohanakrishna, G., Abu-Reesh, I. M., & Al-Raoush, R. I. (2018). Biological anodic oxidation and cathodic reduction reactions for improved bioelectrochemical treatment of petroleum refinery wastewater. *Journal of Cleaner Production*, *190*, 44–52. https://doi.org/10.1016/j.jclepro.2018.04.141

Mohanty, J., Dey, S., Bala, K., & Das, A. P. (2024). Recent advances in microplastic pollution for its sustainable management. In *Sustainable Management of Urban Plastic Waste through Circular Economic Approaches* (pp. 27–48). https://doi.org/10.1201/9781003465447-3

Nag, M., Lahiri, D., Ghosh, S., Sarkar, T., Pati, S., Das, A. P., Ram, D. K., Bhattacharya, D., & Ray, R. R. (2023). Application of microorganisms in biotransformation and bioremediation of environmental contaminant: A review. *Geomicrobiology Journal*. https://doi.org/10.1080/01490451.2023.2261443

Nawaz, M. Z., Xu, C., Qaria, M. A., Zeeshan Haider, S., Rameez Khalid, H., Ahmed Alghamdi, H., Ahmad Khan, I., & Zhu, D. (2023). Genomic and biotechnological potential of a novel oil-degrading strain Enterobacter kobei DH7 isolated from petroleum-contaminated soil. *Chemosphere*, *340*, 139815. https://doi.org/10.1016/j.chemosphere.2023.139815

Okparanma, R. N., Ukoha-Onuoha, E., & Ayotamuno, J. M. (2022). Predicting selected kinetic parameters of hydrocarbon interactions with alum-based water treatment residuals. *Results in Engineering*, *16*, 100726. https://doi.org/10.1016/j.rineng.2022.100726

Ou, Y., Wu, M., Yu, Y., Liu, Z., Zhang, Y., & Yi, N. (2024). Nitrogen utilization efficiency assessment during bioremediation of petroleum-contaminated loess soils: Insights from metagenomic analysis. *Journal of Hazardous Materials*, *478*, 135506. https://doi.org/10.1016/j.jhazmat.2024.135506

Paul, A., Priyadarsini, S., Karakuş, S., Behera, J. K., Gnanasekaran, R., & Das, A. P. (2024). Fate and Accumulation of synthetic microfibers in the environment. In *Sustainable Microbial Technology for Synthetic and Cellulosic Microfiber Bioremediation* (pp. 61–76). https://doi.org/10.1007/978-3-031-62600-5_4

Paul Choudhury, S., Panda, S., Haq, I., & Kalamdhad, A. S. (2022). Enhanced methane production and hydrocarbon removal from petroleum refinery sludge after Pseudomonas putida pretreatment and process scale-up. *Bioresource Technology*, *343*, 126127. https://doi.org/10.1016/j.biortech.2021.126127

Poi, G., Aburto-Medina, A., Mok, P. C., Ball, A. S., & Shahsavari, E. (2017). Large scale bioaugmentation of soil contaminated with petroleum hydrocarbons using a mixed microbial consortium. *Ecological Engineering*, *102*, 64–71. https://doi.org/10.1016/j.ecoleng.2017.01.048

Poi, G., Shahsavari, E., Aburto-Medina, A., Mok, P. C., & Ball, A. S. (2018). Large scale treatment of total petroleum-hydrocarbon contaminated groundwater using bioaugmentation. *Journal of Environmental Management*, *214*, 157–163. https://doi.org/10.1016/j.jenvman.2018.02.079

Priyadarsini, S., & Das, A. P. (2025). Lithium bioleaching: A review on microbial-assisted sustainable technology for lithium bio-circularity. *Journal of Water Process Engineering*, *69*, 106744. https://doi.org/10.1016/j.jwpe.2024.106744

Qv, M., Bao, J., Wang, W., Dai, D., Wu, Q., Li, S., & Zhu, L. (2024). Bentonite addition enhances the biodegradation of petroleum pollutants and bacterial community succession during the aerobic co-composting of waste heavy oil with agricultural wastes. *Journal of Hazardous Materials*, *462*, 132655. https://doi.org/10.1016/j.jhazmat.2023.132655

Raj, K., & Das, A. P. (2023). Lead pollution: Impact on environment and human health and approach for a sustainable solution. *Environmental Chemistry and Ecotoxicology*, *5*, 79–85. https://doi.org/10.1016/J.ENCECO.2023.02.001

Ray, S. K., Dey, S., Das, A. P., & John, A. (2024). Microplastic waste management through greener technologies: A review. In *Sustainable Management of Urban Plastic Waste through Circular Economic Approaches* (pp. 237–261). https://doi.org/10.1201/9781003465447-14

Reunamo, A., Riemann, L., Leskinen, P., & Jørgensen, K. S. (2013). Dominant petroleum hydrocarbon-degrading bacteria in the Archipelago Sea in South-West Finland (Baltic Sea) belong to different taxonomic groups than hydrocarbon degraders in the oceans. *Marine Pollution Bulletin*, *72*(1), 174–180. https://doi.org/10.1016/j.marpolbul.2013.04.006

Safdari, M.-S., Kariminia, H.-R., Rahmati, M., Fazlollahi, F., Polasko, A., Mahendra, S., Wilding, W. V., & Fletcher, T. H. (2018). Development of bioreactors for comparative study of natural attenuation, biostimulation, and bioaugmentation of petroleum-hydrocarbon contaminated soil. *Journal of Hazardous Materials*, *342*, 270–278. https://doi.org/10.1016/j.jhazmat.2017.08.044

Sahoo, P. P., Singh, S., Rout, P. K., Mishra, S., & Das, A. P. (2022). Microbial remediation of plastic pollutants generated from discarded and abandoned marine fishing nets. *Biotechnology and Gnetic Engineering Reviews*. https://doi.org/10.1080/02648725.2022.2152629

Sahu, A., Dey, S., Mallick, K., Priyadarsini, S., Behera, I. D., & Das, A. P. (2024). A review on advanced technology for sustainable management of synthetic microplastic waste. In *Sustainable Management of Urban Plastic Waste through Circular Economic Approaches* (pp. 262–279). https://doi.org/10.1201/9781003465447-15

Sahu, A., Mallick, K., & Das, A. P. (2024). Environmental impact of single-use synthetic face mask and its recycling: A sustainable approach. (pp. 197–212). https://doi.org/10.1007/978-3-031-50840-0_11

Saini, A., Bekele, D. N., Chadalavada, S., Fang, C., & Naidu, R. (2021). Electrokinetic remediation of petroleum hydrocarbon contaminated soil (I). *Environmental Technology & Innovation*, *23*, 101585. https://doi.org/10.1016/j.eti.2021.101585

Salgado, L. D., Marques, A. E. M. L., Kramer, R. D., de Oliveira, F. G., Moretto, S. L., de Lima, B. A., Prodocimo, M. M., Cestari, M. M., de Azevedo, J. C. R., & Silva de Assis, H. C. (2019). Integrated assessment of sediment contaminant levels and biological responses in sentinel fish species Atherinella brasiliensis from a sub-tropical estuary in south Atlantic. *Chemosphere*, *219*, 15–27. https://doi.org/10.1016/j.chemosphere.2018.11.204

Sharma, D., Thakur, A., Sharma, M. K., Bhardwaj, A., Kumar, D., Kumar, A., Kherb, J., Heena, Rana, R., Pandya, N. U., Sharma, A. K., & Om, H. (2025). Electrochemicaland computational insightsinto the utilization of an N-heteroaromatic containing compound2-(4-Methoxy-phenyl)-5-naphthalen-2-yl-[1,3,4]oxadiazole as a promising anticorrosive agent for mild steel in corrosive medium. *Journal of Molecular Structure*, *1321*, 140055. https://doi.org/10.1016/j.molstruc.2024.140055

Sharma, D., Thakur, A., Sharma, M. K., Jakhar, K., Kumar, A., Sharma, A. K., & Om, H. (2023). Synthesis, electrochemical, morphological, computational and corrosion inhibition studies of 3-(5-naphthalen-2-yl-[1,3,4]oxadiazol-2-yl)-pyridine against mild steel in 1 M HCl. *Asian Journal of Chemistry*, *35*(5), 1079–1088.

Sharma, D., Thakur, A., Sharma, M. K., Kumar, A., Jakhar, K., Kumar, S., Sihmar, A., Dahiya, H., Kumar, A., Sharma, A. K., & Om, H. (2023). A convenient synthesis, electrochemical profiling, and morphological studies of a pyridine-based 1,3,4-oxadiazole hybrid: A promising study for corrosion mitigation of mild steel in strongly acidic environment. *Inorganic Chemistry Communications*, *158*(P1), 111554. https://doi.org/10.1016/j.inoche.2023.111554

Siddiqui, Z., Grohmann, E., & Malik, A. (2023). Degradation of alkane hydrocarbons by Priestia megaterium ZS16 and sediments consortia with special reference to toxicity and oxidative stress induced by the sediments in the vicinity of an oil refinery. *Chemosphere*, *317*, 137886. https://doi.org/10.1016/j.chemosphere.2023.137886

Sui, X., Wang, X., Li, Y., & Ji, H. (2021). Remediation of petroleum-contaminated soils with microbial and microbial combined methods: Advances, mechanisms and challenges. *Sustainability (Switzerland)*, *13*(16). https://doi.org/10.3390/su13169267

Sun, Y., Liu, Y., Yue, G., Cao, J., Li, C., & Ma, J. (2023). Vapor-phase biodegradation and natural attenuation of petroleum VOCs in the unsaturated zone: A microcosm study. *Chemosphere*, *336*, 139275. https://doi.org/10.1016/j.chemosphere.2023.139275

Thakur, A., Dagdag, O., Berisha, A., Ebenso, E. E., Kumar, A., Sharma, S., Ganjoo, R., & Assad, H. (2024). Mechanistic insights into the corrosion inhibition of mild steel by eco-benign Asphodelus Tenuifolius aerial extract in acidic environment: Electrochemical and computational analysis. *Surface and Coatings Technology*, *480*, 130568. https://doi.org/10.1016/j.surfcoat.2024.130568

Thakur, A., & Kumar, A. (2024). Unraveling the multifaceted mechanisms and untapped potential of activated carbon in remediation of emerging pollutants: A comprehensive review and critical appraisal of advanced techniques. *Chemosphere*, *346*, 140608. https://doi.org/10.1016/j.chemosphere.2023.140608

Tran, H.-T., Lin, C., Bui, X.-T., Ngo, H.-H., Cheruiyot, N. K., Hoang, H.-G., & Vu, C.-T. (2021). Aerobic composting remediation of petroleum hydrocarbon-contaminated soil. Current and future perspectives. *Science of the Total Environment*, *753*, 142250. https://doi.org/10.1016/j.scitotenv.2020.142250

Ugochukwu, U. C., Ochonogor, A., Jidere, C. M., Agu, C., Nkoloagu, F., Ewoh, J., & Okwu-Delunzu, V. U. (2018). Exposure risks to polycyclic aromatic hydrocarbons by humans and livestock (cattle) due to hydrocarbon spill from petroleum products in Niger-delta wetland. *Environment International*, *115*, 38–47. https://doi.org/10.1016/j.envint.2018.03.010

Upadhyay, M., Hansa, A., Devi, A., Asgari Lajayer, B., Price, G. W., & Sharma, R. (2024). Status and future scope of algal biomass-based remediation for various environmental contaminants. *Journal of Water Process Engineering*, *65*, 105809. https://doi.org/10.1016/j.jwpe.2024.105809

Villeneuve, J.-P., de Mora, S., & Cattini, C. (2004). Determination of organochlorinated compounds and petroleum hydrocarbons in fish-homogenate sample IAEA-406: Results from a worldwide interlaboratory study. *TrAC Trends in Analytical Chemistry*, *23*(7), 501–510. https://doi.org/10.1016/S0165-9936(04)00731-9

Wang, C., Wang, S., Jiao, X., Yang, B., Liang, S., Luo, Z., & Mao, L. (2020). Periodic density as an endpoint of customized plankton community responses to petroleum hydrocarbons: A level of toxic effect should be matched with a suitable time scale. *Ecotoxicology and Environmental Safety*, *201*, 110723. https://doi.org/10.1016/j.ecoenv.2020.110723

Wang, H., Dai, B., Gao, F., Cao, X., Ji, X., Zuo, L., Cheng, C., & Jiang, L. (2024). Petrochemical wastewater treatment in a full-scale biological aerated filter system: Insight into the mechanism of performance deterioration. *Chemical Engineering Journal*, *488*, 151089. https://doi.org/10.1016/j.cej.2024.151089

Wang, J., Zhan, X., Zhou, L., & Lin, Y. (2010). Biological indicators capable of assessing thermal treatment efficiency of hydrocarbon mixture-contaminated soil. *Chemosphere*, *80*(8), 837–844. https://doi.org/10.1016/j.chemosphere.2010.06.009

Wang, J., Zhang, Y., Liu, T., Shi, Y., Ding, Y., Zhang, Y., Xu, W., Zhang, X., Wang, Y., & Li, D. (2023). A biodegradable chitosan-based polymer for sustained nutrient release to stimulate groundwater hydrocarbon-degrading microflora. *Chemosphere*, *344*, 140346. https://doi.org/10.1016/j.chemosphere.2023.140346

Wang, Y., Huang, Y., Xi, P., Qiao, X., Chen, J., & Cai, X. (2021). Interrelated effects of soils and compounds on persulfate oxidation of petroleum hydrocarbons in soils. *Journal of Hazardous Materials*, *408*, 124845. https://doi.org/10.1016/j.jhazmat.2020.124845

Wei, K.-H., Zheng, Y.-M., Sun, Y., Zhao, Z.-Q., Xi, B.-D., & He, X.-S. (2024). Larger aggregate formed by self-assembly process of the mixture surfactants enhance the dissolution and oxidative removal of non-aqueous phase liquid contaminants in aquifer. *Science of the Total Environment*, *912*, 169532. https://doi.org/10.1016/j.scitotenv.2023.169532

Wei, Y., Ding, D., Qu, K., Sun, J., & Cui, Z. (2022). Ecological risk assessment of heavy metal pollutants and total petroleum hydrocarbons in sediments of the Bohai Sea, China. *Marine Pollution Bulletin*, *184*, 114218. https://doi.org/10.1016/j.marpolbul.2022.114218

Widera, B., Tyszkiewicz, N., Truu, J., Rutkowski, P., Młynarz, P., & Pasternak, G. (2024). Relationship between biodiversity and power generated by anodic bacteria enriched from petroleum-contaminated soil at various potentials. *International Biodeterioration & Biodegradation*, *194*, 105849. https://doi.org/10.1016/j.ibiod.2024.105849

Wu, M., Ma, C., Wang, D., Liu, H., Zhu, C., & Xu, H. (2020). Nutrient drip irrigation for refractory hydrocarbon removal and microbial community shift in a historically petroleum-contaminated soil. *Science of the Total Environment*, *713*, 136331. https://doi.org/10.1016/j.scitotenv.2019.136331

Wu, M., Wu, J., Zhang, X., & Ye, X. (2019). Effect of bioaugmentation and biostimulation on hydrocarbon degradation and microbial community composition in petroleum-contaminated loessal soil. *Chemosphere*, *237*, 124456. https://doi.org/10.1016/j.chemosphere.2019.124456

Xia, M., Chakraborty, R., Terry, N., Singh, R. P., & Fu, D. (2020). Promotion of saltgrass growth in a saline petroleum hydrocarbons contaminated soil using a plant growth promoting bacterial consortium. *International Biodeterioration & Biodegradation*, *146*, 104808. https://doi.org/10.1016/j.ibiod.2019.104808

Xue, Y., Chen, L., Xiang, L., Zhou, Y., & Wang, T. (2023). Experimental investigation on electromagnetic induction thermal desorption for remediation of petroleum hydrocarbons contaminated soil. *Journal of Environmental Management*, *328*(December 2022), 117200. https://doi.org/10.1016/j.jenvman.2022.117200

Xue, Z., Zuo, R., Ding, F., Wu, Z., Pan, M., Cai, W., Xu, Y., & Wang, J. (2022). Microwave-induced steam distillation (MISD) remediation in petroleum hydrocarbon-contaminated sites: From process improvement to pilot application. *Environmental Pollution*, *313*, 120059. https://doi.org/10.1016/j.envpol.2022.120059

Yang, S., Zhang, J., Liu, Y., & Feng, W. (2023). Biodegradation of hydrocarbons by Purpureocillium lilacinum and Penicillium chrysogenum from heavy oil sludge and their potential for bioremediation of contaminated soils. *International Biodeterioration & Biodegradation*, *178*, 105566. https://doi.org/10.1016/j.ibiod.2023.105566

Yousaf, U., Ali Khan, A. H., Farooqi, A., Muhammad, Y. S., Barros, R., Tamayo-Ramos, J. A., Iqbal, M., & Yousaf, S. (2022). Interactive effect of biochar and compost with Poaceae and Fabaceae plants on remediation of total petroleum hydrocarbons in crude oil contaminated soil. *Chemosphere*, *286*, 131782. https://doi.org/10.1016/j.chemosphere.2021.131782

Zhang, J., Feng, Y., Hu, T., Xu, X., Zhao, D., Zhao, J., Wang, X., Li, L., Wang, S., Song, C., & Zhao, S. (2024). Antibiotics and polycyclic aromatic hydrocarbons in marine food webs of the Yellow River Estuary: Occurrence, trophic transfer, and human health risks. *Science of the Total Environment*, *943*, 173709. https://doi.org/10.1016/j.scitotenv.2024.173709

Zhen, M., Chen, H., Liu, Q., Song, B., Wang, Y., & Tang, J. (2019). Combination of rhamnolipid and biochar in assisting phytoremediation of petroleum hydrocarbon contaminated soil using Spartina anglica. *Journal of Environmental Sciences*, *85*, 107–118. https://doi.org/10.1016/j.jes.2019.05.013

Zou, Y., Hu, Y., Li, S., Huang, X., Cheng, X., & Pan, W. (2024). Remediation of crude oil contaminated soil through an integrated biological-chemical-biological strategy. *Science of the Total Environment*, *919*, 170756. https://doi.org/10.1016/j.scitotenv.2024.170756

12 Technology and Assessment Emphasizing Fungal Bioremediation of Hydrocarbon-Contaminated Sites

Priya Dubey and Aditi Roy

12.1 INTRODUCTION

The rapid industrialization and increased dependency on petroleum products over the last century have led to significant environmental challenges, one of the most pressing being the contamination of soil and water with hydrocarbons. Hydrocarbon pollution, originating from sources such as oil spills, industrial discharges, and improper waste disposal, poses severe risks to ecosystems and human health (Bala et al., 2024; Dey, Bissoyi, Koteswara Reddy, et al., 2024; Dey, Dash, Ray, et al., 2024). Addressing these environmental hazards necessitates innovative and effective remediation strategies. In recent years, bioremediation has emerged as a promising and sustainable approach to detoxify contaminated environments. Problems in the oil industry, such as leakages, spills, and environmental accidents involving contaminants like fossil fuels and their by-products, occur frequently. The remediation of these incidents has become essential, and environmental control agencies impose penalties when corporate responsibilities are identified. Among the various contaminants, polycyclic aromatic hydrocarbons (PAHs) are particularly concerning due to their hazardous effects on human health, necessitating specific remediation processes. Numerous studies have focused on treating water and soil contaminated with PAHs using methods such as solvent extraction, bioremediation, phytoremediation, chemical oxidation, photocatalytic degradation, electrokinetic remediation, thermal treatment, enzymatic processes, and algae-based treatments targeting specific compounds. Additionally, combined remediation approaches, such as chemical and biological treatments, have also been employed.

The commonly employed physical and chemical methods for environmental remediation are not only costly but also produce hazardous by-products. Studies indicate that more suitable techniques are available for removing these contaminants, with bioremediation standing out as a natural approach. Bioremediation leverages organisms or their by-products to reduce or eliminate the negative

DOI: 10.1201/9781003508496-12

effects of pollutants in the environment. The eco-friendly agents used in bioremediation include microorganisms, enzymes, or plants (Das et al., 2024; Das et al., 2015; Kanungo et al., 2023; Nag et al., 2023; Paul et al., 2023). This method is primarily applied to contaminated matrices such as soil, sludges, and various types of wastewaters, and it can be executed either in-situ or ex-situ using bioreactors. Bioremediation is particularly attractive due to its cost-effectiveness and environmental benefits, especially when applied in-situ. In-situ techniques may involve bioaugmentation through inoculation or biostimulation by adding microbial growth-promoting formulations. Among various bioremediation techniques, fungal bioremediation, or mycoremediation, has garnered significant attention due to fungi's unique metabolic capabilities and adaptability. Fungi, particularly white-rot fungi, are equipped with an extensive enzymatic arsenal capable of degrading complex organic pollutants, including persistent hydrocarbons. Fungi play a crucial role in environmental detoxification, including the purification of wastewater from various toxins like heavy metals, persistent organic pollutants, textile dyes, and pollutants from industries such as leather tanning and petroleum refining. They are also effective in degrading contaminants like PAHs, pharmaceuticals, personal care products, pesticides, and herbicides in diverse environments such as land, freshwater, and marine habitats. Utilizing fungi for remediation purposes is not only cost-effective but also environmentally friendly. Additionally, enzymes secreted by fungi, such as laccases, hold significant promise for enhancing bioremediation. These microorganisms utilize extracellular enzymes, such as lignin peroxidase, manganese peroxidase, and laccase, to break down and assimilate hydrocarbons, transforming them into less harmful substances (Das & Ghosh, 2023; Ghosh et al., 2023). The deployment of fungal bioremediation requires a thorough understanding of both the biological mechanisms involved and the technological applications that facilitate this process. Advances in biotechnology and environmental engineering have enhanced the efficacy of fungal bioremediation. Techniques such as genetic engineering, optimization of growth conditions, and the development of fungal consortia have improved fungi's ability to thrive in contaminated environments and degrade pollutants more efficiently. Furthermore, the integration of monitoring and assessment tools, including molecular biology techniques and remote sensing technologies, has enabled more precise evaluation and management of bioremediation efforts. This chapter delves into the intersection of technology and assessment in the context of fungal bioremediation of hydrocarbon-contaminated sites. It explores the current state of knowledge on fungal mechanisms for hydrocarbon degradation, examines the technological innovations enhancing fungal bioremediation, and discusses the methodologies for assessing the effectiveness and environmental impact of these bioremediation efforts. By providing a comprehensive overview of these topics, the chapter aims to highlight the potential of fungal bioremediation as a viable solution for mitigating hydrocarbon pollution and restoring ecological balance in contaminated areas. As we confront the growing environmental challenges of the 21st century, harnessing the power of fungi through advanced technological and assessment frameworks represents a critical step toward achieving sustainable and effective remediation of hydrocarbon-contaminated sites.

12.2 SIGNIFICANCE OF FUNGI IN BIOREMEDIATION OF HYDROCARBON-CONTAMINATED SITES

Fungi play a significant role in the bioremediation of hydrocarbon-contaminated sites due to their unique enzymatic capabilities and metabolic versatility. Unlike many other organisms, fungi, particularly white-rot fungi, possess the ability to degrade a wide range of hydrocarbon compounds, including complex and recalcitrant ones. Their enzymatic machinery, comprising lignin peroxidase, manganese peroxidase, and laccase among others, enables them to break down hydrocarbons into simpler, less harmful substances (Figure 12.1). This ability makes fungi effective agents for remediation in various environments, from soil to water bodies contaminated with hydrocarbons originating from sources such as oil spills, industrial discharges, and waste disposal (Behera et al., 2020, 2021, 2022; Dey et al., 2023). Furthermore, fungi are adept at thriving in diverse environmental conditions, making them adaptable to a range of contaminated sites. Their role in bioremediation not only offers a sustainable and environmentally friendly approach to mitigating hydrocarbon pollution but also holds promise for addressing some of the most pressing environmental challenges of our time. Specific fungal genera, such as *Pleurotus, Trametes*, and *Phanerochaete*, have demonstrated exceptional abilities to break down and transform these persistent contaminants. The mechanisms behind fungal bioremediation are complex, involving a range of enzymes, such as laccases, peroxidases, and cytochrome P450 monooxygenases, which play crucial roles in the degradation process (Akhtar & Mannan, 2020). Recent advances in analytical chemistry and genetic engineering have provided new insights into the molecular mechanisms underlying fungal biodegradation (Ghosal et al., 2016). Emerging integrative approaches, including metagenomics, meta-transcriptomics, metabolomics, and meta-proteomics, have

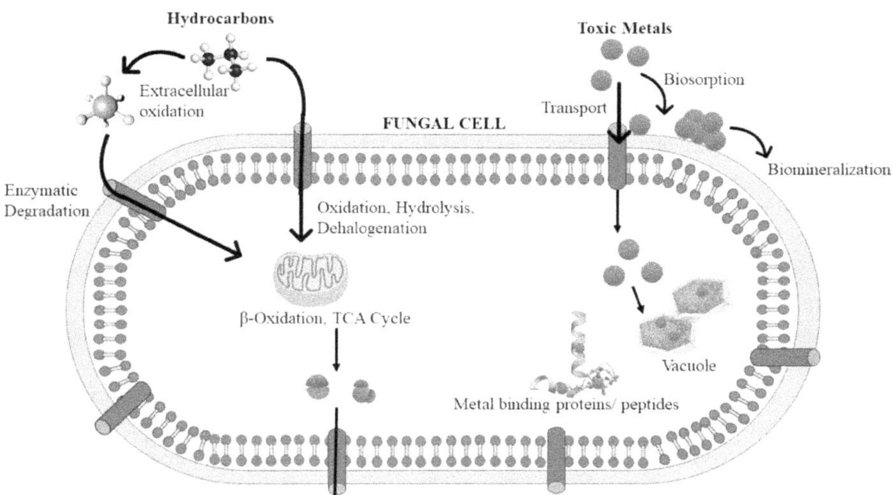

FIGURE 12.1 Diagram of fungal interactions with polycyclic aromatic hydrocarbons (PAHs) and toxic metals.

been used to elucidate the complex interactions between fungi and their associated microbial communities in the remediation of hydrocarbon-contaminated sites (Quintella et al., 2019).

Ulfig (2003) investigated the utilization of keratinolytic fungi as an indicator of hydrocarbon-based pollution stemming from petroleum refining activities at the Czechowice Oil Refinery in Southern Poland over 39 months. Their study specifically targeted the bioremediation of heavily contaminated soil at the Czechowice Oil Refinery site. However, further extensive testing is required to validate the results and determine the practical applicability of keratinolytic fungi in environmental monitoring. Before these fungi can be functionally employed in such monitoring endeavors, it is imperative to establish essential biological databases. The insights gleaned from their study and the potential applications of their findings hold promise for developing more cost-effective and environmentally safe technologies for the natural monitoring of petroleum-contaminated soil remediation efforts (Ulfig, 2003). Another study on the remediation of petroleum-contaminated marine sediments found that a combination of bacteria, fungi, and microalgae was effective in degrading a wide range of hydrocarbon compounds. A study reported the use of white-rot fungi, such as *Trametes versicolor*, in the remediation of soil contaminated with polycyclic aromatic hydrocarbons (PAHs), highlighting their ability to degrade these persistent pollutants (Akhtar & Mannan, 2020). In addition, fungal bioremediation has been successfully applied in the treatment of groundwater contaminated with petroleum hydrocarbons. These case studies underscore the versatility and effectiveness of fungal-based remediation strategies in addressing the environmental challenges posed by hydrocarbon contamination.

12.2.1 Mechanism of Hydrocarbon Degradation

Fungal degradation of hydrocarbons is a complex, multistep process involving the secretion of extracellular enzymes, cellular uptake, and intracellular metabolism. Fungi, particularly those with saprophytic and ligninolytic abilities, play a critical role in breaking down a wide range of hydrocarbons, including aliphatic, aromatic, and polycyclic compounds, which are major components of petroleum and other fossil fuels. The initial phase of hydrocarbon degradation begins with the secretion of oxidative enzymes such as peroxidases (e.g., lignin peroxidase, manganese peroxidase) and laccases. These enzymes catalyze the oxidation of hydrocarbons, facilitating the formation of more reactive intermediates, like alcohols, aldehydes, and ketones. Oxidative mechanisms are particularly effective for breaking down complex, high molecular weight PAHs, which are otherwise recalcitrant to microbial degradation (Bala et al., 2024; Dey, Bissoyi, Koteswara Reddy, et al., 2024b). Once oxidized, hydrocarbons are often transformed into simpler compounds through secondary enzymatic reactions like dehalogenation, hydroxylation, and epoxidation. These processes can lead to the cleavage of complex hydrocarbon rings, converting them into carboxylic acids and other small metabolites that are more easily transported across fungal cell membranes. Inside the fungal cells, these intermediates enter central metabolic pathways such as β-oxidation and the tricarboxylic acid (TCA) cycle, where they are further catabolized to produce acetyl-CoA, NADH, and $FADH_2$, contributing to cellular respiration and energy production.

In addition to enzymatic degradation, fungi can adapt to the presence of toxic hydrocarbons by modifying their cell membranes, which enhances permeability and facilitates hydrocarbon uptake. Fungi can also employ active transport systems to import smaller metabolites generated during extracellular degradation. Furthermore, many fungi exhibit an ability to sequester toxic intermediates, such as halogenated hydrocarbons, in vacuoles or bind them with metal-binding proteins, reducing their toxicity to the fungal cell.

12.3 ROLE OF ENZYMES AND METABOLITES

Many transformative interactions between fungi and various pollutants rely on the secretion of diverse extracellular substances and metabolites (Kirtzel et al., 2020). Fungi degrade petroleum hydrocarbons by producing enzymes such as laccases, tyrosinases, manganese peroxidases, cytochrome P450 monooxygenases, and reductive dehalogenases, while also influencing metal speciation through the secretion of other metabolites like organic acids, amino acids, siderophores, and extracellular proteins. Several fungal species use petroleum hydrocarbons as a carbon and energy source, incorporating them into fungal biomass. Fungal taxa like *Amorphoteca, Neosartorya, Talaromyces, Aspergillus, Fusarium, Paecilomyces, Sporobolomyces, Cephalosporium, Penicillium*, and *Graphium* have been identified as potent petroleum hydrocarbon degraders (Das & Chandran, 2011; Varjani & Patel, 2017).

Some species oxidize pollutants, such as phenols and aromatic amines, via extracellular enzyme production, including laccases (Martínková et al., 2016), manganese peroxidases (Zhang, 2016), and lignin peroxidases (Falade et al., 2017). Additionally, fungal cell membranes are permeable to various organic pollutants, which are degraded by intracellular enzymes like cytochrome P450, reductive dehalogenases (Stella, 2020), and nitroreductases (Singh et al., 2020). These pollutants are then broken down into simpler organic compounds, followed by further metabolism through pathways such as β-oxidation and entry into the TCA cycle (Varjani & Patel, 2017). The degradation process involves multiple enzymatic steps, and the biodegradability of hydrocarbons depends on their chemical structure and other factors affecting fungal growth, metabolism, and pollutant chemistry. Generally, the biodegradation efficiency is maximum in linear alkanes followed by branched alkanes, small aromatics, and lastly cyclic alkanes (Das & Chandran, 2011; Varjani & Patel, 2017).

In co-contaminated soils, petroleum hydrocarbons can serve as a carbon and energy source for certain fungal species, while toxic metals can have notable effects on fungal populations and their activity. Despite the toxicity of many metal species, numerous fungi are capable of thriving in contaminated environments, though shifts in species composition may occur (Fomina et al., 2005). The primary mechanisms enabling survival involve alterations in metal mobility, which result from various fungal tolerance and resistance strategies (Gadd, 2010). Fungi possess several mechanisms that mitigate metal toxicity and influence metal mobility, including the production of metal-binding proteins, organic and inorganic precipitation, active transport systems, and intracellular compartmentalization. Additionally, fungal cell walls, along with their pigments and polysaccharides, exhibit significant

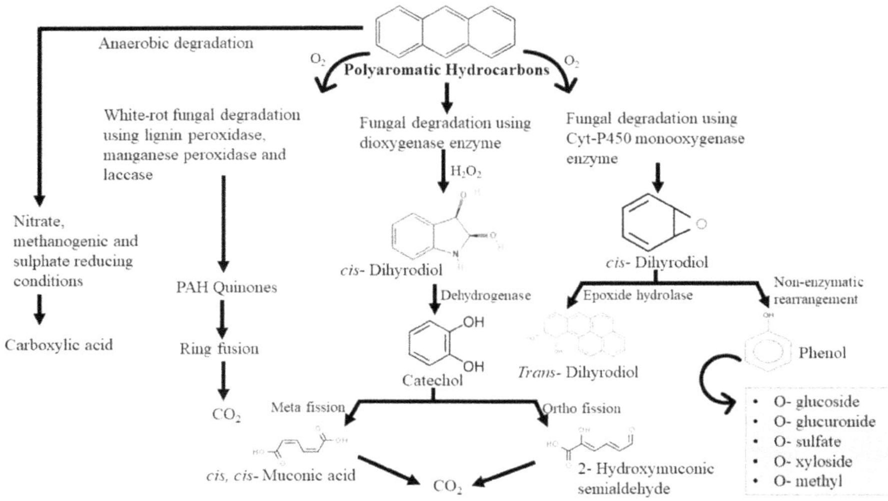

FIGURE 12.2 Fungal biodegradation of PAHs.

metal-binding capacities. The secretion of fungal metabolites, such as metal-binding peptides, polysaccharides, amino acids, and organic acids, plays a crucial role in metal and mineral transformations, contributing to both the mobilization and immobilization of metal species (Gadd, 2007; Gadd et al., 2014). Additionally, fungal phenolic polymers and melanin contain numerous potential metal-binding sites, which feature oxygen-containing groups like carboxyl, phenolic and alcoholic hydroxyl, carbonyl, and methoxyl groups (Figure 12.2).

Metal immobilization is particularly significant for bioremediation strategies, and fungi can mediate the precipitation of metals in the form of insoluble oxalates, oxides, carbonates, and phosphates (Gadd et al., 2014; Liang & Gadd, 2017). For instance, phosphate release from organic or inorganic phosphate hydrolysis has been shown to be an effective method for immobilizing metals such as Zn, Pb, La, and U, which precipitate on and around fungal hyphae (Liang & Gadd, 2017). Urease-positive fungi, like *Neurospora crassa*, *Pestalotiopsis* sp., and *Myrothecium gramineum*, are promising candidates for toxic metal immobilization due to their ability to hydrolyze urea. When grown in urea-containing media, these fungi produce ammonia and free carbonate, leading to the precipitation of metals as carbonates, including $BaCO_3$, $CdCO_3$, $CoCO_3$, $Cu_2(OH)_2CO_3$, $La_2(CO_3)_3$, and $NiCO_3$ (Liang & Gadd, 2017; Li et al., 2020).

Additionally, fungi can produce various metal oxalates through interactions with metals and metal-bearing minerals, including those of Ca, Cd, Co, Cu, Mg, Mn, Sr, Zn, Ni, and Pb (Gadd et al., 2014). Extracellular proteins, amino acids, and polysaccharides also play key roles in toxic metal immobilization. For example, extracellular nickel precipitation has been linked to the removal of extracellular proteins (Li et al., 2020), with studies showing that extracellular proteins may act as templates for biomineral formation, influencing the size and structure of the resultant minerals (Li et al., 2020).

12.4 MATRIX FOR HYDROCARBON REMEDIATION

The matrix studied are soil, water, wastewater, sludge, and including keywords like metabolites, genes, phytoremediation, and contaminants (Figure 12.3). Fungal bioremediation of hydrocarbons has been extensively studied using various matrices such as soil, water, wastewater, sludge, and through processes like phytoremediation. Soil has emerged as one of the most commonly explored matrices due to its complex microbial interactions and the persistence of hydrocarbons in terrestrial environments. Water and wastewater matrices are critical in understanding fungal adaptability and efficiency in aquatic ecosystems, where hydrocarbons can disperse widely. Sludge, often a by-product of industrial processes, serves as a rich medium for fungal growth, providing both nutrients and pollutants for degradation studies. Phytoremediation, in conjunction with fungal activity, highlights the symbiotic relationship between plants and fungi in hydrocarbon degradation, enhancing the bioremediation process. Additionally, studies focusing on metabolites and genes involved in fungal bioremediation have provided insights into the enzymatic pathways and molecular mechanisms fungi employ to degrade hydrocarbons. Understanding contaminants at the molecular level has furthered the development of more targeted and efficient fungal bioremediation strategies. These matrices collectively form the foundation for advancing fungal bioremediation applications in hydrocarbon-contaminated environments.

Figure 12.3a shows the accumulated number of articles published and cited in the past 20 years for fungal bioremediation in the soil matrix. Fungal remediation of hydrocarbons in soil has garnered significant research attention, with a total of 360 publications addressing the topic. The highest number of publications and citations occurred in 2021, marking it as a peak year for advancements in this field, generating around 1,507 citations and 28 publications. This surge reflects an increased global focus on sustainable solutions to soil pollution, particularly through the use of fungi, which are known for their robust enzymatic capabilities in breaking down complex hydrocarbon compounds. Following 2021, the number of publications remained strong in 2022 and 2023, indicating sustained interest and ongoing developments in fungal bioremediation techniques. A total of 1,508 citations and 26 publications were generated in 2022, and 977 citations, and 27 publications in 2023. Additionally, 2019 saw notable contributions to the research, laying the groundwork for the subsequent surge in scholarly activity. The total publications for the topic were around 25, generating around 1,446 citations. The trend suggests a growing recognition of the potential of fungal species in mitigating hydrocarbon pollution in soil, emphasizing their role in environmental cleanup and the importance of optimizing fungal bioremediation strategies.

Fungal bioremediation of hydrocarbons in water and wastewater has been the focus of specialized research, with 88 publications addressing its application in water environments and 17 in wastewater treatment (Figure 12.3b and c). For hydrocarbon bioremediation in water, the year 2022 saw the highest citation count, reflecting significant advancements in fungal capabilities to degrade pollutants in aquatic systems. This was followed by continued interest in 2023 and 2021, highlighting ongoing research into optimizing fungal species and conditions for water purification.

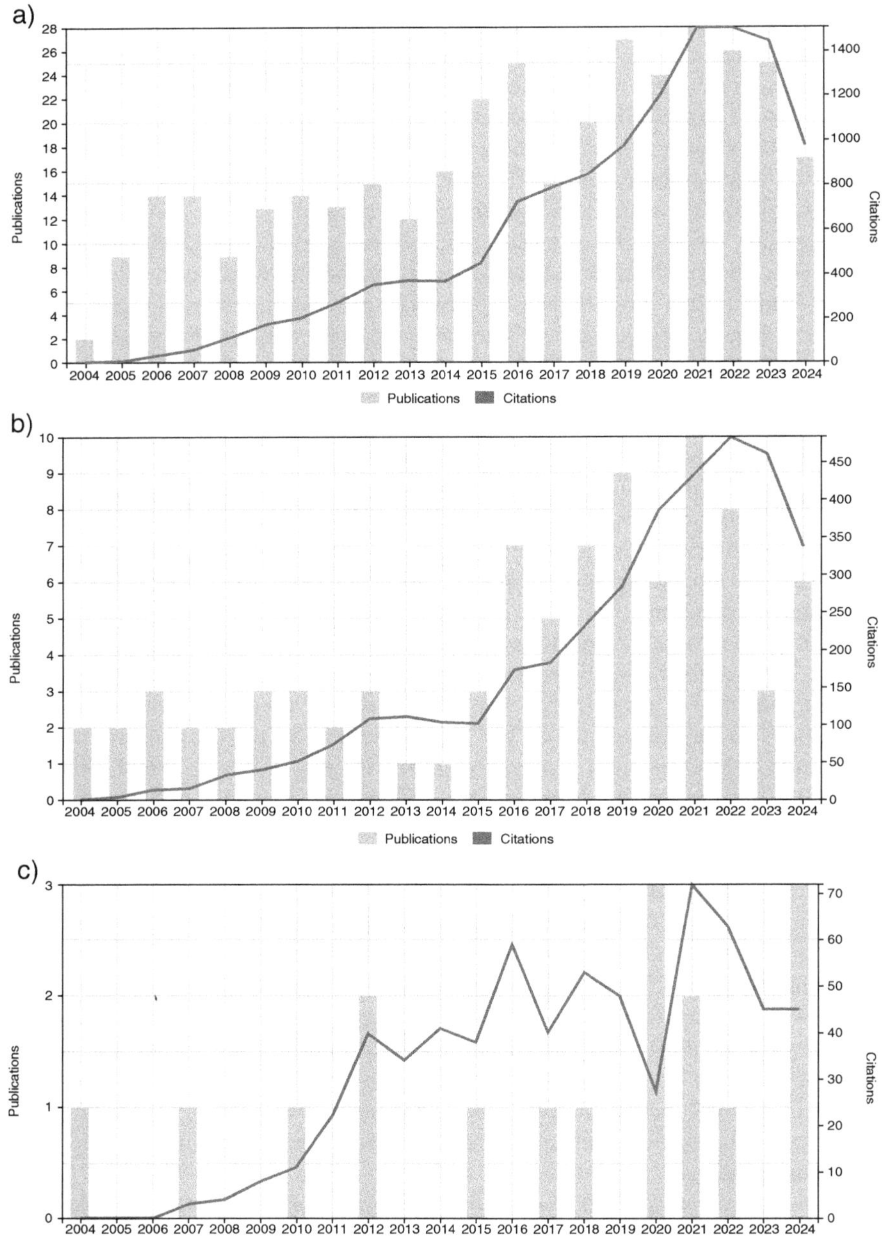

FIGURE 12.3 Accumulated numbers of citation and publication from year 2004 to 2024 for (a) soil, (b) water, (c) wastewater, (d) sludge, (e) enzyme and metabolites, (f) genes, (g) contaminants, and (h) phytoremediation.

(Continued)

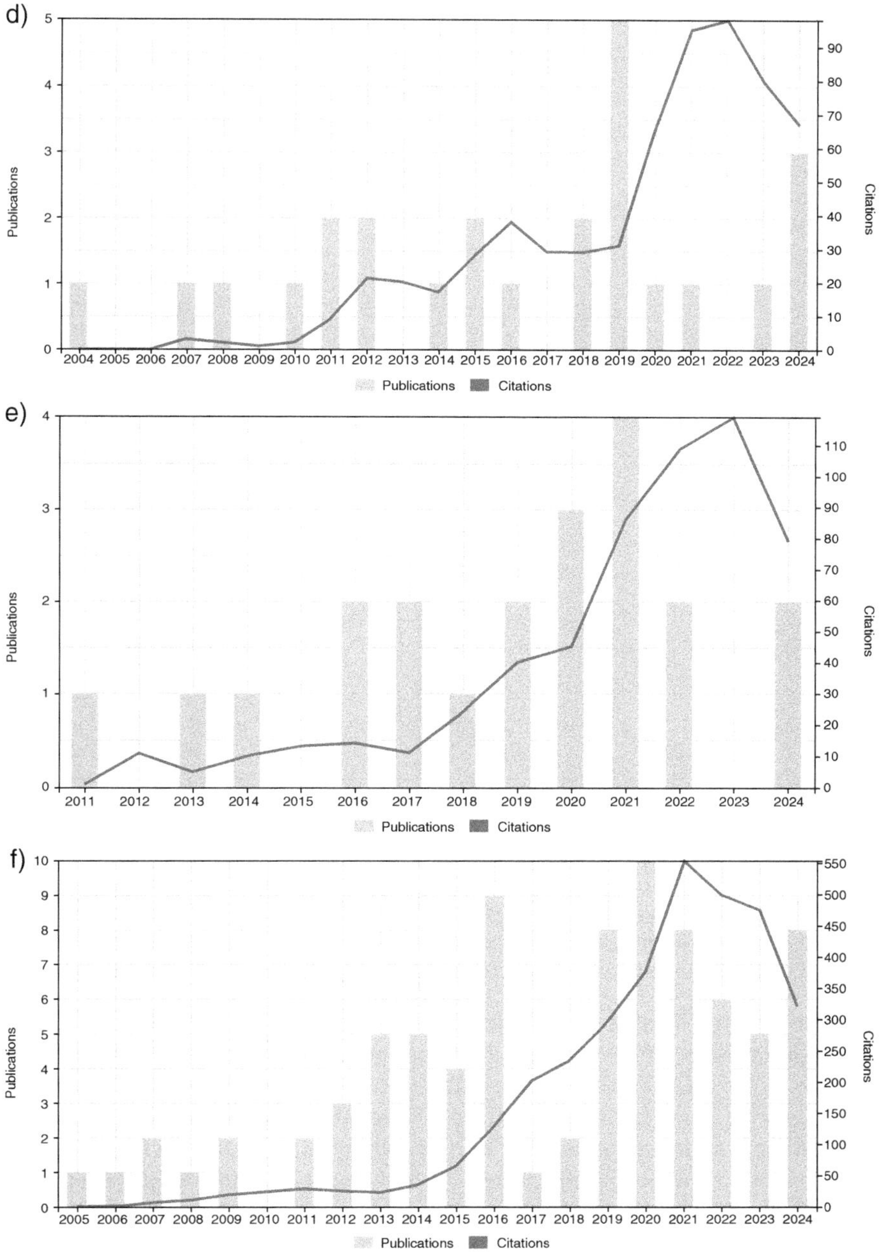

FIGURE 12.3 (*Continued*) Accumulated numbers of citation and publication from year 2004 to 2024 for (a) soil, (b) water, (c) wastewater, (d) sludge, (e) enzyme and metabolites, (f) genes, (g) contaminants, and (h) phytoremediation.

(Continued)

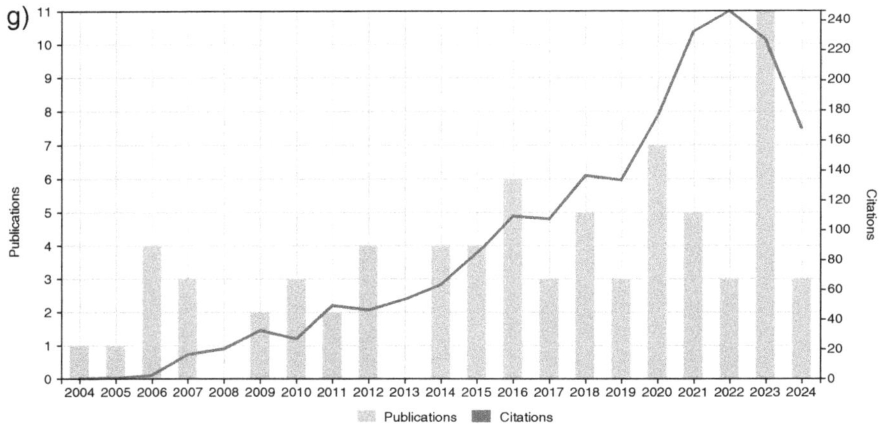

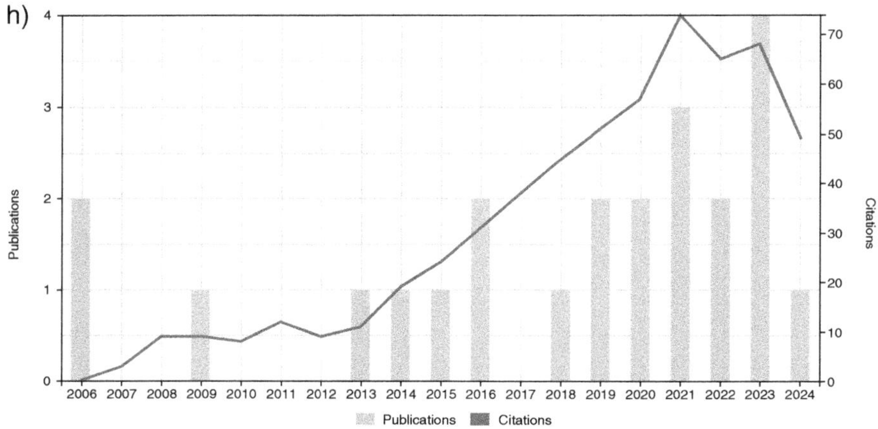

FIGURE 12.3 (*Continued*) Accumulated numbers of citation and publication from year 2004 to 2024 for (a) soil, (b) water, (c) wastewater, (d) sludge, (e) enzyme and metabolites, (f) genes, (g) contaminants, and (h) phytoremediation.

The number of publications, however, was maximum in 2021, publishing around ten articles followed by 2022 and 2023. In wastewater, where conditions are more complex due to the presence of diverse contaminants, the peak in citations occurred in 2021, signaling important breakthroughs in fungal-based treatments for hydrocarbon degradation in such environments. The research trends indicate growing recognition of fungi as an effective biological tool for both water and wastewater remediation, driven by their ability to adapt to varying conditions and break down hydrocarbons, offering sustainable solutions for environmental pollution management.

Fungal remediation of hydrocarbons in sludge has been explored in 25 publications, reflecting a niche yet important area of research in environmental biotechnology (Figure 12.3d). The highest citation count occurred in 2022 (98), indicating key advancements in understanding how fungal species can degrade hydrocarbons in sludge, a complex and nutrient-rich matrix often associated with industrial and wastewater treatment processes. The research momentum continued in 2021 (95) and 2023 (80), highlighting the sustained interest in utilizing fungi for sludge bioremediation. These studies emphasize the potential of fungi to efficiently break down persistent organic pollutants in sludge, where other biological methods may fall short due to the dense and contaminated nature of the material. Fungi's enzymatic diversity and adaptability to harsh conditions make them a promising tool for treating hydrocarbon-contaminated sludge, contributing to sustainable waste management practices.

Fungal remediation of hydrocarbons with a focus on enzymes and metabolites has been explored in 21 publications, underscoring the significance of fungal biochemistry in environmental cleanup (Figure 12.3e). The year 2023 saw the highest citation count (119), reflecting substantial advancements in understanding the enzymatic pathways and metabolites involved in fungal degradation of hydrocarbons. This was followed by 2022 (109) and 2021 (86), indicating a growing body of research aimed at identifying key enzymes, such as lignin peroxidases, laccases, and manganese peroxidases, that enable fungi to break down complex hydrocarbon molecules. In 2024, citations (81) further solidified the role of fungal metabolites, which act as intermediates or by-products during the degradation process, providing insights into the molecular mechanisms behind bioremediation. These studies collectively emphasize the potential of harnessing fungal enzymes and metabolites to enhance hydrocarbon degradation efficiency, opening pathways for more targeted and eco-friendly bioremediation strategies.

Fungal remediation of hydrocarbons, particularly focusing on the genes involved in the degradation process, has gained increasing attention, with a total of 83 publications shedding light on this critical area of bioremediation (Figure 12.3f). The year 2021 stands out as the most impactful year in terms of citations (553), indicating significant advances in identifying and characterizing the genes that play a vital role in the fungal degradation of hydrocarbons. These genes are responsible for encoding key enzymes, such as monooxygenases, dioxygenases, laccases, and peroxidases, which are crucial for breaking down complex hydrocarbon molecules into simpler, less harmful compounds. Following 2021, the research continued to gain traction with substantial citations in 2022 (499) and 2023 (475), reinforcing the importance of genetic research in fungal bioremediation. In terms of publication trends, 2020 saw the highest number of studies (10), marking a pivotal year for foundational research in this field. This was followed by eight publications in 2021, six in 2022, and five in 2023, indicating a steady stream of research focused on unraveling the genetic basis of fungal hydrocarbon degradation. The research during these years emphasized the identification of specific gene clusters involved in hydrocarbon metabolism, regulation of gene expression in response to pollutant stress, and the potential for genetic manipulation to enhance the bioremediation capabilities of fungi. The insights gained from these studies have opened new avenues for optimizing fungal strains through genetic engineering, allowing for the development of more robust and efficient bioremediation agents. Researchers have also explored the horizontal gene transfer between fungal species, which could lead to the evolution of new metabolic pathways for hydrocarbon degradation. These genetic studies are crucial not only for understanding the natural mechanisms of fungal remediation but also for designing biotechnological interventions to combat environmental pollution, especially in areas contaminated with petroleum hydrocarbons, PAHs, and other recalcitrant organic pollutants. Thus, the genetic approach to fungal bioremediation holds immense promise for creating sustainable, scalable solutions to address the global issue of hydrocarbon contamination in diverse environments (Das et al., 2024; Dey, Dash, & Das, 2024; Satapathy et al., 2024).

Fungal remediation of hydrocarbons, with a focus on the specific contaminants involved, has been extensively studied in 74 publications (Figure 12.3g). The research peaked in terms of citations in 2022 (246), highlighting key advancements in understanding how fungi degrade various hydrocarbon pollutants, including petroleum hydrocarbons, PAHs, alkanes, and other recalcitrant organic compounds. The significant citations in 2021 (232) further emphasized the ability of fungi to tackle complex hydrocarbon structures that are otherwise resistant to microbial breakdown. By 2023, citations (227) remained high, reflecting continued interest in optimizing fungal strains for the degradation of diverse contaminants under different environmental conditions. The studies have focused on the metabolic pathway fungi employ to break down these hydrocarbons, often mediated by enzymes such as laccases, peroxidases, and cytochrome P450s. These enzymatic systems enable fungi to attack the carbon bonds in hydrocarbon pollutants, leading to their mineralization or transformation into less harmful by-products. Moreover, research has delved into the ability of fungi to degrade highly toxic and persistent contaminants like PAHs, which are often found in oil spills and industrial waste. This knowledge has contributed to the development of more effective bioremediation strategies, tailored to address specific contaminants in soil, water, and sludge environments. Overall, fungal remediation offers a promising, eco-friendly approach to mitigating the environmental impact of hydrocarbon contamination.

Fungal remediation of hydrocarbons in combination with phytoremediation has gained increasing attention, with 23 publications exploring this synergistic approach (Figure 12.3h). The highest number of citations occurred in 2021 (74), highlighting key breakthroughs in understanding how fungi and plants work together to enhance hydrocarbon degradation in contaminated environments. This was followed by continued research in 2023 (68) and 2022 (65), reflecting sustained interest in optimizing the combined use of fungi and plants for bioremediation. In fungal–phytoremediation systems, plants and fungi form symbiotic relationships that enhance the degradation of hydrocarbons. Fungi, particularly mycorrhizal fungi, colonize plant roots, extending their hyphal networks into the surrounding soil, which increases the bioavailability of hydrocarbons for both fungi and plants. The fungi produce enzymes such as laccases, peroxidases, and oxidases, which break down complex hydrocarbons, while plants help stabilize the soil, improve aeration, and secrete root exudates that stimulate fungal activity. This dual approach has proven especially effective in tackling persistent organic pollutants like PAHs and aliphatic hydrocarbons in contaminated soils and water bodies. The combined remediation process leverages the natural capabilities of both fungi and plants, making it a sustainable and eco-friendly solution to mitigating hydrocarbon pollution.

12.5 USE OF MYCOREMEDIATION FOR OIL CONTAMINATION

This section provides a detailed technological evaluation of oil bioremediation, focusing on oil as the primary contaminant within the matrix and highlighting the bioremediation agents with the greatest potential for advancing future technologies. Soils and fungi were selected for their promising prospects, as they rank second in terms of both articles and patents. For instance, patents involving fungi for soil bioremediation following oil spills often feature Aspergillus or Penicillium due to their widespread availability. These patents originate from the US, Great Britain, China, the European Community, and Brazil. An example is the patent "Delivery Systems for Mycotechnologies, Mycofiltration, and Mycoremediation" from Cisco Technology, which incorporates fungal spore masses or hyphal fragments into landscaping cloths, fiber substrates, paper products, hydroseeding equipment, and agricultural tools. The fungi used may include saprophytic types, gourmet and medicinal mushrooms, mycorrhizal fungi, entomopathogenic fungi, parasitic fungi, and fungi imperfecti. These fungi play a crucial role as keystone species in the bioremediation of toxic wastes and contaminated sites (Stamets, 2002).

The invention titled "Process for Preparing Solid Fungus to Degrade Petroleum," developed by the Institute of Applied Ecology and the Chinese Academy of Sciences, details a method for creating a solid microbial preparation specifically designed to break down petroleum. This preparation involves a mixture of various fungi, including Mucor, Penicillium, Aspergillus, Trichoderma, and white-rot fungi. These fungi are cultured using techniques such as slant culture, shake-flask culture, or solid culture, and then combined in precise proportions. The solid microbial preparation is intended to enhance the degradation of petroleum by leveraging the diverse enzymatic capabilities of these fungal species to target and break down complex hydrocarbon compounds present in petroleum (Li et al., 2001).

Another noteworthy invention, "Process of Biodegradable Attainment of Products for Application in the Remediation of Ground, Contaminated Waters, and Environments and Method of Functioning," from the Brazilian Federal University of Bahia, outlines an innovative approach for in-situ bioremediation. This process is designed to produce biodegradable products that are used to remediate contaminated soils and mangrove ecosystems affected by a variety of pollutants, including spilled oil, residual oil, fuel oils, polishes, fats, aromatic hydrocarbons, naphthalene, and polycyclic hydrocarbons. The bioremediation process utilizes Aspergillus species in conjunction with by-products from biodiesel production, such as crude glycerin and castor cake. This combination facilitates the breakdown of contaminants in mangrove sediments, which are characterized by their sandy and saline conditions. This approach aims to restore polluted mangrove environments by enhancing the natural degradation processes through the application of these biodegradable products (Quintella et al., 2019). The invention titled "Electrically-Assisted Microbial Remediation Method for Oil-Polluted Soil," developed by the Chinese Northeast Forestry University, describes an ex-situ remediation process with the following steps: (1) introduction of the oil-contaminated soil; (2) addition of a nutrient solution containing nitrogen and phosphorus; (3) incorporation of a microbial blend that includes bacteria and fungi such as *Fusarium subglutinans, Pseudallescheria boydii,*

Eupenicillium crustaceum, Isaria farinosa, Aspergillus terreus, and *Aspergillus protuberus*; (4) inclusion of organic additives like surfactants; (5) installation of a wire mesh system with a resistance wire heating plate, a parallel temperature controller, and an external power plug; (6) placement of the mixture into a reaction tank with the heating plate at the bottom and activation of the power supply (Jia et al., 2013).

The patents related to oil bioremediation of soils highlight various approaches using a combination of bioremediation agents. The US patent "Composition and Methods of Use" by Janet Angel involves a mix of microbes, enzymes, emulsifiers, and nutrients to remove or degrade hydrocarbons in contaminated areas, utilizing bacteria, fungi, algae, or their combinations (Angel, 2014). Another patent, "Biocatalytic Composition for Treatment of Substrates," filed by Elio Fabio Bortoli and recognized in the European Community, the US, and Mexico, describes a biocatalytic blend used for agricultural, zootechnical, and environmental recovery. This composition includes components with coenzymatic and enzymatic activities, such as various vitamins, fats, proteins, and a range of microorganisms, including Bacillus species and Aspergillus species (Bortoli, 2015). Additionally, the invention "Composite Inoculant for Deep Oil-Contaminated Soil Bioremediation Processes," by Chinese Xi'an Huanuo Environment Protection, focuses on a composite inoculant consisting of fungi and bacteria, including *Aspergillus niger* and *Penicillium chrysogenum*, designed to achieve over 90% degradation of petroleum hydrocarbons in deep soil (Zhang, 2016). Overall, the trend indicates that effective oil bioremediation often involves a strategic combination of fungi and bacteria to enhance degradation (Table 12.1).

TABLE 12.1
Biodegradation of Hydrocarbons through Different Fungal Species

Fungus Species	Hydrocarbons	References
Keratinolytic fungi	As an indicator of soil petroleum hydrocarbon contamination and associated bioremediation contamination	Ulfig (2003)
Phanerochaete chrysosporium	Significant reduction of TPH (total petroleum hydrocarbon)	Borazjani et al. (2005)
Autochthonous filamentous fungi	Bioremediation of a soil-contaminated hydrocarbon	D'Annibale et al. (2006)
Basidiomycete fungi	Impact of basidiomycete fungi on the wettability of soil	Hallett et al. (2006)
Filamentous fungi	Removal of weathered hydrocarbons from polluted soil	Perez-Armendariz et al. (2010)
Fungi isolated from soil and the marine environment	Bioremediation of industrial effluents	Saritha et al. (2010)
Aspergillus niger	Bioremediation of endosulfan-contaminated soil	Bhalerao (2012)

(*Continued*)

TABLE 12.1 (*Continued*)
Biodegradation of Hydrocarbons through Different Fungal Species

Fungus Species	Hydrocarbons	References
Acromonium sp., *Alternaria* sp., *Aspergillus terreus* and *Penicillium* sp.	Petroleum removal	Mohsenzadeh et al. (2012)
Aspergillus niger, Aspergillus terreus, Rhizopus sp., and *Penicillium* sp.	Bioremediation of oil-contaminated soil	Lotfinasabasl et al. (2012)
Aspergillus flavus	Bioremediation of oil contamination	Maruthi et al. (2013)
Aspergillus and *Rhizopus* sp.	Mycoremediation of used engine oil-contaminated soil	Thenmozhi et al. (2013)
Fungi belonging to different genera like *Aspergillus, Curvularia, Drechslera, Fusarium* etc.	Bioremediation of PAHs-contaminated environments	Balaji et al. (2014)
Aspergillus niger, Aspergillus fumigatus, Fusarium solani, and *Penicillium funiculosum*	Biodegradation of petroleum hydrocarbon	Al-Jawhari (2014)
Bionectria ochroleuca	Degradation of more than 70% of the C12–C28 hydrocarbons, and with 100% degradation of C12 and C28 hydrocarbons	Kota et al. (2014)
Penicillium sp.	Biodegradation of petrol	Vanishree et al. (2014)
Phanerochaete velutina	Bioremediation of PAH-contaminated soil	Winquist et al. (2014)
Aspergillus niger, Candida glabrata, Candida krusei, and Saccharomyces cerevisiae	Crude oil remediation	Burghal et al. (2016)
Acremonium sp. P0997 and *Bacillus subtilis*	Bioremediation of fluoranthene-contaminated sand	Ma et al. (2016)
Aspergillus niger and *Lichtheimia ramose*	Reduction of oil spill	Moustafa (2016)
Oenothera picensis	Bioremediation of a copper-contaminated soil	Cornejo et al. (2017)
Combined bacteria-white-rot fungi	Bioremediation of petroleum-contaminated soil	Liu et al. (2017)
Pleurotus pulmonarius	Bioremediation of crude oil-contaminated soil	Stanley et al. (2017)
Penicillium sp.	Decane	Govarthanan et al. (2017)
Aspergillus sp.	N-hexadecane	Al-Hawash et al. (2018)
Fusarium sp.	N-octadecane	Hidayat and Tachibana (2013)
Phomopsis liquidambari	Phenanthrene	Fu et al. (2018)
Irpex lacteus	Anthracene	Drevinskas et al. (2016)
Pleurotus ostreatus	Anthracene	Drevinskas et al. (2016)

(*Continued*)

TABLE 12.1 (*Continued*)
Biodegradation of Hydrocarbons through Different Fungal Species

Fungus Species	Hydrocarbons	References
Ganoderma lucidum	Pyrene	Agarwal et al. (2018)
Polyporus sp.	Chrysene	Hadibarata et al. (2009)
Acremonium sp.	Naphthalene, fluorine, phenanthrene, anthracene, and fluoranthene	Ma et al. (2016)
Pleurotus ostreatus	Pyrene, benzo[*a*]anthracene, chrysene, benzo[*k*]fluoranthene, benzo[*a*]pyrene, dibenzo[*a,h*] anthracene, and benzo[*ghi*] perylene	Baldrian et al. (2000)
Fusarium flocciferum, Trichoderma sp., *Trametes versicolor, and Pleurotus* sp.	Benzo[*a*]anthracene, benzo[*a*] fluoranthene, benzo[*a*]pyrene, chrysene, phenanthrene	Al-Turki (2009)
Pleurotus ostreatus	Crude oil	Anacletus et al. (2017)

12.6 SWOT ANALYSIS

A Strengths, Weaknesses, Opportunities, and Threats (SWOT) analysis is a valuable tool for evaluating the bioremediation technology, offering insights into its positioning and guiding policy and strategic decisions for academic researchers, companies, and government bodies. This analysis involves assessing the Strengths, Weaknesses, Opportunities, and Threats associated with the technology. Strengths may include the technology's proven effectiveness in degrading pollutants, its scalability, and the environmental benefits of using natural processes. Weaknesses might involve limitations such as high costs, slow processing times, or technical challenges in applying the technology to different types of contaminants or environments. Opportunities in the SWOT analysis could highlight the potential for technological advancements, the growing demand for sustainable environmental solutions, and the possibility of integrating bioremediation with other technologies for enhanced effectiveness. Threats may address factors such as regulatory hurdles, competition from alternative remediation technologies, or challenges in obtaining funding and public support.

Based on a review of relevant articles and patents, along with a critical evaluation by the authors, a comprehensive SWOT analysis of bioremediation technology at an international level was conducted. This analysis is detailed in a matrix (Table 12.2), which systematically categorizes and presents the key factors influencing the technology's adoption and development. The matrix serves as a strategic tool, helping stakeholders understand the current landscape, identify areas for improvement, and make informed decisions about advancing bioremediation technology and addressing potential challenges.

TABLE 12.2
SWOT Analysis

Strengths:

- **Biodegradability**: Fungi have a remarkable ability to break down complex hydrocarbons into simpler, less harmful substances.
- **Versatility**: Fungi can thrive in various environmental conditions, including extreme pH levels and temperatures, making them suitable for diverse bioremediation applications.
- **Cost-Effectiveness**: Fungal bioremediation can be more cost-effective compared to traditional remediation methods, such as chemical treatment or excavation.
- **Sustainability**: Fungal bioremediation is environmentally friendly and does not produce harmful by-products, making it a sustainable approach for hydrocarbon cleanup.
- **Selective Degradation**: Fungi can selectively target specific hydrocarbon contaminants, minimizing damage to the surrounding ecosystem.
- Technologies can be utilized across various bioremediation scenarios and for a broad spectrum of pollutants, facilitating entry into multiple markets;
- Adherence to environmental regulations to uphold sustainability;
- Biodegradable secondary products and residues;
- Strong capability to degrade hazardous compounds swiftly and effectively compared to alternative methods;
- Ability to function aerobically or anaerobically;
- Tolerance to elevated temperatures and salinity levels;
- Potential for synergistic partnerships with other remediation agents to enhance outcomes;

Weakness:

- **Time-Consuming**: Fungal bioremediation can be a slow process, particularly for complex hydrocarbon compounds, requiring weeks to months for significant remediation.
- **Specificity**: Some fungi may have limited effectiveness against certain types of hydrocarbons, requiring the selection of appropriate fungal strains for optimal results.
- **Environmental Factors**: The effectiveness of fungal bioremediation can be influenced by environmental factors such as temperature, pH, and nutrient availability, which may not always be conducive to optimal fungal growth.
- **Risk of Contamination**: Introducing non-native fungal species for bioremediation purposes may carry the risk of introducing invasive species, potentially disrupting local ecosystems.
- Expertise-focused technologies;
- Tailored technologies designed for specific customers that may not perform adequately in different settings;
- Ensuring optimal microorganism conditions during transportation

Opportunities:

- **Research and Development**: Continued research into fungal species and their enzymatic capabilities could lead to the discovery of more efficient and versatile fungi for hydrocarbon bioremediation.
- **Integration with Other Technologies**: Fungal bioremediation can be integrated with other remediation techniques, such as phytoremediation or microbial consortia, to enhance overall remediation efficiency.
- **Market Demand**: Growing environmental concerns and stricter regulations regarding hydrocarbon pollution create a significant market opportunity for fungal bioremediation technologies.

(*Continued*)

TABLE 12.2 (*Continued*)
SWOT Analysis

- **Public Awareness**: Increasing public awareness of the environmental benefits of fungal bioremediation could drive demand and adoption of this technology.
- Global recognition of increasingly eco-friendly technologies;
- Restoration of arable land for agricultural cultivation;
- Emerging innovative technologies creating fresh market prospects;
- Numerous technologies in the public domain across multiple nations, accessible solely due to filings in select priority countries;
- Limited exploration of genetic engineering in this context;
- Escalating global emphasis on environmental regulations and standards, particularly favoring eco-friendly technologies;
- Multinational corporations showing interest in bioremediation technologies;
- Heightened global consciousness regarding environmental sustainability.

Threats:

- **Competition from Alternative Technologies**: Fungal bioremediation faces competition from alternative remediation technologies, such as chemical treatments or thermal desorption, which may be perceived as faster or more effective in certain contexts.
- **Regulatory Challenges**: Regulatory approval processes for the use of fungal bioremediation in large-scale cleanup projects may be complex and time-consuming, delaying widespread adoption.
- **Resistance and Adaptation**: Hydrocarbon-degrading fungi may develop resistance or adaptability to certain contaminants over time, reducing their effectiveness in long-term remediation efforts.
- **Public Perception**: Public perception of fungal bioremediation, including concerns about its safety and effectiveness, could hinder its acceptance and implementation in some communities.
- Global legal frameworks and impediments to international trade;
- Microbial technologies are still nascent and could encounter future hurdles in import/export;
- Scarce availability of raw materials for inputs in the market leading to price hikes;
- Limited internationalization efforts by patent holders, resulting in technologies developed in one country being utilized without royalty payments in another;
- Market remains largely untapped;
- Annual rise in bioremediation patent filings suggests impending intense market competition.

12.7 CONCLUSION AND FUTURE SCOPE

The analysis of bioremediation technologies reveals distinct trends based on technological readiness levels and regional patent activity. Technologies with lower technological readiness levels were primarily identified through academic articles, while those with medium technological readiness levels were documented through patents. A significant observation is that most bioremediation patents originate from China and the US. However, while Chinese patents are predominantly applicable within China, many US patents are designed for international export, reflecting a broader global reach and commercial intent for these technologies. The focus of bioremediation patents is predominantly on water and soil remediation, with a relatively minor emphasis on sludge treatment. Among the various agents used in these technologies, bacteria, fungi, and enzymes are the most commonly employed, followed by algae, plants, and protozoa. A notable recent trend in bioremediation is the application of mutation and genetic engineering processes, though this is still emerging within the field. One of the significant weaknesses identified in bioremediation technology is the regulatory and logistical barriers to importing or exporting microorganism-based technologies. These barriers can limit the global dissemination and application of effective bioremediation solutions. A promising trend in oil bioremediation of soils is the increasing use of combined bioremediation agents, such as fungi and bacteria, within a single process. This combination approach enhances the robustness of the bioremediation process, making it more adaptable to varying environmental conditions. This trend presents a valuable research opportunity, as integrating multiple agents could lead to more efficient and versatile bioremediation strategies capable of addressing diverse contamination scenarios.

REFERENCES

Agarwal, S., Gupta, K.K., Chaturvedi, V.K., Kushwaha, A., Chaurasia, P.K., Singh, M.P. (2018). Chapter 12: The potential application of peroxidase enzyme for the treatment of industry wastes. In: Bharati, S.L., Chaurasia, P.K. (Eds.), *Research Advancements in Pharmaceutical, Nutritional, and Industrial Enzymology* (pp. 278–293). IGI Global, Hershey, PA. https://doi.org/10.4018/978-1-5225-5237-6.ch012

Al-Jawhari, I.F.H. (2014). Ability of some soil fungi in biodegradation of petroleum hydrocarbon. *Journal of Applied & Environmental Microbiology*, 2(2), 46–52.

Akhtar, N., Mannan, M.A.U. (2020). Mycoremediation: expunging environmental pollutants. *Biotechnology Reports*, 26, e00452. https://doi.org/10.1016/j.btre.2020.e00452

Al-Hawash, A.B., Zhang, J., Li, S., Liu, J., Ghalib, H.B., Zhang, X., Ma, F. (2018). Biodegradation of n-hexadecane by Aspergillus sp. RFC-1 and its mechanism. *Ecotoxicology and Environmental Safety*, 164, 398–408. https://doi.org/10.1016/j.ecoenv.2018.08.049

Al-Turki, A.I. (2009). Microbial polycyclic aromatic hydrocarbons degradation in soil. *Research Journal of Environmental Toxicology*, 3(1), 1–8. https://doi.org/10.3923/rjet.2009.1.8

Anacletus, F., Nwauche, K., Ighorodje-Monago, C. (2017) Mineral and heavy metal composition of crude oil polluted soil amended with non-ionic surfactant (Triton X-100) and white rot fungus (Pleurotus ostratus). *Journal of Environmental & Analytical Toxicology*, 7, 449–451. https://doi.org/10.4172/2161-0525.1000449

Angel, J. Inventor. Angel Janet. Assignee. (2014). Composition and Methods of Use. PCT patent WO2014152350.

Bala, K., Dogra, A., Lata, P., Das, A.P. (2024). Discussing sources and biological-chemical remediation approaches for microplastic pollution. In *Sustainable Microbial Technology for Synthetic and Cellulosic Microfiber Bioremediation*, 37–59. https://doi.org/10.1007/978-3-031-62600-5_3

Balaji, V., Arulazhagan, P., Ebenezer, P. (2014). Enzymatic bioremediation of polyaromatic hydrocarbons by fungal consortia enriched from petroleum contaminated soil and oil seeds. *Journal of Environmental Biology*, 35(3), 521–529.

Baldrian, P., in der Wiesche, C., Gabriel, J., Nerud, F., Zadražil, F. (2000). Influence of cadmium and mercury on activities of ligninolytic enzymes and degradation of polycyclic aromatic hydrocarbons by Pleurotus ostreatus in soil. *Applied and Environmental Microbiology*, 66(6), 2471–2478. https://doi.org/10.1128/aem.66.6.2471-2478.2000

Behera, I.D., Basak, G., Kumar, R.R., Sen, R., & Meikap, B.C. (2020). Treatment of petroleum refinery sludge by petroleum degrading bacterium Stenotrophomonas pavanii IRB19 as an efficient novel technology. *Journal of Environmental Science and Health - Part A Toxic/Hazardous Substances and Environmental Engineering*, 56(2). https://doi.org/10.1080/10934529.2020.1866924

Behera, I.D., Nayak, M., Biswas, S., Meikap, B.C., Sen, R. (2021). Enhanced biodegradation of total petroleum hydrocarbons by implementing a novel two-step bioaugmentation strategy using indigenous bacterial consortium. *Journal of Environmental Management*, 292, 112746. ISSN 0301–4797, https://doi.org/10.1016/j.jenvman.2021.112746

Behera, I.D., Nayak, M., Mishra, A., Meikap, B.C., Sen, R. (2022). Strategic implementation of integrated bioaugmentation and biostimulation for efficient mitigation of petroleum hydrocarbon pollutants from terrestrial and aquatic environment. *Marine Pollution Bulletin*, 177, 113492. ISSN 0025-326X, https://doi.org/10.1016/j.marpolbul.2022.113492

Bhalerao, T.S. (2012). Bioremediation of endosulfan-contaminated soil by using bioaugmentation treatment of fungal inoculant Aspergillus niger. *Turkish Journal of Biology*, 36, 561–567.

Borazjani, H., Wiltcher, D., Diehl, S. (2005). Bioremediation of polychlorinated biphenyl and petroleum contaminated soil. In: Lyon, W.G., Hong, J., Reddy, R.K. (Eds.), *Proceeding of Environmental science and Technology*. pp. 502–507. ISBN: 0-9768853-5-2. New Orleans, LA, USA: American Science Press.

Bortoli, E.F. Inventor. Bortoli Elio Fabio. Assignee. (2015). Biocatalytic Composition for Treatment of Substrates. European Patent EP3104988.

Burghal, A.A., Abu-Mejdad, N.M.J.A., Al-Tamimi, W.H. (2016). Mycodegradation of crude oil by fungal species isolated from petroleum contaminated soil. *International Journal of Innovative Research in Science, Engineering and Technology*, 5(2), 1517–1524.

Cornejo, P., Meier, S., García, S., Ferrol, N., Durán, P., Borie, F., Seguel, A. (2017). Contribution of inoculation with arbuscular mycorrhizal fungi to the bioremediation of a copper contaminated soil using Oenothera picensis. *Journal of Soil Science and Plant Nutrition*, 17(1), 14–21.

D'Annibale, A., Rosetto, F., Leonardi, V., Federici, F., Petruccioli, M. (2006). Role of autochthonous filamentous Fungi in bioremediation of a soil historically contaminated with aromatic hydrocarbons. *Applied and Environmental Microbiology*, 72(1), 28–36.

Das, A.P., Ghosh, S. (2023). Microorganisms in manganese biomining processes. In: *Manganese Mining Microorganisms* (pp. 55–61). https://doi.org/10.1016/B978-0-12-822147-1.00013-3

Das, A.P., Ghosh, S., Mohanty, S., & Sukla, L.B. (2015). Advances in manganese pollution and its bioremediation. In: Sukla, L.B., Pradhan, N., Panda, S., Mishra, B.K. (Eds.), *Environmental Microbe Biotechnology* (1st ed., pp. 313–328). Springer, Cham. https://doi.org/10.1007/978-3-319-19018-1_16

Das, A., Dey, S., Bala, K., Gnanasekaran, R., Das, A.P. (2024). Microbial bioremediation technology for sustainable treatment and management of synthetic microfiber waste. In *Sustainable Microbial Technology for Synthetic and Cellulosic Microfiber Bioremediation*, 221–238. https://doi.org/10.1007/978-3-031-62600-5_12

Das, N., Chandran, P. (2011). Microbial degradation of petroleum hydrocarbon contaminants: an overview. *Biotechnology Research International*, 2011(1), 941810.

Dey, S., Bissoyi, A., Koteswara Reddy, G., Das, A.P. (2024). Synthetic microfibers: an urban emerging pollutant demanding for sustainable environmental solutions. In: *Sustainable Microbial Technology for Synthetic and Cellulosic Microfiber Bioremediation* (pp. 107–121). https://doi.org/10.1007/978-3-031-62600-5_7

Dey, S., Das, A., Mallick, K., Sahu, A., Das, A.P. (2023). Environmental petroleum waste: pollution, toxicity, sustainable remediation. 159–175. https://doi.org/10.1007/978-3-031-48220-5_7

Dey, S., Dash, A., Das, A.P. (2024). Bioremediation as a sustainable and circular solution for mining waste management and bioenergy generation. In: *Sustainable Management of Mining Waste and Tailings: A Circular Economy Approach* (pp. 107–119). https://doi.org/10.1201/9781003442455-6

Dey, S., Dash, N., Ray, S.K., Priyadarsini, S., Chhotaray, C., Das, A.P. (2024). Occurrence and quantification of tire wear plastics pollutants from highway city dust of Bhubaneswar City of Odisha. In: *Sustainable Microbial Technology for Synthetic and Cellulosic Microfiber Bioremediation* (pp. 77–89). https://doi.org/10.1007/978-3-031-62600-5_5

Drevinskas, T., Mickienė, R., Maruška, A., Stankevičius, M., Tiso, N., Mikašauskaitė, J., Ragažinskienė, O., Levišauskas, D., Bartkuvienė, V., Snieškienė, V., Stankevičienė, A., Polcaro, C., Galli, E., Donati, E., Tekorius, T., Kornyšova, O., Kaškonienė, V. (2016). Downscaling the in vitro test of fungal bioremediation of polycyclic aromatic hydrocarbons: methodological approach. *Analytical and Bioanalytical Chemistry*, 408(4), 1043–1053. https://doi.org/10.1007/s00216-015-9191-3

Falade, A.O., Nwodo, U.U., Iweriebor, B.C., Green, E., Mabinya, L.V., Okoh, A.I. (2017). Lignin peroxidase functionalities and prospective applications. *Microbiology Open*, 6(1), e00394.

Fomina, M.A., Alexander, I.J., Colpaert, J.V., Gadd, G.M. (2005). Solubilization of toxic metal minerals and metal tolerance of mycorrhizal fungi. *Soil Biology and Biochemistry*, 37(5), 851–866.

Fu, W., Xu, M., Sun, K., Hu, L., Cao, W., Dai, C., Jia, Y. (2018). Biodegradation of phenanthrene by endophytic fungus Phomopsis liquidambari in vitro and in vivo. *Chemosphere*, 203, 160–169. https://doi.org/10.1016/j.chemosphere.2018.03.164

Gadd, G.M. (2010). Metals, minerals and microbes: geomicrobiology and bioremediation. *Microbiology*, 156(3), 609–643. https://doi.org/10.1099/mic.0.037143-0

Gadd, G.M., Bahri-Esfahani, J., Li, Q., Rhee, Y.J., Wei, Z., Fomina, M., Liang, X. (2014). Oxalate production by fungi: significance in geomycology, biodeterioration and bioremediation. *Fungal Biology Reviews*, 28(2–3), 36–55. https://doi.org/10.1016/j.fbr.2014.05.001

Ghosal, D., Ghosh, S., Dutta, T.K., Ahn, Y. (2016). Current state of knowledge in microbial degradation of polycyclic aromatic hydrocarbons (PAHs): a review. *Frontiers in Microbiology*, 7, 1369. https://doi.org/10.3389/fmicb.2016.01369

Ghosh, S., Sinha, J.K., Ghosh, S., Vashisth, K., Han, S., Bhaskar, R. (2023). Microplastics as an emerging threat to the global environment and human health. *Sustainability*, 15(14), 10821. https://doi.org/10.3390/SU151410821

Govarthanan, M., Fuzisawa, S., Hosogai, T., Chang, Y.-C. (2017). Biodegradation of aliphatic and aromatic hydrocarbons using the filamentous fungus Penicillium sp. CHY-2 and characterization of its manganese peroxidase activity. *RSC Advances*, 7(34), 20716–20723. https://doi.org/10.1039/C6RA28687A

Hadibarata, T., Tachibana, S., Itoh, K. (2009). Biodegradation of chrysene, an aromatic hydrocarbon, by Polyporus sp. S133 in liquid medium. *Journal of Hazardous* Materials, 164(2), 911–917. https://doi.org/10.1016/j.jhazmat.2008.08.081

Hallett, P.D., White, N.A., Ritz, K. (2006). Impact of basidiomycete fungi on the wettability of soil contaminated with a hydrophobic polycyclic aromatic hydrocarbon. *Biologia,* Bratislava, 61(Suppl.19), S334–S338. Section Botany. https://doi.org/10.2478/s11756-006-0184-8

Hidayat, A., Tachibana, S. (2013). Crude oil and n-octadecane degradation under saline conditions by Fusarium sp., F092. *Journal of Environmental Science and Technology*, 6(1), 29–40.

Jia, H., Song, X., Yan, S., Qu, L., Wang, Q. Inventors. Northeast Forestry University. Assignee. (2013). Electrically-assistant Microbial Remediation Method of Oil Polluted Soil. China patent CN104056857B.

Kanungo, J., Sahoo, T., Bal, M., Behera, I.D. (2023). Performance of bioremediation strategy in waste lubricating oil pollutants: a review. *Geomicrobiology Journal*. https://doi.org/10.1080/01490451.2023.2245395

Kirtzel, J., Ueberschaar, N., Deckert-Gaudig, T., Krause, K., Deckert, V., Gadd, G.M., Kothe, E. (2020). Organic acids, siderophores, enzymes and mechanical pressure for black slate bioweathering with the basidiomycete Schizophyllum commune. *Environmental Microbiology*, 22(4), 1535–1546. https://doi.org/10.1111/1462-2920.14749

Kota, M.F., Hussaini, A.A.S.A., Zulkharnain, A., Roslan, H.A. (2014). Bioremediation of crude oil by fungal genera. *Asian Journal of Plant Biology*, 2(1), 11–18.

Li, P., Zhang, H., Yao, D. Inventors. Institute of Applied Ecology, Chinese Academy of Sciences. Assignees. (2001). Process for Preparing Solid Fungus to Degrade Petroleum. China patent CN1382758.

Li, Q., Liu, J., Gadd, G.M. (2020). Fungal bioremediation of soil co-contaminated with petroleum hydrocarbons and toxic metals. *Applied Microbiology and Biotechnology*, 104(21), 8999–9008.

Liang, X., Gadd, G.M. (2017). Metal and metalloid biorecovery using fungi. *Microbial Biotechnology*, 10(5), 1199–1205. https://doi.org/10.1111/1751-7915.12767

Liu, B., Liu, J., Ju, M., Li, X., Wang, P. (2017). Bacteria-white-rot fungi joint remediation of petroleum-contaminated soil based on sustained release of laccase. *RSC Advances*, 7, 39075.

Lotfinasabasl, S., Gunale, V.R., Rajurkar, N.S. (2012). Assessment of petroleum hydrocarbon degradation from soil and tar ball by fungi. *Bioscience Discovery*, 3(2), 186–192.

Ma, X.-K., Ding, N., Peterson, E.C., Daug, A.J. (2016). Heavy metals species affect fungal-bacterial synergism during the bioremediation of fluoranthene. *Applied Microbiology and Biotechnology*, 100, 7741–7750.

Martínková, L., Kotik, M., Marková, E., Homolka, L. (2016). Biodegradation of phenolic compounds by Basidiomycota and its phenol oxidases: a review. *Chemosphere*, 149, 373–382.

Maruthi, Y.A., Hossain, K., Thakre, S. (2013). Aspergillus flavus: a potential bioremediator for oil contaminated soils. *European Journal of Sustainable Development*, 2(1), 57–66.

Mohsenzadeh, F., Chehregani Rad, A., Akbari, M. (2012). Evaluation of oil removal efficiency and enzymatic activity in some fungal strains for bioremediation of petroleum-polluted soils. *Journal of Environmental Health Science and Engineering*, 9, 26.

Moustafa, A.M. (2016). Bioremediation of oil spill in Kingdom of Saudi Arabia by using fungi isolated from polluted soils. *International Journal of Current Microbiology and Applied Sciences*, *5*, 680–691. http://dx.doi.org/10.20546/ijcmas.2016.505.069

Nag, M., Lahiri, D., Ghosh, S., Sarkar, T., Pati, S., Das, A.P., Ram, D.K., Bhattacharya, D., Ray, R.R. (2023). Application of microorganisms in biotransformation and bioremediation of environmental contaminant: a review. *Geomicrobiology Journal*, 1–18. https://doi.org/10.1080/01490451.2023.2261443

Paul, A., Dey, S., Ram, D.K., Das, A.P. (2023). Hexavalent chromium pollution and its sustainable management through bioremediation. *Geomicrobiology Journal*, 1–11. https://doi.org/10.1080/01490451.2023.2218377

Perez-Armendariz, B., Martinez-Carrera, D., Calixto-Mosqueda, M., Alba, J., Rodriguez-Vazquez, R. (2010). Filamentous fungi remove weathered hydrocarbons from polluted soil of tropical Mexico. *Revista Internacional de Contaminacion Ambiental*, 26(3), 193–199.

Quintella, C.M., Mata, A.M., Lima, L.C. (2019). Overview of bioremediation with technology assessment and emphasis on fungal bioremediation of oil contaminated soils. *Journal of Environmental Management*, 241, 156–166. https://doi.org/10.1016/j.jenvman.2019.04.019

Saritha, V., Maruthi, Y.A., Mukkanti, K. (2010). Potential fungi for bioremediation of industrial effluents. *BioResources*, 5(1), 8–22.

Satapathy, S., Kar, D., Priyadarsini, S., Das, A.P., Kuanar, A. (2024). Application of microorganisms in emerging waste management strategies and remediation of synthetic and cellulosic microfibers. In: *Sustainable Microbial Technology for Synthetic and Cellulosic Microfiber Bioremediation* (pp. 183–206). https://doi.org/10.1007/978-3-031-62600-5_10

Singh, R.K., Tripathi, R., Ranjan, A., Srivastava, A.K. (2020). Fungi as potential candidates for bioremediation. In: Sigh, P., Kumar, A., Borthakur, A. (Eds.), *Abatement of Environmental Pollutants: Trends and Strategies*, (pp. 177–191). New York: Elsevier.

Stamets, P. Inventor. Cisco Technology. Assignee. (2002). Delivery Systems for Mycotechnologies, Mycofiltration, and Mycoremediation. PCT patent WO200265836.

Stanley, H.O., Offorbuike, O.M., Stanley, C.N. (2017). Bioremediation of crude oil contaminated soil using Pleurotus pulmonarius, a white-rot fungus. *IOSR Journal of Environmental Science, Toxicology and Food Technology*, 11(4), 122–128.

Stella, T. (2020). Mycoremediation of contaminated soils. In: *Advanced Nano-Bio Technologies for Water and Soil Treatment* (pp. 445–465). Cham: Springer International Publishing. https://doi.org/10.1007/978-3-030-29840-1_22

Thenmozhi, R., Arumugam, K., Nagasathya, A., Thajuddin, N., Paneerselvam, A. (2013). Studies on Mycoremediation of used engine oil contaminated soil samples. *Advances in Applied Science Research*, 4(2), 110–118.

Ulfig, K. (2003). Studies of keratinolytic and keratinophilic fungi in sewage sludge by means of a multi-temperature hair baiting method. *Polish Journal of Environmental Studies*, 12(4), 461–466.

Vanishree, M., Thatheyus, A.J., Ramya, D. (2014). Biodegradation of petrol using the fungus Penicillium sp. *Science International*. https://doi.org/10.17311/sciintl.2014.26.31

Varjani, S.J., Patel, R.K. (2017). Fungi: a remedy to eliminate environmental pollutants. *Mycoremediation and Environmental Sustainability*, 1, 53–67.

Winquist, E., Björklöf, K., Schultz, E., Räsänen, M., Salonen, K., Anasonye, F., Cajtham, T., Steffen, K.T., Jørgensen, K.S., Tuomela, M. (2014). Bioremediation of PAH-contaminated soil with fungi- from laboratory to field scale. *International Biodeterioration & Biodegradation*, 86, 238–247.

Zhang, F. Inventor. Xi An Huanuo Environment Protection. Assignee. (2016). Composite Inoculant for Deep Oil Contaminated Soil Bioremediation Process. China patent CN106244479.

13 Bioremediation for Hydrocarbon Waste Present in Soaps

Anamika Basu

13.1 USE OF SOAPS AND HAND DISINFECTANTS DURING THE COVID-19 PANDEMIC

The use of soaps and hand disinfectants became essential during the COVID-19 pandemic as part of personal hygiene and infection control measures to reduce the spread of the virus. Soap is highly effective in removing the virus from hands because it breaks down the lipid (fatty) membrane that encases SARS-CoV-2 virus, making it inactive. Regular handwashing with soap and water for at least 20 seconds was recommended as a primary method of disinfection. Soap and water are widely available and are effective against various pathogens, not just viruses, making them a universal hygiene solution. Soap is effective in cleaning visibly dirty hands, which is crucial since many disinfectants work best on clean skin. Furthermore, alcohol-based sanitizers with at least 60% alcohol content were recommended when soap and water were not available. Alcohol kills viruses by denaturing their proteins and dissolving their lipid membranes.

The CDC and WHO recommended washing hands frequently, especially after being in public, touching surfaces, coughing, or sneezing. Proper handwashing was emphasized as the gold standard. The pandemic led to increased awareness of hand hygiene, with many people adopting handwashing and sanitizing practices more regularly. Public places installed more hand sanitizer dispensers, and individuals began carrying sanitizers as part of their daily routines (Mohanty et al., 2024; Sahu et al., 2024).

13.2 INGREDIENTS OF SOAPS AND HAND DISINFECTANTS

13.2.1 Ingredients in Soaps

Soaps are composed of a variety of ingredients that help cleanse, moisturize, and create lather. The core ingredients of soap include both basic ingredients and additional ingredients. Basic ingredients contains fats and oils. Soaps are made through a process called saponification, where fats or oils (like palm oil, coconut oil, or olive oil) react with a strong alkali. These fats and oils can be animal-based (like tallow) or plant-based. Lye (sodium hydroxide or potassium hydroxide) is present in soaps. This is the alkali used in saponification. Sodium hydroxide is used for making solid soaps, while potassium hydroxide is used for making liquid soaps. Water is used in

DOI: 10.1201/9781003508496-13

the soap-making process to dissolve the alkali and create the lye solution. Additional ingredients of soaps contain fragrances, colorants, emollients and moisturizers, preservatives, and exfoliants, etc. Essential oils or synthetic fragrances are added to impart pleasant smells. Dyes or natural colorants (like clays or plant extracts) give soap its appearance. Emollients and moisturizers are ingredients like glycerine, shea butter, and aloe vera are added to keep the skin moisturized and prevent dryness (Bhattacharjee et al., 2021). Antimicrobial preservatives, like sodium benzoate, are sometimes added to extend the shelf life of soaps. Some soaps may contain small particles like oats, seeds, or synthetic beads for exfoliation.

13.2.2 Ingredients in Hand Disinfectants (Hand Sanitizers)

Hand sanitizers, particularly alcohol-based ones, contain specific ingredients to ensure their effectiveness in killing germs. The core and additional ingredients are included in hand sanitizers. Ethanol (ethyl alcohol) or isopropyl alcohol are the primary active ingredients in alcohol-based hand sanitizers. They kill germs by denaturing their proteins and dissolving their lipid membranes. Hand sanitizers must contain at least 60% alcohol to be effective.

Inactive ingredients contain water, humectants, emollients, thickening agents, neutralizers, fragrances, and antimicrobial additives, etc. Water is used to dilute the alcohol and provide moisture. Humectants (e.g., glycerin) are added to retain moisture in the skin and prevent it from drying out after alcohol use. Emollients are ingredients like aloe vera, tocopherol (vitamin E), or moisturizers help soothe and hydrate the skin. Thickening agents (e.g., carbomer) give the sanitizer its gel-like consistency and make it easier to apply. Neutralizers (e.g., triethanolamine) help adjust the pH of the formula to be skin-friendly. Some hand sanitizers include fragrances to mask the smell of alcohol. Occasionally, non-alcohol-based sanitizers may contain other antimicrobial agents like benzalkonium chloride, but these are less commonly used than alcohol-based sanitizers.

Summary of Key Ingredients Present in Soap and Hand Sanitizer

Product	Key Ingredients
Soap	Fats and oils, sodium/potassium hydroxide (lye), water, fragrances, emollients
Hand sanitizer	Ethyl alcohol/isopropyl alcohol, water, glycerin, thickening agents, aloe vera, fragrance

These ingredients work together to cleanse hands, eliminate germs, and maintain skin health.

13.3 EMPHASIZING THE ROLE OF PARABENS IN HAND SANITIZERS AND HUMAN HEALTH

Parabens are a class of synthetic chemicals commonly used as preservatives in cosmetics, pharmaceuticals, and personal care products, including some soaps. They are not typically used as antibacterial agents but are rather included to prevent the growth of harmful bacteria, mold, and fungi in products, thereby extending their shelf life. The most commonly used parabens in personal care products include methylparaben, ethylparaben, propylparaben, and butylparaben. Parabens inhibit the growth of microbes in products that contain water. They are effective against a broad range of microorganisms, which help prevent contamination and spoilage. Parabens work by disrupting the cell membranes of microorganisms, making it difficult for bacteria and fungi to thrive in the product (Mpayipheli et al., 2024). However, they do not serve the same function as dedicated antibacterial agents like triclosan (TCS) or alcohol, which are included specifically to kill bacteria on the skin (Priyadarshini et al., 2024; Tripathy et al., 2024).

13.4 PARABEN RESIDUES IN WASTEWATER AND SURFACE WATER DURING COVID-19

With heightened hygiene practices during the pandemic, there was a surge in the use of products containing parabens (e.g., hand sanitizers, soaps, and disinfectants). This contributed to increased paraben residues in wastewater and surface water. The higher frequency of washing and the use of these products led to more significant discharges into the water system. Parabens enter the wastewater system primarily through household sewage, where they are washed down the drain during bathing, cleaning, or washing hands. In wastewater treatment plants (WWTPs), some parabens are degraded, but a portion of them can still survive the treatment process and get released into surface water bodies like rivers and lakes.

Studies conducted during and after the pandemic detected parabens and their metabolites in both wastewater and surface water. Surface water pollution can have implications for aquatic ecosystems, as parabens have been found to disrupt endocrine function in aquatic organisms. They can also accumulate in sediments, potentially leading to long-term environmental effects. The persistence of parabens in the environment poses risks to both aquatic life and humans (Bolujoko et al., 2022). The substances can act as endocrine disruptors, mimicking hormones and interfering with natural biological processes. As parabens are not fully removed in conventional wastewater treatment, their accumulation in the environment became a concern during the pandemic due to elevated concentrations linked to increased usage of hygiene products (Mishra & Das, 2023).

During COVID-19, regulatory bodies increased monitoring efforts to assess the environmental impact of heightened paraben use. Some studies called for improved wastewater treatment technologies that focus on removing endocrine-disrupting chemicals (EDCs), including parabens (Wahab et al., 2023). The pandemic has highlighted the need for sustainable practices in product formulation and better management of wastewater to reduce the release of paraben residues into the environment. There are ongoing efforts to find alternatives to parabens in personal care products to minimize their environmental impact.

Parabens used as preservatives in soaps (Pham et al., 2024). Their primary role is to preserve the product from microbial contamination rather than act as an antibacterial agent on the skin. They ensure that the soap remains safe and effective over time.

Ingredients like TCS, benzalkonium chloride, and para chloro meta xylenol (PCMX) are specifically used to provide antibacterial properties on the skin. They kill or inhibit bacteria directly on the skin's surface during use. Over the years, parabens have come under scrutiny due to potential health risks such as hormonal disruption and skin irritation. Some studies suggest that parabens can mimic estrogen in the body and may be linked to hormonal disruptions (Lin et al., 2024). This raised concerns about their potential link to breast cancer, though scientific consensus has not conclusively proven this link. For some individuals, parabens can cause allergic reactions or skin irritation, especially in those with sensitive skin (Tran-Lam et al., 2023). Due to health and environmental concerns (Karthikraj et al., 2017), many manufacturers have moved toward using paraben-free formulations. They are increasingly using alternative preservatives, such as phenoxyethanol, sodium benzoate, and potassium sorbate. These alternatives still help prevent microbial growth but are considered less controversial.

13.5 STUDY OF RISK ASSESSMENT OF PARABENS IN A TRANSCRIPTOMICS-BASED IN VITRO SYSTEM

The data from the NCBI database (Barrett et al., 2012), the gene expression profiling by microarray with GEO accession ID GSE233332 (Seidel et al., 2023), is analyzed with GEO2R software (Clough et al., 2024). Top differentially expressed genes (DEGs) are analyzed. Among the GSE233332 dataset, ten samples with sample IDs GSM7425743 and GSM7425744 (both selected as control), GSM7425700 and GSM7425701 (both induced with methylparaben), GSM7425675 and GSM7425676 (both induced with ethylparaben), GSM7425727 and GSM7425728 (both induced with propylparaben), GSM7425657 and GSM7425658 (both induced with butylparaben) are selected for the identification of DEGs. Gene symbol, gene title, gene ontology, and disease correlation are described in Table 13.1. Differential gene expression induced by methylparaben, ethylparaben, propylparaben, and butylparaben is designated as MP, EP, PP, and BP, respectively, in Table 13.1. The Omics Discovery Index (OmicsDI) provides a knowledge discovery framework across heterogeneous omics data (genomics, proteomics, transcriptomics, and metabolomics) (Perez-Riverol et al., 2017).

Some representative gene expression patterns are shown in Figure 13.1. Here, the expression values for genes such as NAV1, PLEKHA2, KCNIP3, FAM84A, NR2E1, TDGF1, NRDC, and FBXL20 are graphically compared to control cells, as well as cells that are induced by methyl, ethyl, propyl, and butyl parabens.

In the presence of methylparaben, RCSD1 gene expression is increased, which is related to acute lymphoblastic leukemia. Moreover, ethylparaben is responsible for increased expression of the DNA binding-histone cluster 1 (HIST1) gene. This specific gene is connected to Paget's disease, breast cancer, and schizophrenia. Small nuclear RNA (SNORD) expression is highly increased in the presence of ethylparaben compared to other paraben compounds. Genes, namely, BRD8 and MACF1 are differentially expressed due to the presence of propylparaben. The expression of the pseudogene ID2B, which has a correlation with a specific virus entry and attachment with the human host cell, is exhibited when induced by butylparaben.

TABLE 13.1
Name and Biological Functions of Differentially Expressed Genes after Induction with Parabens

Sl No.	Gene Symbol	Gene Title	Comparison of Gene Expression with Respect to Control				GO Function	Biological Activities/ Disease Correlation from OmicsDI Dataset
			MP	EP	PP	BP		
1.	NAV1	Neuron navigator 1	↑	↑	↓	↓	Function/activity not available	Regulates learning, memory
2.	PLEKHA2	Pleckstrin homology domain containing A2	↑	↑ ↑	↑ ↑	↑ ↑ ↑	PDZ domain binding	Related to prolymphocytic leukemia
3.	BRD8	Bromodomain containing 8	–	–	↑ ↑	–	Thyroid hormone receptor activity	Role in colorectal cancer cells
4.	TNSF13B	Tumor necrosis factor Superfamily member 13B	–	–	↑	↑ ↑	Cytokine activity	Modulation of osteoclastogenesis, susceptibility of rheumatoid arthritis, systemic lupus erythromatus
5.	ZADH2	Zinc-binding alcohol dehydrogenase domain containing 2	–	–	↑ ↑	↑	Zinc ion binding	Links DNA hypomethylation in blood and Alzheimer's disease
6.	HIST1H2BC	Histone cluster 1, H2bc	–	↑ ↑	–	–	DNA binding	Related to Paget's disease, triple-negative breast cancers, and schizophrenia
7.	EPM2AIP1	EPM2A interacting protein	↓	–	↑ ↑	↑ ↑	RNA polymerase II TF activity	Participates in primary constitutional MLH1 epimutation

(Continued)

TABLE 13.1 (*Continued*)
Name and Biological Functions of Differentially Expressed Genes after Induction with Parabens

Sl No.	Gene Symbol	Gene Title	Comparison of Gene Expression with Respect to Control				GO Function	Biological Activities/ Disease Correlation from OmicsDI Dataset
			MP	EP	PP	BP		
8.	KCNIP3	Potassium voltage-gated channel-interacting protein 3	↑	↑ ↑ ↑	↑	↑ ↑	RNA polymerase II core promoter proximal region sequence-specific DNA binding, calcium ion binding, potassium channel regulator activity	Together with PSEN2 involved in the modulation of amyloid-beta formation
9.	CDS2	CDP diacylglycerol synthase 2	↑	↑ ↑	–	–	Phosphatidate cytidyl transferase activity	Increased expression in asthma and COPD patients
10.	NR2E1	Nuclear receptor subfamily 2 group E member 1	↑ ↑	↑ ↑ ↑	↑ ↑	↑	Histone deacetylase binding	Inhibits growth and progression of triple-negative cancer
11.	PDGFB	Platelet-derived growth factor unit B	↑ ↑ ↑	–	↑ ↑	↑ ↑ ↑	Ras guanyl nucleotide exchange factor activity	Related to non-small cell lung carcinoma, pulmonary hypertension, neuroblastoma
12.	SNORD73A// RPS3A		–	↑ ↑ ↑	–	–	Small nuclear RNA, C/D box 73A, ribosomal S3A poly (A) RNA binding	ADP-ribosylated protein confirmed tyrosine as ADPr-acceptor amino acid in RPS3A
13.	RCSD1	RCSD1 domain containing 1	↑	↓	↓	–	Actin filament binding	Acute lymphoblastic leukemia

(*Continued*)

TABLE 13.1 (*Continued*)
Name and Biological Functions of Differentially Expressed Genes after Induction with Parabens

Sl No.	Gene Symbol	Gene Title	Comparison of Gene Expression with Respect to Control				GO Function	Biological Activities/ Disease Correlation from OmicsDI Dataset
			MP	EP	PP	BP		
14.	ID2B	Inhibitor of DNA binding 2B (pseudogene)	–	↓	–	↑	Ion channel binding	Respiratory syncytial virus (RSV) attachment and entry (according to Reactome Interacting pathway)
15.	UTP18	UTP18, small subunit processome component	↑ ↑	↑ ↑	↓	–	Poly (A) RNA binding	Promotes adenoma carcinogenesis
16.	TDGF1	Teratocarcinoma-derived growth factor 1	↑	↑ ↑	↑	↑	Growth factor activity	Related to pluripotency and oncogenes, enriched in human epithelial cells
17.	SLC7A2	Solute carrier family 7 member 2	↑ ↑ ↑	↑ ↑	–	–	L-Amino acid transmembrane transporter activity	Colorectal cancer
18.	FAM 84 A	Family with sequence similarity 84 member A	↑ ↑	↑ ↑ ↑	↑	↑	Function/activity not available	Promotes tumor development in papillary thyroid cancer
19.	PODN	Podocan	↑ ↑	↑ ↑ ↑	↑ ↑	–	Collagen binding	A prognostic biomarker in osteosarcoma, and breast cancer
20.	GNG8	G protein subunit gamma 8	–	↑ ↑ ↑	↑ ↑	↑	Signal transducer activity /// GTPase activity	Cervical cancer, chronic lymphocytic leukemia

(*Continued*)

TABLE 13.1 (*Continued*)
Name and Biological Functions of Differentially Expressed Genes after Induction with Parabens

Sl No.	Gene Symbol	Gene Title	Comparison of Gene Expression with Respect to Control				GO Function	Biological Activities/ Disease Correlation from OmicsDI Dataset
			MP	EP	PP	BP		
21.	MACF1	Microtubule-actin cross-linking factor 1	–	↓	↑ ↑ ↑	–	Cadherin binding is involved in cell–cell adhesion, ATPase activity	Osteoblast differentiation, aging-related osteoporosis
22.	MORF4L2-AS1		–	↑ ↑	↑ ↑ ↑	↑ ↑	MORF4L2 antisense RNA 1	Esophageal squamous cell carcinoma
23.	NRDC	Nardilysin convertor	↓	↓	↑ ↑	↑ ↑	Epidermal growth factor binding	Pancreas cancer cell growth, involvement in gastric cancer, and megakaryocyte cell lysis
24.	LPCAT4	Lysophosphatidyl choline acyltransferase 4	–	–	↑ ↑ ↑	↑	1-Acylglycerol-3-phosphate *O*-acyl transferase activity	Colorectal cancer, pancreatic ductal adenocarcinoma
25.	SNRK	SNF-related kinase	–	↑ ↑ ↑	↑ ↑ ↑	↑ ↑	ATP binding/MAP kinase kinase kinase activity	Promotes angiogenesis, suppressor of adipocyte inflammation
26.	LRRCC66	Leucine-rich repeat-containing 66	–	↑	↑ ↑	↑	Protein kinase inhibitor activity	Polymorphism in the gene correlates with sperm motility in idiopathic infertility
27.	FBXL20	F-box and leucine-rich repeat protein 20	↑ ↑	↑ ↑	↑ ↑ ↑	↑ ↑	Function/activity not available	Prostate cancer metastasis, metastatic ovarian carcinoma, triple-negative breast cancer

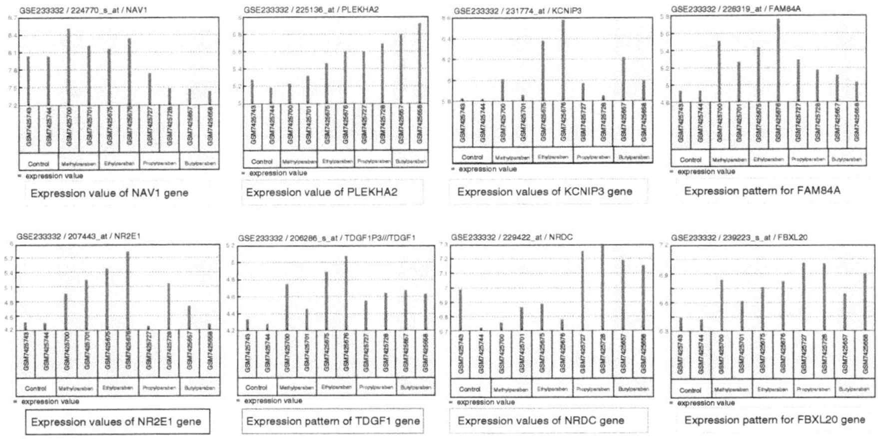

FIGURE 13.1 Gene expression patterns for differentially expressed genes.

13.6 ANALYSIS OF INGREDIENTS PRESENT IN PERSONAL CARE PRODUCTS, E.G., HAND WASH

Personal care products are manufactured with 10,500 unique chemical ingredients, some of which are known or suspected carcinogens, toxic to the reproductive system, or known to disrupt the endocrine system. Although some companies make products that are safe to use, others use dangerous ingredients, like coal tar and formaldehyde, both of which are human carcinogens, and lead acetate, a developmental toxin. EWG launched Skin Deep (https://www.ewg.org/skindeep/) in 2004, to fill the gap by educating the public about the ingredients in their cosmetics and personal care products. The core of Skin Deep is an electronic product database that contains ingredients from over 100,000 products. Skin Deep makes it easier for shoppers to understand potential hazards and health concerns related to ingredients in cosmetics and personal care products. EWG's ratings represent the best available information about the safety of personal care product ingredients.

Skin Deep currently contains information on 8,892 personal care product ingredients, culled from ingredient labels on products and from the scientific and industry literature on personal care products. A standardized name for each ingredient in the Skin Deep database is assigned, generally taken as the International Nomenclature for Cosmetic Ingredients (INCI) standard, with some exceptions where alternate names are more easily recognized by consumers. Each of these ingredient names is associated with a unique ingredient identification number in our database. The database currently contains 152,580 unique chemicals. EWG created a core, integrated database of chemical hazards, regulatory status, and study availability by pooling the data of nearly 60 databases and sources from government agencies, industry panels, academic institutions, or other credible bodies. Collectively, these data sources detail more than 1,535 unique chemical classifications. EWG uses these databases (https://www.ewg.org/skindeep/) to assess potential health hazards and data gaps for cosmetic ingredients.

TABLE 13.2
Comparative Studies of Ingredients Present in Hand Washes

Sl. No.	Ph Balanced Handwash	Handwash 2
1	Water	Water
2	Ammonium lauryl sulfate	Sodium laureth sulfate
3	Sodium laureth sulfate	Sodium chloride
4	Glycol distearate	Glycol distearate
5	Glycerin	Cocamide mea
6	Sodium chloride	Citric acid
7	Cocamide mea	Fragrance
8	Fragrance	Acrylates copolymer
9	Salicylic acid	Sodium benzoate
10	Tetrasodium edta	PPG-9
11	Citric acid	Sodium hydroxide
12	Methylchloroisothiazolinone	Tetrasodium edta
13	Methylisothiazolinone	Terpineol
14	Citronellol	Thymol
15	Hexyl cinnamal	Sodium carbonate
16	Limonene	Sodium sulfate
17	Benzyl salicylate	Benzyl salicylate
18	Butylphenyl	Hexyl cinnamal
19	Lilial	Linalool
20		FD&C Red No. 4 (CI 14700)
21		D&C Red No. 33 (CI 17200)

Among the ingredients present in Table 13.2, the following ingredients contain a high hazard score according to the Skin Deep database (https://www.ewg.org/skindeep/).

Methylisothiazolinone (MIT) is a widely used preservative; it has been associated with allergic reactions (Reeder et al., 2023; Geier et al., 2012). Lab studies on the brain cells of mammals also suggest that MIT may be neurotoxic.

Salicylic acid is a naturally occurring and synthetically produced beta-hydroxy acid used in skin treatment products. The word "fragrance" or "parfum" on the product label represents an undisclosed mixture of various scent chemicals and ingredients used as fragrance dispersants such as diethyl phthalate (DEP). Fragrance mixes have been associated with allergies, dermatitis, respiratory distress, and potential effects on the reproductive system. Lilial (butylphenyl methylpropanal) is a synthetic scent ingredient associated with allergies and contact dermatitis.

13.7 STUDY ON MIT AS POLLUTANT

MIT is a synthetic chemical commonly used as a preservative in a wide range of personal care products (like shampoos, soaps, and lotions) (Kim et al., 2019), cleaning products, and industrial applications. It is effective at inhibiting the growth of bacteria, fungi, and mold. However, MIT has raised significant environmental and health concerns due to its classification as a pollutant and its toxicity to aquatic ecosystems (Tang et al., 2024).

13.7.1 Environmental Impact of MIT

MIT is highly toxic to aquatic organisms, including fish, algae, and other marine life. Even at low concentrations, it can cause harm, disrupting the balance of aquatic ecosystems (Burnett et al., 2010). It tends to persist in water systems, where it breaks down slowly, leading to prolonged exposure to marine and freshwater species. This makes MIT a concern when it is washed down drains from household products or released from industrial processes.

While MIT is not known to bioaccumulate significantly in larger animals, its persistence in aquatic environments can cause a continual cycle of exposure, potentially harming sensitive organisms (Dey et al., 2023; Dey, Bissoyi, et al., 2024; Ray et al., 2024). Since MIT is commonly found in a variety of consumer products, it enters wastewater systems when these products are washed off or disposed of. MIT may not be entirely removed during wastewater treatment, allowing it to flow into natural water bodies and contribute to water pollution. This issue has led to increasing regulatory scrutiny in some regions where wastewater regulations aim to limit the release of harmful chemicals into water systems.

MIT degrades relatively slowly in water, and under certain environmental conditions, it can break down into byproducts that may also pose environmental risks. Its incomplete degradation in some treatment facilities makes it problematic for long-term environmental health. In addition to its environmental impact, MIT is known for causing skin sensitization and allergic reactions in humans (Paul et al., 2023, 2024). Prolonged exposure, particularly in personal care products, has led to increased cases of contact dermatitis, especially among individuals with sensitive skin.

Due to its environmental toxicity and potential health risks, several regulatory bodies have imposed restrictions on the use of MIT. The European Union has limited the concentration of MIT in cosmetic products and banned its use in leave-on products (like lotions) due to concerns about skin sensitization. In the United States, MIT use is regulated but not banned. However, there is increasing pressure from environmental groups to reduce its usage in both personal care and cleaning products. Due to environmental and health concerns, many manufacturers are moving away from using MIT as a preservative. Phenoxyethanol, sodium benzoate, and potassium sorbate are commonly used alternatives that are considered less toxic and more environmentally friendly.

13.7.2 Potential Bioremediation Strategies for MIT

Microbial bioremediation involves using bacteria, fungi, or other microorganisms capable of breaking down MIT into less harmful substances. This approach is effective because many microorganisms can metabolize a wide range of chemical pollutants, including synthetic compounds like MIT. Certain bacterial strains may be capable of degrading MIT through enzymatic processes. These bacteria can be isolated from contaminated environments and studied for their ability to break down

MIT (Kumar et al., 2022). Some Pseudomonas species are known to degrade various organic pollutants, and they might be adapted to break down MIT as well. Fungi, especially white-rot fungi, have shown the ability to degrade persistent organic compounds (Das & Ghosh, 2022; Ghosh et al., 2023). Their enzymatic systems, including lignin-degrading enzymes, could potentially be applied to MIT degradation.

Phytoremediation involves using plants to absorb, sequester, or detoxify pollutants from the environment. While this method is primarily used for heavy metals and organic pollutants, its application to MIT pollution could involve plants that are capable of absorbing the compound from water or soil. Aquatic plants such as water hyacinth (*Eichhornia crassipes*) or duckweed (*Lemna minor*) are commonly used in wastewater treatment systems due to their ability to absorb pollutants from water (Adki et al., 2014). These plants may be tested for their ability to uptake and store MIT, though more research is required to determine their effectiveness.

Enzymatic bioremediation is another method for the bioremediation of organic waste. In this method, enzymes specifically tailored to break down MIT are either introduced directly into the environment or produced by engineered microorganisms. Enzymatic bioremediation offers targeted degradation of pollutants. Researchers are investigating enzymes that can degrade biocides like MIT. This could involve isolating natural enzymes from microbes (Nowak et al., 2020) that naturally encounter MIT in contaminated environments or developing engineered enzymes with enhanced degradation capabilities.

Bioaugmentation involves adding specific strains of microorganisms that are known to degrade MIT into contaminated environments. These microbes can be isolated from sites where MIT pollution is present or genetically engineered to possess enhanced MIT-degrading capabilities. This approach could be applied in industrial WWTPs, where bioengineered microbial communities are used to specifically target MIT and other similar pollutants.

13.7.3 Challenges in Bioremediation of MIT

MIT is chemically resistant and not easily biodegradable, making it more challenging to break down compared to some other pollutants. Microorganisms and plants may struggle to metabolize it, requiring specific adaptations or engineered solutions. Due to its antimicrobial properties, MIT can be toxic to the very microorganisms needed to degrade it. This can hinder bioremediation efforts, as only highly resistant species may survive and thrive in MIT-polluted environments.

Slow Degradation Rates: Even when microbial strains capable of degrading MIT are identified, the degradation process may be slow, requiring optimization of environmental conditions such as pH, temperature, and nutrient availability.

13.7.4 Applications of Bioremediation for MIT

WWTPs: One of the primary entry points for MIT into the environment is through wastewater systems. By incorporating bioremediation techniques, such as microbial bioaugmentation, WWTPs could break down MIT before it is discharged into natural water bodies.

Constructed Wetlands: Constructed wetlands can be used to treat MIT-contaminated water by combining phytoremediation and microbial degradation. These systems mimic natural wetlands and provide an environment where microorganisms and plants work together to remove pollutants.

13.7.5 Research and Development Needs

Isolation of MIT-Degrading Microorganisms: Continued research is needed to identify and cultivate microbial strains capable of efficiently breaking down MIT in various environments.

Genetic Engineering: Advances in genetic engineering could enhance the biodegradation capabilities of microorganisms, making them more resilient to MIT's toxicity and accelerating the degradation process.

Optimization of Bioremediation Systems: Bioremediation systems, whether involving microbial cultures or constructed wetlands, need to be optimized for factors like temperature, pH, and oxygen levels to enhance MIT breakdown.

13.8 METHYLCHLOROISOTHIAZOLINONE AS WASTE WATER POLLUTANT

Methylchloroisothiazolinone (MCI) is a synthetic preservative commonly used in personal care products, cosmetics, household cleaners, and industrial applications. Like its counterpart MIT, MCI is highly effective at preventing the growth of bacteria, fungi, and mold, but it is also classified as an environmental pollutant due to its toxicity to aquatic life and persistence in wastewater.

13.8.1 Aquatic Toxicity

Toxic to Aquatic Organisms: MCI is highly toxic to aquatic life, including fish, algae, and other microorganisms in both freshwater and marine environments. Even at low concentrations, MCI can cause serious harm to these organisms, potentially disrupting entire aquatic ecosystems.

Bioavailability: MCI tends to persist in water environments and does not break down easily, leading to prolonged exposure to aquatic species.

Ecological Imbalance: The presence of MCI in water bodies can lead to reduced biodiversity and altered structure of aquatic communities. Sensitive species may be particularly vulnerable to its toxic effects, while resistant species may proliferate, altering the natural balance of ecosystems.

13.8.2 Presence in Wastewater

Entry through Wastewater: MCI enters wastewater systems primarily through the use of products like shampoos, lotions, cleaning agents, and industrial products that contain the preservative (Rafoth et al., 2007). When these products are washed off or disposed of, MCI is carried into the sewage system.

Incomplete Removal: Conventional WWTPs are not always equipped to completely remove synthetic organic chemicals like MCI. While some fraction of MCI may degrade or be removed during treatment, a significant portion can still pass through the system and enter natural water bodies like rivers, lakes, and oceans.

Persistent in Water: MCI is not easily biodegradable, making it resistant to break down in both wastewater treatment processes and natural aquatic environments. This persistence contributes to long-term environmental pollution.

13.8.3 Effects on Wastewater Treatment Systems

Impact on Microbial Communities: Wastewater treatment systems rely on microbial communities to break down organic pollutants. Due to MCI's antimicrobial properties, it can disrupt these communities, reducing the efficiency of biological treatment processes.

Inhibition of Biodegradation: The antimicrobial nature of MCI may inhibit the activity of bacteria and other microorganisms in treatment plants, leading to incomplete removal of organic pollutants from wastewater. This can increase the overall pollution load discharged into the environment.

13.8.3.1 Environmental and Health Concerns

13.8.3.1.1 Human Health Impacts

Skin Sensitization: MCI is a known skin sensitizer and can cause allergic reactions, particularly in individuals with sensitive skin. It is one of the leading causes of allergic contact dermatitis (Carbajo et al., 2015), especially when used in leave-on products like lotions and cosmetics.

Cumulative Exposure: While wastewater pollution primarily affects aquatic life, cumulative exposure to MCI through various products can also pose risks to human health.

13.8.3.1.2 Regulation and Restrictions

European Union: The European Union has placed restrictions on the use of MCI in personal care products due to its allergenic properties. It is now banned in leave-on products (like creams and lotions) but is still permitted in rinse-off products (like shampoos and soaps) at regulated concentrations.

United States and Other Regions: While MCI use is regulated in many countries, including the United States, its presence in consumer products and its release into wastewater remain a concern.

13.8.3.2 Mitigating MCI Pollution

13.8.3.2.1 Advanced Wastewater Treatment Technologies

Activated Carbon Filtration: This method can be used to remove synthetic organic chemicals like MCI from wastewater. Activated carbon adsorbs these pollutants, preventing them from entering natural water systems.

Ozonation and Advanced Oxidation Processes (AOPs): These processes can degrade organic pollutants, including MCI, by breaking down their chemical structure using ozone or other reactive species (Huang et al., 2022a, b). AOPs can be more effective at eliminating persistent chemicals than conventional treatment methods.

Membrane Filtration: Membrane technologies like reverse osmosis and ultrafiltration can also help remove MCI and similar pollutants from wastewater by physically separating them from water.

13.8.3.2.2 Bioremediation

Microbial Degradation: Some microorganisms can degrade isothiazolinone compounds like MCI, though the process can be slow and not entirely efficient. Bioaugmentation (the introduction of specialized microorganisms) could enhance this process in wastewater treatment systems.

Constructed Wetlands: These engineered ecosystems use plants and microorganisms to remove pollutants from water naturally. While primarily used for organic and nutrient removal, they can be adapted to target synthetic chemicals like MCI.

13.8.3.2.3 Consumer Awareness and Product Formulation

Paraben-Free and MCI-Free Products: With growing awareness of the environmental and health risks posed by MCI, many manufacturers have begun formulating personal care products without it, using alternative preservatives like phenoxyethanol or sodium benzoate, which are considered safer for both humans and the environment.

Eco-Friendly Product Labeling: Consumers can look for eco-friendly certifications on products that indicate reduced environmental impact and the absence of harmful chemicals like MCI.

13.8.3.3 Alternatives to MCI

Due to environmental concerns, many companies are moving away from MCI in favor of safer preservatives such as phenoxyethanol, sodium benzoate, potassium sorbate, and different essential oils. Among them, phenoxyethanol is a less toxic preservative commonly used as an alternative to MCI in personal care products. Sodium benzoate and potassium sorbate are natural preservatives that are commonly used in "green" or eco-friendly products. Furthermore, in some products, essential oils with antimicrobial properties (like tea tree oil or lavender oil) are used as preservatives.

13.9 EFFECT OF SALICYLIC ACID IN WASTE WATER

Salicylic acid, a widely used ingredient in cosmetics, pharmaceuticals, and personal care products, such as acne treatments, shampoos, and exfoliants, can enter the environment through wastewater. Although it is biodegradable, its increasing use raises concerns about its potential impact on aquatic ecosystems and wastewater treatment processes.

13.9.1 Sources of Salicylic Acid in Wastewater

Salicylic acid enters wastewater systems primarily through personal care products as commonly found in acne treatments, exfoliants, and dandruff shampoos. After use, it is washed off and enters wastewater. Similarly, it is used in topical medications and can also be a metabolic byproduct of aspirin (acetylsalicylic acid), which breaks down into salicylic acid in the body. Not only is salicylic acid used in some manufacturing processes, but residues may enter wastewater (Camacho-Muñoz et al., 2014).

13.9.2 Behavior of Salicylic Acid in WWTPs

Salicylic acid is considered biodegradable, meaning it can be broken down by microorganisms in WWTPs. However, the rate of degradation can depend on the concentration, microbial community, and environmental conditions in the treatment plant. Studies suggest that a significant portion of salicylic acid is removed during conventional biological wastewater treatment processes (Khamis et al., 2011; Karunanayake et al., 2017) though not always completely. While most salicylic acid may be broken down in WWTPs, some portion may still pass through treatment systems, especially when concentrations are high or the plant's microbial populations are insufficient to fully degrade it. Secondary treatment processes, which involve microbial breakdown, are generally more effective in removing salicylic acid than primary treatment processes, which rely on physical filtration and sedimentation.

13.9.3 Environmental Impact of Salicylic Acid in Water Bodies

Compared to other pollutants, salicylic acid has relatively low acute toxicity to aquatic life. However, chronic exposure or accumulation in environments where wastewater is not adequately treated can still pose risks to sensitive species, especially aquatic invertebrates, algae, and microorganisms (Martín et al., 2012). Salicylic acid, like many pharmaceutical compounds, may disrupt the microbial balance in aquatic ecosystems, potentially affecting the decomposition of organic matter and the cycling of nutrients in aquatic environments. Although salicylic acid itself is not considered an endocrine disruptor, it can act as a pharmaceutical contaminant, and in combination with other pharmaceutical residues, it could contribute to broader concerns about the presence of EDCs in water bodies. Pharmaceuticals in general, including salicylic acid, may have cumulative effects on wildlife, especially when combined with other contaminants present in wastewater.

Salicylic acid is unlikely to bioaccumulate in aquatic organisms due to its relatively low lipophilicity (fat solubility) and high biodegradability. However, prolonged exposure to environments with continuous inputs of wastewater could lead to sublethal effects in organisms.

Since WWTPs rely on microbial communities to break down organic and pharmaceutical contaminants, high concentrations of salicylic acid (along with other pharmaceuticals) may inhibit the activity of certain microorganisms. This could reduce the overall efficiency of the treatment process, especially in systems that are not designed to handle pharmaceutical residues. Salicylic acid, like other pharmaceuticals, is classified as an emerging contaminant. Although it is biodegradable, its continuous discharge into wastewater systems raises concerns about long-term impacts on aquatic environments and the potential for low-level contamination over time.

13.9.4 Mitigation and Removal

13.9.4.1 Advanced Treatment Technologies

Activated Carbon Filtration: Activated carbon can adsorb salicylic acid and other pharmaceuticals, providing an additional method for removing it from wastewater (Vedenyapina et al., 2023). This is particularly useful for WWTPs that handle high loads of pharmaceutical waste. Ozonation and AOPs break down organic contaminants like salicylic acid by generating reactive species (such as hydroxyl radicals) that degrade the chemical structure of the pollutant. AOPs are more effective than conventional biological treatment at removing resistant organic compounds. Minimizing the overuse of personal care products containing salicylic acid, and improving the disposal of pharmaceuticals, can reduce the overall load of salicylic acid in wastewater. Green Chemistry Alternatives is another method to reduce salicylic acid in waste water Manufacturers could consider formulating products with biodegradable ingredients that break down more easily during wastewater treatment and have minimal environmental impact.

13.10 ROLE OF DEP IN WASTE WATER POLLUTION

DEP is a type of phthalate ester commonly used as a plasticizer in the manufacturing of plastics, personal care products (like perfumes and lotions), cosmetics, and various industrial applications. DEP enters the environment primarily through wastewater, raising concerns about its potential impact on aquatic ecosystems, wastewater treatment processes, and human health (Gao & Wen, 2016).

13.10.1 Sources of DEP in Wastewater

DEP is often used in personal care products like perfumes, lotions, and deodorants to help stabilize fragrances. It is also found in nail polish, hairspray, and skin care products. DEP is used as a plasticizer in various industrial processes, such as the production of flexible plastics. It can enter wastewater systems through industrial discharges. After the use of products containing DEP, it is washed off and enters wastewater systems via household drains. DEP may leach from plastic products into water sources, including wastewater, when plastics degrade or are discarded.

13.10.2 Biodegradability and Removal of DEP in WWTPs

DEP is considered moderately biodegradable in aerobic conditions (where oxygen is present), meaning that some portion of it can be broken down by microorganisms in WWTPs. However, it can be persistent in anaerobic conditions (low-oxygen environments) commonly found in certain treatment processes or natural water bodies. While some DEP is degraded in conventional biological treatment processes, a portion often remains in treated effluent, allowing it to enter rivers, lakes, and oceans. Studies have found phthalates, including DEP, in treated wastewater and sediments (Dargnat et al., 2009). DEP can bind to particles and settle in sediments at the bottom of water bodies, where it may persist for long periods. This can create a reservoir of contamination that continues to affect aquatic life. While DEP itself has a low potential for bioaccumulation compared to other phthalates, it can still enter the food chain, particularly in aquatic organisms like fish, invertebrates, and plants. Over time, this could lead to low-level exposure in higher organisms, including humans.

13.10.3 Environmental Impact of DEP in Water Bodies

DEP is considered less toxic than other more heavily regulated phthalates, such as di(2-ethylhexyl) phthalate (DEHP). However, it can still have adverse effects on aquatic organisms, particularly with chronic exposure or in environments with high concentrations of phthalates. DEP has been shown to cause reproductive and developmental effects in fish and invertebrates. It can interfere with hormonal systems, leading to endocrine disruption in wildlife [33], affecting reproduction and growth. DEP is part of a broader group of chemicals known as EDCs (Saini et al., 2016). While not as potent as other phthalates, DEP can still interfere with the normal functioning of hormone systems in aquatic organisms, potentially affecting their reproduction, growth, and development. Long-term exposure to phthalates in water can lead to reproductive issues in fish, amphibians, and other species. For example, exposure to EDCs like DEP may affect the production of sex hormones, causing feminization or masculinization in fish populations. The presence of DEP in water bodies can affect the biodiversity and health of aquatic ecosystems. Sensitive species may be more vulnerable to the effects of phthalate pollution, leading to changes in species composition and ecosystem functioning over time (Olujimi et al., 2012).

Humans can be exposed to DEP through direct contact with personal care products, ingestion of contaminated water, or consumption of seafood contaminated with phthalates. While DEP is less toxic than other phthalates, chronic exposure still poses potential risks. Prolonged exposure to DEP has been linked to reproductive toxicity, developmental effects, and potential endocrine disruption in humans. While direct risks from low levels in water may be minimal, the cumulative exposure through various environmental sources raises concern (Das et al., 2015; Dey, Dash, et al., 2024). DEP is often found in combination with other phthalates and chemicals in wastewater, which may result in additive or synergistic effects. Even at low concentrations, the combined impact of various chemicals may pose greater risks to both aquatic life and human health than DEP alone.

13.10.4 Challenges in Wastewater Treatment for DEP

Conventional wastewater treatment processes, such as primary sedimentation and biological treatments, are not fully effective at removing DEP and other phthalates. As a result, DEP can pass through treatment plants and enter surface waters. Advanced treatment technologies, such as activated carbon adsorption, ozonation, and AOPs, may be needed to more effectively remove DEP from wastewater (Li et al., 2020). The presence of DEP in wastewater can also affect microbial communities in treatment plants. As DEP degrades, its byproducts may impact the performance of microorganisms responsible for breaking down organic matter in wastewater, potentially reducing the efficiency of the treatment process. Activated carbon can adsorb DEP and other organic pollutants, making it an effective method for removing phthalates from wastewater. Ozonation and AOPs processes generate reactive oxygen species that can break down DEP and other resistant organic compounds. Ozonation is particularly effective in degrading phthalates that persist in conventional treatment processes. Techniques like reverse osmosis and ultrafiltration can physically remove phthalates from wastewater, though these methods are expensive and not widely implemented in all treatment plants.

Bioremediation through microbial degradation is frequently used for phthalate degradation. Certain bacteria and fungi have been identified that can degrade phthalates, including DEP, through enzymatic processes. However, the biodegradation of DEP is typically slow, and bioaugmentation (the introduction of specialized microbial strains) may be needed to enhance the process in WWTPs (Liang et al., 2007). The degradation of phthalates, including DEP, by bacteria and fungi occurs through enzymatic processes. Several microbial species have been identified that can metabolize phthalates, breaking them down into less harmful byproducts. These microorganisms produce enzymes that catalyze the breakdown of the ester bonds in phthalates, facilitating their biodegradation.

Bacteria-mediated phthalate degradation conducted with the help of Pseudomonas, Rhodococcus, Acinetobacter, Bacillus, Sphingomonas, Aspergillus, and Penicillium species (Ren et al., 2018; Hu et al., 2021).

Pseudomonas putida and *Pseudomonas fluorescens* are commonly known for their ability to degrade phthalates, including DEP. Pseudomonas species produce esterases and hydrolases, which hydrolyze the ester bonds in DEP, converting it into phthalic acid and ethanol. Phthalic acid is further metabolized through the phthalate degradation pathway, where it is converted into protocatechuic acid, and then into intermediates of the TCA (tricarboxylic acid) cycle. *Rhodococcus erythropolis* and *Rhodococcus ruber* have been shown to degrade various phthalates, including DEP (Lu et al., 2009). These bacteria utilize monooxygenases and dioxygenases to initiate the oxidative cleavage of the phthalate ring, leading to further breakdown into simpler organic compounds. Rhodococcus species possess genes that code for enzymes involved in the breakdown of phthalates into intermediate compounds that can enter central metabolic pathways.

Acinetobacter baumannii and *Acinetobacter lwoffii* can degrade phthalates, including DEP, by producing carboxylesterases, which break down the ester bond in phthalates. DEP is hydrolyzed into phthalic acid and ethanol, with phthalic acid being metabolized into intermediates that are eventually mineralized into carbon dioxide and water (Perpetuo et al., 2020). *Bacillus subtilis* and *Bacillus megaterium* are also capable of degrading phthalates, including DEP. Bacillus species produce esterases that hydrolyze the ester bonds in phthalates. Phthalic acid produced during this process is further degraded via oxidative pathways. These bacteria metabolize phthalic acid into simpler organic acids and eventually into carbon dioxide and water. *Sphingomonas paucimobilis* is another bacterial species capable of degrading DEP and other phthalates. Sphingomonas species utilize phthalate dioxygenase to cleave the aromatic ring of phthalic acid, leading to further breakdown into compounds that enter the TCA cycle.

13.10.5 Fungi for Phthalate Degradation

Aspergillus niger and *Aspergillus oryzae* are well-known for their ability to degrade phthalates, including DEP (Wang et al., 2017). These fungi produce esterases and lipases that hydrolyze phthalates into phthalic acid (Mohanty et al., 2017). Phthalic acid is then further metabolized through fungal metabolic pathways. The aromatic ring of phthalic acid is cleaved by dioxygenases, leading to the formation of intermediates that are funneled into the TCA cycle.

Penicillium chrysogenum and *Penicillium simplicissimum* are able to degrade phthalates through enzymatic processes. These fungi secrete esterases and oxidases that break down DEP and other phthalates into phthalic acid, which is then metabolized into simpler compounds. Penicillium species are widely studied for their potential use in bioremediation of environments contaminated with phthalates. *Phanerochaete chrysosporium* (white-rot fungus), a ligninolytic fungus, is known for its ability to degrade a wide range of pollutants, including phthalates. *Phanerochaete chrysosporium* produces lignin peroxidase and manganese peroxidase, which break down complex organic molecules like phthalates (Savinova et al., 2022). It can mineralize phthalic acid completely into carbon dioxide and water. Due to its ability to degrade a wide range of organic pollutants, this fungus is commonly used in bioremediation efforts for various contaminants, including phthalates.

Trichoderma viride and *Trichoderma harzianum* have been shown to degrade phthalates, including DEP. These fungi produce esterases and oxidases that cleave the ester bonds in phthalates, producing phthalic acid and other degradation products. Trichoderma species are studied for their potential use in the biodegradation of industrial pollutants and wastewater treatment. The first step in phthalate degradation involves the hydrolysis of the ester bond by esterases or hydrolases, producing phthalic acid and alcohol (in the case of DEP, ethanol). After the formation of phthalic acid, phthalate dioxygenase initiates the oxidative cleavage of the aromatic ring, breaking it down into simpler organic acids, which can then enter the central metabolic pathways of microorganisms. The degradation intermediates formed from phthalic acid eventually feed into the TCA cycle, where they are completely mineralized into carbon dioxide and water.

Bioaugmentation is a method where the addition of specialized bacteria or fungi that degrade phthalates to contaminated environments can enhance the biodegradation process (Naveen et al., 2022). Some plants, in association with phthalate-degrading microbes, can be used for the bioremediation of contaminated soils and water systems. Fungi such as Aspergillus and Penicillium species can be applied in composting systems to degrade phthalates in organic waste.

Various bacteria and fungi play essential roles in the biodegradation of phthalates, including DEP, by producing enzymes that break down the chemical structure of these pollutants. These microorganisms hold significant potential for bioremediation efforts, especially in environments contaminated with phthalates. Advanced techniques such as bioaugmentation and bioreactors utilizing these microbes could further enhance the degradation of phthalates in wastewater treatment systems and polluted sites.

Regulatory bodies have imposed restrictions on the use of certain phthalates, but DEP remains less tightly regulated in many regions. However, growing awareness of its environmental impact has prompted some manufacturers to phase out DEP in favor of safer alternatives. Consumers are increasingly seeking phthalate-free personal care products. This shift can reduce the amount of DEP entering wastewater systems from household sources.

13.11 DISCUSSION

In conclusion, both soaps and hand disinfectants played crucial roles in curbing the spread of COVID-19 by maintaining hand hygiene, which is one of the most effective preventive measures against viral transmission. While parabens are used in soaps primarily as preservatives, they are not classified as antibacterial agents. They protect the product from contamination rather than providing an antibacterial effect on the skin. Dedicated antibacterial ingredients serve a more active role in killing or inhibiting bacteria on the skin itself. During the pandemic, the use of personal care and hygiene products, including disinfectants, hand sanitizers, and pharmaceuticals, increased significantly. Parabens, which are commonly used as preservatives in cosmetics, pharmaceuticals, and personal care products, have been a subject of concern due to their potential impact on the environment and human health. In summary, the COVID-19 pandemic led to increased levels of paraben residues in wastewater and surface water, highlighting the need for improved water treatment methods and a shift toward more sustainable, environmentally friendly products. The DEGs extracted from the microarray study show clinical correlation with various cancers and neurodegenerative disorders.

MIT is a widely used preservative but poses significant environmental risks, particularly to aquatic ecosystems due to its high toxicity and persistence. Its role as a pollutant has prompted regulatory action and a shift toward safer, more sustainable alternatives in personal care and cleaning products. Bioremediation is an eco-friendly approach that uses microorganisms or plants to detoxify or remove pollutants from contaminated environments, including water, soil, and air. For pollutants like MIT, which can persist in aquatic systems and be toxic to living organisms, bioremediation presents a promising solution to mitigate environmental damage. However, due to the chemical complexity of MIT and its resistance to degradation, the bioremediation process for this compound is still under research and development. Bioremediation for MIT pollution holds promise but is still in the experimental phase due to MIT's persistence and antimicrobial nature. The key strategies involve microbial degradation, phytoremediation, bioaugmentation, and the potential use of engineered enzymes. However, challenges such as the compound's resistance to degradation and toxicity to microbes need to be addressed for bioremediation to be a viable large-scale solution.

MCI is an effective preservative but a problematic pollutant in wastewater due to its persistence and high toxicity to aquatic organisms. Its presence in wastewater and subsequent release into the environment can lead to significant ecological damage. Mitigating MCI pollution requires a combination of advanced wastewater treatment technologies, bioremediation strategies, and a shift toward safer alternatives in consumer products.

Salicylic acid in wastewater primarily originates from personal care products and pharmaceuticals. Although it is biodegradable and considered to have low toxicity to aquatic organisms, its persistence in wastewater systems can contribute to chronic, low-level pollution. Advanced treatment methods like activated carbon filtration or AOPs can improve the removal of salicylic acid from wastewater. To mitigate its environmental impact, both consumers and industries can take steps to reduce unnecessary waste and explore more sustainable alternatives.

DEP is a significant wastewater pollutant due to its widespread use in personal care products and industrial applications. While it is moderately biodegradable, it is often not fully removed during conventional wastewater treatment, leading to its presence in aquatic ecosystems where it can cause toxicity and endocrine disruption. Advanced treatment technologies such as bioaugmentation and bioreactors and regulatory measures can help mitigate its environmental impact, but reducing its use in consumer products remains an important step in minimizing DEP pollution.

REFERENCES

Adki, V. S., Jadhav, J. P., & Bapat, V. A. (2014). At the cross roads of environmental pollutants and phytoremediation: a promising bio remedial approach. *Journal of Plant Biochemistry and Biotechnology*, *23*, 125–140.

Barrett, T., Wilhite, S. E., Ledoux, P., Evangelista, C., Kim, I. F., Tomashevsky, M., … & Soboleva, A. (2012). NCBI GEO: archive for functional genomics data sets—update. *Nucleic Acids Research*, *41*(D1), D991–D995.

Bhattacharjee, J., Mishra, S., & Das, A. P. (2021). Recent advances in sensor-based detection of toxic dyes for bioremediation application: a review. *Applied Biochemistry and Biotechnology*, *2021*, 1–20. https://doi.org/10.1007/S12010-021-03767-7

Bolujoko, N. B., Ogunlaja, O. O., Alfred, M. O., Okewole, D. M., Ogunlaja, A., Olukanni, O. D., … & Unuabonah, E. I. (2022). Occurrence and human exposure assessment of parabens in water sources in Osun State, Nigeria. *Science of the Total Environment*, *814*, 152448.

Burnett, C. L., Bergfeld, W. F., Belsito, D. V., Klaassen, C. D., Marks, J. G., Shank, R. C., … & Andersen, F. A. (2010). Final report of the safety assessment of methylisothiazolinone. *International Journal of Toxicology*, *29*(4_suppl), 187S–213S.

Camacho-Muñoz, D., Martín, J., Santos, J. L., Aparicio, I., & Alonso, E. (2014). Concentration evolution of pharmaceutically active compounds in raw urban and industrial wastewater. *Chemosphere*, *111*, 70–79.

Carbajo, J. B., Perdigón-Melón, J. A., Petre, A. L., Rosal, R., Letón, P., & García-Calvo, E. (2015). Personal care product preservatives: risk assessment and mixture toxicities with an industrial wastewater. *Water Research*, *72*, 174–185.

Clough, E., Barrett, T., Wilhite, S. E., Ledoux, P., Evangelista, C., Kim, I. F., … & Soboleva, A. (2024). NCBI GEO: archive for gene expression and epigenomics data sets: 23-year update. *Nucleic Acids Research*, *52*(D1), D138–D144.

Dargnat, C., Teil, M. J., Chevreuil, M., & Blanchard, M. (2009). Phthalate removal throughout wastewater treatment plant: case study of Marne Aval station (France). *Science of the Total Environment*, *407*(4), 1235–1244.

Das, A., & Ghosh, S. (2022). Role of microorganisms in extenuation of mining and industrial wastes. *Geomicrobiology Journal*, *39*, 173–175. https://doi.org/10.1080/01490451.2022.2038953

Das, A. P., Ghosh, S., Mohanty, S., & Sukla, L. B. (2015). Consequences of manganese compounds: a review. *Toxicological & Environmental Chemistry*, *96*(7), 981–997. https://doi.org/10.1080/02772248.2015.1005428

Dey, S., Bissoyi, A., Koteswara Reddy, G., & Das, A. P. (2024). Synthetic microfibers: an urban emerging pollutant demanding for sustainable environmental solutions. In *Sustainable Microbial Technology for Synthetic and Cellulosic Microfiber Bioremediation* (pp. 107–121). https://doi.org/10.1007/978-3-031-62600-5_7

Dey, S., Dash, A., & Das, A. P. (2024). Bioremediation as a sustainable and circular solution for mining waste management and bioenergy generation. In *Sustainable Management of Mining Waste and Tailings: A Circular Economy Approach* (pp. 107–119).

Dey, S., Tripathy, B., Kumar, M. S., & Das, A. P. (2023). Ecotoxicological consequences of manganese mining pollutants and their biological remediation. *Environmental Chemistry and Ecotoxicology*, *5*, 55–61. https://doi.org/10.1016/J.ENCECO.2023.01.001

Gao, D. W., & Wen, Z. D. (2016). Phthalate esters in the environment: a critical review of their occurrence, biodegradation, and removal during wastewater treatment processes. *Science of the Total Environment*, *541*, 986–1001.

Geier, J., Lessmann, H., Schnuch, A., & Uter, W. (2012). Recent increase in allergic reactions to methylchloroisothiazolinone/methylisothiazolinone: is methylisothiazolinone the culprit? *Contact Dermatitis*, *67*(6), 334–341.

Ghosh, S., Sinha, J. K., Ghosh, S., Vashisth, K., Han, S., & Bhaskar, R. (2023). Microplastics as an emerging threat to the global environment and human health. *Sustainability 2023*, *15*(14), 10821. https://doi.org/10.3390/SU151410821

Hu, R., Zhao, H., Xu, X., Wang, Z., Yu, K., Shu, L., ... & Wang, C. (2021). Bacteria-driven phthalic acid ester biodegradation: current status and emerging opportunities. *Environment International*, *154*, 106560.

Huang, N., Shao, W. T., Wang, Q., Wang, W. L., Wu, Q. Y., & Hu, H. Y. (2022a). Degradation of chloromethylisothiazolinone antimicrobial by vacuum-ultraviolet/ultraviolet irradiation: reactive species, degradation pathway and toxicity evaluation. *Chemosphere*, *302*, 134821.

Huang, N., Shao, W. T., Wang, W. L., Wang, Q., Chen, Z. Q., Wu, Q. Y., & Hu, H. Y. (2022b). Removal of methylisothiazolinone biocide from wastewater by VUV/UV advanced oxidation process: kinetics, mechanisms and toxicity. *Journal of Environmental Management*, *315*, 115107.

Karthikraj, R., Vasu, A. K., Balakrishna, K., Sinha, R. K., & Kannan, K. (2017). Occurrence and fate of parabens and their metabolites in five sewage treatment plants in India. *Science of the Total Environment*, *593*, 592–598.

Karunanayake, A. G., Todd, O. A., Crowley, M. L., Ricchetti, L. B., Pittman Jr, C. U., Anderson, R., & Mlsna, T. E. (2017). Rapid removal of salicylic acid, 4-nitroaniline, benzoic acid and phthalic acid from wastewater using magnetized fast pyrolysis biochar from waste Douglas fir. *Chemical Engineering Journal*, *319*, 75–88.

Khamis, M., Karaman, R., Ayyash, F., Qtait, A., Deeb, O., & Manssra, A. (2011). Efficiency of advanced membrane wastewater treatment plant towards removal of aspirin, salicylic acid, paracetamol and p-aminophenol. *Journal of Environmental Science and Engineering*, *5*(2), 121–137.

Kim, M. K., Kim, K. B., Lee, J. Y., Kwack, S. J., Kwon, Y. C., Kang, J. S., ... & Lee, B. M. (2019). Risk assessment of 5-chloro-2-methylisothiazol-3 (2H)-one/2-methylisothiazol-3 (2H)-one (CMIT/MIT) used as a preservative in cosmetics. *Toxicological Research*, *35*, 103–117.

Kumar, V., Agrawal, S., Bhat, S. A., Américo-Pinheiro, J. H. P., Shahi, S. K., & Kumar, S. (2022). Environmental impact, health hazards, and plant-microbes synergism in remediation of emerging contaminants. *Cleaner Chemical Engineering*, *2*, 100030.

Li, X., Wan, J., Wang, Y., Yan, Z., Chi, H., & Ding, S. (2020). Mechanism of accurate recognition and catalysis of diethyl phthalate (DEP) in wastewater by novel MIL100 molecularly imprinted materials. *Applied Catalysis B: Environmental*, *266*, 118591.

Liang, D. W., Zhang, T., & Fang, H. H. (2007). Anaerobic degradation of dimethyl phthalate in wastewater in a UASB reactor. *Water Research*, *41*(13), 2879–2884.

Lin, Y., Zhao, Y., Liu, Y., Lan, Y., Zhu, J., Cai, Y., … & Xue, J. (2024). Occurrence and bioaccumulation of parabens and their metabolites in the biota from a subtropical freshwater river ecosystem: implications for human exposure. *Environmental Research*, *240*, 117530.

Lu, Y., Tang, F., Wang, Y., Zhao, J., Zeng, X., Luo, Q., & Wang, L. (2009). Biodegradation of dimethyl phthalate, diethyl phthalate and di-n-butyl phthalate by Rhodococcus sp. L4 isolated from activated sludge. *Journal of Hazardous Materials*, *168*(2–3), 938–943.

Martín, J., Camacho-Muñoz, D., Santos, J. L., Aparicio, I., & Alonso, E. (2012). Occurrence of pharmaceutical compounds in wastewater and sludge from wastewater treatment plants: removal and ecotoxicological impact of wastewater discharges and sludge disposal. *Journal of Hazardous Materials*, *239*, 40–47.

Mishra, S., & Das, A. P. (2023). Microbial remediation of synthetic microfiber contaminated wastewater. In M. P. Shah (Ed.), *Microbial Technologies in Industrial Wastewater Treatment* (pp. 337–350). Springer, Singapore. https://doi.org/10.1007/978-981-99-2435-6_17

Mohanty, M., Mohanty, J., Dey, S., Dutta, K., Shah, M. P., & Das, A. P. (2024). The face mask: a tale from protection to pollution and demanding sustainable solution. *Emerging Contaminants*, *10*(2), 100298. https://doi.org/10.1016/J.EMCON.2023.100298

Mohanty, S., Ghosh, S., Nayak, S., & Das, A. P. (2017). Bioleaching of manganese by Aspergillus sp. isolated from mining deposits. *Chemosphere*, *172*, 302–309. https://doi.org/10.1016/J.CHEMOSPHERE.2016.12.136

Mpayipheli, N., Mpupa, A., Madala, N. E., & Nomngongo, P. N. (2024). Paraben residues in wastewater and surface water: a case study of KwaZulu Natal and Gauteng provinces (South Africa) during the COVID-19 pandemic. *Frontiers in Environmental Science*, *12*, 1418375.

Naveen, K. V., Saravanakumar, K., Zhang, X., Sathiyaseelan, A., & Wang, M. H. (2022). Impact of environmental phthalate on human health and their bioremediation strategies using fungal cell factory: a review. *Environmental Research*, *214*, 113781.

Nowak, M., Zawadzka, K., & Lisowska, K. (2020). Occurrence of methylisothiazolinone in water and soil samples in Poland and its biodegradation by *Phanerochaete chrysosporium*. *Chemosphere*, *254*, 126723.

Olujimi, O. O., Fatoki, O. S., Odendaal, J. P., & Daso, A. P. (2012). Chemical monitoring and temporal variation in levels of endocrine disrupting chemicals (priority phenols and phthalate esters) from selected wastewater treatment plant and freshwater systems in Republic of South Africa. *Microchemical Journal*, *101*, 11–23.

Paul, A., Dey, S., Ram, D. K., & Das, A. P. (2023). Hexavalent chromium pollution and its sustainable management through bioremediation. *Geomicrobiology Journal*, 1–11. https://doi.org/10.1080/01490451.2023.2218377

Paul, A., Priyadarsini, S., Karakuş, S., Behera, J. K., Gnanasekaran, R., & Das, A. P. (2024). Fate and accumulation of synthetic microfibers in the environment. In *Sustainable Microbial Technology for Synthetic and Cellulosic Microfiber Bioremediation* (pp. 61–76). https://doi.org/10.1007/978-3-031-62600-5_4

Perez-Riverol, Y., Bai, M., da Veiga Leprevost, F., Squizzato, S., Park, Y. M., Haug, K., … & Hermjakob, H. (2017). Discovering and linking public omics data sets using the Omics Discovery Index. *Nature Biotechnology*, *35*(5), 406–409.

Perpetuo, E. A., da Silva, E. C. N., Karolski, B., & do Nascimento, C. A. O. (2020). Biodegradation of diethyl-phthalate (DEP) by halotolerant bacteria isolated from an estuarine environment. *Biodegradation*, *31*(4), 331–340.

Pham, P. T., Quan, T. C., Le, Q. T., Bui, M. Q., Tran, A. H., Phung, A. T. T., … & Tran, T. M. (2024). Quantification of parabens in marine fish samples by a rapid, simple, effective sample preparation method. *Environmental Science and Pollution Research*, *31*(11), 16571–16582.

Priyadarshini, B., Santhosh Kumar, M., & Das, A. P. (2024). Microbial degradation of synthetic plastics released from COVID-19 generated single use disposable face masks. In *Environmental Science and Engineering, Part F2827* (pp. 1–16). https://doi.org/10.1007/978-3-031-50840-0_1

Rafoth, A., Gabriel, S., Sacher, F., & Brauch, H. J. (2007). Analysis of isothiazolinones in environmental waters by gas chromatography–mass spectrometry. *Journal of Chromatography A, 1164*(1–2), 74–81.

Ray, S. K., Dey, S., Das, A. P., & John, A. (2024). Microplastic waste management through greener technologies: a review. In *Sustainable Management of Urban Plastic Waste through Circular Economic Approaches* (pp. 237–261). https://doi.org/10.1201/9781003465447-14

Reeder, M. J., Warshaw, E., Aravamuthan, S., Belsito, D. V., Geier, J., Wilkinson, M., … & Uter, W. (2023). Trends in the prevalence of methylchloroisothiazolinone/methylisothiazolinone contact allergy in North America and Europe. *JAMA Dermatology, 159*(3), 267–274.

Ren, L., Lin, Z., Liu, H., & Hu, H. (2018). Bacteria-mediated phthalic acid esters degradation and related molecular mechanisms. *Applied Microbiology and Biotechnology, 102*, 1085–1096.

Sahu, A., Mallick, K., & Das, A. P. (2024). Environmental impact of single-use synthetic face mask and its recycling: a sustainable approach. (pp. 197–212). https://doi.org/10.1007/978-3-031-50840-0_11

Saini, G., Pant, S., Singh, S. O., Kazmi, A. A., & Alam, T. (2016). A comparative study of occurrence and fate of endocrine disruptors: diethyl phthalate and dibutyl phthalate in ASP-and SBR-based wastewater treatment plants. *Environmental Monitoring and Assessment, 188*, 1–13.

Savinova, O. S., Shabaev, A. V., Glazunova, O. A., Eremin, S. A., & Fedorova, T. V. (2022). Biodestruction of phthalic acid esters by white rot fungi. *Applied Biochemistry and Microbiology, 58*(5), 598–612.

Seidel, F., Kappenberg, F., Fayyaz, S., Scholtz-Illigens, A., Cherianidou, A., Derksen, K., … & Hengstler, J. G. (2023). Risk assessment of parabens in a transcriptomics-based in vitro test. *Chemico-Biological Interactions, 384*, 110699.

Tang, L., Liu, M., Li, J., Zhou, B., Lam, P. K., Hu, C., & Chen, L. (2024). Isothiazolinone disrupts reproductive endocrinology by targeting the G-protein-coupled receptor signaling. *Environmental Science & Technology, 58*(2), 1076–1087.

Tran-Lam, T. T., Phung, A. T. T., Pham, P. T., Bui, M. Q., Dao, Y. H., & Le, G. T. (2023). Occurrence, biomagnification, and risk assessment of parabens and their metabolites in marine fish: the case study of Vietnam. *Chemosphere, 344*, 140221.

Tripathy, B., Dey, S., Dubey, N. K., Lo, W.-C., & Das, A. P. (2024). An omics-based approach for managing the risk of COVID-19-generated wastes. (pp. 51–66). https://doi.org/10.1007/978-3-031-50840-0_4

Vedenyapina, M. D., Kulayshin, S. A., Kurmysheva, A. Y., Sandulyak, D. A., Rakishev, A. K., Weichgrebe, D., & Nair, R. R. (2023). Adsorption of salicylic acid on modified active carbon from an aqueous medium. *Russian Chemical Bulletin, 72*(5), 1099–1106.

Wahab, R. A., Omar, T. F. T., Nurulnadia, M. Y., & Rozulan, N. N. A. (2023). Occurrence, distribution, and risk assessment of parabens in the surface water of Terengganu River, Malaysia. *Marine Pollution Bulletin, 192*, 115036.

Wang, J. W., Xu, W., Zhong, T. H., He, G. Y., & Luo, Z. H. (2017). Degradation of dimethyl phthalate esters by a filamentous fungus Aspergillus versicolor isolated from deep-sea sediments. *Botanica Marina, 60*(3), 351–359.

14 Recent Advances in Microbial Degradation of Petroleum Hydrocarbon Contaminants and Bioremediation

Suman Jagatee, Sujata Priyadarshini, Mukta Mayee Kumbhar, Alok Prasad Das, Akalabya Bissoyi, Chiranjibi Chhotaray, and Selcan Karakus

14.1 INTRODUCTION

Hydrocarbon contamination, generated by a variety of petrochemical industry uses, is currently a major environmental concern. The accidental discharge of petroleum products raises serious environmental concerns. Hydrocarbon molecules, whether saturated or aromatic, are broadly distributed throughout the ecosystem. Such compounds have a low degree of chemical reactivity. For eras, it was assumed that hydrocarbons could only be destroyed physiologically in the presence of air or O_2. However, over the last three decades, a large number of bacteria with the potential to degrade hydrocarbon compounds directly under a stringent O_2 shortage supply have been overlooked (Abdel-Shafy and Mansour, 2018). Hydrocarbons are primarily made of hydrogen and carbon. They lack functional groups and are mostly polar, which permits relatively mild chemical interactions at room temperature. The primary factors determining the chemical reactivity of hydrocarbons are the kind, quantity, and arrangement of their aromatic (π) bonds. As a result, these substances are divided into two groups based on the sorts of bonds they have: aliphatic and aromatic hydrocarbons. Additionally, it is simple to identify the three types of aliphatic hydrocarbons: nonaromatic rings, branched trains, and straight chains. The monocyclic, bicyclic, and polycyclic molecules that make up the aromatic hydrocarbons are categorized according to their structures. Alkylbenzenes, where the alkyl groups replace several H_2 atoms from the benzene ring, are important among these classes. Because hydrocarbons are fuels and initiator constituents of a large range of biochemical preparations, their availability

DOI: 10.1201/9781003508496-14

is fundamentally important to our modern society. The majority of hydrocarbons, if not all of them, are found in nature and may be readily synthesized from there. These substances are the result of long-term geochemical interactions between the metabolism of living things and extremely ancient, entombed biomass. Therefore, there are other ways besides human activity-related pollution and the environment responsible for global contamination of crude oil. On the other hand, it may cause them to accumulate more in the environment (Widdel and Rabus, 2001). Petroleum is a mixture of hydrocarbons, including asphaltenes, resins, aromatic hydrocarbons, and alkanes. Polycyclic aromatic hydrocarbons are often noxious, persistent, and carcinogenic. Some petroleum compounds can be biologically broken down in the marine environment, but others, like methyl tertiary butyl ether, may not (Abdel-Shafy and Mansour, 2018). These substances pose a significant environmental concern since they are not biodegradable. Because they disrupt the ecosystems around them, petroleum hydrocarbons (PHs) have been dubbed the primary environmental contaminants. These contaminants include a varied amount of salts, PAHs, organic compounds, phenols, alcohols, acids, radionuclides, and heavy metals, which may be present in groundwater, petroleum-polluted soil, and wasterwater as well (Abdel-Shafy and Kamel, 2016). Mixtures of hydrocarbons, on the other hand, are categorized as organic pollutants with neurotoxic and carcinogenic properties. First and foremost, it is forbidden to burn or remove petroleum wastes from hazardous landfills; in addition, it is expensive since large volumes of these contaminants are discharged into the ecosystem and are challenging to clean up. The majority of mechanical and chemical methods used to break down or eradicate petroleum hydrocarbons (PHs) from contaminated zones are not only expensive but also have a number of negative effects (Moursy and Abdel-Shafy, 1984). Conversely, the most suitable method for eliminating PHs from the contaminated areas is bioremediation (Das et al., 2024; Pradhan et al., 2023). They can totally remove these wastes and are also quite affordable. These procedures essentially fall under the biodegradation method. They speak about the complete breakdown of these wastes' CO_2 from organic contaminants, inorganic materials, water, and protein cells. On the other hand, they also discuss how bioagents—effective microorganisms—transform complex organic pollutants into simpler ones. PH pollutants are effectively broken down by a number of indigenous microorganisms in soil and water (Das and Chandran, 2011). In this way, the variety of hydrocarbons and the microbiological population in soil are closely linked. For the bioremediation of hydrocarbons, the choice of microorganisms is therefore crucial and significant. The destiny of PHs is mostly determined by how they react to soil microorganisms and how they interact with the soil matrix. The capacity of the microorganisms and the chemical makeup of these pollutants to degrade are correlated with the biodegradation ratio of these pollutants (Dey et al., 2023; Lahiri et al., 2021). For PHs to be bioremediated, polluted soils must have certain unique characteristics as well as the right environmental conditions, including the activity of the microorganisms living there. Additionally, the metabolites produced by the bacteria in these soils are not inhibited, and the soil contains active native hydrocarbons that the microbes may use. Molecular markers that regulate the effectiveness of the microorganisms' activity are produced chemically, biochemically, and microbiologically during the biological breakdown of PHs. The environmental degradation of petro-hydrocarbons is assessed and analyzed using various methodologies, including respirometry, metagenomics, phospholipid fatty acid profiling, rapid plate reader techniques (microtiter), accessible protocols that elucidate the functions and adaptations of microorganisms over time and space, and molecular techniques (Chikere et al., 2011).

PHs and associated contaminants can be bioremediated, although this is a complex and demanding process. The changes that these pollutants undergo, whether quantitative or qualitative, mostly depend on the type and amount of these pollutants that exist in the surroundings. This includes variations in temperature, O_2 that is either free or dissolved, and the physical and chemical dispersion of the petroleum. In addition to the native components of the microorganism groups, turbulent circumstances also influence the bioremediation process. The ideal temperature range for petroleum degradation was found to be 20°C–35°C. The initial stage in microbial breakdown of PHs is to target its aliphatic or low-molecular-weight aromatic portions. High-molecular-weight crude oil is classified as a recalcitrant species; hence, its bioremediation values are much lower. On the other hand, in ideal circumstances, their eradication has been shown in a few studies as quite significant. Temperature increases often result in an increase in the biodegradation value. As a result, in the extremely low-temperature habitat, PHs degrade very slowly. In the aquatic environment, these chemicals also biodegrade poorly since there aren't enough nutrients, such as phosphorus and nitrogen. Salinity and pressure, on the other hand, are thought to be important variables in estuaries and deep waters.

14.2 AEROBIC DEGRADATION OF POLLUTANTS

The genetic and molecular underpinnings of the breakdown of organic pollutants are made possible by the growing amount of genomic data on bacteria. Lessons may be drawn from the recent genomic investigations of two of the largest known bacterial genomes—*Rhodococcus* sp. strain RHA1 and *Burkholderia xenovorans* LB400—which are among the most persistent of these pollutants: aromatic chemicals. These investigations have advanced our knowledge of the development of huge bacterial genomes, non-catabolic physiological adaptation to organic chemicals, and bacterial catabolism. First, in terms of general structure, the metabolic pathways of isolates that differ phylogenetically are quite similar. In *Pseudomonas*, different chemicals are channeled into central aromatic routes through peripheral aromatic channels. These pathways are genetically structured differently depending on the species (Mishra et al., 2021). Some routes, such as non-oxygenolytic ring-cleavage strategies in aerobic aromatic degradation processes, are more prevalent than previously thought, according to a comparative genomics study. There are a few examples of true redundancy in species with high homologous enzyme numbers, according to functional genomics research. The cryptic aromatic catabolism of certain terpenoids and steroids may be the cause of the high concentration of ring-cleaving dioxygenases in rhodococcal isolates. Nonetheless, it is more plausible that the extensive gene repertoires of potent pollutant degraders like LB400 and RHA1 evolved mostly through earlier pathways, indicating the ancient origin of catabolic. However, it is more probable that the extensive gene repertoires of potent chemical degraders like LB400 and RHA1 evolved mainly through earlier methods. It is remarkable that this is the case in such phylogenetically diverse animals, which further supports the catabolic ability's ancient origins (McLeod and Eltis, 2008).

14.3 ANAEROBIC DEGRADATION OF POLLUTANTS

Recalcitrant organic contaminants can be mineralized by anaerobic microbial processes, which are highly significant for the environment and entail interesting new biochemical processes (Jugder et al., 2015). In particular, it has long been disputed whether hydrocarbons and halogenated compounds can be broken down in the absence of O_2. However, in recent decades, the isolation of previously undiscovered anaerobic hydrocarbon-degrading and reductively dehalogenating bacteria has offered the strongest evidence for these processes occurring in nature. Although much of the earlier research on this topic focused on chlorinated compounds, more recent investigations have shown that bromine and iodine moieties in aromatic insecticides undergo reductive dehalogenation. It has been demonstrated that other processes, such as abiotic reduction caused by soil minerals that are organically triggered (Tor et al., 2000) deactivate relatively durable herbicides based on aniline faster than those seen in aerobic settings. Since most of these bacteria lack genetic systems that are easily applied, progress in our knowledge of their molecular makeup has been somewhat modest, despite the discovery of several novel biochemical events that enable the corresponding metabolic pathways. But a fresh and exciting approach to gaining molecular understanding of these novel metabolic features is now available thanks to the growing use of genomics in environmental microbiology. In the past several years, the full genomes of many bacteria that can degrade organic pollutants anaerobically have been sequenced. *Aromatoleum aromaticum* strain EbN1 has the first genome of an anaerobic hydrocarbon degrader that uses toluene or ethylbenzene as substrates with a genome size of around 4.7 Mb.

About 200 gene clusters, including multiple paralogs, were found in the genome sequence to code for a sophisticated catabolic network that breaks down aromatic chemicals both aerobically and anaerobically (Behera et al., 2020). The present comprehensive research on the control of pathways and enzyme structures is based on the genome sequence. The entire genomes of anaerobic hydrocarbon-degrading bacteria are found in *Geobacter metallireducens* and *Dechloromonas aromatica*. Through halorespiration, these bacteria may break down halogenated hydrocarbons anaerobically. Several paralogous reductive dehalogenase genes are found in the full 1.4 Mb genomes of *Dehalococcoides ethenogenes* strain 195, *Dehalococcoides* sp. strain CBDB1, and the 5.7 Mb genome of *Desulfitobacterium hafniense* strain Y51. The evolution of reductive dehalogenation and other niche adaptation schemes can be better understood thanks to genome sequencing. Some species, including *Desulfitobacterium chlororespirans*, which was first evaluated for halorespiration on chlorophenols, have recently been shown to be able to use specific brominated compounds as electron acceptors for growth. These substances include, for instance, the herbicide bromoxynil and its main metabolite.

14.4 PH MICROBIAL DEGRADATION

The kind and amount of hydrocarbons present affect the complicated process of PH degradation. These hydrocarbons fall into four groups: resins, asphaltenes, aromatics, and saturates. They are challenging to eliminate or break down due to factors including their restricted availability to microbes (Dey et al., 2024). The most efficient natural process for eliminating petroleum hydrocarbon contaminants from the ecosystem is microbial degradation. Petroleum-derived biodegraded aromatic hydrocarbons have been found in marine sediments, according to recent studies. Prior to any noticeable degradation of the crude oil's n-alkane profile, it was revealed that the microbial degradation of alkyl aromatics was widespread in the marine ecosystem. These hydrocarbons are mostly biodegraded by bacteria, yeast, and fungi. Biodegradation efficiency varies for different microorganisms; for soil fungus, the biodegradation efficiency is approximately 80%; for soil bacteria, it is around 50%; and for marine bacteria, it is almost 100%. N-alkanes with chain lengths between C10 and C40 can serve as the sole "C" source for *Acinetobacter* sp. In both aquatic and terrestrial settings, protozoa and algae play an important role in the microbial ecosystem (Singh et al., 2023) (Figure 14.1).

14.5 ELEMENTS AFFECTING THE DEGRADATION OF PHs

Temperature, the availability of nutrients, and the makeup of the pollutants all have an impact on how quickly PHs biodegrade. Delays in biodegradation are caused by temperature's effects on the physiology, variety, and chemistry of pollutants as well as microbial flora. The maximum rates of deterioration take place near about at 40°C in soil habitats, 30°C in freshwater ecosystem, and 20°C in aquatic surroundings (Dey, Bissoyi, et al., 2024; Mallick et al., 2024). Microbial action and the characteristics of oil spills are also impacted by ambient temperature. The effective biological degradation of hydrocarbon contaminants requires nutrients, mostly iron, phosphorus, and nitrogen. Certain nutrients, however, may turn into limiting elements and interfere with the processes of biodegradation. The carbon supply rises and the accessibility of N_2 and P becomes the limiting factor for degradation of oil when large-scale oil spills happen in freshwater and marine ecosystems (Dey, Dash, Ray, et al., 2024). This is particularly noticeable in maritime areas since saltwater has minimal amounts of phosphorus and nitrogen. Since freshwater wetlands usually lack nutrients, it is necessary to supplement them in order to stimulate the degradation of oil pollutants. Biodegradation activity can be inhibited by excessive nutrient concentrations. It has been shown that high nutrient levels have a detrimental effect on the degradation of hydrocarbons, especially on aromatics. Fowl dung, has been demonstrated to improve biodegradation in polluted soil, whereas fertilizers have been proven to be useful for crude oil bioremediation in subarctic intertidal sediments. Through improved microbial activity and increased bioavailability, photooxidation has been shown to boost the biodegradability of PHs Das and Chandran (2011).

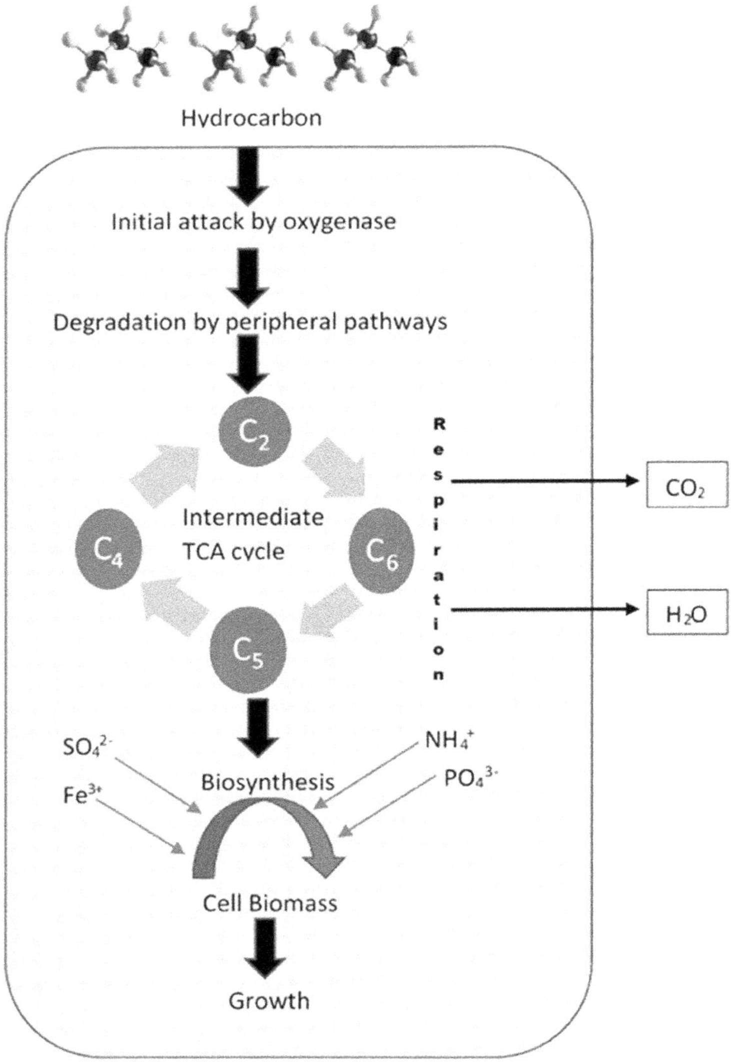

FIGURE 14.1 Microbial degradation of hydrocarbon.

14.6 ENZYMES PARTICIPATING IN THE DEGRADATION OF HYDROCARBONS

Heme-thiolate monooxygenases, acknowledged as cytochrome P450 alkane hydroxylases, are crucial to the microbial breakdown of many substances, including gasoline additives, oil, and chlorinated hydrocarbons. Enzyme systems are needed to transfer O_2 into the substrate to start degradation, depending on the length of the chain. Higher eukaryotes often have many P450 families, each of which is made up of diverse P450 forms. These isoforms may ensemble to contribute to the metabolic conversion of a particular substrate. P450 multiplicity in microorganisms is limited to a few species. PHs are shown to be biodegraded by cytochrome P450 enzyme systems. Multiple microsomal forms of cytochrome P450 facilitate the capacity of some yeast sp. to utilize n-alkanes and other aliphatic hydrocarbons as their exclusive source of carbon and energy. These enzymes, known as cytochrome P450, were extracted from yeast sp., including *Candida apicola, Candida tropicalis,* and *Candida maltose.* In their discussion of alkane oxygenase systems, Das and Chandran (2011) highlighted the enzymes such as cytochrome P450, soluble di-iron methane monooxygenases, membrane-bound copper-containing methane monooxygenases, and integral-membrane di-iron alkane hydroxylases (e.g., alkB) that play an active role in the aerobic degradation of alkanes.

14.7 HYDROCARBON ABSORPTION BY BIOSURFACTANTS

A diverse range of microorganisms creates a heterogeneous set of chemical molecules that are surface active, known as biosurfactants. Surfactants improve the solubilization and elimination of impurities. Because surfactants increase the bioavailability of contaminants, they also improve biodegradation (Das and Chandran, 2011). It has been claimed that oil sludge can be bioremediated utilizing biosurfactants. This study employed a microbial consortium made up of a single strain of *Rhodococcus erythropolis* and two isolates of *Pseudomonas aeruginosa* from soil polluted with oily sludge. Ninety percent of the hydrocarbons were broken down by the collaboration in 6 weeks in liquid culture. Two distinct field trials were conducted to evaluate the consortium's capacity to break down the hydrocarbons found in sludge. Furthermore, the impact of a combination of nutrients and an unrefined biosurfactant preparation on the process's efficiency was estimated. A consortium member generated the biosurfactant, which was found to be a combination of eleven rhamnolipid congeners. In 5 weeks, the partnership damaged 91% of the hydrocarbon content of soil polluted with 1% (v/v) crude oil sludge. The hydrocarbon content was reduced by 91%–95% in 4 weeks when the consortium and individual additives were used separately. The crude biosurfactant preparation was shown to be a more potent facilitator of degradation. However, when both additives were applied in tandem with the consortium, hydrocarbon depletion was achieved to a level of greater than 98%. The information supported the hydrocarbon remediation method's usage of a basic biosurfactant. *Pseudomonads* are the most well-known bacteria that can utilize hydrocarbons as sources of energy and carbon to create biosurfactants. *P. aeruginosa*, one of the *Pseudomonas* species, is well-researched for its ability to produce glycolipid-type biosurfactants. However, certain other species, such as *Pseudomonas putida* and *Pseudomonas chlororaphis,* have also been shown to produce glycolipid-type biosurfactants. The oil surface area is increased by biosurfactants, making more oil accessible for bacterial use (Nikolopoulou and Kalogerakis, 2009). Biosurfactants can operate as emulsifying agents by reducing surface tension and forming micelles, where the microbial cell surface encloses the microdroplets, which are subsequently internalized and broken down.

14.8 PH BIODEGRADATION BY IMMOBILIZED CELLS

Immobilized cells have been utilized to bioremediate a variety of hazardous chemicals, easing separation and recovery operations and making the application more reusable. Das and Chandran (2011) employed *Pseudomonas* sp. to break down petroleum in aqueous solution, which resulted in enhanced cell-hydrocarbon droplet contact and rhamnolipid production. This resulted in quicker hydrocarbon absorption into cells and enhanced interaction with solubilized hydrocarbon droplets. In various culture salinities, immobilizing microbial cells boosted the rate of degradation in crude oil when compared with free living cells (Singh Jadaun et al., 2022). Immobilization can be done in batches or continuously, using packed-bed reactors with local microorganisms as the entrapment matrix. After 32 days, immobilized systems proved to be the most successful in removing diesel. In an experiment using immobilized bacteria in alginate beads, the microbial consortium's biodegradation activity did not diminish with repeated application. Immobilization of cells is a feasible method for bioremediation of hydrocarbon-contaminated areas.

14.9 COMMERCIAL BIOREMEDIATION AGENTS

To lessen the impact of the discharge as bioremediation agents, the United States Environmental Protection Agency (US EPA) disregarded nutrition additives, enzyme additions, and microbiological cultures that greatly accelerate biodegradation. Based on two main methods for oil spill bioremediation, bioaugmentation and biostimulation agents are the two categories into which bioremediation agents are divided. Particularly in the early 1990s, when bioremediation was hailed as "the ultimate solution" to oil spills, vendors offered and marketed a range of bioremediation options. In order to meet the requirements of the Clean Water Act, the Oil Pollution Act of 1990, the National Contingency Plan (NCP), and the National Hazardous Substances Pollution Contingency Plan (NCP), the US EPA created a list of 15 bioremediation solutions. However, the list was changed, and there are now just nine bioremediation agents. Research revealed that while bioremediation products could work well in the lab, their effectiveness in real-world applications is far lower. This is due to the fact that complex real-world variables, including nutritional mass transit constraints, biological interactions, regional variability, and climatic impacts, are sometimes difficult for laboratory research to replicate. As a result, field studies and applications serve as the most accurate instances of bioremediation product efficacy (Kanungo et al., 2023; Pradhan et al., 2023). Compared to microbiological treatments, few nutrient additions have been developed and expressly offered as commercial bioremediation agents for oil spill cleanup. It's most likely because ordinary fertilizers are widely accessible, reasonably priced, and, when applied correctly, have been proven to be effective. In light of the drawbacks of conventional fertilizers—such as their quick erosion by tide and wave action—a number of organic nutrient products, including oleophilic nutrient products, have lately been tested and put on the market as bioremediation agents. This category also includes the four agents mentioned on the NCP Product Schedule: Inipol EAP22, Oil Spill Eater II (OSE II), BIOREN 1, and BIOREN 2. As per its application, it is arguably the most well-known bioremediation agent for oil spill recovery and is classified as a nutrition supplement on the NCP Product Schedule. This nutritional product is a microemulsion that contains urea, phosphorus, 2-butoxy-1-ethanol, and oleic acid to impart hydrophobicity to the substance. Inipol EAP22 is said to have the following benefits: (1) it lowers oil viscosity and interfacial tension; (2) it releases phosphorus and nitrogen in a controlled manner, facilitating oil biodegradation; and (3) it shows good biodegradability and no toxicity to plants or animals. The NCP Schedule also lists Oil Spill Eater II (Oil Spill Eater International, Corp.) as a nutritional supplement. This product contains N, P, C, and vitamins for rapid spontaneous bacterial colonization and is classified as an enzyme/nutrient supplement. A field experiment was conducted to assess the effectiveness of OSEII for promoting hydrocarbon degradation in a fuel-contaminated zone at a bioventing site in a Marine Corps Air Ground Combat Center (MCAGCC) in California. Another field study was also carried out in an estuary setting by researchers from the European EUREKA BIOREN initiative to assess the efficacy of two bioremediation products (BIOREN 1 and 2). The two nutritional products were obtained as granular forms from fish meals, utilizing protein as the carbon source and urea and superphosphate as sources of N and P. The presence of a biosurfactant in BIOREN 1 was the main distinction between the two formulations. According to the findings, BIOREN 1's biosurfactant content was the most active component that increased the rates at which oil degraded, but BIOREN 2—which lacked biosurfactant—was ineffective in this regard. The biosurfactant may have increased the hydrocarbons' bioavailability to microbial assault.

14.10 BIODEGRADATION PATHWAYS AND RELATED GENES

14.10.1 Biodegradation Pathways

The biodegradation of n-alkanes occurs via four primary pathways: the Finnerty process, the diterminal oxidation pathway, the subterminal oxidation pathway, and the terminal oxidation pathway. A typical breakdown mechanism for hexadecane is alkane hydroxylases, which oxidize it to 1-hexadecanol. Alcohol dehydrogenases next oxidize it to 1-hexadecanal, and aldehyde dehydrogenases eventually convert it to hexadecanoic acid. β-oxidation is initiated by this fatty acid. *P. aeruginosa* was the first organism to exhibit subterminal oxidation, and Gordonia sp. strain TY-5 later demonstrated this phenomenon. In contrast to the terminal process, alkane hydroxylases oxidize hexadecane to 2-hexadecanol, 2-hexadecanal, and finally tetradecyl acetate via Baeyer–Villiger monooxygenase (BVMO). The subsequent procedures are the same as for the terminal pathway after esterase further oxidizes the tetradecyl acetate to the matching alcohol. First identified in a bacterium, the diterminal oxidation pathway is also present in many strains of yeast. Because of the presence of ω-hydroxylase, which changes hexadecanoic acid into the more valuable 16-hydroxyhexadecanoic acid, it is special. Hexadecanedioic acid enters β-oxidation as a result of alcohol and aldehyde dehydrogenases, further oxidizing this 16-hydroxyhexadecanoic acid. According to Ji et al. (2013), *Acinetobacter* sp. strain HO1 N exhibits the Finnerty route, in which dioxygenase catalyzes the conversion of hexadecane into 1-hydroperoxyhexadecane and hexadecaneperoxoic acid. Next, hexadecaneperoxoic acid undergoes further oxidation to yield hexadecanoic acid, which proceeds into β-oxidation.

14.10.2 Related Genes

Numerous degrading pathway genes are involved in alkane breakdown in microorganisms, which is controlled by a number of activities. including recognition, transport, and oxidative stress response. Taking into account every gene that is involved is essential while building and refining an alkane-degrader. As synthetic biology and metabolic engineering techniques have advanced, a large number of genes involved in the breakdown of alkanes have been found and studied. AlkB, alkM, ladA, and the gene encoding the CYP153C1 protein are some of the crucial genes involved in the process of alkane degradation. Prokaryotes have alkane hydroxylase coenzymes encoded by rubA and rubB and a transcriptional encoding the CYP153C1 protein encoded by alkR. Alkane degradation typically involves the CYP52 gene family, particularly in yeast. Alcohols are broken down by laoA and laoB, which encode alcohol dehydrogenase and its coenzyme, respectively, to break down primary long-chain alcohols (Panasia and Philipp, 2018). HFD1–HFD4 also acts as a laoC in the alkane degradation pathway. On a cluster of genes related to secondary alcohol degradation, sadC breaks down secondary alcohol, while sadD, sadA, and sadB encode a BVMO and two esterases that catalyze secondary alcohol to the corresponding primary alcohols (Iwama et al., 2014). Two significant factors that affect the breakdown of alkanes are the identification and movement of alkanes as well as the oxidative stress response.

As per Mounier et al. (2018), two genes aupA and aupB encoding for outer membrane and inner membrane proteins enable the prokaryotes to absorb alkanes. Whereas in eukaryotes, the ABC1transporter, encoded by the gene ABC1, is involved in the transport of alkanes. Certain genes that encode fatty acid transport proteins, including fadL (van den Berg, 2005) and exfadLO (Martínez et al., 2013) in prokaryotes and ScFAT1 and YlFAT1 (Dulermo et al., 2014) in eukaryotes, may have an indirect effect on alkane breakdown by influencing the buildup of fatty acids. The alkane-responsive biosensor encoded by the three genes Yas1p, Yas2p, and Yas3p is instructive in terms of how to build *S. cerevisiae* that is capable of breaking down alkanes (Dabirian et al., 2020). Hazardous compounds such as H_2O_2, O_2, and organic sulfonates may be generated during the breakdown of alkanes, which can lead to oxidative damage to cells. Related genes that affect the endoplasmic reticulum membrane environment and are involved in the synthesis of Alk proteins include tauD, ssuD, oxyR, and P24 (Iwama et al., 2018).

14.10.3 Alkane Hydroxylases

Alkane hydroxylases catalyze the first reaction in alkane degradation by adding oxygen atoms, which is thought to be the rate-limiting step (Ji et al., 2013). As of now, dioxygenases and several alkane hydroxylases such as cytochrome P450, flavoprotein, AlmA, LadA, AlkB, and AlkM have all been found. Of these, studies on the structure and modification of certain alkane hydroxylases, i.e., integral-membrane and cytochrome P450 alkane hydroxylases, are more developed and detailed than those of flavoprotein and the dioxygenase family, which is detailed below.

14.10.3.1 Integral-Membrane

According to Wang et al. (2022), the two main enzyme systems were AlkB and AlkM, which were found in *P. putida* Gpol and *Acinetobacter calcoaceticus* ADP1, respectively. AlkM has been reported to oxidize long-chain n-alkanes (C16–C30), whereas AlkB oxidizes alkanes with shorter chain lengths (C5–C12). This is the difference in function between the two. The structure of AlkB has been examined more thoroughly than that of AlkM. Six alpha-helical transmembrane segments, each containing eight histidines, were identified in a prior work on AlkB. These eight histidines are crucial to the AlkB structure. The system's electron-transfer components include two rubredoxins and one rubredoxin reductase in addition to AlkB. Additionally, rubA and rubB, two additional rubredoxins and rubredoxin reductases from *Alcanivorax borkumensis*, *Acinetobacter* sp. M-1, were discovered by the researchers. Researchers discovered a new gene called alkW1, which is made up of a rubredoxin and AlkB, in the Dietzia strain DQ12–45-1b. Long-chain alkanes can be oxidized by the AlkB-fused rubredoxin, according to the study. Furthermore, the alkB gene promoter (PalkB), which influences alkB expression, was found and examined. In terms of DNA sequence homology, gene alkM is distinct from gene alkB. Since AlkM also requires rubredoxin and rubredoxin reductase, it is still acknowledged to be an integral-membrane alkane hydroxylase. The positional connection between the AlkM gene and the coenzyme gene is also different from AlkB to AlkM. AlkB is the reverse of alkM, and it is not associated with the genes in *Acinetobacter* sp. strain ADP1 that encode for rubredoxin and rubredoxin reductase. Furthermore, the identification of transcriptional regulators has been completed. AlkM's transcriptional regulation depends on AlkR. The AlkR gene is situated next to the AlkM gene. Genes alkMa and alkMb, as well as alkRa and alkRb, found in *Acinetobacter* sp. strain M-1, are comparable to strain ADP1 in terms of sequence homology and positional connection.

14.10.3.2 Cytochrome P450

The cytochrome P450 family of heme-thiolate proteins includes a large number of alkane hydroxylases (Maier et al., 2001). Class I and Class II of cytochrome P450, which are divided into components, are now the most researched (Van Beilen and Funhoff, 2005). The crucial Class I enzymes CYP153 were discovered in *A. calcoaceticus* EB104 (Maier et al., 2001). CYP153 enzymes, such CYP153C1, which was cloned from *Novosphingobiu aromaticivorans* DSM12444, have the ability to oxidize n-alkanes (C6–C11). Wang et al. (2022) state that knowing these enzymes will enable us to use synthetic biology and metabolic engineering techniques to change them in the way we want. Integral-membrane alkane hydroxylases and the P450cam system (CYP101) found in *P. putida* ATCCI7453 are comparable. Putidaredoxin, putidaredoxin reductase, and cytochrome P450 alkane hydroxylases are part of the P450cam system. It is the job of the two coenzymes to move electrons from NADH to P450cam. The amino acid residues in the active site of P450cam have been changed by substitution. The activity of the mutations in alkane degradation is increased. P450 BM-3 and the CYP52 family have been investigated in Class II. The diterminal oxidation pathway, present in fungi like *C. tropicalis* ATCC 20336, depends on the CYP52 family for oxidation. The CYP52 family in *C. tropicalis* ATCC 20336 oxidizes n-alkanes to produce more useful by-products, including fatty acids and dicarboxylic acids (DCAs). The source of P450 BM-3 is *Bacillus megaterium* 14581. This enzyme's structure is unique in that it combines the reductase and hydroxylase domains onto a single polypeptide chain, making it the most active P450 enzyme. Furthermore, alterations of P450 BM-3 have been investigated, including rational evolution, which enhances n-alkane deprivation.

14.10.3.3 Flavoprotein

Reaction involving NAD(P)H, substrate, and cofactor flavin is catalyzed by flavoprotein alkane hydroxylases where they transport electrons through redox processes (Ji et al., 2013). LadA and AlmA are currently the two flavoprotein hydroxylases that have been extensively researched. From *Geobacillus thermodenitrificans* NG80–2, LadA was isolated (Tani et al., 2005). Studies revealed that it can use the terminal oxidation route to oxidize long-chain n-alkanes from C15 to at least C36. LadA is a member of the SsuD subfamily, which has a barrel structure for triosephosphate isomerase (TIM) (Li et al., 2008). The TIM barrel's flavoprotein hydroxylase for the FMN was shown to be responsible. To increase LadA's activity, certain alterations, including site-directed and random mutagenesis, have been used (Dong et al., 2012). Studies have demonstrated that *Pseudomonas fluorescens* strain KOB21 may grow more quickly when hexadecane is added when the LadA mutations are expressed. Long-chain alkanes can be oxidized by AlmA, which was initially isolated from *Acinetobacter* sp. Strain DSM17874 (Throne-Holst et al., 2007). Additionally, it was discovered in *P. aeruginosa* SJTD-1, *Alcanivorax hongdengensis* A-11–3 (Wang and Shao, 2010), *Acinetobacter pittii* SW-1, and *Acinetobacter oleivorans* DR1 (Park et al., 2017). Only a small number of the degrading mechanisms in which AlmA participates have been described, nevertheless. One dioxygenase was discovered in *Acinetobacter* sp. strain M-1. Using hexadecane as an example, dioxygenase can oxidize it in the Finnerty manner to 1-hydroperoxy hexadecane. This enzyme needs both Cu^{2+} and flavinadenine dinucleotide to function. Alkanes ranging from C10 to C30 are catalyzed by the enzyme. In the next 25 years, dioxygenases occurred in the aerobic deprivation route of naphthalene and phenanthrene, despite the fact that there isn't a true replication of the study that can fully confirm that dioxygenases operate on n-alkanes (Mawad et al., 2020). It leads scientists to believe that while the potential exists, further study is necessary to confirm it.

14.11 ENGINEERED MICROBIAL CHASSIS

Wang et al. (2023) found that certain microbial species such as *Mycobacterium, Rhodococcus, Pseudomonas, Dietzia, Acinetobacter, Aspergillus, Fusarium, Penicillium, Ochrobactrum,* and *Brevibacterium* are capable of expressing alkane hydroxylases. However, due to their unclear genetic background, they are challenging to modify. *Escherichia coli* is the most often used model bacterium for a microbial chassis. Some nonmodel microorganisms may also be altered as a result of the development of synthetic biology technology; these organisms may possess traits that model organisms do not, including a complete degradation system. We draw attention to the developments in the study of certain coli form and bacilli species, which can be modified as potential degradation chassis.

14.12 GENETICALLY MODIFIED BACTERIA

There has been a lot of interest in using genetically modified microorganisms (GEMs) in bioremediation to enhance the breakdown of hazardous pollutants in experimental settings.

There have been instances of several microbes breaking down environmental pollutants. Below is a discussion of several significant uses of mutation techniques to boost bacterial bioremediation of the pollutants (Bala et al., 2024). The mutated organisms show greater degradative capability. Nevertheless, ecological problems, as well as legislative limits, provide significant challenges to the field testing of GEM. These issues must be addressed before GEM can offer a successful cleanup method at a cheaper cost. Several researchers have explored genetically altered bacteria to examine bioremediation processes, strains, stress responses, endpoint analysis, and toxicity (Das & Ghosh, 2022; Das et al., 2024). The pollutants that have been evaluated for the remediation process include chlorinated nonpolar aromatic hydrocarbons. Effective in situ bioremediation with genetically modified bacteria requires a mix of field engineering designs, biochemical processes, and knowledge of microbiology and ecology.

14.12.1 *E. coli*

One significant model microbe is *E. Coli*. Currently, *E. coli* is used to express alkane hydroxylase and biooxidize alkanes across the complete cell. This method not only eliminates the need for laborious stages in the purification process but also addresses the issue of the enzymes' decreased activity. Two facets of the *E. coli* study can be distinguished. On the one hand, alkane hydroxylases are expressed in *E. coli* to induce alkane degradation. When the alk genes were attempted to be inserted into *E. coli* back in 1993, Alkane hydroxylase was overexpressed in a distinct cytoplasmic membrane subfraction as a result. While more alkane hydroxylase can be expressed via recombinant *E. coli.* (AlkB) 2 years later, a research found that AlkB's specific activity was five or six times lower in *E. coli* than it was in *Pseudomonas oleovorans* (Staijen et al., 1999). Recombinant *E. coli* was not used in either study, nevertheless, to break down alkane (Igeño et al., 2019; Park et al., 2019). According to recent research, it is possible to create primary alcohols and carboxylic acids with the *E. coli* GEC137 pCEc47 J strain (Grant et al., 2011). This is the first instance of n-alkane (n-dodecane) degradation in a recombinant organism employing AlkB. The next year, the pCom8-GPo1 AlkB plasmid was transformed into *E. coli* DH5a, and the *P. putida* GPo1 AlkB gene was produced in a PCom8 expression vector (Luo et al., 2015). According to the findings, the diesel oil degradation ratio increased from 31% to 50% in 24 hours when recombinant *E. coli* was cultured with the petroleum biodegradation bacterial population (Luo et al., 2015). However, other research concentrated on oxidation, which is a process that may transform alkanes into more useful compounds. These investigations have demonstrated that medium-chain n-alkanes at low cost may be converted to useful medium-chain-diols and -DCAs via diterminal oxidation. Another work examined the biocatalytic conversion of fatty acid esters to hydroxy fatty acid esters utilizing recombinant *E. coli*, which generates the AlkBGT enzymes (Van Nuland et al., 2017). Furthermore, the strategies for increasing the yield and product output of HFAs and related chemicals in *E. coli* were meticulously demonstrated in a recent work. These results imply that mutated *E. coli* can be used to degrade alkanes and produce valuable by-products.

14.12.2 *Pseudomonas* sp.

Pseudomonas species may produce biosurfactants using crude oil as the carbon source and are generally found throughout nature. Pseudomonas sp. strains have been isolated and identified in large numbers recently (Table 14.2). Numerous genes linked to degradation were discovered as a result of these investigations. For instance, detailed explanations were provided for a gene cluster that encodes a suspected alcohol dehydrogenase (PA0364/LaoA), a likely inner membrane protein (PA0365/LaoB), and a possible aldehyde dehydrogenase (PA0366/LaoC) (Panasia et al., 2018). Later on, variations of *Pseudomonas* sp. were also progressively created. After being given the NAH7 plasmid, *P. putida* KT2440 was able to break down naphthalene. In addition, this change lessened the stress on cells caused by this harmful substance (Fernandez et al., 2012). This is an illustration of pseudomonas rational modification. Additional studies focus on *Pseudomonas* irrational alterations. *Pseudomonas pseudoalcaligenes* CECT5344 was cultivated utilizing furfuryl alcohol, furfural, and furoic acid as carbon sources, and the evolved strain was obtained as the strain adapted (Igeno et al., 2019). The original strain had a several-day lag period, but the evolved strain had no lengthy lag phases (Igeno et al., 2019). These studies paved the way for the development of communities capable of efficient alkane degradation, as well as a framework for future genetic engineering of wild *Pseudomonas*.

14.12.3 *Bacillus* sp.

Bacillus species have robust resistance to exogenous pathogens. Bacillus sp. can therefore be used as a chassis for the degradation of alkanes. They can be found in animal intestines, water, soil, and the air (Mohapatra et al., 2017). Numerous Bacillus sp. bacteria have recently been isolated and identified (Table 14.2). Currently, target gene strengthening and promoter replacement are the two major biological techniques used to change strains. According to Qiu et al. (2014), the modified strain generated 2,149 mg/L of lichenysin, nearly about a 17-fold increase over the wild strain, using the promoter of the lichenysin biosynthetic operon as a substitute.

By overexpressing two regulatory genes either separately or in tandem (Zhang et al., 2017), inserting a potent constitutive promoter upstream of the operon (Dang et al., 2019), and activating two competence-stimulating pheromones to stimulate operon transcription (Jung et al., 2012), engineered bacteria increase the yield of their target products. In addition, Wu et al. (2019) developed a methodical engineering approach to improve biosynthesis. The final surfactant titer reached 12.8 g/L in the metabolically modified strain, resulting in 65.0 mmol/mol sucrose (42% of the theoretical yield). However, no modification research has been conducted on Bacillus sp.'s capacity to degrade.

14.13 MICROBIAL COMMUNITY

A community of microorganisms can establish a commensal and co-metabolic system, enhance the range of degradable substrates, and be more successful at breaking down PHs than a single bacterium. The degradation effects of various combinations, such as the pure bacterial community (Bidja et al., 2020), pure fungal community (Hamad et al., 2021), bacterial–fungal community (Ghorbannezhad et al., 2018), and pure protozoa community, have been compared. The environment has yielded the isolation and identification of several microbial species. These certainly offer a wealth of theoretical and material information for a microbial community which can build a system of commensalism and co-metabolism, enhance the range of degradable substrates, and be more successful at breaking down PHs than a single bacterium (Pradhan et al., 2024; Priyadarshini & Das 2024). The degradation effects of various combinations, such as the pure bacterial community (Bidja et al., 2019), pure fungal community (Hamad et al., 2021), bacterial–fungal community (Ghorbannezhad et al., 2018), and pure protozoa community, have been compared. Numerous microorganisms have been isolated and identified from the environment. These certainly offer a wealth of material and a theoretical foundation for our future investigation. However, it is very tough to influence microorganisms to carry out certain degrading activities; this can only be achieved by fully understanding the cell–cell interactions in communities and the endogenous genes of each microbe in microbial communities (Das et al., 2023; Jagatee et al., 2024; Mallick et al., 2023). Because of this, scientists are starting to consider using modified strains and a rational strain allocation to create microbial communities for the breakdown of PHs. "Alkane-degrader and Alkane-degrader" or "Alkane-degrader and Helper" are two possible descriptions of the interactions between the microorganisms in this chapter (Figure 14.2).

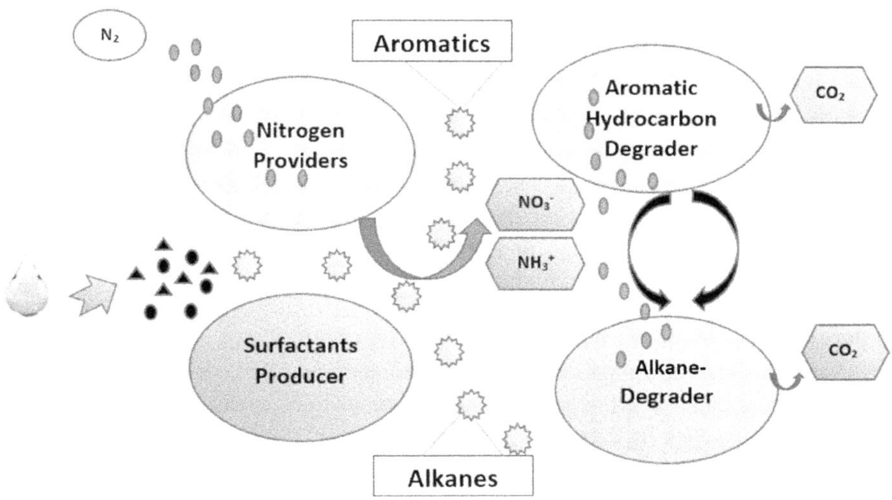

FIGURE 14.2 Degradation of petroleum hydrocarbon contaminants.

14.14 ALKANE-DEGRADER

Each member of a community known as "Alkane-degrader and Alkane-degrader" is capable of degrading alkanes, although the kind of alkanes they break down vary. After each member completes a particularly difficult alkane degradation, the complex PHs are fully degraded. A recent study examined a co-culture of fungus, *Talaromyces* sp., and bacteria, *Acinetobacter baumannii*. In this community, bacteria are better at degrading other substances like aromatics and iso alkanes, while fungi are more adept at breaking down n-alkanes. This community's crude oil can degrade up to 80% overall (Zhang et al., 2021). Some researchers often utilize synthetic bacteria to build a microbial community in addition to using wild bacteria to establish an artificial community. In one study, the degradation of C10–C20 or C22–C32 n-alkanes was accelerated by an *Acinetobacter venetianus* strain RAG-1 mutant with an alkMa or alkMb deletion. Next, alkMa/almA and alkMb were assembled into a focused and tunable community. In comparison to the single wild-type strain, this community degraded light crude oil approximately 89%, viscous crude oil 90%, and high waxy crude oil by about 61% in 10 days. The microbial community, as indicated by "Alkane-degrader and Alkane-degrader," is clearly useful in the breakdown of complex alkane components.

14.15 ALKANE-DEGRADER AND HELPER

Within a community dubbed "Alkane-degrader and Helper," some individuals break down alkanes while other members perform various tasks to assist the former in breaking down alkanes more effectively. For instance, by making the alkanes more soluble, surfactant makers can aid in the degradation of alkanes by an "Alkane-degrader." Nitrogen suppliers can aid in polluted soil by feeding the alkane-degrader with nitrogen.

14.16 SURFACTANTS PRODUCERS

Surfactants are secreted by most "Helpers" to aid "Alkane-degraders" (Souza et al., 2014). But how is this going to benefit the microbes? There is still much to be answered on this issue. The standard response to this query is that biosurfactants can boost solubility, which can lead to increased bacterial proliferation and improved biodegradation efficiency (Xia et al., 2019). Nonetheless, a more thorough examination of this process has been conducted, accounting for several cellular alterations that occur, including the genome, metabolic activity, and cell surface features. These discovered genomic alterations on exposure to surfactants and concentrated on providing a detailed account of the interactions between hydrocarbons and bacteria (Ray et al., 2024). The findings suggested that the advantages of using surfactants could be connected to profound changes in the omics level as well as the surface characteristics of the cell (Behera et al., 2022). However, further study is still needed to fully understand the mechanism of profound change at the omics levels. It should be noted that surfactants occasionally have no progressive effect on the breakdown of PHs and may even be detrimental. Anionic surfactants react with divalent ions like Ca^{2+} or get permanently absorbed in the soil during the remediation of soil contamination. Petroleum biodegradation will be negatively impacted, leading to a significant loss of surfactant or a rise in the quantity of organic carbon in the soil. Furthermore, it has been documented that certain surfactants can promote or prevent the development of microorganisms (Bailey et al., 2012). Bioemulsifiers, which have higher molecular weights, like polymeric and particulate biosurfactants, and biosurfactants, which have lower molecular weights, like lipopeptides, rhamno lipids, and sophorolipids, are the two categories of biosurfactants based on the changes in their chemical compositions. Wang and his colleagues also mentioned the distinction in function between the two kinds, namely that low-molecular-weight biosurfactants can assist bacteria in increasing the hydrophobic substrates' surface area and in boosting the hydrophobic compounds' bioavailability through efficient surface tension reduction (Gregson et al., 2019). High-molecular-weight bioemulsifiers serve this purpose by covering and preventing the coalescence of oil droplets. Alkane-degraders can deteriorate more quickly when biosurfactants are present. Researchers have been interested in isolating and identifying surfactant-producing microorganisms from the environment. Given the tight relationship between surfactant synthesis and alkane degradation, most microorganisms capable of producing surfactants can also degrade crude oil. For the future growth of microbial communities, researchers have produced useful tools and theories. Manufacturers of surfactants have been used in these communities to speed up the process of breakdown. These communities may consist of a single bacterium or bacterial communities, such as single cultures that may produce biosurfactants. For all alkane-degraders, the rate of alkane hydrocarbon disintegration improves dramatically when the "Helper" joins the community; the biggest improvement occurs at around 55.4%.

14.17 NITROGEN PROVIDERS

Nitrogen plays a crucial role in the rehabilitation of polluted soil, with appropriate nitrogen biostimulation promoting the decomposition of aromatic hydrocarbons. However, excessive nitrogen stimulation reduces microbial degradation efficiency. In a community, alkane-degraders and nitrogen-fixing bacteria collaborate to deliver carbon and nitrogen to each other. Nitrogen deficiency significantly reduces cell attachment to soil particles. When uric acid is combined with crude oil, it provides the necessary nitrogen supply for alkane-degraders. Gao et al. (2021) found the nitrogen metabolic route in bioremediation of diesel-contaminated soil using metagenomics analyses, finding that organic ammonium form of nitrogen is preferred for increased bioremediation. This work contributes to our understanding of nitrogen's role in the breakdown of crude oil. Various nitrogen-producing bacteria have been isolated and identified, with a nitrogen-fixing bacteria related to *Azospirillum* from an oil production mixture at Yumen Oilfield. Additionally, a genome for new nitrogen-fixing bacteria capable of using aromatic hydrocarbons was constructed. This project will broaden the scope of bioremediation techniques in a nitrogen-stressed environment. Nitrogen suppliers have recently been employed in communities to accelerate the deterioration process (Thevenieau et al., 2007). For instance, after 40 days of treatment, two more oil-degrading microbes (*Acinetobacter sp.* K-6 + *Rhodococcus sp.* Y2-2) helped one nitrogen-fixing microbe, *Azotobacter vinelandii* KCTC2426, break down the diesel from the soil, eliminating 83.1% of it (Chaudhary et al., 2019). By boosting fungal abundance, accelerating the activity of plastic-degrading enzymes, and forming or interacting with fungal communities that break down plastic, nitrogen-fixing bacteria helped the microbiological degradation of poly (butylene succinate-co-adipate) in another study. Future attempts to create a microbial community with a higher rate of disintegration might be greatly inspired by these discoveries (Yin et al., 2022).

14.18 CONCLUSION

PH cleanup in the subterranean environment is a real-world issue. Improved comprehension of the biodegradation mechanism, which relies on it uses indigenous microbes to mineralize or transform organic contaminants, is highly significant from an ecological point of view. Once significant quantities of oil have been removed from the environment utilizing a range of chemical and physical methods, the process of microbial breakdown aids in clearing the surrounding region after oil spills. The ability of microbes to decompose and use various hydrocarbons as a source of carbon and energy is made feasible by their enzymatic systems. Utilizing genetically modified (GM) microorganisms is a cutting-edge area of study with far-reaching consequences. The use of genetically engineered microorganisms has a lot of potential advantages. Though native species frequently function well enough, our poor knowledge of different phytoremediation mechanisms, such as the control of enzyme systems that digest contaminants, prevents us from fully using the potential of wild species. Because of this, the need for GM bacteria may be questioned in many situations. Thus, it is possible to draw the conclusion that microbial degradation is a crucial element of the cleaning plan for PH remediation based on the current review.

REFERENCES

Abdel-Shafy, H. I., & Kamel, A. H. (2016). Groundwater in Egypt issue: Resources, location, amount, contamination, protection, renewal, future overview. *Egyptian Journal of Chemistry, 59*(3), 321–362.

Abdel-Shafy, H. I., & Mansour, M. S. (2018). Microbial degradation of hydrocarbons in the environment: an overview. In *Microbial Action on Hydrocarbons* (pp. 353–386).

Bailey, K. L., Tilton, F., Jansik, D. P., Ergas, S. J., Marshall, M. J., Miracle, A. L., & Wellman, D. M. (2012). Growth inhibition and stimulation of Shewanella oneidensis MR-1 by surfactants and calcium polysulfide. *Ecotoxicology and Environmental Safety, 80*, 195–202.

Bala, K., Dogra, A., Lata, P., & Das, A. P. (2024). Discussing sources and biological-chemical remediation approaches for microplastic pollution. In *Sustainable Microbial Technology for Synthetic and Cellulosic Microfiber Bioremediation* (pp. 37–59). https://doi.org/10.1007/978-3-031-62600-5_3

Behera, I. D., Basak, G., Kumar, R. R., Sen, R., & Meikap, B. C. (2020). Treatment of petroleum refinery sludge by petroleum degrading bacterium Stenotrophomonas pavanii IRB19 as an efficient novel technology. *Journal of Environmental Science and Health - Part A Toxic/Hazardous Substances and Environmental Engineering, 56*(2). https://doi.org/10.1080/10934529.2020.1866924

Behera, I. D., Nayak, M., Mishra, A., Meikap, B. C., and Sen, R. (2022). *Strategic implementation of integrated bioaugmentation and biostimulation for efficient mitigation of petroleum hydrocarbon pollutants from terrestrial and aquatic environment. Marine Pollution Bulletin, 177,* 113492.

Bidja Abena, M. T., Chen, G., Chen, Z., Zheng, X., Li, S., Li, T., & Zhong, W. (2020). Microbial diversity changes and enrichment of potential petroleum hydrocarbon degraders in crude oil-, diesel-, and gasoline-contaminated soil. *3 Biotech*, *10*(2), 42.

Chaudhary, D. K., Bajagain, R., Jeong, S. W., & Kim, J. (2019). Development of a bacterial consortium comprising oil-degraders and diazotrophic bacteria for elimination of exogenous nitrogen requirement in bioremediation of diesel-contaminated soil. *World Journal of Microbiology and Biotechnology*, *35*, 1–11.

Chikere, C. B., Okpokwasili, G. C., & Chikere, B. O. (2011). Monitoring of microbial hydrocarbon remediation in the soil. *3 Biotech*, *1*(3), 117–138.

Dabirian, Y., Skrekas, C., David, F., & Siewers, V. (2020). Does co-expression of Yarrowia lipolytica genes encoding Yas1p, Yas2p and Yas3p make a potential alkane-responsive biosensor in Saccharomyces cerevisiae? *PLoS One*, *15*(12), e0239882.

Dang, Y., Zhao, F., Liu, X., Fan, X., Huang, R., Gao, W., ... & Yang, C. (2019). Enhanced production of antifungal lipopeptide iturin A by Bacillus amyloliquefaciens LL3 through metabolic engineering and culture conditions optimization. *Microbial Cell Factories*, *18*, 1–14.

Das, A. P., Dutta, K., Khatun, R., Behera, I. D., Singh, S., & Mishra, S. (2023). Microfiber pollution and its microbial mitigation: a review on current trends and future prospects. *Journal of the Taiwan Institute of Chemical Engineers*, 105104. https://doi.org/10.1016/J.JTICE.2023.105104

Das, A., & Ghosh, S. (2022). Role of microorganisms in extenuation of mining and industrial wastes. *Geomicrobiology Journal*, *39*, 173–175. https://doi.org/10.1080/01490451.2022.2038953

Das, A., Dey, S., Bala, K., Gnanasekaran, R., & Das, A. P. (2024). Microbial bioremediation technology for sustainable treatment and management of synthetic microfiber waste. In *Sustainable Microbial Technology for Synthetic and Cellulosic Microfiber Bioremediation* (pp. 221–238). https://doi.org/10.1007/978-3-031-62600-5_12

Das, N., & Chandran, P. (2011). Microbial degradation of petroleum hydrocarbon contaminants: an overview. *Biotechnology Research International*, *2011*(1), 941810.

Dey, S., Bissoyi, A., Koteswara Reddy, G., & Das, A. P. (2024). Synthetic microfibers: an urban emerging pollutant demanding for sustainable environmental solutions. In *Sustainable Microbial Technology for Synthetic and Cellulosic Microfiber Bioremediation* (pp. 107–121). https://doi.org/10.1007/978-3-031-62600-5_7

Dey, S., Das, A., Mallick, K., Sahu, A., & Das, A. P. (2023). Environmental petroleum waste: pollution, toxicity, sustainable remediation. (pp. 159–175). https://doi.org/10.1007/978-3-031-48220-5_7

Dey, S., Dash, N., Ray, S. K., Priyadarsini, S., Chhotaray, C., & Das, A. P. (2024). Occurrence and quantification of tire wear plastics pollutants from highway city dust of Bhubaneswar City of Odisha. In *Sustainable Microbial Technology for Synthetic and Cellulosic Microfiber Bioremediation* (pp. 77–89). https://doi.org/10.1007/978-3-031-62600-5_5

Dong, Y., Yan, J., Du, H., Chen, M., Ma, T., & Feng, L. (2012). Engineering of LadA for enhanced hexadecane oxidation using random-and site-directed mutagenesis. *Applied Microbiology and Biotechnology*, *94*, 1019–1029.

Dulermo, R., Gamboa-Meléndez, H., Dulermo, T., Thevenieau, F., & Nicaud, J. M. (2014). The fatty acid transport protein Fat1p is involved in the export of fatty acids from lipid bodies in Yarrowia lipolytica. *FEMS Yeast Research*, *14*(6), 883–896.

Fernández, M., Niqui-Arroyo, J. L., Conde, S., Ramos, J. L., & Duque, E. (2012). Enhanced tolerance to naphthalene and enhanced rhizoremediation performance for Pseudomonas putida KT2440 via the NAH7 catabolic plasmid. *Applied and Environmental Microbiology*, *78*(15), 5104–5110.

Gao, Y., Du, J., Bahar, M. M., Wang, H., Subashchandrabose, S., Duan, L., … & Naidu, R. (2021). Metagenomics analysis identifies nitrogen metabolic pathway in bioremediation of diesel contaminated soil. *Chemosphere*, *271*, 129566.

Ghorbannezhad, H., Moghimi, H., & Dastgheib, S. M. M. (2018). Evaluation of heavy petroleum degradation using bacterial-fungal mixed cultures. *Ecotoxicology and Environmental Safety*, *164*, 434–439.

Grant, C., Woodley, J. M., & Baganz, F. (2011). Whole-cell bio-oxidation of n-dodecane using the alkane hydroxylase system of P. putida GPo1 expressed in E. coli. *Enzyme and Microbial Technology*, *48*(6–7), 480-486.

Gregson, B. H., Metodieva, G., Metodiev, M. V., & McKew, B. A. (2019). Differential protein expression during growth on linear versus branched alkanes in the obligate marine hydrocarbon-degrading bacterium Alcanivorax borkumensis SK2T. *Environmental Microbiology*, *21*(7), 2347–2359.

Hamad, A. A., Moubasher, H. A., Moustafa, Y. M., Mohamed, N. H., & Abd-El Rhim, E. H. (2021). Petroleum hydrocarbon bioremediation using native fungal isolates and consortia. *The Scientific World Journal*, *2021*(1), 6641533.

Igeño, M. I., Macias, D., & Blasco, R. (2019). A case of adaptive laboratory evolution (ALE): biodegradation of furfural by Pseudomonas pseudoalcaligenes CECT 5344. *Genes*, *10*(7), 499.

Igeño, M. I., Macias, D., & Blasco, R. (2019). A case of adaptive laboratory evolution (ALE): biodegradation of furfural by Pseudomonas pseudoalcaligenes CECT 5344. *Genes*, *10*(7), 499.

Iwama, R., Hara, M., Mizuike, A., Horiuchi, H., & Fukuda, R. (2018). Osh6p, a homologue of the oxysterol-binding protein, is involved in production of functional cytochrome P450 belonging to CYP52 family in n-alkane-assimilating yeast Yarrowia lipolytica. *Biochemical and Biophysical Research Communications*, *499*(4), 836–842.

Iwama, R., Kobayashi, S., Ohta, A., Horiuchi, H., & Fukuda, R. (2014). Fatty aldehyde dehydrogenase multigene family involved in the assimilation of n-alkanes in Yarrowia lipolytica. *Journal of Biological Chemistry*, *289*, 33275–33286.

Jagatee, S., Priyadarshini, S., Rath, C. C., & Das, A. P. (2024). Synthetic microfiber: an enduring environmental problem linked to sustainable development. *Environmental Science and Engineering*, *Part F2333*, 93–112. https://doi.org/10.1007/978-3-031-51792-1_6

Ji, Y., Mao, G., Wang, Y., & Bartlam, M. (2013). Structural insights into diversity and n-alkane biodegradation mechanisms of alkane hydroxylases. *Frontiers in Microbiology*, *4*, 58.

Jugder, B. E., Ertan, H., Lee, M., Manefield, M., & Marquis, C. P. (2015). Reductive dehalogenases come of age in biological destruction of organohalides. *Trends in Biotechnology*, *33*(10), 595–610.

Jung, J., Yu, K. O., Ramzi, A. B., Choe, S. H., Kim, S. W., & Han, S. O. (2012). Improvement of surfactin production in Bacillus subtilis using synthetic wastewater by overexpression of specific extracellular signaling peptides, comX and phrC. *Biotechnology and Bioengineering*, *109*(9), 2349–2356.

Kanungo, J., Sahoo, T., Bal, M., & Behera, I. D. (2023). Performance of bioremediation strategy in waste lubricating oil pollutants: a review. *Geomicrobiology Journal*. https://doi.org/10.1080/01490451.2023.2245395

Lahiri, D., Nag, M., Dey, A., Sarkar, T., Joshi, S., Pandit, S., Das, A. P., Pati, S., Pattanaik, S., Tilak, V. K., & Ray, R. R. (2021). Biofilm mediated degradation of petroleum products. *Geomicrobiology Journal*, 1–10. https://doi.org/10.1080/01490451.2021.1968979

Li, L., Liu, X., Yang, W., Xu, F., Wang, W., Feng, L., & Rao, Z. (2008). Crystal structure of long-chain alkane monooxygenase (LadA) in complex with coenzyme FMN: unveiling the long-chain alkane hydroxylase. *Journal of Molecular Biology*, *376*(2), 453–465.

Luo, Q., He, Y., Hou, D. Y., Zhang, J. G., & Shen, X. R. (2015). GPo1 alkB gene expression for improvement of the degradation of diesel oil by a bacterial consortium. *Brazilian Journal of Microbiology*, *46*, 649–657.

Maier, T., Förster, H. H., Asperger, O., & Hahn, U. (2001). Molecular characterization of the 56-kDa CYP153 from Acinetobacter sp. EB104. *Biochemical and Biophysical Research Communications, 286*(3), 652–658.

Mallick, K., Sahu, A., Dubey, N. K., & Das, A. P. (2023). Harvesting marine plastic pollutants-derived renewable energy: a comprehensive review on applied energy and sustainable approach. *Journal of Environmental Management, 348*, 119371. https://doi.org/10.1016/J.JENVMAN.2023.119371

Mallick, K., Sahu, S., Sahu, A., Dey, S., & Das, A. P. (2024). A critical review of marine microfiber pollution routes, toxicity, and its sustainable remediation. In *Environmental Science and Engineering, Part F2333* (pp. 189–211). https://doi.org/10.1007/978-3-031-51792-1_10

Martínez, E., Estupiñán, M., Pastor, F. J., Busquets, M., Díaz, P., & Manresa, A. (2013). Functional characterization of ExFadLO, an outer membrane protein required for exporting oxygenated long-chain fatty acids in Pseudomonas aeruginosa. *Biochimie, 95*(2), 290–298.

Mawad, A. M., Abdel-Mageed, W. S., & Hesham, A. E. L. (2020). Quantification of naphthalene dioxygenase (NahAC) and catechol dioxygenase (C23O) catabolic genes produced by phenanthrene-degrading Pseudomonas fluorescens AH-40. *Current Genomics, 21*(2), 111–118.

McLeod, M. P., & Eltis, L. D. (2008). Genomic insights into the aerobic pathways for degradation of organic pollutants. In Díaz, E. (Ed.), *Microbial Biodegradation: Genomics and Molecular Biology* (pp. 1–23). Norfolk: Caister Academic Press.

Mishra, S., Singh, R., & Arora, N. K. (2021). *Metabolic diversity of Pseudomonas species in petroleum hydrocarbon degradation: Peripheral aromatic pathways and genetic organization. Environmental Microbiology Reports, 13*(6), 1045–1058.

Mohapatra, B., Bapuji, M., & Sree, A. (2017). *Biodegradation of petroleum hydrocarbons by Bacillus species: A review. International Journal of Current Microbiology and Applied Sciences, 6*(7), 2807–2821.

Mounier, J., Hakil, F., Branchu, P., Naïtali, M., Goulas, P., Sivadon, P., & Grimaud, R. (2018). AupA and AupB are outer and inner membrane proteins involved in alkane uptake in Marinobacter hydrocarbonoclasticus SP17. *MBio, 9*(3), 10–1128.

Moursy, A. S., & Abdel-Shafy, H. I. (1984). *Petroleum hydrocarbon pollution: Limitations of mechanical and chemical remediation methods. Water Research, 18*(7), 881–886.

Nikolopoulou, M., & Kalogerakis, N. (2009). Biostimulation strategies for fresh and chronically polluted marine environments with petroleum hydrocarbons. *Journal of Chemical Technology & Biotechnology: International Research in Process, Environmental & Clean Technology, 84*(6), 802–807.

Panasia, G., & Philipp, B. (2018). LaoABCR, a novel system for oxidation of long-chain alcohols derived from SDS and alkane degradation in Pseudomonas aeruginosa. *Applied and Environmental Microbiology, 84*(13), e00626–18.

Park, C., Shin, B., Jung, J., Lee, Y., & Park, W. (2017). Metabolic and stress responses of Acinetobacter oleivorans DR 1 during long-chain alkane degradation. *Microbial Biotechnology, 10*(6), 1809–1823.

Park, H. G., Kim, V., Kim, H., Lee, R., Cho, M. A., Park, S. W., … & Kim, D. (2019). CYP52A23 from Candida albicans and its substrate preference for fatty acid hydroxylation. *Archives of Biochemistry and Biophysics, 671*, 27–34.

Pradhan, G., Dey, S., Priyadarsini, S., Kumar, M. S., & Das, A. P. (2024). Characterization and quantification of microplastics pollutants in sediment samples from Daya River of Odisha State in India for their appropriate management. In *Environmental Science and Engineering, Part F2333* (pp. 75–92). https://doi.org/10.1007/978-3-031-51792-1_5

Pradhan, G., Tripathy, B., Ram, D. K., Digal, A. K., & Das, A. P. (2023). Bauxite mining waste pollution and its sustainable management through bioremediation. *Geomicrobiology Journal*, 1–10. https://doi.org/10.1080/01490451.2023.2235353

Priyadarshini, B., & Das, A. P. (2024). The adverse impact of synthetic microplastics released from disposable face masks pollutants on aquatic microbiota, aquatic species and humans: a review. In *Environmental Science and Engineering, Part F2827* (pp. 67–81). https://doi.org/10.1007/978-3-031-50840-0_5

Qiu, Y., Xiao, F., Wei, X., Wen, Z., & Chen, S. (2014). Improvement of lichenysin production in Bacillus licheniformis by replacement of native promoter of lichenysin biosynthesis operon and medium optimization. *Applied Microbiology and Biotechnology*, *98*, 8895–8903.

Ray, S. K., Dey, S., Das, A. P., & John, A. (2024). Microplastic waste management through Greener Technologies: a review. In *Sustainable Management of Urban Plastic Waste through Circular Economic Approaches* (pp. 237–261). https://doi.org/10.1201/9781003465447-14

Singh Jadaun, J., Bansal, S., Sonthalia, A., Rai, A. K., & Singh, S. P. (2022). Biodegradation of plastics for sustainable environment. *Bioresource Technology*, *347*, 126697. https://doi.org/10.1016/J.BIORTECH.2022.126697

Souza, E. C., Vessoni-Penna, T. C., & de Souza Oliveira, R. P. (2014). Biosurfactant-enhanced hydrocarbon bioremediation: an overview. *International Biodeterioration & Biodegradation*, *89*, 88–94.

Staijen, I. E., Marcionelli, R., & Witholt, B. (1999). The PalkBFGHJKL promoter is under carbon catabolite repression control in Pseudomonas oleovorans but not in Escherichia coli alk1 recombinants. *Journal of Bacteriology*, *181*, 1610–1616.

Tani, A., Ishige, T., Sakai, Y., & Kato, N. (2005). *Isolation of LadA, a novel long-chain alkane monooxygenase from Geobacillus thermodenitrificans NG80–2, involved in petroleum hydrocarbon degradation. Journal of Bacteriology, 183*(19), 6694–6702.

Thevenieau, F., Le Dall, M. T., Nthangeni, B., Mauersberger, S., Marchal, R., & Nicaud, J. M. (2007). Characterization of Yarrowia lipolytica mutants affected in hydrophobic substrate utilization. *Fungal Genetics and Biology*, *44*(6), 531–542.

Throne-Holst, M., Wentzel, A., Ellingsen, T. E., Kotlar, H. K., & Zotchev, S. B. (2007). Identification of novel genes involved in long-chain n-alkane degradation by Acinetobacter sp. strain DSM 17874. *Applied and Environmental Microbiology*, *73*(10), 3327–3332.

Tor, J. M., Xu, C., Stucki, J. M., Wander, M. M., & Sims, G. K. (2000). Trifluralin degradation under microbiologically induced nitrate and Fe (III) reducing conditions. *Environmental Science & Technology*, *34*(15), 3148–3152.

Van Beilen, J. B., & Funhoff, E. G. (2005). Expanding the alkane oxygenase toolbox: new enzymes and applications. *Current Opinion in Biotechnology*, *16*(3), 308–314.

van den Berg, B. (2005). The FadL family: unusual transporters for unusual substrates. *Current Opinion in Structural Biology*, *15*(4), 401–407.

Van Nuland, Y. M., de Vogel, F. A., Scott, E. L., Eggink, G., & Weusthuis, R. A. (2017). Biocatalytic, one-pot diterminal oxidation and esterification of n-alkanes for production of α, ω-diol and α, ω-dicarboxylic acid esters. *Metabolic Engineering*, *44*, 134–142.

Wang, W., & Shao, Z. (2014). The long-chain alkane metabolism network of Alcanivorax dieselolei. *Nature Communications*, *5*(1), 5755.

Wang, W., Shao, Z., & Liu, J. (2022). *Functional diversity of alkane hydroxylases in petroleum hydrocarbon degradation: AlkB and AlkM systems in Pseudomonas putida GPo1 and Acinetobacter calcoaceticus ADP1. Applied and Environmental Microbiology, 88*(3), 2145–21.

Widdel, F., & Rabus, R. (2001). Anaerobic biodegradation of saturated and aromatic hydrcarbons. *Current Opinion in Biotechnology*, *12*(3), 259–276.

Wu, Q., Zhi, Y., & Xu, Y. (2019). Systematically engineering the biosynthesis of a green biosurfactant surfactin by Bacillus subtilis 168. *Metabolic Engineering*, *52*, 87–97.

Xia, M., Fu, D., Chakraborty, R., Singh, R. P., & Terry, N. (2019). Enhanced crude oil depletion by constructed bacterial consortium comprising bioemulsifier producer and petroleum hydrocarbon degraders. *Bioresource Technology*, *282*, 456–463.

Yin, C. F., Xu, Y., Li, T., & Zhou, N. Y. (2022). Wide distribution of the sad gene cluster for subterminal oxidation in alkane utilizers. *Environmental Microbiology*, *24*(12), 6307–6319.

Zhang, X., Kong, D., Liu, X., Xie, H., Lou, X., & Zeng, C. (2021). Combined microbial degradation of crude oil under alkaline conditions by Acinetobacter baumannii and Talaromyces sp. *Chemosphere*, *273*, 129666.

Zhang, Z., Ding, Z. T., Zhong, J., Zhou, J. Y., Shu, D., Luo, D., ... & Tan, H. (2017). Improvement of iturin A production in Bacillus subtilis ZK0 by overexpression of the comA and sigA genes. *Letters in Applied Microbiology*, *64*(6), 452–458.

15 Recycle of Hydrocarbon Content from Biomedical Waste (BMW) Approaching Zero Waste

G. Koteswara Reddy, K. Naga Sai Abhiram, Alok Prasad Das, Akbar John, Chiranjibi Chhotaray, and Selcan Karakus

15.1 INTRODUCTION

Biomedical waste (BMW) represents a unique challenge in waste management due to its hazardous nature and significant environmental impact. Generated by healthcare facilities, such as hospitals, clinics, and laboratories, this waste includes a mixture of infectious, chemical, and nonbiodegradable components (Behera et al., 2020, 2021, 2022). Among these, plastic materials derived from hydrocarbons form a substantial proportion, contributing to environmental pollution if not properly managed. Coronavirus disease 2019 (COVID-19) is a highly contagious disease caused by Severe Acute Respiratory Syndrome Coronavirus 2 (SARS-CoV-2). As per the reports of WHO (World Health Organization, 2023) on 11 January 2023, there have been 6,691,495 deaths out of 660,378,145 confirmed cases globally (WHO, 2023). This has had beneficial effects on the environment and climate as well. Especially, this change was evident in the cities as the movement was restricted in the name of lockdown. On the other side, increased cases of COVID-19 have increased the production of BMW (M. Mohanty et al., 2024; Sahu, Mallick, et al., 2024). The increased use of the personal protective equipment (PPE) kits and their improper disposal has created a huge amount of hospital waste, which has a huge negative impact on the environment (Rume & Islam, 2020). Imposing lockdown has been the greatest challenge for waste management throughout the world. COVID-19 has had a serious effect on the composition and quality of BMW, and the amount of waste generated and contagious wastes in landfills have also increased rapidly (Lambodaran et al., 2020). Besides this, increased use of use-and-throw masks and gloves has also created a potentially infectious waste, which also has a great concern (Dey, Bissoyi, et al., 2024; Ray et al., 2024).

This BMW is the place for various microbes that can easily penetrate a healthy human body and can cause various diseases, which may sometimes be even fatal. The hydrocarbon content in BMW, primarily found in single-use plastics, syringes,

 DOI: 10.1201/9781003508496-15

IV bags, and PPE can be repurposed through advanced recycling technologies. The microbes may be bacteria, viruses, parasites, or even animal and human waste, which can become a life-threatening problem to human beings (Kanungo et al., 2023; Mohanta et al., 2023). Safety measures should be at a high level, especially for the healthcare and social workers who work for the public concern. Though there are other wastes which are causing a serious threat to the environment, the major problem of BMW is that it causes immediate and sometimes even severe health effects. To overcome these challenges, WHO has published a book called "Blue Book" in 2014 for the safe disposal of BMW. In this review article, we focus on the sources, effects, and control methods of BMW containing hydrocarbons (Ngoc et al., 2022).

15.2 SOURCES AND EFFECTS OF BMW CONTAINING HYDROCARBONS

Apart from the effect of the pandemic, there are so many other sources as well from which BMW is generated. Even the production of waste from these other sources increased during the COVID-19 pandemic, as all of them were linked with the pandemic (Ngoc et al., 2022). The different waste generated from different sources are described in Figure 15.1 and Table 15.1.

The waste from these sources was also a serious concern for mankind, and this waste should be disposed of properly, as it is linked directly or indirectly to the COVID-19 pandemic.

BMW can be in both liquid and solid form. In the case of COVID-19, the liquid waste includes the body fluids of the patients, especially the suctioned fluid from the patient's lungs which are affected by the virus. Solid waste includes the PPE of both the patient and the doctor or healthcare worker (Mallick et al., 2023). These are

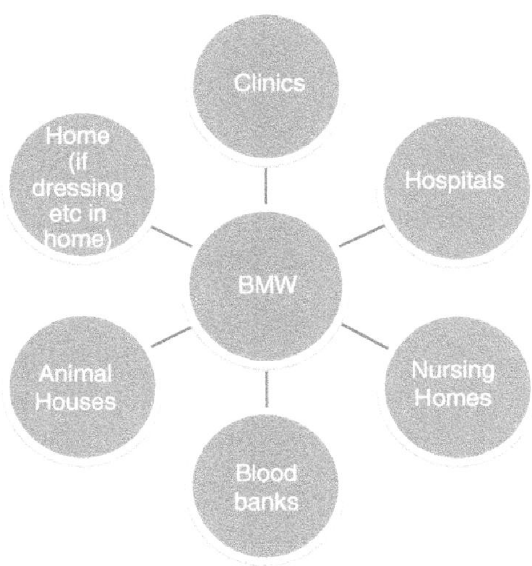

FIGURE 15.1 Sources of biomedical waste (BMW).

TABLE 15.1
Main Sources of Hydrocarbons with their Examples and Origin

S.No.	Source	Examples	Origin
1	Plastics	Syringes, catheters, IV bags, gloves, masks, gowns, and face shields	Polyethylene (PE), polypropylene (PP), and polyvinyl chloride (PVC)
2	Rubber-based products	Surgical gloves, tubing, and seals	Latex, neoprene, and nitrile
3	Hydrocarbon-based fluids and chemicals	Disinfectants, solvents, and some laboratory reagents	Alcohols, ketones, and phenols
4	Synthetic textiles	Surgical drapes, disposable gowns, and wipes	Polyester and nylon
5	Biomedical equipment components	Casings, tubing, connectors, and nonmetallic components	Thermoplastics and composites derived from hydrocarbons
6	Residual pharmaceuticals	Hydrocarbon-based creams, gels, and petroleum derivatives like paraffin	Derived from refined petroleum products

the primary sources of BMW during the pandemic. The secondary sources include ambulance services, local clinics, small healthcare centers, close contact with the patient, paramedics, etc. Household waste, transport services, education centers, etc. are the tertiary sources to produce BMW (Bala et al., 2024; Satapathy et al., 2024, Das, Dey, et al., 2024). The hydrocarbons in BMW primarily originate from petroleum-basedmaterials. These hydrocarbons form the backbone of many synthetic materials widely used in healthcare settings.

Hydrocarbon-containing materials are nonbiodegradable and can persist in the environment if improperly disposed of. Recycling and recovering hydrocarbons from these materials through methods like pyrolysis or chemical recycling offer opportunities to reduce environmental impact and promote sustainability (Ghosh et al., 2023). BMWs are classified into harmful and non-harmful wastes (Jalal et al., 2021). All household kitchen waste falls under non-harmful waste, whereas harmful waste is further classified as shown in Table 15.2. Countries all over the world are suffering from BMW. 16,649.48 tons/day of BMW was produced globally during the COVID-19 pandemic (Jalal et al., 2021). The following countries are the most BMW generators in their respective continents, as shown in Figure 15.2 (Singh et al., 2022).

Research indicates that BMW has rapidly increased during the COVID-19 pandemic. This increase may cause adverse effects on the environment and human health as well. Improper disposal of BMW from COVID-19 can make the pandemic more challenging (Das, Ray, et al., 2024). According to the studies, 5.2 million people worldwide died from diseases caused by improperly disposed medical waste, among which 4 million were children (Malekahmadi et al., 2014). In the case of coronavirus, since it can be attached to surfaces, lack of proper care leads to the contagious COVID-19 viral infection, as depicted in Figure 15.3.

TABLE 15.2
Types of Biomedical Waste and Their Sources/Examples

S. No.	Types of Biomedical Waste	Source/Examples	References
1	Infectious waste	Nasal swabs, excreta, body fluids, masks, gloves, and PPE kits	Ackerman (2020)
2	Pharma waste	Expired drugs	Udupa et al. (2013)
3	Pathological wastes	Body tissues like organs and blood	Ghosh and Das (2018)
4	Sharp objects	Knives, blades, needles, scalpels, etc.	Moray et al. (2020)
5	Radioactive waste	Unused liquids of radiotherapy	Padmanabhan and Barik (2019)
6	Genotoxic waste	Cytotoxic drugs used in cancer	Ghafuri and Nabizadeh (2017)
7	Chemical waste	Lab reagents and disinfectants	Meerambika Behera et al. (2021)

FIGURE 15.2 Amount of biomedical waste generated by the countries per day.

15.3 BIOMEDICAL WASTE MANAGEMENT

Biomedical waste management (BMWM) is critical for ensuring public health, environmental safety, and resource recovery. Effective recycling of hydrocarbon-rich materials from BMW involves a structured approach, integrating segregation, safe handling, and advanced recycling technologies to minimize environmental impact and move toward a zero-waste economy (Dey, Dash, & Das, 2024; J. Mohanty et al., 2024; Paul et al., 2024; Sahu, Dey, et al., 2024). Managing BMW is the biggest challenge all over the world. The major reason behind this problem is lack of social awareness, responsibility, and lack of novel solutions (Jiang et al., 2022).

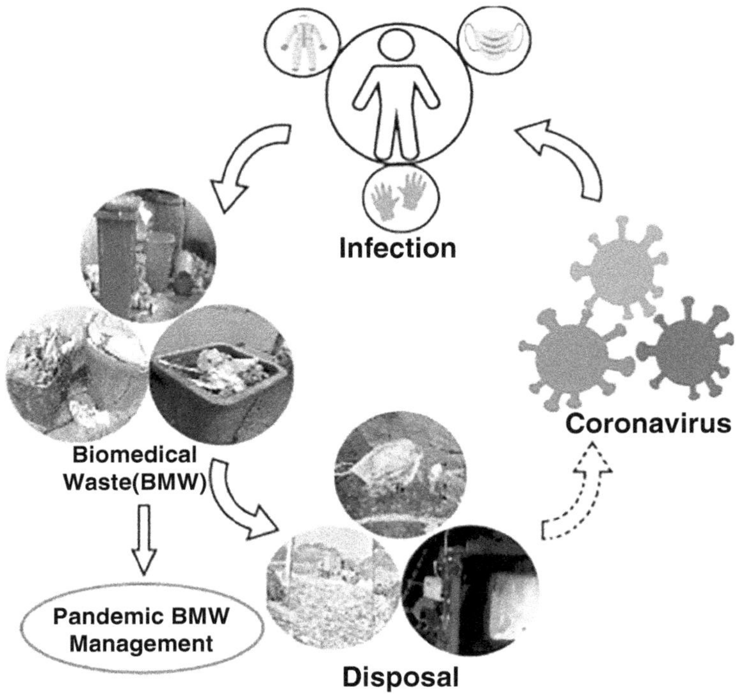

FIGURE 15.3 Improper disposal leads to infection.

Green technologies can be used in BMWM to reduce the toxic human impacts on nature. Especially in the case of COVID-19 pandemic, these green technologies are very useful in the management of BMW generated due to the virus (Dey, Das, et al., 2024; Mishra et al., 2024a; Paul et al., 2023).

15.4 WASTE SEGREGATION AND CLASSIFICATION

The very first step in the management of any kind of waste is segregation. Segregation refers to the separation of waste on a particular basis at the level of source. This will reduce the chance of infection and the handling cost. Separation of BMW is done with the help of color-coded bins or bags which play a key role in BMWM (Sharad Chand et al., 2021). Segregation is followed by the collection process which should be done with proper care. After segregation, collection, storage, and then transportation of these waste should be done with proper and personal safety equipment such as gloves, masks, boots, and gowns, as depicted in Figure 15.4.

Various methods such as autoclaving, pyrolysis, incineration, encapsulation, hydroclaving, and chemical disinfecting can be used in the management process of BMW (Singh et al., 2021). Plastic waste can be managed by improving the concept of 4R's – reduce, reuse, recycle, and recovery. Informing the world about the fact that the COVID-19-causing-corona virus survives for a long time on plastic surfaces may reduce the usage of plastic products (Corpet, 2021). Production of washable

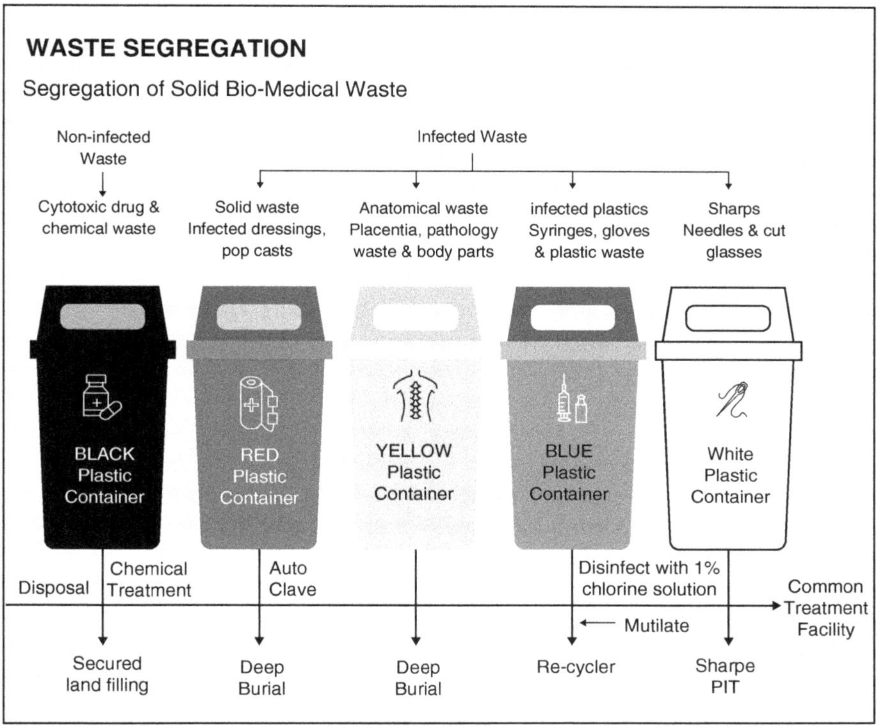

FIGURE 15.4 Separation of biomedical waste with the help of color-coded bins.

and reusable face masks will reduce a lot of biomedical plastic waste as the usage of use-and-throw masks is a major concern to the environment and human health (Das et al., 2015; Das & Singh, 2011, Khan et al., 2020). Used plastic gloves and face masks may become a major threat to the aquatic bodies as huge amounts of BMW are disposed of into the water bodies (Hasan et al., 2021). Unused and Expired Medicines should be collected and disposed of properly (Aluko et al., 2022). Though the studies show that the fungal degradation of plastic can be done by a marine fungus called *Zalerion maritimum*, which helps in the degradation of 43% polyethylene, in-situ degradation of plastic particles is limited. Water bodies should be chemically disinfected to eradicate the spread of harmful contagious diseases. Proper sanitization methods should be followed while handling the contagious waste.

15.5 RECYCLING TECHNOLOGIES FOR HYDROCARBON CONTENT

Recycling hydrocarbon-rich BMW, like plastics and rubber, involves specialized technologies to recover and repurpose valuable resources. These processes aim to convert waste into usable materials, fuels, or new products while reducing environmental harm. Here's a closer look at the key recycling methods:

15.5.1 Pyrolysis

Pyrolysis is a highly efficient process for breaking down plastic and rubber materials present in the environment (Bhattacharjee et al., 2021; Mishra, Dash, & Das, 2022; Mishra et al., 2024b; Mishra & Das, 2021). Waste is heated in an oxygen-free chamber at temperatures ranging from 300°C to 700°C. The heat breaks down the long hydrocarbon chains in plastics and rubber into smaller molecules. Pyrolysis is effective in managing mixed or contaminated plastics that cannot be recycled mechanically (Khatun et al., 2023; Mishra et al., 2023; Mishra & Das, 2023; Pradhan et al., 2023). However, it does require significant energy input to maintain high temperatures (Sangkham 2020). Uses: It can be refined into fuel like diesel and can power the pyrolysis system itself, making the process partially self-sustaining.

15.5.2 Gasification

Gasification converts waste into a fuel source by partially burning it in a controlled environment with limited oxygen. Waste is heated to over 800°C, breaking hydrocarbons into a gas mixture called syngas (hydrogen and carbon monoxide). Gasification efficiently handles waste that is not suitable for direct recycling. It also has the added benefit of producing energy, making it a dual-purpose process. However, the initial setup cost for gasification plants can be high (Ma et al., 2021). Uses: It can generate electricity or serve as a raw material for producing synthetic fuels and chemicals.

15.5.3 Mechanical Recycling

This is a more traditional method of recycling plastics and rubber, focusing on physical rather than chemical processes. Items are cleaned, sorted, shredded, and melted into pellets or other forms that can be reused to create new products. It's cost-effective and effective for clean and uncontaminated plastics (Mishra et al., 2019, 2021; Mishra, Dash, Al-Tawaha, et al., 2022; Mishra & Das, 2022; Sahoo et al., 2022). However, it struggles with highly mixed or soiled waste (Penta Floras et al., 2022). Uses: Lower-grade plastics for noncritical applications, like construction materials or packaging.

15.5.4 Chemical Recycling

Chemical recycling is a more advanced process that can break down plastics into their basic building blocks. Plastics are treated with heat, catalysts, or chemicals to convert them into monomers or simpler hydrocarbons (Priyadarshini & Das, 2024a, 2024b). This method is especially useful for dealing with degraded or mixed plastics that mechanical recycling can't handle. However, it is expensive and requires specialized facilities. Uses: Monomers can be used to make high-quality new plastics, and hydrocarbons can be processed into fuels or industrial raw materials.

15.6 COMBINING METHODS FOR BETTER RECYCLING

No single method can handle all types of hydrocarbon-rich BMW effectively. By combining methods: Mechanical recycling can handle clean, sorted plastics; Pyrolysis and gasification can manage contaminated or mixed materials; and Chemical recycling can be used for high-value recovery of degraded plastics. These methods not only divert waste from landfills but also create new opportunities for resource recovery and energy production. By integrating these technologies, we can transform BMW into valuable materials while protecting the environment and moving closer to a zero-waste future (Das, Dey et al., 2024).

15.7 CONCLUSION

Recycling hydrocarbon content from BMW represents a transformative approach to managing the growing challenge of hazardous healthcare waste. By leveraging advanced technologies like pyrolysis, gasification, and chemical recycling, it is possible to extract valuable resources from waste while minimizing environmental harm. This process not only reduces dependency on fossil fuels but also aligns with the principles of a circular economy and zero-waste initiatives. Implementing such strategies requires robust infrastructure, effective waste segregation, and strict adherence to regulatory standards. While challenges like high costs and contamination exist, fostering public–private partnerships, increasing awareness, and advancing research can overcome these barriers. Ultimately, the recycling of hydrocarbons from BMW is a vital step toward sustainable waste management, conserving natural resources, and protecting public health. It paves the way for a cleaner environment and a future where waste is no longer a burden but a valuable resource.

REFERENCES

Ackerman, F. (2000). Waste management and climate change. *Local Environment*, 5(2), 223–229. https://doi.org/10.1080/13549830050009373

Aluko, O. O., Imbianozor, G. T., Jideama, C. O., Ogundele, O. V., Fapetu, T. E., Afolabi, O. T., & Odewade, O. L. (2022). The perception and disposal practices of unused and expired medicines by households in an urban municipality, southwest Nigeria: A comparative cross-sectional study. *Waste Management*, 140, 121–132. https://doi.org/10.1016/j.wasman.2022.08.028

Bala, K., Dogra, A., Lata, P., & Das, A. P. (2024). Discussing sources and biological-chemical remediation approaches for microplastic pollution. In *Sustainable Microbial Technology for Synthetic and Cellulosic Microfiber Bioremediation*, 37–59. https://doi.org/10.1007/978-3-031-62600-5_3

Behera, I. D., Basak, G., Kumar, R. R., Sen, R., & Meikap, B. C. (2020). Treatment of petroleumrefinery sludge by petroleum degrading bacterium Stenotrophomonas pavanii IRB19 as an efficient novel technology. *Journal of Environmental Science and Health, Part A*, 1–13. https://doi.org/10.1080/10934529.2020.1866924

Behera, I. D., Nayak, M., Biswas, S., Meikap, B. C., & Sen, R. (2021). Enhanced biodegradation of total petroleum hydrocarbons by implementing a novel two-step bioaugmentation strategy using indigenous bacterial consortium. *Journal of Environmental Management*, 292, 112746. https://doi.org/10.1016/j.jenvman.2021.112746

Behera, I. D., Nayak, M., Mishra, A., Meikap, B. C., & Sen, R. (2022). Strategic implementation of integrated bioaugmentation and biostimulation for efficient mitigation of petroleum hydrocarbon pollutants from terrestrial and aquatic environment. *Marine Pollution Bulletin*, 177, 113492. https://doi.org/10.1016/j.marpolbul.2022.113492

Bhattacharjee, J., Mishra, S., & Das, A. P. (2021). Recent advances in sensor-based detection of toxic dyes for bioremediation application: A review. *Applied Biochemistry and Biotechnology*, 1–20. https://doi.org/10.1007/S12010-021-03767-7

Chand, S., Shastry, C. S., Hiremath, S., Joel, J. J., Krishnabhat, C. H., & Mateti, U. V. (2021). Updates on biomedical waste management during COVID-19: The Indian scenario. *Clinical Epidemiology and Global Health*, 11, 100715. https://doi.org/10.1016/j.cegh.2021.100715

Corpet, D. E. (2021). Why does SARS-CoV-2 survive longer on plastic than on paper? *Medical Hypotheses*, 146, 110429. https://doi.org/10.1016/j.mehy.2020.110429

Das, A., Dey, S., Bala, K., Gnanasekaran, R., & Das, A. P. (2024). Microbial bioremediation technology for sustainable treatment and management of synthetic microfiber waste. In *Sustainable Microbial Technology for Synthetic and Cellulosic Microfiber Bioremediation*, 221–238. https://doi.org/10.1007/978-3-031-62600-5_12

Das, A., Dey, S., & Das, A. P. (2024). Ecotoxicological impacts of synthetic microfiber pollutants and development of sustainable mitigation strategies. In *Environmental Chemistry and Ecotoxicology*. ISSN 2590-1826, https://doi.org/10.1016/j.enceco.2024.12.003

Das, A., Ray, S. K., Mohanty, M., Mohanty, J., Dey, S., & Das, A. P. (2024). Ecotoxicity of microplastic wastes and their sustainable management: A review. *Environmental Chemistry and Ecotoxicology*, 6, 144–152. https://doi.org/10.1016/J.ENCECO.2024.05.003

Das, A. P., Ghosh, S., Mohanty, S., & Sukla, L. B. (2015). Consequences of manganese compounds: A review. *Toxicological & Environmental Chemistry*, 96(7), 981–997. https://doi.org/10.1080/02772248.2015.1005428

Dey, S., Bissoyi, A., Koteswara Reddy, G., & Das, A. P. (2024). Synthetic microfibers: An urban emerging pollutant demanding for sustainable environmental solutions. In *Sustainable Microbial Technology for Synthetic and Cellulosic Microfiber Bioremediation*, 107–121. https://doi.org/10.1007/978-3-031-62600-5_7

Dey, S., Dash, A., & Das, A. P. (2024). Bioremediation as a sustainable and circular solution for mining waste management and bioenergy generation. In *Sustainable Management of Mining Waste and Tailings: A Circular Economy Approach*, 107–119. https://doi.org/10.1201/9781003442455-6

ey, S., Dash, N., Ray, S. K., Priyadarsini, S., Chhotaray, C., & Das, A. P. (2024). Occurrence and quantification of tire wear plastics pollutants from highway city dust of Bhubaneswar City of Odisha. In *Sustainable Microbial Technology for Synthetic and Cellulosic Microfiber Bioremediation*, 77–89. https://doi.org/10.1007/978-3-031-62600-5_5

Ghafuri, Y., & Nabizadeh, R. (2017). Composition and quantity of cytotoxic waste from oncology wards: A survey of environmental characterization and source management of medical cytotoxic waste. *Biosciences Biotechnology Research Communications*, 10(3), 438–444.

Ghosh, S., & Das, A. P. (2018). Metagenomic insights into the microbial diversity in manganese-contaminated mine tailings and their role in biogeochemical cycling of manganese. *Scientific Reports*, 8(1), 1–12. https://doi.org/10.1038/s41598-018-26311-w

Ghosh, S., Sinha, J. K., Ghosh, S., Vashisth, K., Han, S., & Bhaskar, R. (2023). Microplastics as an emerging threat to the global environment and human health. *Sustainability*, 15(14), 10821. https://doi.org/10.3390/SU151410821

Hasan, N. A., Heal, R. D., Bashar, A., & Haque, M. M. (2021). Face masks: Protecting the wearer but neglecting the aquatic environment? A perspective from Bangladesh. *Environmental Challenges*, 4, 100126. https://doi.org/10.1016/j.envc.2021.100126

Jalal, S. M., Akhter, F., Abdelhafez, A. I., & Alrajeh, A. M. (2021). Assessment of knowledge, practice, and attitude about biomedical waste management among healthcare professionals during COVID-19 crises in Al-Ahsa. *Healthcare (Basel)*, 9(6), 747. https://doi.org/10.3390/healthcare9060747

Jiang, Q., Xu, Z., Ye, G., Pahlow, M., Hu, M., & Qu, S. (2022). A systematic scoping review of environmental and socio-economic effects of COVID-19 on the global ocean-human system. *Science of the Total Environment*, 849, 157925. https://doi.org/10.1016/j.scitotenv.2022.157925

Kanungo, J., Sahoo, S., Bal, M., & Behera, I. D. (2023). Performance of bioremediation strategy in waste lubricating oil pollutants: A review. *Geomicrobiology Journal*, 41, 360–373.

Khan, S., Syed, A., Ahmad, R., Rather, T. A., Ajaz, M., & Jan, F. (2010). Radioactive waste management in a hospital. *International Journal of Health Sciences*, 4(1), 39–46.

Khatun, R., Singh, S., Dubey, N. K., & Prasad Das, A. (2023). A review on marine-based phytochemicals and their application in biomedical research. In *Recent Frontiers of Phytochemicals: Applications in Food, Pharmacy, Cosmetics, and Biotechnology*, 383–395. Elsevier. https://doi.org/10.1016/B978-0-443-19143-5.00013-X

Lambodaran, G., Ganesh, S. S., Shyam, S., & Prabhakar, M. (2020). Biomedical waste: A review. *Medico-Legal Update*, 20(4), 1–12.

Ma, X., Hosany, A., & Budihardjo, M. (2021). Advancing gasification techniques for handling biomedical waste during pandemics. *Science of the Total Environment*, 780, 146523. https://doi.org/10.1016/j.scitotenv.2021.146523

Malekahmadi, F., Yunesian, M., & Yaghmaeian, K. (2014). Analysis of the healthcare waste management status in Tehran hospitals. *Journal of Environmental Health Science and Engineering*, 12, 116. https://doi.org/10.1186/s40201-014-0116-4

Mallick, K., Sahu, A., Dubey, N. K., & Das, A. P. (2023). Harvesting marine plastic pollutants-derived renewable energy: A comprehensive review on applied energy and sustainable approach. *Journal of Environmental Management*, 348, 119371. https://doi.org/10.1016/J.JENVMAN.2023.119371

Mishra, S., & Das, A. P. (2021). Current treatment technologies for removal of microplastic and microfiber pollutants from wastewater. In *Wastewater Treatment: Cutting-Edge Molecular Tools, Techniques and Applied Aspects*, 237–251. https://doi.org/10.1016/B978-0-12-821881-5.00011-8

Mishra, S., & Das, A. P. (2022). Treatment of the wastewater polluted with synthetic microfiber released from washing machine. In B. B. Das, H. Hettiarachchi, P. K. Sahu, & S. Nanda (Eds.), *Lecture Notes in Civil Engineering* (Vol. 207, pp. 109–117). Springer Science and Business Media Deutschland GmbH. https://doi.org/10.1007/978-981-16-7509-6_9

Mishra, S., & Das, A. P. (2023). Microbial remediation of synthetic microfiber contaminated wastewater. In M. P. Shah (Ed.), *Microbial Technologies in Industrial Wastewater Treatment* (pp. 337–350). Springer, Singapore. https://doi.org/10.1007/978-981-99-2435-6_17

Mishra, S., Dash, D., Al-Tawaha, A. R. M. S., & Das, A. P. (2022). A review on heavy metal ion adsorption on synthetic microfiber surface in aquatic environments. *Applied Biochemistry and Biotechnology*, 194(10), 4639–4654. https://doi.org/10.1007/s12010-022-04029-w

Mishra, S., Dash, D., & Das, A. P. (2022). Detection, characterization and possible biofragmentation of synthetic microfibers released from domestic laundering wastewater as an emerging source of marine pollution. *Marine Pollution Bulletin*, 185, 114254. https://doi.org/10.1016/J.MARPOLBUL.2022.114254

Mishra, S., Dash, D., & Das, A. P. (2023). Aquatic microbial diversity on plastisphere: Colonization and potential role in microplastic biodegradation. *Geomicrobiology Journal*, 1–12. https://doi.org/10.1080/01490451.2023.2209750

Mishra, S., Dash, D., & Das, A. P. (2024a). Aquatic microbial diversity on plastisphere: Colonization and potential role in microplastic biodegradation. *Geomicrobiology Journal*, 41(4), 312–323. https://doi.org/10.1080/01490451.2023.2209750

Mishra, S., Dash, D., & Das, A. P. (2024b). Isolation and molecular characterization of polyester-degrading bacterial strains from municipal wastewater and their application in microplastic degradation in contaminated wastewater. *Geomicrobiology Journal*. https://doi.org/10.1080/01490451.2024.2376101

Mishra, S., Rath, C. C., & Das, A. P. (2019). Marine microfiber pollution: A review on present status and future challenges. *Marine Pollution Bulletin*, 140, 188–197. https://doi.org/10.1016/J.MARPOLBUL.2019.01.039

Mishra, S., Rout, P. K., & Das, A. P. (2021). Emerging microfiber pollution and its remediation. In R. Prasad (Ed.), *Environmental and Microbial Biotechnology* (pp. 247–266). Springer, Singapore. https://doi.org/10.1007/978-981-15-5499-5_9

Mohanta, S., Pradhan, B., & Behera, I. D. (2023). Impact and remediation of petroleum hydrocarbon pollutants on agricultural land: A review. *Geomicrobiology Journal*, 1–5. https://doi.org/10.1080/01490451.2023.2243925

Mohanty, J., Dey, S., Bala, K., & Das, A. P. (2024). Recent advances in microplastic pollution for its sustainable management. In *Sustainable Management of Urban Plastic Waste through Circular Economic Approaches* (pp. 27–48). https://doi.org/10.1201/9781003465447-3

Mohanty, M., Mohanty, J., Dey, S., Dutta, K., Shah, M. P., & Das, A. P. (2024). The face mask: A tale from protection to pollution and demanding sustainable solution. *Emerging Contaminants*, 10(2), 100298. https://doi.org/10.1016/J.EMCON.2023.100298

Moray, K. V., Manjunath, K., Martina Shalini, A. J., Pricilla, S. R. A., John, S. M., & Prasad, J. H. (2020). The insulin sharps disposal study: Evaluation of a structured patient education initiative in an urban community health centre in India. *Journal of Family Medicine and Primary Care*, 9(12), 6164–6170. https://doi.org/10.4103/jfmpc.jfmpc_1231_20

Ngoc, S.-M., Nguyen, M.-A., Nguyen, T.-L., Thi, H., Dao, T. L., Bui, T., Hoang, V. T., & Chu, D.-T. (2022). COVID-19 and environmental health: A systematic analysis for the global burden of biomedical waste by this epidemic. *Case Studies in Chemical and Environmental Engineering*, 6, 100245. https://doi.org/10.1016/j.cscee.2022.100245

Paço, A., Duarte, K., Costa, J. P., Santos, P. S. M., Pereira, R., Pereira, M. E., Freitas, A. C., Duarte, A. C., & Rocha-Santos, T. A. P. (2017). Biodegradation of polyethylene microplastics by the marine fungus Zalerion maritimum. *Science of the Total Environment*, 586, 10–15. https://doi.org/10.1016/j.scitotenv.2017.02.017

Padmanabhan, K. K., & Barik, D. (2019). Health hazards of medical waste and its disposal. In *Energy from toxic organic waste for heat and power generation* (pp. 99–118). Woodhead Publishing.

Paul, A., Dey, S., Ram, D. K., & Das, A. P. (2023). Hexavalent chromium pollution and its sustainable management through bioremediation. *Geomicrobiology Journal*, 1–11. https://doi.org/10.1080/01490451.2023.2218377

Paul, A., Priyadarsini, S., Karakuş, S., Behera, J. K., Gnanasekaran, R., & Das, A. P. (2024). Fate and accumulation of synthetic microfibers in the environment. In *Sustainable Microbial Technology for Synthetic and Cellulosic Microfiber Bioremediation* (pp. 61–76). https://doi.org/10.1007/978-3-031-62600-5_4

Pradhan, G., Tripathy, B., Ram, D. K., Digal, A. K., & Das, A. P. (2023). Bauxite mining waste pollution and its sustainable management through bioremediation. *Geomicrobiology Journal*, 1–10. https://doi.org/10.1080/01490451.2023.2235353

Priyadarshini, B., Santhosh Kumar, M., & Das, A. P. (2024). Microbial degradation of synthetic plastics released from COVID-19 generated single use disposable face masks. In *Environmental Science and Engineering, Part F2827* (pp. 1–16). https://doi.org/10.1007/978-3-031-50840-0_1

Ray, S. K., Dey, S., Das, A. P., & John, A. (2024). Microplastic waste management through greener technologies: A review. In *Sustainable Management of Urban Plastic Waste through Circular Economic Approaches* (pp. 237–261). https://doi.org/10.1201/9781003465447-14

Rume, T., & Islam, S. M. D. (2020). Environmental effects of COVID-19 pandemic and potential strategies of sustainability. *Heliyon*, 6(9), e04965. https://doi.org/10.1016/j.heliyon.2020.e04965

Sahoo, P. P., Singh, S., Rout, P. K., Mishra, S., & Das, A. P. (2022). Microbial remediation of plastic pollutants generated from discarded and abandoned marine fishing nets. *Biotechnology and Gnetic Engineering Reviews*. https://doi.org/10.1080/02648725.2022.2152629

Sahu, A., Dey, S., Mallick, K., Priyadarsini, S., Behera, I. D., & Das, A. P. (2024). A review on advanced technology for sustainable management of synthetic microplastic waste. In *Sustainable Management of Urban Plastic Waste through Circular Economic Approaches* (pp. 262–279). https://doi.org/10.1201/9781003465447-15

Sajgure, A., Kulkarni, A., Joshi, A., Sajgure, V., Pathak, V., Melinkeri, R., ... & Date, G. (2023). Safety and efficacy of mycophenolate in COVID-19: a nonrandomised prospective study in western India. *The Lancet Regional Health-Southeast Asia*, 11.

Sahu, A., Mallick, K., & Das, A. P. (2024). Environmental impact of single-use synthetic face mask and its recycling: A sustainable approach. (pp. 197–212). https://doi.org/10.1007/978-3-031-50840-0_11

Sangkham, S. (2020). Face mask and medical waste disposal during the novel COVID-19 pandemic in Asia. *Case Studies in Chemical and Environmental Engineering*, 2, 100052. https://doi.org/10.1016/j.cscee.2020.100052

Satapathy, S., Kar, D., Priyadarsini, S., Das, A. P., & Kuanar, A. (2024). Application of microorganisms in emerging waste management strategies and remediation of synthetic and cellulosic microfibers. In *Sustainable Microbial Technology for Synthetic and Cellulosic Microfiber Bioremediation* (pp. 183–206). https://doi.org/10.1007/978-3-031-62600-5_10

Singh, N., Ogunseitan, O. A., & Tang, Y. (2021). Medical waste: Current challenges and future opportunities for sustainable management. *Critical Reviews in Environmental Science and Technology*, 52, 2000–2022. https://doi.org/10.1080/10643389.2021.1885325

Singh, N., Ogunseitan, O. A., & Tang, Y. (2022). Medical waste: Current challenges and future opportunities for sustainable management. *Critical Reviews in Environmental Science and Technology*, 52(11), 2000–2022. https://doi.org/10.1080/10643389.2021.1885325

Udupa, N., Muragundi, P. M., Naik, A. N., & Janodia, M. (2013). PHP154: Disposal of date expired and unused medicines in India: A conceptual framework. *Value in Health*, 16(3), A269. https://doi.org/10.1016/j.jval.2013.03.1309

16 Recent Advances in the Combustion of Petroleum Waste for Sustainable Energy Generation

Subhashree Parida, Sudeshna Dey, Harjeet Nath, Akbar John, Alok Prasad Das, and Akalabya Bissoyi

16.1 INTRODUCTION

Although it has been employed for numerous decades, the thermal treatment of waste, with or without concomitant energy recovery, continues to be a subject of considerable contention. Historical waste incineration facilities possessed minimal to nonexistent emissions control mechanisms and, as a result, played a significant role in exacerbating air pollution, thereby inducing detrimental health outcomes. Petroleum waste, a byproduct of the petroleum refining and extraction processes, contains hydrocarbons, different types of polymers, and other toxic substances that pose a threat to ecosystems (Khatun et al., 2023a; Pradhan et al., 2023). Petroleum waste is usually responsibly disposed of and recycled in developing nations, but it is often carelessly dumped on land or even into the ocean, especially in aquatic environments where plastics break down over time and have negative effects on wildlife, limited options for reducing petroleum waste, and marine life is threatened (Das et al., 2023). Petroleum waste and plastic waste are interconnected environmental waste that demands urgent attention due to their significant ecological impact.

The sheer volume of petroleum waste generated globally is staggering. Since the extensive production of synthetic polymers commenced in the mid-20th century, it is estimated that over 8.3 billion metric tons of petroleum have been synthesized, with a substantial fraction of this material ultimately being disposed of as waste (Das & Ghosh, 2023b). According to contemporary assessments, approximately 60% of all plastics ever synthesized have become entrenched in landfills or the natural environment, as these synthetic materials are typically engineered for durability and exhibit significant resistance to biodegradation (Ghosh et al., 2023). Petroleum waste can take hundreds to thousands of years to decompose, leading to a buildup that far exceeds the capacity of existing waste management systems (Kumar et al., 2023). In 2023 alone, global petroleum production reached nearly

 DOI: 10.1201/9781003508496-16

450 million metric tons, with a substantial portion of this output being single-use plastics. The demand for single-use plastics has been driven by consumer convenience, particularly in the food, beverage, and healthcare industries, but this convenience has come at a significant environmental cost. Once discarded, these items often end up in landfills or, worse, in natural environments such as rivers, lakes, and oceans, where they pose long-term ecological threats (Nayanathara Thathsarani Pilapitiya & Ratnayake, 2024). Micro-petroleum particles present in the air as well as in water may have a hazardous direct influence on human beings through direct exposures by intake or breathing, which can cause a range of health impacts (Anand et al., 2023; Paul et al., 2023). Phthalates are micropollutants that can pose serious health risks, such as breast cancer, blood infections that can lead to cell death and inflammation, and damage to the liver, kidneys, and intestines (Bal et al., 2017; Browne et al., 2013). Additionally, phthalates can reduce the amount of oxidative stress molecules in the liver (Das, Dey, & Das, 2024). Other side effects of phthalates include early puberty, reduced male reproductivity, decreased sperm fertility level, and reproductive and genital defects (Dey et al., 2023). A chemical present in plastics called Bisphenol A (BPA) causes serious damaging impacts on female reproductive hormones. DNA and protein interruption are the effects of other chemicals that can be harmful in microplastics (MPs) once they enter our bodies. Inhalation of air contaminated with particles leads to the deposition of these particles in our lung tissues, which ultimately paves the way for lung inflammation (Das & Ghosh, 2023c,d).

MPs tend to be transported by wind and water, as well as other processes, both natural and human, and are washed to the sea in extremely large quantities, settling at the seafloor and also drifting in levels of water, endangering aquatic as well as marine life (Pradhan et al., 2023). Creatures often mistake petroleum waste for food and ingest it, obstructing their digestive systems (Dey, Bissoyi, Koteswara Reddy, et al., 2024; Ghosh, Tripathy, et al., 2023). Animals can also get entangled in marine detritus, including petroleum waste, leading to starvation, choking, lacerations, disease, reduced reproductive success, and even death (John & Das, 2024). Petroleum waste that gets trapped by coastlines poses an ecological setback and has negative effects on tourism. financial losses are connected to reduced tourism income, negative impacts on tourism, and damage to the marine ecosystem (Paul et al., 2023). Deposited shoreline petroleum has an effect that is certainly unfavorable on delivery infrastructure, power generation, fishing, and aquaculture. There have been cases where large jeopardized tortoises had been found having suffocated due to the unwanted swallowing of plastics along with seaweed (Das & Ghosh, 2023e).

Petroleum accumulation in the water environment harms the pivotal value, natural rawness, and health of ecosystems, as stated by over 395 marine fish species have been found to have ingested environmental debris, including 220 commercially important species. Sea turtles often consume minute pollutants of varying sizes. According to Reichert et al. (2019), marine litter has affected many marine vertebrates and mammals. Several seabird species are advertised by entanglement and consumption. The cons of marine discharge have also touched the nests of many organisms. Seabirds comprise nearly 60% of the total, followed by pisces with 20%, marine creatures with 18%, and sea turtles with 2%, flame retardants, and phthalates incorporated in ethene-based waste have potentially toxic effects on pisces, mollusks, and mammals (Jagatee et al., 2024; Sahoo et al., 2022a).

In 1950, petroleum production was approximately 2 million tons, which experienced a substantial escalation to 368 million tons by 2019. Hence, petroleum consumption has surged nearly 180-fold from 1950 to 2018. It is projected that the global production of petroleum will reach an estimated 400.3 million tons in 2022. In 2023, 4.86 billion tons of petroleum will be generated, of which 453 million tons may be recycled, 589 million tons will be combusted with energy recovery, and 299 million tons will be landfilled (Mishra & Das, 2022a). Large petroleum materials undergo deterioration through their engagement with the natural environment, primarily due to a range of physical, chemical, and biological phenomena, including mechanical degradation, biodegradation, thermal influences, ultraviolet degradation, photodegradation, mechanical forces (for instance, friction), turbulence, among other factors (Das & Ghosh, 2022). Petroleum also has plastic materials, which cause a serious impact on human health. Lambert and Wagner (2018) categorized plastics into macroplastics as less than 5 mm, mesoplastics as 1–5 mm, MPs as less than 1 mm but higher than 0.1 μm, and nanoplastics (NPs) as less than 0.1 μm. Small-sized synthetic polymer fragments—macroplastics, MPs, and NPs—are inexorably integrated into natural ecosystems as they originate from either primary or secondary sources that infiltrate the ecosystem directly via personal care products (e.g., shampoos, detergents, cosmetic formulations, and paints) or undergo degradation from larger plastic entities owing to their inherent adaptability and extensive range (Kumar et al., 2023). Anticipating future developments, projections indicate that the global accumulation of petroleum waste may reach an alarming 27 billion tons by the year 2050, surpassing the anticipated population surge (Ghosh, kumar et al., 2018). The incineration of petroleum waste to generate renewable energy constitutes a significant component of nonhazardous waste management technologies, which encompass a diverse array of waste management strategies ranked from the most to the least environmentally sustainable.

Pyrolysis is frequently examined and theoretically modeled as a recycling technique due to its economic efficiency, which minimizes environmental pollution and represents a significantly enhanced approach for the retrieval of chemicals from petroleum waste, subsequently transforming them into a viable energy resource (Mallick et al., 2024; Mishra et al., 2019a). Cold plasma techniques, on the other hand, can help to recover other chemicals and materials, particularly from petroleum waste, which are then added back into the manufacturing cycle and can solve issues related to nonrecyclable and low-grade waste (Mishra & Das, 2023c). They can be installed in tourist/hilly areas of the country where discharging carbon-based wastes is a worry, and significant transportation costs can be saved vis-à-vis energy can be co-generated. Gasification entails higher temperatures, which promote a diverse spectrum of materials, while processing also safeguards a cleaner technology. Gasification depicts higher temperatures, allowing a broader range of materials selected and processed while also ensuring a cleaner technology by eliminating practically all subproducts and impurities (Mishra et al., 2023; Tripathy et al., 2025). Gasification promotes efficient results for a variety of leftovers, minimizing waste while also providing valuable assets such as renewable energy generation.

It is crucial to promptly tackle the challenge of improperly managed plastics, rather than permitting their continuation in the ecosystem; simultaneously, it is vital to promote feasible transformations of waste into energy (WtE) and waste into products (WtP). Consequently, the implementation of a sustainable framework for petroleum waste management necessitates the coordinated application of both nontechnical/noninstrumental and technical/instrumental methodologies. The nontechnical/noninstrumental strategies primarily encompass the enhancement of public awareness, the fortification of existing legislative frameworks by governmental entities, and the improvement of petroleum waste collection processes, along with the formulation of effective collection rates and strategies. Conversely, the post-collection management of petroleum waste entails the transformation of such waste into WtP and WtE through the adoption of cutting-edge technologies, while also engaging a diverse array of societal stakeholders—including manufacturers, consumers, and waste collection personnel—to implement effective petroleum waste control initiatives. Nevertheless, it is crucial to augment current waste management competencies and to integrate contemporary methodologies, alongside the adoption of sustainable practices that promote reduced production and consumption, as vital factors in the mitigation of petroleum waste generation (Ghosh, Tripathy, et al., 2016, 2023). The realization of these initiatives necessitates the proactive involvement of policymakers and all relevant stakeholders, including consumers, manufacturers, and both formal and informal waste management operatives, as well as a heightened awareness among the general public (Lahiri et al., 2021).

The primary aim of this chapter is to furnish a comprehensive evaluation of the detrimental effects on natural ecosystems, elucidate existing knowledge deficiencies, and explore policy initiatives within transdisciplinary frameworks via life cycle assessment, circular economy principles, and sustainability paradigms. The principal objectives of this chapter encompass (1) elucidating the potential consequences of plastics on ecosystem services and climate change dynamics, (2) underscoring the significance of effective petroleum waste management, (3) delineating interconnections between the petroleum pollution, waste management strategies, and sustainable development objectives, and (4) examining life cycle assessment methodologies of recycling of petroleum waste.

16.2 PETROLEUM WASTE AS AN ENVIRONMENTAL POLLUTANT

It is undeniable that plastics have significantly aided the advancement of human civilization; nevertheless, the proliferation of petroleum waste (macro-, micro-, and NPs) within the environment and its incorporation into biological systems have emerged as a critical concern. The production of petroleum has increased significantly in recent years, resulting in a surge in demand from 420 million tons in 2020 to over 600 million tons in 2024 (Huang et al., 2024). This rise in petroleum production has a significant impact on the environment and all living beings. Petroleum has become a part and parcel of commonality, but the continuous usage and production of plastics, especially in poor countries, is a cause of concern due to the potential lack of resources to manage petroleum waste effectively (Dey et al., 2026). The effects of petroleum waste are increasing, resulting in water, soil, and air pollution, and posing a severe hazard to our ecosystems (Das et al., 2015a). However, petroleum waste is rapidly becoming a major concern due to its detrimental impact on aquatic life. According to estimates, around 8.3 million tons of waste make their way into the waters annually. The unhindered presence of waste particles in the aquatic ecosystem for years is a significant issue (Yusuf et al., 2022). Land-based sources of garbage, such as littering, petroleum trash disposal, and landfills, as well as groundwater runoff, combined sewer overflows, and industrial discharge, all contribute to the infiltration of garbage into aquatic bodies.

When introduced into aquatic ecosystems, petroleum microparticles significantly augment the toxicity levels encountered by diverse organisms. These environmental pollutants permeate the biological systems of fauna or are absorbed by marine organisms, ultimately resulting in human exposure. Various hazardous environmental pollutants, such as manganese, chromium, polychlorinated biphenyls (PCBs), and dichlorodiphenyltrichloroethane (DDTs) from industries, can lead to the spread of pollutants into the environment. These pollutants are frequently found in waste since they are nonpolar molecules that repel water molecules. The fact that more petroleum waste will be found in water by 2030 indicates an urgent need for action to address this threat. Additionally, skin absorption can make humans sick when they come into contact with pollutants used in petroleum industries. To give plastics specific qualities, synthetic chemicals like Bisphenol A, phthalates, and fire retardants are usually added throughout the process. However, health hazards can result from monomeric building blocks (such as Bisphenol A), additives (such as plasticizers), or a combination of both (e.g., antimicrobial polycarbonate). The major sources of human exposure to BPA are food consumption and gaseous inhalation. BPA is linked to various diseases such as breast and prostate cancer, metabolic disorders, and pain. In females, it may cause obesity, endometrial hyperplasia, irregular periods, infertility, and PCOS, among other health problems (Dey, Dash, & Das, 2024). MPs have the potential to exert biological effects on living organisms by influencing their growth, reproductive capabilities, and long-term toxicological responses. MPs originating from polyvinyl chloride (PVC) disrupt growth processes, induce oxidative stress, interfere with oviposition, and affect the activity of biological enzymes. The biological ramifications experienced by lugworms resulting from the assimilation of MPs and additives within their gastrointestinal tissues. Furthermore, insects such as Culex mosquitoes exhibit detrimental interactions with MPs.

16.3 PETROLEUM POLLUTION AND MANAGEMENT

Petroleum waste is generated at an unparalleled rate and is subsequently disseminated into the environment at a concerning velocity, a phenomenon attributable to the expansion of global industrial activities and increases in population density. Each year, around 450 million tons of petroleum are produced, with a significant proportion ultimately infiltrating and contaminating the natural ecosystem (Singh, Mishra, & Das, 2020; Mishra, Rath, & Das, 2019; Das et al., 2024). Plastics present in petrochemical waste demonstrate an inherent resistance to biodegradation; instead, they disintegrate into MPs, diminutive particles that persist in the environment for centuries. Aquatic organisms frequently misidentify petroleum materials as nutritional sources, leading to ingestion that precipitates malnutrition, internal injuries, and mortality (Sahu, Mallick & Das, 2024; Moharana et al., 2024). The consequences of petroleum pollution on human health are increasingly concerning. Addressing petroleum pollution requires a comprehensive strategy that includes policy reform, technological advancements, and transformations in societal practices. A viable framework for the management of petroleum waste is the circular economy, which aims to extend the utility of materials for as long as practicable (Das, Nayak, & Behera, 2024; Ghosh, Sinha, et al., 2023; Mallick et al., 2024; Pradhan et al., 2024). Within the circular economy paradigm, products are designed for reuse, repair, or recycling, thereby minimizing waste production and conserving vital resources (Mallick et al., 2024). The characteristics and mechanisms governing the degradation of petroleum waste are exceptionally complex, necessitating rigorous examination regarding feasibility, reactor design, and the scaling process (Das et al., 2024; Dey et al., 2024; Mishra et al., 2024; Sahu et al., 2024).

The recovered garbage is repurposed by incorporating the traits and characteristics of the original material. Petroleum waste may be physically recycled into secondary raw materials or products while maintaining their chemical structure. It also covers operations that recycle petroleum waste through mechanical processes such as grinding, washing, separating, drying, re-granulating, and compounding, which may result in petroleum products of varied grades (Mallick et al., 2023; J. Mohanty et al., 2025; Sahoo et al., 2022b). This means that the same or comparable items may be manufactured several times using recycled polymers, resulting in a closed-loop system. It is the most successful method for recycling polyethylene (PE) garbage. Re-extraction of energy and usable compounds from waste is achieved through pyrolysis, which involves the thermal decomposition of the plastic. Pyrolysis enables simultaneous polymer decomposition and classification at different temperatures. The deterioration of catalysts and heat degradation appear similar (Zhao et al., 2024; Ma et al., 2023). The main stand-apart factor in catalyst degradation during pyrolysis operations involves the addition of catalysts to promote conversion, and fuel efficiency, increase selectivity, and reduce pyrolysis temperature and residence time. The investigation into the pyrolysis of petroleum waste derived from both industrial sources and domestic sources was conducted utilizing a microwave reactor constructed from quartz, operating at a frequency of 25 GHz within a nitrogen atmosphere, employing carbon as a microwave adsorbent. The resultant pyrolytic products included bio-oil and biochar (Behera, 2020, 2021, 2022). The liquid byproduct exhibited considerable viscosity, containing a notable fraction of aromatic compounds that may serve as potential solvents and precursors. The analysis of the surface morphology and the amorphous characteristics of the solid byproduct indicated its applicability for the reduction of tar (Bhue et al., 2026).

Gasification refers to the transformation of petroleum materials into gaseous fuels. Additionally, it is characterized as converting the energy content inherent in solid fuels into chemical energy in the form of gaseous fuel. This process changes solid fuel into syngas containing hydrogen and carbon monoxide. Petroleum waste can be thermally treated to yield carbonaceous materials through various methodologies, including anoxic pyrolysis, catalytic carbonization, pressure carbonization, flash Joule heating, microwave processing, molten salt techniques, and hydrothermal processes (Bala et al., 2024; Dey, Dash, Ray, et al., 2024). These methodologies are frequently employed to attain the targeted products, which may encompass carbonaceous solids such as carbon nanotubes (CNTs), graphene, hierarchically porous carbon (HPC), and carbon nanosheets.

16.4 RENEWABLE ENERGY-EFFICIENT RECYCLING OF PETROLEUM WASTES

Petroleum waste recycling refers to any method of reusing rather than discarding petroleum trash. When petroleum waste is reused, it is considered to be downcycled; however, the recycled substance does not preserve its original qualities. Carbon-based waste is considered upcycled if it can be reused while retaining its originality (Mishra et al., 2026). Identifying methodologies for recycling and upcycling petroleum waste can reduce refuse in landfills and aquatic environments, the emission of greenhouse gases, and the dependency on finite resources. Due to their distinct advantages and limitations, petroleum waste recycling and upcycling processes must be used for individual petroleum wastes (Dey, Bissoyi, Koteswara Reddy, et al., 2024; Mishra et al., 2024). For the recycling of plastics, four basic approaches are used in global waste management schemes. Primary recycling (closed-loop), secondary recycling (mechanical), tertiary recycling (chemical), and quaternary recycling. The landfilling of petroleum trash is the only remaining evident alternative to these procedures. Different processes can be used for recycling plastics, including mechanical (secondary) recycling to recover plastics, and primary recycling for pre-consumer leftovers. Energy recovery from petroleum waste can be in the form of heat, syngas, bio-oil, etc. is one of the technologically challenging procedures of reusing petroleum waste, and it currently has good scope. Figure 16.1 represents the thermochemical recycling of petroleum wastes.

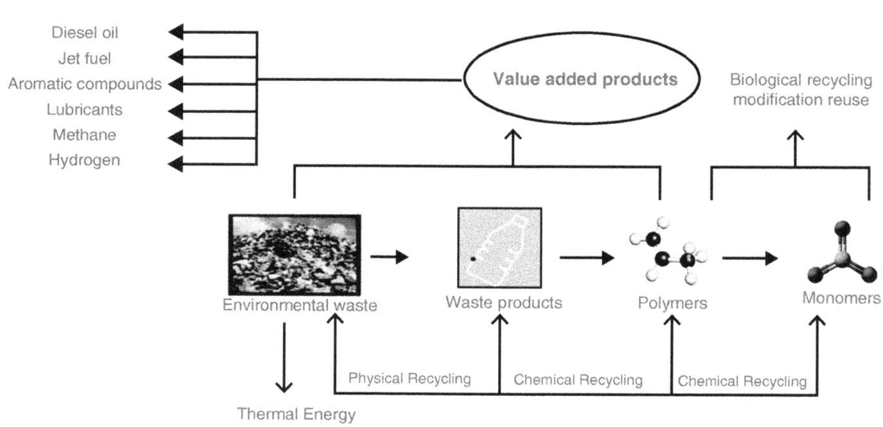

FIGURE 16.1 Represents the thermochemical recycling of petroleum wastes.

16.5 WASTE TO ENERGY USING THERMOS/ THERMOS-CHEMICAL CONVERSION

Thermochemical conversion is an energy-intensive technique that uses the heat degradation of chemicals in microalgal biomass to produce other products (Das et al., 2015b; Das & Ghosh, 2018; Ghosh et al., 2016). Thermochemical conversion processes may be divided into four categories based on temperature, pressure, and the presence of oxygen. These include direct combustion, gasification, pyrolysis, and thermal liquefaction.

16.5.1 Combustion

Combustion is a type of thermochemical conversion method, in which carbonaceous material or biomasses are burned in a supply of excess oxygen (atmospheric oxygen), resulting in the formation of heat and light as energy (John & Das, 2024). Noncatalytic combustion requires relatively high temperatures compared to catalytic combustion. In an ideal combustion process, the amount of fuel and oxidant is balanced to achieve complete combustion. However, achieving complete combustion is difficult in practice because chemical equilibrium is not always reached. This results in the formation of unburned byproducts. Smoke produced during this process is toxic and usually contains unburned or partially burned materials. Nitrogen oxide (NO_x) is parallel to any combustion reaction because atmospheric air contains 78% nitrogen. Out of all the above thermochemical processes, combustion is the fastest controlled chemical reaction known to humans. In ancient times, this thermochemical reaction was used for cooking food and to obtain heat and light. In the present scenario, combustion is used in rocket engines, where hydrogen reacts with oxygen to form water vapor. The quality of combustion depends upon the dynamics of combustion machines like burners and internal combustion engines.

16.5.2 Gasification

Gasification is a method that uses high temperatures to convert a solid or liquid material into a gas. This is achieved through a chemical reaction, known as thermochemical conversion, which breaks down the material into its constituent parts. In this process, biomasses (like wood, sawdust, cow dung, etc.) or any carbonaceous materials (like coal and plastics/MPs) are converted into valuable gaseous products at an elevated temperature approximately from 900°C to 1,200°C. Gasifying agents like air, carbon dioxide (CO_2), steam (H_2O), pure oxygen (O_2), or their mixture are used in a sub-stoichiometric environment where the equivalent ratio (ER) is kept between 0.2 and 0.4. The crucial air-to-fuel ratio for syngas/producer gas production ranges from 1.5:1 to 1.7:1 (Das et al., 2015c). Gasification proves highly advantageous for waste management compared to alternative methods like landfilling and combustion, as it can process a diverse range of waste materials such as biomass, paper, and plastics, yielding multiple useful products (Das et al., 2011, 2015c; Das & Ghosh, 2023f; Ghosh & Das, 2018). Gasification encompasses several processes, including drying, pyrolysis, partial oxidation, cracking, and reduction. The heat liberated during partial oxidation aids in reducing the initial moisture content in the feedstock. Maintaining feedstock moisture content within the range of 9%–22% is crucial for effective gasification. Pyrolysis occurs within a temperature range of 200°C–700°C, generating gases such as H_2, CO, CO_2, and CH_4, along with moisture, hydrocarbons, and char. The char produced during pyrolysis serves as a valuable input in gasification. Any tar generated during intermediary processes undergoes thermal cracking, leading to the release of non-condensable gases and light hydrocarbons.

Gasification is a promising method for converting carbon-based materials into green energy in the form of syngas. The most commonly utilized feedstock for gasification includes wood, agro-waste, and the organic component of municipal solid waste (MSW).

One effective method for deriving energy and power from biomass involves employing the ranking cycle in conjunction with a steam boiler. However, this approach lacks significant energy efficiency, typically ranging from 65% to 25%, as reported. Gasification, as a process, promotes the elimination of nitrogen and sulfur compounds that would otherwise produce SO_2 and NO_2 during burning. Syngas is produced as a byproduct of gasification, along with other gases such as CO, CO_2, CH_4, and H_2, as well as traces of hydrocarbons and gasification agents. Char formed during gasification is considered a byproduct of some other output. The primary contaminants include ash, tar, H_2S, HCl, and NH_3. Syngas composition and properties are determined by a variety of factors, including feedstock type, gasifier type, and operational parameters. Temperature, however, is the most important factor influencing the nature of syngas. The generation of CO and H_2 increases with rising temperature; at the same time, CO_2 & CH_4 production decrease. The combustion/oxidation reaction is the principal source of energy during the gasification process. Figure 16.2 demonstrates the Operating Flow of a Plasma Gasifier. Petroleum garbage becomes more available as time passes and lifestyles change, owing to its widespread use in our daily lives. Various ways are utilized to generate energy from waste plastics, including gasification. Petroleum gasification produces nearly identical results as biomass gasification (Mastellone et al., 2010).

16.5.3 Pyrolysis

Pyrolysis, also referred to as thermolysis, is a widely employed thermal decomposition process that occurs at elevated temperatures and is irreversible. The term "pyrolysis" originates from two Ancient Greek words: "pyro," meaning fire, and "lysis," meaning separation. Thus, pyrolysis entails separation by fire or heat. In this process, raw materials undergo thermal decomposition under oxygen-limited or inert-gas environments compared to stoichiometry, leading to the formation of solid product charcoal, liquid product bio-oil, and gaseous fuel. The rate of pyrolysis is directly proportional to temperature (Li et al., 2023). The chemical composition of the raw material changes during pyrolysis. A simple analogy for pyrolysis is cooking food: as the temperature of the food rises, the molecular vibrations of food molecules increase, causing large complex molecules to break down into smaller, simpler molecules that are easier to digest. Pyrolysis typically occurs at temperatures ranging from 400°C to 800°C and serves as the initial step for various other processes such as

TABLE 16.1
Range of Operating Parameters for Pyrolysis Processes

Types of Pyrolysis	Pyrolysis Temperature (°C)	Residence Time (Seconds)	Heating Rate (°C/s)	Particle Size (mm)	Major Product
Conventional pyrolysis	300–700 Low to med	550–950 Long	0.1–1 Low	5–50	Gas, char Bio-oil (tar)
Fast pyrolysis	600–1,000 Med to high	0.5–10 Short	10 to 200 High	<1	Bio-oil (thinner) Gases, char
Flesh Pyrolysis	800–1,000 High	<0.5 Very short	>1,000 Very high	<0.2 (Dust)	Gases Bio-oil

Maschio, Koufopanos, and Lucchesi (1992).

gasification and combustion. By adjusting temperature and residence time, the resulting product distribution can be altered. At low pyrolysis temperatures, the yield of liquid and solid products is higher, whereas at high pyrolysis temperatures, gas yield predominates over liquid and solid products. Consequently, the choice of pyrolysis temperature depends on the desired product composition. The outcome of pyrolysis is influenced by various factors including temperature, time, raw material types, feedstock size, reactors used, and catalytic or noncatalytic conditions. For maximum liquid product yield, a low temperature, short residence time, and high heating rate are required for pyrolysis. Conversely, a low temperature may be preferred if the aim is char production. Pyrolysis can be categorized into three subcategories based on their operating conditions, including pyrolysis temperature, rate of heating, particle size, and solid residence time. Table 16.1 lists the various operating parameters and their corresponding ranges. Fast and flesh pyrolysis necessarily comes under the scenario because the utilization of liquid fuel is easy and safe.

16.5.3.1 Slow Pyrolysis

Slow pyrolysis, also known as conventional pyrolysis, is conducted at relatively low temperatures ranging from approximately 300°C to 700°C for an extended duration typically lasting between 450 and 550 seconds. The rate of heating is also very low ~0.1°C/s compared to other pyrolysis processes. The required particle size for slow pyrolysis is between 5 and 50 mm. This type of pyrolysis is very useful for char formation (Jahirul et al., 2012). Additionally, slow pyrolysis poses certain limitations that render it less conducive to the production of high-quality bio-oil. The prolonged residence time and limited heat transfer characteristics necessitate additional energy input for the decomposition of primary products during slow pyrolysis.

16.5.3.2 Flash Pyrolysis

The process of flash pyrolysis is conducted within a specific temperature range of 800°C–1,000°C, more or less similar to fast pyrolysis, for a time period of less than 0.5 seconds. The rate of heating is very high, more than 1,000°C/s. The required particle size for this process is very fine (less than 0.2 mm) because the time needed for the process is significantly lower in comparison to fast pyrolysis. In different paper processes, perimeters vary little but are almost the same. Flash pyrolysis, also known as ultrafast pyrolysis, is a technique increasingly recognized for its versatility in converting not only biomass but also petroleum waste into valuable components. As the utilization of plastics continues to rise across various sectors, so does the accumulation of petroleum waste (Das et al., 2023; Das & Ghosh, 2023a; Khatun et al., 2023b; Mallick et al., 2023b; M. Mohanty et al., 2024; Nag et al., 2023). Traditional recycling methods for petroleum are often hindered by high labor costs and water pollution during processing, prompting researchers to explore alternative energy recovery approaches. One such method involves employing pyrolysis to decompose waste plastic, comprised of long-chain polymer molecules, into smaller and less complex molecules. This decomposition process yields three primary products: oil, gas, and char, all of which hold value across multiple industries, including manufacturing and refineries. Pyrolysis is particularly favored among researchers due to its ability to generate a substantial amount of liquid oil, reaching up to 80% by mass, at a moderate temperature of around 500°C (Grammelis et al., 2009). Moreover, the process parameters can be adjusted according to specific requirements, offering flexibility and customization. The oil produced through pyrolysis can be directly utilized for boilers, furnaces, and diesel engines, requiring no further treatment, thus presenting a promising avenue for energy recovery from petroleum waste. So, we can consider pyrolysis as a green technology because no waste is generated during the entire process, and no pollution takes place. Gas generated during pyrolysis can add up to the overall energy required for pyrolysis plants.

Exactly, the composition of various plastics varies, and these differences influence the ratios of oil, gas, and ash obtained under the same pyrolysis conditions. Proximate analysis serves as a valuable tool for identifying and quantifying the chemical properties of petroleum compounds (Bal et al., 2022). Parameters such as moisture content, volatile matter, fixed carbon, and ash content are measured through proximate analysis. Notably, the percentages of volatile matter and ash content are significant factors affecting the yield of liquid oil during the pyrolysis process. Higher percentages of volatile matter typically led to higher liquid oil yields (Agarwal, 2007). Therefore, understanding the proximate analysis results allows for the optimization of pyrolysis conditions to achieve desired product yields. Table 16.2 summarizes the proximate analysis of different petroleum compounds.

16.5.4 Liquefaction

Liquefaction, also known as hydrothermal liquefaction (HTL), is a thermal depolymerization process utilized to convert macromolecules such as plastics into valuable crude oil under moderate temperatures and high pressure. Depolymerization reactions are facilitated by homogeneous and/or heterogeneous catalysts to enhance the quality of the crude oil. HTL is particularly effective under supercritical conditions (with critical temperature $T_c = 374°C$ and critical pressure $P_c = 22$ MPa). Polymer depolymerization occurs rapidly at temperatures exceeding 400°C in a supercritical fluid. Moreover, at supercritical conditions, the dielectric constant of water decreases, rendering it an effective solvent for nonpolar organic compounds. Consequently, organic

TABLE 16.2
Proximate Analysis Data for Various Polymer Materials

Type of Polymer	Moisture (wt%)	Fixed Carbon (wt%)	Volatile (wt%)	Ash (wt%)
Polyethylene	0.45	07.76	92.29	0.02
terephthalate (PET)	0.61	13.21	87.56	0.00
High-density	0.00	0.02	99.89	0.18
polyethylene	0.00	0.04	97.58	1.40
Polyvinyl chloride	0.81	6.33	93.75	0.00
(PVC)	0.74	5.19	94.82	0.00
Low-density	0.32	0.00	99.75	0.00
polyethylene	–	–	99.69	0.40
Polypropylene (PP)	0.19	1.26	95.11	3.55
	0.18	0.16	97.85	1.99
Polystyrene (PS)	0.26	0.13	99.63	0.00
	0.30	0.20	99.50	0.00

chemical reactions can proceed in a single fluid phase without interphase heat and mass-transfer limitations. HTL is a kind of energy recovery technique from waste biomass/plastic. Generally, we throw away plastics after using them; as a result, petroleum waste increases with time. These waste plastics slowly degrade and convert into micro-molecules from macromolecules (Das, Dey, Bala, et al., 2024). Pollutants generated from the degradation of macromolecules (waste plastic) are very dangerous compared to visible waste petroleum. The micro-molecule or micropollutant generated during degradation slowly enters our food chain. With the help of HTL or any other similar technique, we can stop micro-petroleum contamination in the environment by treating macromolecules at the source, and at the same time, we can recover some energy from it. A few limitations of HTL are the high processing cost because it is carried out under high pressure, and decomposition is better in near/supercritical fluid in this audit. According tothe sequence of decomposition time at the same temperature is polyethylene (PE > polypropylene (PP) > polystyrene (PS). Understanding the chemistry and the circumstances in which various polymers break down in the HTL process is crucial since various polymers produce different products or yields under multiple circumstances. For example, HTL process under supercritical conditions, PE at 425°C–530°C yields 80%–90% oil whereas PP at 450°C for 1 hour yields 92% oil. Liquefaction of high-impact polystyrene (HIPS) at 350°C and 490°C gives 89.5% and 77% oil recovery, respectively. HTL decomposition of polyethylene terephthalate (PET) gives the main product of terephthalic acid and ethylene glycol. The main output of HIPS liquefaction at a temperature range of 350°C to 490°C is 1,3-diphenyl propane and ethylbenzene. The key product of the HTL process is a long-chain hydrocarbon. Similarly, Bai et al. (2019) studied HTL on various other plastics like PE terephthalate and polycarbonate at 400°C for 15 minutes. At 400°C, polycarbonate fully liquefies into oil products, whereas PET does not undergo liquefaction under the same conditions. Conversely, certain polymers like PP may not form oil at lower temperatures. Research on HTL is primarily focused on two key factors: oil yield and product yield from different plastics. Most studies have concentrated on polyolefins, with fewer investigations conducted on other plastics such as polycarbonate and polyester, and over a wider temperature range. Hence, it is imperative to study a myriad of polymers (plastics) under a spectrum of HTL conditions and compare their conversion rates, energy recovery potential, and oil composition. This comprehensive approach is crucial for developing effective strategies for the treatment and utilization of mixed petroleum waste streams through HTL processes.

16.6 CIRCULAR ECONOMY AS AN APPROACH FOR ADDRESSING PETROLEUM WASTE

The escalating apprehensions regarding environmental degradation concomitant with the proliferation of petroleum waste within the natural ecosystem have catalyzed comprehensive investigations into sustainable substitutes (Kanungo, Sahoo, Bal, et al., 2023; Kanungo, Sahoo, Swain, et al., 2023). The circular economy represents a principled, solution-oriented approach that integrates loops within the material flow across the value chain, diverging from the traditional continuous linear flow (take-make-use-dispose) that spans from production through utilization to the end-of-life phase. This model effectively diminishes material and resource consumption while ensuring the prolonged retention of resources within the economy (Ray et al., 2024). At present, a significant proportion of petroleum materials adhere to a linear economic model, exacerbating the issue of single-use plastics. This prevailing utilization pattern serves as the primary impetus behind the persistent phenomenon of petroleum pollution (Yaqoob et al., 2024). Consequently, the circular economy emerges as an innovative concept that reconfigures a linear process into a circular framework, prioritizing reduction, reuse, recycling, arrangement, assembly, and circulation. Thus, this framework is predicated upon three foundational principles: (1) mitigate petroleum waste and pollution via strategic product design, (2) sustain resources and products in active use, and (3) regenerate and conserve natural systems. Beyond merely preserving materials within the system at their highest value for extended durations, this framework aspires to remediate historical environmental degradation and restore ecological equilibrium.

16.7 CONCLUSION

Managing plastic waste has become a complicated procedure that is linked to environmental implications, carbon emissions, and economic viability. While the advancement of compostable and biodegradable petroleum packaging is of paramount importance, precise disposal protocols must be delineated to prevent the contamination of recyclable plastics until they fully supplant petroleum-derived packaging. Similarly, establishing bio-PET, a bio-based petroleum alternative to petroleum-based plastic, demands weighing both environmental and economic considerations. Collaboration among industry, stakeholders, and the government is critical to establishing a sustainable recovery of petroleum packaging waste and diverting it away from landfills and inefficient recycling techniques. In comparison to other forms of nonbiodegradable waste, including metallic debris or electronic waste, the economic valuation of waste plastics is inadequate to justify extensive resource recovery initiatives. The ongoing issue of cross-border waste disposal necessitates a robust political commitment and coordinated efforts between exporting and receiving nations to eradicate unethical practices, such as the misrepresentation of plastic waste destined for recycling. Any initiatives aimed at closing the recycling loop, whether through the recovery of materials or the reformation of monomers, are considered to be the most ecologically sustainable approach. The difficulty in recycling is to recover high-purity monomers that can generate polymers of equivalent grade to those made from virgin monomers. Chemically, this needs thorough sorting and classification of various waste polymers, as well as management of their additives. Consequently, by eliminating challenging-to-recycle plastics from the marketplace, we can strive for a more sustainable and eco-friendlier environment by enhancing the worth of our resources and diminishing the waste we generate, thereby yielding significant economic, social, and ecological benefits.

REFERENCES

Agarwal, A. K. (2007). Biofuels (alcohols and biodiesel) applications as fuels for internal combustion engines. *Progress in Energy and Combustion Science*, *33*(3), 233–271. https://doi.org/10.1016/J.PECS.2006.08.003

Anand, U., Dey, S., Bontempi, E., Ducoli, S., Vethaak, A. D., Dey, A., & Federici, S. (2023). Biotechnological methods to remove microplastics: a review. *Environmental Chemistry Letters*, *21*(3), 1787–1810. https://doi.org/10.1007/S10311-022-01552-4

Bal, B., Das, B. P., Das, A. P., & Akbar John, B. (2022). Molecular Identification and phylogenetic analysis of Indian mangrove horseshoe crab. *International Horseshoe Crab Conservation and Research Efforts: 2007-2020*, 29–39. https://doi.org/10.1007/978-3-030-82315-3_3

Bal, B., Nayak, S., & Das, A. P. (2017). Recent advances in molecular techniques for the diagnosis of foodborne diseases. *Nanotechnology Applications in Food: Flavor, Stability, Nutrition and Safety*, 267–285. https://doi.org/10.1016/B978-0-12-811942-6.00013-3

Bala, K., Dogra, A., Lata, P., & Das, A. P. (2024). Discussing sources and biological-chemical remediation approaches for microplastic pollution. In *Sustainable Microbial Technology for Synthetic and Cellulosic Microfiber Bioremediation*, 37–59. https://doi.org/10.1007/978-3-031-62600-5_3

Behera, I. D., Basak, G., Kumar, R. R., Sen, R., & Meikap, B. C. (2020). Treatment of petroleum refinery sludge by petroleum degrading bacterium Stenotrophomonas pavanii IRB19 as an efficient novel technology. *Journal of Environmental Science and Health, Part A*, *56*(2), 226–239.

Behera, I. D., Nayak, M., Biswas, S., Meikap, B. C., & Sen, R. (2021). Enhanced biodegradation of total petroleum hydrocarbons by implementing a novel two-step bioaugmentation strategy using indigenous bacterial consortium. *Journal of Environmental Management*, *292*, 112746.

Behera, I. D., Nayak, M., Mishra, A., Meikap, B. C., & Sen, R. (2022). Strategic implementation of integrated bioaugmentation and biostimulation for efficient mitigation of petroleum hydrocarbon pollutants from terrestrial and aquatic environment. *Marine Pollution Bulletin*, *177*, 113492.

Bhue, D., Dey, S., & Das, A. P. (2026). Circular economy solutions for managing discarded plastic pollution. In *Bioremediation as a Greener Solution for Marine Plastic Pollutants* (pp. 301–314). Elsevier.

Browne, M. A., Niven, S. J., Galloway, T. S., Rowland, S. J., & Thompson, R. C. (2013). Microplastic moves pollutants and additives to worms, reducing functions linked to health and biodiversity. *Current Biology*, *23*(23), 2388–2392. https://doi.org/10.1016/J.CUB.2013.10.012

Das, A. P., & Ghosh, S. (2018). Bioleaching of manganese from mining waste materials. *Materials Today: Proceedings*, *5*(1), 2381–2390. https://doi.org/10.1016/J.MATPR.2017.11.459

Das, A. P., & Ghosh, S. (2022). Role of microorganisms in extenuation of mining and industrial wastes. *Geomicrobiology Journal*, *39*(3–5), 173–175. https://doi.org/10.1080/01490451.2022.2038953

Das, A. P., & Ghosh, S. (2023a). Evolution of biomining technology. In *Manganese Mining Microorganisms*, 5–19. https://doi.org/10.1016/B978-0-12-822147-1.00012-1

Das, A. P., & Ghosh, S. (2023b). Manganese and its application. In *Manganese Mining Microorganisms*, 29–40. https://doi.org/10.1016/B978-0-12-822147-1.00007-8

Das, A. P., & Ghosh, S. (2023c). Metagenomic insights into unculturable microbial diversity. In *Manganese Mining Microorganisms*, 63–76. https://doi.org/10.1016/B978-0-12-822147-1.00002-9

Das, A. P., & Ghosh, S. (2023d). Microorganisms in manganese biomining processes. In *Manganese Mining Microorganisms*, 55–61. https://doi.org/10.1016/B978-0-12-822147-1.00013-3

Das, A. P., & Ghosh, S. (2023e). Mixed culture bioleaching: an insight into manganese biomining process and efficacy. In *Manganese Mining Microorganisms*, 129–141. https://doi.org/10.1016/B978-0-12-822147-1.00001-7

Das, A. P., & Ghosh, S. (2023f). Role of metabolites and metabolomics for understanding of manganese bioleaching. In *Manganese Mining Microorganisms*, 117–127. https://doi.org/10.1016/B978-0-12-822147-1.00011-X

Das, A. P., Dutta, K., Khatun, R., Behera, I. D., Singh, S., & Mishra, S. (2023). Microfiber pollution and its microbial mitigation: a review on current trends and future prospects. *Journal of the Taiwan Institute of Chemical Engineers*, 105104. https://doi.org/10.1016/J.JTICE.2023.105104

Das, A. P., Ghosh, S., Mohanty, S., & Sukla, L. B. (2015a). Advances in manganese pollution and its bioremediation. In L. B. Sukla, N. Pradhan, S. Panda, & B. K. Mishra (Eds.), *Environmental Microbe Biotechnology* (1st ed.), 313–328. Springer, Cham. https://doi.org/10.1007/978-3-319-19018-1_16

Das, A. P., Ghosh, S., Mohanty, S., & Sukla, L. B. (2015b). Consequences of manganese compounds: a review. *Toxicological & Environmental Chemistry*, *96*(7), 981–997. https://doi.org/10.1080/02772248.2015.1005428

Das, A. P., Nayak, B., & Behera, I. D. (2024). Applications of microorganisms in advanced geomicrobiological and biotechnological exploration. *Geomicrobiology Journal*, *41*(4), 309–311. https://doi.org/10.1080/01490451.2024.2327059

Das, A., Dey, S., & Das, A. P. (2024). Ecotoxicological impacts of synthetic microfiber pollutants and development of sustainable mitigation strategies. *Environmental Chemistry and Ecotoxicology*, *7*, 201–210. https://doi.org/10.1016/J.ENCECO.2024.12.003

Das, A., Dey, S., Bala, K., Gnanasekaran, R., & Das, A. P. (2024). Microbial bioremediation technology for sustainable treatment and management of synthetic microfiber waste. In *Sustainable Microbial Technology for Synthetic and Cellulosic Microfiber Bioremediation*, 221–238. https://doi.org/10.1007/978-3-031-62600-5_12

Das, A., Ray, S. K., Mohanty, M., Mohanty, J., Dey, S., & Das, A. P. (2024). Ecotoxicity of microplastic wastes and their sustainable management: a review. *Environmental Chemistry and Ecotoxicology*, *6*, 144–152. https://doi.org/10.1016/J.ENCECO.2024.05.003

Dey, S., Bissoyi, A., Koteswara Reddy, G., & Das, A. P. (2024). Synthetic microfibers: an urban emerging pollutant demanding for sustainable environment°al solutions. In *Sustainable*

Microbial Technology for Synthetic and Cellulosic Microfiber Bioremediation, 107–121. https://doi.org/10.1007/978-3-031-62600-5_7

Dey, S., Dash, A., & Das, A. P. (2024). Bioremediation as a sustainable and circular solution for mining waste management and bioenergy generation. In *Sustainable Management of Mining Waste and Tailings*, 107–119. https://doi.org/10.1201/9781003442455-6

Dey, S., Dash, N., Ray, S. K., Priyadarsini, S., Chhotaray, C., & Das, A. P. (2024). Occurrence and quantification of tire wear plastics pollutants from highway city dust of Bhubaneswar City of Odisha. In *Sustainable Microbial Technology for Synthetic and Cellulosic Microfiber Bioremediation*, 77–89. https://doi.org/10.1007/978-3-031-62600-5_5

Dey, S., Kar, D., Kuanar, A., & Das, A. P. (2026). Microbial-Assisted Sustainable Management of Microplastic Waste in the Environment-A Circular Economy. *Industrial Biotechnology for Sustainability and Circular Economy*, 171.

Dey, S., Tripathy, B., Kumar, M. S., & Das, A. P. (2023). Ecotoxicological consequences of manganese mining pollutants and their biological remediation. *Environmental Chemistry and Ecotoxicology*, *5*, 55–61. https://doi.org/10.1016/J.ENCECO.2023.01.001

Ghosh, S., & Das, A. P. (2018). Metagenomic insights into the microbial diversity in manganese-contaminated mine tailings and their role in biogeochemical cycling of manganese. *Scientific Reports*, *8*(1), 1–12. https://doi.org/10.1038/s41598-018-26311-w

Ghosh, S., Kumar, M. S., Bal, B., & Das, A. P. (2018). Application of bioengineering in revamping human health. In *Synthetic Biology: Omics Tools and Their Applications*, 21–37. https://doi.org/10.1007/978-981-10-8693-9_2

Ghosh, S., Mohanty, S., Akcil, A., Sukla, L. B., & Das, A. P. (2016). A greener approach for resource recycling: manganese bioleaching. *Chemosphere*, *154*, 628–639. https://doi.org/10.1016/J.CHEMOSPHERE.2016.04.028

Ghosh, S., Sinha, J. K., Ghosh, S., Vashisth, K., Han, S., & Bhaskar, R. (2023). Microplastics as an emerging threat to the global environment and human health. *Sustainability*, *15*(14), 10821. https://doi.org/10.3390/SU151410821

Ghosh, S., Tripathy, B., Dey, S., & Das, A. P. (2023). Comparative investigation of fungal and bacterial manganese biomining mechanisms. In *Manganese Mining Microorganisms*, 99–115. https://doi.org/10.1016/b978-0-12-822147-1.00016-9

Huang, J., Ye, L., Wang, J., Deng, Y., Du, B., Liu, W., & Su, G. (2024). A new approach to monitoring typical organophosphorus compounds (OPs) in environmental media: from database building to suspect screening. *Environment International*, *189*, 108802. https://doi.org/10.1016/J.ENVINT.2024.108802

Jagatee, S., Priyadarshini, S., Rath, C. C., & Das, A. P. (2024). Synthetic microfiber: an enduring environmental problem linked to sustainable development. In *Environmental Science and Engineering*, *Part F2333*, 93–112. https://doi.org/10.1007/978-3-031-51792-1_6

Jahirul, M. I., Rasul, M. G., Chowdhury, A. A., & Ashwath, N. (2012). Biofuels production through biomass pyrolysis: a technological review. *Energies*, *5*(12), 4952–5001. MDPI AG. https://doi.org/10.3390/en5124952

John, A., & Das, A. P. (2024). Government strategies for achieving circular economy for plastic waste management in Southeast Asia. In *Sustainable Management of Urban Plastic Waste through Circular Economic Approaches*, 72–82. https://doi.org/10.1201/9781003465447-6

Kanungo, J., Sahoo, T., Bal, M., & Behera, I. D. (2023). Performance of bioremediation strategy in waste lubricating oil pollutants: a review. *Geomicrobiology Journal*. https://doi.org/10.1080/01490451.2023.2245395

Kanungo, J., Sahoo, T., Swain, L. P., & Behera, I. D. (2023). Toxicity of persistent hydrocarbon pollutants, sources and sustainable remediation process. 39–65. https://doi.org/10.1007/978-3-031-48220-5_2

Khatun, R., Singh, S., Dubey, N. K., & Prasad Das, A. (2023b). A review on marine-based phytochemicals and their application in biomedical research. *Recent Frontiers of*

Phytochemicals: Applications in Food, Pharmacy, Cosmetics, and Biotechnology, 383–395. https://doi.org/10.1016/B978-0-443-19143-5.00013-X

Kumar, M., Bolan, S., Padhye, L. P., Konarova, M., Foong, S. Y., Lam, S. S., Wagland, S., Cao, R., Li, Y., Batalha, N., Ahmed, M., Pandey, A., Siddique, K. H. M., Wang, H., Rinklebe, J., & Bolan, N. (2023). Retrieving back plastic wastes for conversion to value added petrochemicals: opportunities, challenges and outlooks. *Applied Energy*, *345*, 121307. https://doi.org/10.1016/J.APENERGY.2023.121307

Lahiri, D., Nag, M., Dey, A., Sarkar, T., Joshi, S., Pandit, S., Das, A. P., Pati, S., Pattanaik, S., Tilak, V. K., & Ray, R. R. (2021). Biofilm mediated degradation of petroleum products. *Geomicrobiology Journal*, *39*(3–5), 389–398. https://doi.org/10.1080/01490451.2021.1968979

Lambert, S., & Wagner, M. (2018). Microplastics are contaminants of emerging concern in freshwater environments: an overview. *Handbook of Environmental Chemistry*, *58*, 1–23. https://doi.org/10.1007/978-3-319-61615-5_1

Li, J., Yu, D., Pan, L., Xu, X., Wang, X., & Wang, Y. (2023). Recent advances in plastic waste pyrolysis for liquid fuel production: critical factors and machine learning applications. *Applied Energy*, *346*, 121350. https://doi.org/10.1016/J.APENERGY.2023.121350

Ma, Y. B., Xie, Z. Y., Hamid, N., Tang, Q. P., Deng, J. Y., Luo, L., & Pei, D. S. (2023). Recent advances in micro (nano) plastics in the environment: distribution, health risks, challenges and future prospects. *Aquatic Toxicology*, *261*. https://doi.org/10.1016/j.aquatox.2023.106597

Mallick, K., Sahu, A., Dubey, N. K., & Das, A. P. (2023). Harvesting marine plastic pollutants-derived renewable energy: a comprehensive review on applied energy and sustainable approach. *Journal of Environmental Management*, *348*, 119371. https://doi.org/10.1016/J.JENVMAN.2023.119371

Mallick, K., Sahu, S., Sahu, A., Dey, S., & Das, A. P. (2024). A critical review of marine microfiber pollution routes, toxicity, and its sustainable remediation. In *Environmental Science and Engineering*, *Part F2333*, 189–211. https://doi.org/10.1007/978-3-031-51792-1_10

Maschio, G., Koufopanos, C., & Lucchesi, A. (1992). Pyrolysis, a promising route for biomass utilization. *Bioresource Technology*, *42*, 219–231.

Mastellone, M. L., Zaccariello, L., & Arena, U. (2010). Co-gasification of coal, plastic waste and wood in a bubbling fluidized bed reactor. *Fuel*, *89*(10). https://doi.org/10.1016/j.fuel.2010.05.019

Mishra, S., Das, A. P. (2023). Microbial Remediation of Synthetic Microfiber Contaminated Wastewater. In: Shah, M.P. (eds) *Microbial Technologies in Industrial Wastewater Treatment*. Springer, Singapore. https://doi.org/10.1007/978-981-99-2435-6_17

Mishra, S., Dey, S., & Das, A. P. (2026). Microbial Bioremediation Technology for Conversion of Microfiber Waste to Value-Added Products. In *Bioconversion of Environmental Waste to Value Added Products: A Sustainable Approach* (pp. 131–156). Cham: Springer Nature Switzerland.

Mishra, S., Ghosh, S., van Hullebusch, E. D., Singh, S., & Das, A. P. (2023). A critical review on the recovery of base and critical elements from electronic waste-contaminated streams using microbial biotechnology. *Applied Biochemistry and Biotechnology*, *195*(12), 7859–7888.

Mishra, S., Rout, P. K., & Das, A. P. (2019). Solar photovoltaic panels as next generation waste: A review. *Biointerface Research in Applied Chemistry*, *9*(6), 4539.

Mohanty, M., Mohanty, J., Priyadarsini, S., Rodríguez-Díaz, J. M., Yadav, S., Cabral-Pinto, M. M., ... & Das, A. P. (2025). Environmental hydrocarbon pollutants: Sources, transport, and effect on human health. In *Environmental hydrocarbon pollution and zero waste approach towards a sustainable waste management* (pp. 21-44). Cham: Springer Nature Switzerland.

Moharana, S., Dey, S., Priyadarsini, S., Kumar, M. S., & Das, A. P. (2024). Identification and characterization of microplastic pollutants from the marine sediments of Paradeep Coast of Bay of Bengal, India for their sustainable management. In *Environmental Science and Engineering, Part F2333*, 57–73. https://doi.org/10.1007/978-3-031-51792-1_4

NayanatharaThathsarani Pilapitiya, P. G. C., & Ratnayake, A. S. (2024). The world of plastic waste: a review. *Cleaner Materials, 11*, 100220. https://doi.org/10.1016/J.CLEMA.2024.100220

Paul, A., Dey, S., Ram, D. K., & Das, A. P. (2023). Hexavalent chromium pollution and its sustainable management through bioremediation. *Geomicrobiology Journal*, 1–11. https://doi.org/10.1080/01490451.2023.2218377

Pradhan, G., Dey, S., Priyadarsini, S., Kumar, M. S., & Das, A. P. (2024). Characterization and quantification of microplastics pollutants in sediment samples from Daya River of Odisha State in India for their appropriate management. In *Environmental Science and Engineering, Part F2333*, 75–92. https://doi.org/10.1007/978-3-031-51792-1_5

Pradhan, G., Tripathy, B., Ram, D. K., Digal, A. K., & Das, A. P. (2023). Bauxite mining waste pollution and its sustainable management through bioremediation. *Geomicrobiology Journal*, 1–10. https://doi.org/10.1080/01490451.2023.2235353

Ray, S. K., Dey, S., Das, A. P., & John, A. (2024). Microplastic waste management through greener technologies: a review. In *Sustainable Management of Urban Plastic Waste Through Circular Economic Approaches*, 237–261. https://doi.org/10.1201/9781003465447-14

Reichert, J., Arnold, A. L., Hoogenboom, M. O., Schubert, P., & Wilke, T. (2019). Impacts of microplastics on growth and health of hermatypic corals are species-specific. *Environmental Pollution, 254*, 113074. https://doi.org/10.1016/J.ENVPOL.2019.113074

Sahoo, P. P., Singh, S., Rout, P. K., Mishra, S., & Das, A. P. (2022a). Microbial remediation of plastic pollutants generated from discarded and abandoned marine fishing nets. *Biotechnology and Gnetic Engineering Reviews*. https://doi.org/10.1080/02648725.2022.2152629

Sahu, A., Dey, S., Mallick, K., Priyadarsini, S., Behera, I. D., & Das, A. P. (2024). A review on advanced technology for sustainable management of synthetic microplastic waste. In *Sustainable Management of Urban Plastic Waste through Circular Economic Approaches*, 262–279. https://doi.org/10.1201/9781003465447-15

ahu, A., Mallick, K., & Das, A. P. (2024). Environmental impact of single-use synthetic face mask and its recycling: A sustainable approach. *Impact of COVID-19 Waste on Environmental Pollution and its Sustainable Management: COVID-19 Waste and its Management*, 197–212.

Singh, R. P., Mishra, S., & Das, A. P. (2020). Synthetic microfibers: Pollution toxicity and remediation. *Chemosphere, 257*, 127199.

Singh, R. P., Mishra, S., & Das, A. P. (2020). Synthetic microfibers: Pollution toxicity and remediation. *Chemosphere, 257*, 127199.

Tripathy, I. P., Dey, S., Behera, I. D., & Das, A. P. (2025). Nano-based Remediation Strategies for Microplastic Pollution in the Environment. In *Emerging Micropollutants: Status in the Environment and Management through Bioremediation* (pp. 711–739). Cham: Springer Nature Switzerland.

Yaqoob, H., Tan, E. S., Ali, H. M., Ong, H. C., Jamil, M. A., & Farooq, M. U. (2024). Sustainable energy generation from plastic waste: an in-depth review of diesel engine application. *Environmental Technology & Innovation, 34*, 103467. https://doi.org/10.1016/J.ETI.2023.103467

Yusuf, A., Sodiq, A., Giwa, A., Eke, J., Pikuda, O., Eniola, J. O., Ajiwokewu, B., Sambudi, N. S., & Bilad, M. R. (2022). Updated review on microplastics in water, their occurrence, detection, measurement, environmental pollution, and the need for regulatory standards. *Environmental Pollution, 292*, 118421. https://doi.org/10.1016/J.ENVPOL.2021.118421

Zhao, X., Wan, C., Pan, Y., Fan, Y., & Liu, X. (2024). Pyrolysis behavior of sewage sludge coexisted with microplastics: kinetics, mechanism, and product characteristics. *Journal of Environmental Management, 370*, 123030. https://doi.org/10.1016/J.JENVMAN.2024.123030

17 Microbial Wastewater Treatment and Its Benefits over Physical and Chemical Processes

Raj Kumar, Kiran Bala, Arushi Dogra, Sarfaraz Ali, Alok Prasad Das, and Kumari Aditi

17.1 INTRODUCTION

Globally, wastewater treatment represents one of the most critical environmental challenges. Emphasis is laid on economic and efficient technology for the treatment of wastewater to minimize pollution produced by pollutants, increasing the water quality. There are various options for treating wastewater including constructed wetlands (CW) as an economical alternative having low operational expenditures, while needing a huge land area and long treatment times. New technologies are currently being considered to increase the overall efficiency of CW systems. It has to do with the study of structural configurations, choice of substrates, plant combinations (biotic elements), aeration, and incorporation of CW treatment system into other systems working in tandem (Zhang et al., 2022). Conventional activated sludge (CAS) wastewater treatment plants (WWTPs) have the purpose of ammonia removal and nitrate abatement through microbial communities. These still emit notable amounts of micro-pollutants (MPs) into the water body and are an important environmental burden. The moving bed biofilm reactor (MBBR) technology processes are cheaper, smaller, and efficient solution than the classic activated sludge reactor. It provides benefits like easy installation, lowest sludge production, and high treatment efficiency (Li et al., 2023). This chapter highlights advances in wastewater treatment and recycling, serving as a good corridor for regulatory actions aimed at environmental protection while also bridging key scientific gaps. Wastewater treatment prevents the release of contaminants into natural ecosystems such as seas, rivers, and lakes, since wastewater often contains microorganisms harmful to human health that can cause zoonoses.

Some of the methods used for treatment of types or categories of wastewater are physicochemical/non-biological, which turn out to be inefficient at times, consuming a lot of energy. Microalgae are not only ubiquitous but also able to flourish in water bodies containing the nutrients necessary for growth. This property is of particular interest when treating wastewater, as they can adsorb nitrogen, phosphorus,

DOI: 10.1201/9781003508496-17

and other trace elements (Al-Jabri et al., 2020). The growing importance of microbial fermentations following the biorefinery concept and their use for both sustainable production and environmental remediation has raised more attention toward microorganism-based industrial biotechnologies. Considerable investigations have been performed into the potential of microalgae for the production of various renewable resources, such as biohydrogen, transportation fuels, food supplements, and high-value bioproducts. These resources are generated by solar energy and CO_2 without using the land suitable for agriculture (Del Rio-Chanona et al., 2019). This capability makes microalgae a potentially excellent candidate for greenhouse gas capture and contributes to combating climate change on a global scale. Anthropocentric activities and the irrational use of pharmaceuticals (medicaments) including antibiotics have increased their levels in wastewater and other aquatic environments. This especially requires the development of effective bioremediation methods given that over 3,000 pharmaceutical compounds have recently been detected in environmental samples, and some as low as micrograms to nanograms per liter are present in wastewater, posing a contamination burden on water sources (Chandel et al., 2022). Other key causes of rising environmental pollution challenges include population growth, urbanization, and industrialization. These variables pose formidable challenges to maintaining ecosystem services. One of the biggest problems, particularly in developing countries like India, is the release of untreated wastewater into some selected bodies of water. Factors such as growing populations, industrialization, and overuse of fertilizers and pesticides aggravate this issue, leading to water shortages and elevated contamination levels in these watersheds (Renuka et al., 2015).

17.2 LIST OF WATER-PURIFYING MICROBES

During operation, the biofilm was dominated by Euryarchaeota and Proteobacteria, leading to improvements in water filtration system efficiency (Li et al., 2023). *Escherichia coli* and *Pseudomonas* are two bacteria used to clean water. In addition, various microorganisms (algae and fungi) are employed in water treatment to purify or render potable drinking water. Bacteria such as rhizospheric and endophytic bacteria are helpful for the removal of pollutants through floating treatment wetlands (Krzeminski et al., 2019). The floating treatment wetland (FTW) microbial community degrades nitrogen, phosphate, and toxic metals through easily accessible basic nutrients and converts metal ions to mobilized form, thus promoting plant uptake of pollutants under the shared root zone (Shahid et al., 2020). Microalgae, the ultimate chemical weapons masters, excel at treating contaminated water because they utilize nitrogen, phosphate, and trace elements present in wastewaters for their growth while simultaneously remediating pollutants (Al-Jabri et al., 2020). Algal–bacterial consortia containing *Scenedesmus almeriensis* adsorb ciprofloxacin, tetracycline, sulfadiazine, and sulphamethoxazole from wastewater. The consortia employed algae for providing oxygen to bacteria, which in turn degraded contaminants of wastewater (Chandel et al., 2022).

Microalgae range from *Botryococcus braunii* and other ectoine-accumulating marine species to phylogenetically diverse freshwater cyanobacteria (*Phormidium* sp.), green algae (*Chlamydomonas reinhardtii* and *Chlorella vulgaris*), and chlorophyte ciliate symbioses for silicon removal as well as nitrate uptake by the eukaryote *Scenedesmus rubescens/Auxenochlorella protothecoides*. Among microalgae, *Chlorella* species have been extensively studied for nutrient removal and biomass production, showing broad spectrum and potential success depending on culture conditions and wastewater source (Renuka et al., 2015). These are being used by researchers to improve sewage treatment and produce biofuels, bioplastics, and fertilizers, while reducing greenhouse gas emissions compared to traditional fertilizer production. Microbial colonies in water tanks release butyric acid and hydrogen sulfide at certain stages, indicating active pollutant degradation that aids urban and rural sewage management. Aerobic and anaerobic bacteria decompose wastewater organic matter, while protozoa and rotifers contribute to sedimentation during wastewater treatment. Heterotrophs, nitrifiers, denitrifiers, and phosphate accumulators are four key bacterial groups on which effluent treatment plants rely on, depending on water pollution levels. These systems provide habitats for active microorganisms based on the effluent treatment technologies employed and pollutants present (Taher et al., 2024).

The enzymes produced by microorganisms, such as ligninase, lipases, proteases, and cellulase, are essential for wastewater treatment. Lipases and lysozymes in microbes contribute to dewatering of sludge, while chitinase converts shellfish waste into *N*-acetyl glucosamine (Sahal et al., 2023). For instance, *Pseudomonas* and *Clostridium* species (bacteria), *Bacillus cloacae* (soil bacterium) and *Aspergillus* and *Penicillium* spp.(fungi) are proven degraders in bioremediation processes that utilize their biofilm formation ability. Meanwhile *Penicillium funiculosum* has been shown to produce peroxidase (POD) activity more efficiently. With the water quality improvement, microalgae contribute to nutrient removal in biofilm-based wastewater treatment systems. *Enterobacter*, *Pseudomonas*, *Psychrobacter*, *Deinococcus*, *Cupriavidus*, *Bacillus*, and *Staphylococcus* are mercury (Hg)-resistant bacteria with mere operons that transform poisonous Hg into harmless Hg. Also, *Shewanella oneidensis*, *Geobacter sulfurreducens*, *Geobacter metallireducens*, and *Klebsiella pneumoniae* can remove Hg using various processes (Ghosh et al., 2024) (Figure 17.1).

17.3 SPECIFIC FUNCTIONS OF MICROBES IN BREAKING DOWN DIFFERENT TYPES OF POLLUTANTS

In wastewater treatment processes, microorganisms play a crucial role in breaking down organic substances into water and carbon dioxide, which eventually removes contaminants from the water. Anaerobic digestion (in the absence of oxygen) involves microorganisms that decompose organic materials in oxygen-deprived environments. This process produces biogas, a combination of CO_2 and methane, as a by-product while remediating the wastewater from pollutants. For instance, some microbes carry out anaerobic digestion of organic matter to produce biogas and lower the organic load in wastewater. This process removes a substantial proportion of organic contaminants in wastewater (Singh & Shyu, 2024).

To further mitigate the environmental effects of such wastewater, microbes generate enzymes that convert organic substrates (such as proteins, lipids, and

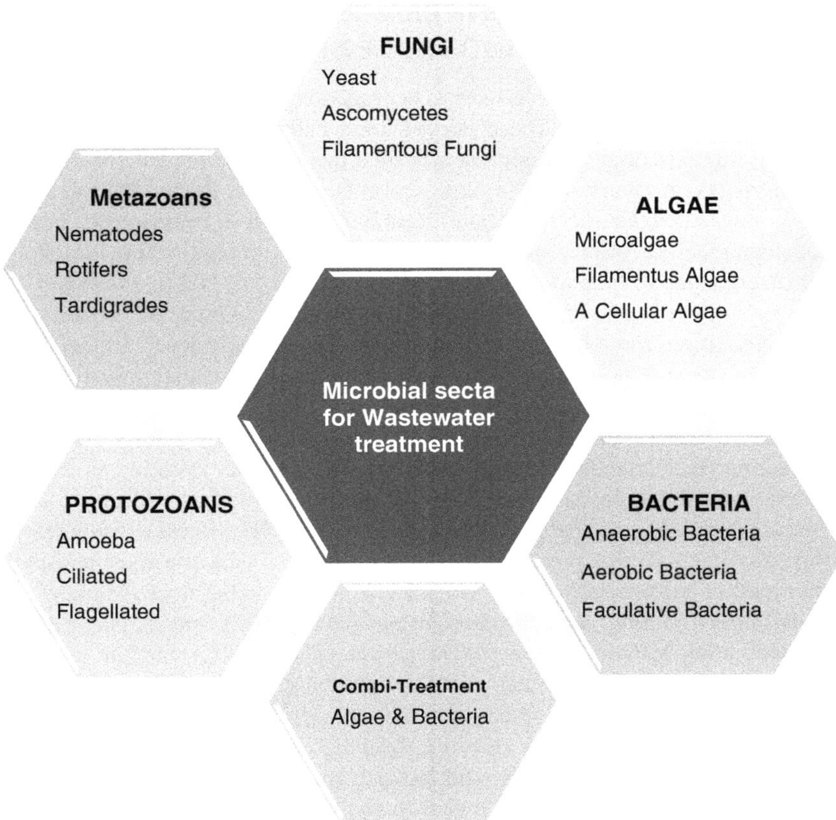

FIGURE 17.1 Microbes utilized in wastewater treatment.

carbohydrates) into less harmful forms. Various microbial enzymes, such as proteases, lipases, and amylases, are utilized in the wastewater treatment to purify water resources by catalyzing the decomposition of recalcitrant pollutants into nontoxic molecules. Heterotrophs are essential for the biodegradation of organic pollutants in wastewater, as they metabolize these substances as a carbon source (Yadav et al., 2024). Nitrifiers and denitrifiers play a central role in the nitrogen cycle as they are responsible for converting ammonia to nitrate and subsequently to N_2, enabling the removal of nitrogen-based contaminants from wastewater. Autotrophic bacteria such as cyanobacteria (blue-green algae) and purple sulfur bacteria, commonly found in wastewater treatment, use sunlight or inorganic compounds for photosynthesis while breaking down pollutants into less noxious concentrations. White-rot fungi serve as a model for optimal fungal species in nutrient-unbalanced substrate, effectively cleaving complex organic contaminants like lignin (benzene derivatives) and nonylphenol ethoxylates (NPAs). Rotifers and protozoa are also critical players in wastewater treatment as they consume bacteria and other microbes, ensuring that there is a balance within the microbiome and optimizing system performance.

17.4 MICROBIAL TREATMENTS BASED ON POLLUTANT KINDS AND PROCESSES

Microbial enzymes from multiple sources degrade toxic wastewater pollutants under mild operational conditions. These enzymes are capable of degradation or transformation of a wide range of toxic components, including pathogenic bacteria, suspended particles, inorganic salts, organic nutrients, minerals, and contaminants such as heavy metals. Enzymatic wastewater treatment costs less compared to physical/chemical approaches. Wastewater is purified by the microbiological process of using microbial enzymes to degrade contaminants (Yadav et al., 2024). Microbial engineering enhances the efficiency of organic and inorganic wastewater contaminants removal. The use of microbial consortia developed for wastewater treatment can target a broader range of pollutants, with greater adaptability to changing environmental scenarios than the cases where pure cultures are used.

Microalgae excel at removing nitrates and phosphates from sewage water while also reducing their chemical oxygen demand (COD) levels, resulting in improved water quality and decreased pollutant levels. Heterotrophic bacteria play a critical role in the enhanced biological phosphorus removal (EBPR) process at water resource recovery facilities. In this process, the efficient removal of wastewater phosphorus is performed by accumulating polyphosphate in active sludge flocs (Mathew et al., 2022). Bacteria and fungi, *e.g.*, Saccharomyces and white-rot fungus, play a key role in bioremediating MPs from industrial wastewaters. Innovative enzyme and hybrid materials featuring complementary growth approaches enable highly efficient separation of MPs from industrial wastewater effluents, serving as superior alternative to conventional treatment solutions (Islam & Mahdi, 2022). Silver nanoparticles synthesized by bacteria such as *E. coli* exhibit strong antibacterial activity against waterborne pathogens, inhibiting microbial growth and decomposing hazardous pollutants into safer forms in contaminated aquatic environments. This silver nanoparticle-based bioremediation proves to be cost-effective, environmentally friendly, and a promising solution for water pollution (Singh & Tiwari, 2021).

Microbes break down or utilize nitrogen, phosphate, potassium, iron, calcium salts, lipids, carbohydrates, and proteins from wastewater. They convert hazardous pollutants to environmentally friendly compounds, cleaning wastewater and preserving the ecological balance (Kumar et al., 2023). Effective microbes (EM) technology, a consortium of beneficial microorganisms, provides comprehensive wastewater treatment. These bacteria can function both aerobically and anaerobically, benefiting humans, animals, and nature. EM induce the synthesis of useful molecules, such as highly active hydrolytic enzymes and antioxidants in wastewater treatment systems, enabling a large-scale effective purification from several types of pollutants in wastewater treatment facilities. EM also scavenge heavy metals. Research on Çankırı wastewater found that EM treatment lowered strontium content to the lower detection limit. This shows that EM removes heavy metals from wastewater, reduces odors, accelerates organic matter sedimentation, and lowers COD, biological oxygen demand (BOD), and suspended solid particles (Yalçın et al., 2023). Autochthonous bacteria like *Pseudomonas fluorescens*, *Bacillus* sp., *Acinetobacter* sp., and *Rhodococcus* sp. decontaminate wastewater by biosorption, bioaccumulation, and enzyme-mediated bioremediation. Bacterial consortia have proved successful in wastewater treatment because cocultivated bacterial strains degrade contaminants synergistically increasing nutrient uptake (Bhtnagar & Kalra, 2022). Oxygenases and other microbial oxidoreductases increase the reactivity or water solubility of organic compounds, especially chlorinated aliphatics and other halogenated organic compounds found in herbicides, insecticides, and pesticides. Anaerobic biodegradation, where bacteria create methane gas instead of CO_2 in the absence of oxygen, helps clean the environment (Sangwan & Dukare, 2018).

17.5 RECENT ADVANCEMENTS IN MICROBIAL TREATMENT METHODS

Microbial wastewater treatment technologies are also noteworthy for their efficiency and practicality with respect to reducing pollutants/toxins from different substrates that, in return, have a better associated environmental–economical–health value: anaerobic process produces biogas (a mixture of methane and carbon dioxide) while converting the wastewater (Singh & Shyu, 2024). A biotechnological approach leveraging metagenomics, microbial community analysis, functional gene analysis, and synthetic biology optimizes biological wastewater treatment systems. The efficient microbial treatment in combination with new technologies, such as nexus-based water–energy recovery and decentralized sewage treatment systems, provides highly proficient ways toward microbiome sustainability. Microbial fuel cells (MFCs) can simultaneously purify wastewater and recover energy, which will greatly reduce the energy needs for wastewater treatment. As the microbial fingerprint is engineered for specific MFC applications, this so-called microbial nexus can be harnessed to improve anodes and even cathode performance leading to efficiency gains in power generation and wastewater treatment. There is a vast scope of technological incorporation in the wastewater treatment field where biosynthesized nanoparticles are being applied for water purification using bio-nanotechnology. In general, nanomaterials have a high surface-to-volume ratio, which allows them to adsorb contaminants from wastewater more effectively. These advancements significantly enhance the treatment modalities (Taher et al., 2024). In wastewater treatment, biofilms shield the bacteria that degrade dangerous pollutants, improving water quality before discharge into a natural habitat. In addition, molecular methods developed to analyze biofilms and their structure can be used to determine the biochemical characteristics of planktonic and sessile cultures at optimum conditions, resulting in efficient wastewater remediation (Sharma & Dhiman, 2023). Biofiltration of organic and inorganic impurities existing in industrial effluents, as well as wastewater systems, is accomplished by biofilters, which are packed with various types of micro, macro, and extracellular polymeric substrates. Microorganisms like *Nitrosomonas*, *Nitrobacter*, *Microalgae,* and *Pseudomonas* are used for the treatment of aliphatic hydrocarbons (aldehyde), aromatic hydrocarbons (ester), chlorinated hydrocarbons, amines/alcohols/amide/terpenes/phosphenate/heteromine/hydroxylamine/methylamine/ketones, nitrile and inorganic compounds (Patel et al., 2022). These technologies are based on the optical and magnetic polarizability, catalysis, and antibacterial activity of silver nanoparticles synthesized by bacteria such as *E. coli* for wastewater treatment. These silver nanoparticles can effectively and safely reduce hazardous pollutants to negligible levels by constraining the microbial growth in wastewater (Singh & Tiwari, 2021).

New hybrid biofilm reactor and membrane bioreactors utilizing anaerobic processes excel at MPs removal from wastewater bioreactors, particularly when treating organic matter-rich effluents (Naz et al., 2021). Recent advancements in microbial treatment such as immobilization of bacteria and microorganisms in matrices of polyvinyl alcohol (PVA) or polyethylene glycol (PEG) or activated sludge systems achieved better stability and improved removal performance than previous generations of media, with >99% nitrification–denitrification and ~85% organic matter removal.

Metagenomics, metatranscriptomics, and metaproteomics approaches have been employed to unravel the uncultivable microbial diversity in wastewater treatment systems, identifying active bioprocesses that can be optimized for better efficiency. Analysis of treatment plant ecology enables the prediction of microbial responses toward xenobiotic and human-induced community shifts leading to proactive management strategies (Suneja & Sharma, 2020).

17.6 ADVANTAGES OF MICROBIAL TREATMENT METHODS

Plants and microorganisms coworking in a floating treatment wetland system enhance the decolorization of textile wastewater with reactive dye, without chemical needs. This also makes for less dangerous sludge production. This treatment system can be even more effective if these bacteria are removed by pollutant-degrading and plant-growth-promoting strains such as *Bacillus cereus* or *Bacillus subtilis.* As a result, it may greatly decrease three key parameters, including BOD, COD, and dye concentration in treated wastewater (Tusief et al., 2022).

Microbial electrochemical technologies offer a sustainable solution to wastewater treatment by combining abiotic-electrochemical reactions with the biodegradation of organic substrates in wastewaters, using microbiologically catalyzed redox processes as the primary means for electron transfer via metal or carbon electrodes, generating energy while treating wastewater. Selective promotion of organotrophic metabolism in activated sludge may improve the management and performance of WWTPs. This enables the use of activated sludge oxygenation for degrading organic waste constituents in wastewater by stimulating microorganisms to digest biodegradable contaminants using a small amount of ozone (ozone + O_2) or pure oxygen. These microorganisms can be awakened by creating the right conditions for them to break down these contaminants through biochemical pathways not typically found in nature (Ohki & Rich, 2020).

Microbial methods increase the sludge thickening degree and hence improve wastewater treatment plant performance. Moreover, the operation model treats and disposes of wastewater in an eco-friendly manner by removing organic waste from water via biodigestion (which takes place during bacteria cycling) as well as through bacterial-plant-nutrient synergy that incorporates algae growth into treatment system architecture. Furthermore, microbial clusters produce extracellular polymeric substances (EPS), which prevent the destruction of biomass structure by harmful compounds, thus enhancing the treatment process globally (Dadrasnia et al., 2017).

Microbial treatment technologies (e.g., activated sludge and facultative lagoons), due to the low operation cost and capital required, are also suitable for domestic wastewater and industrial applications. They rely on microorganisms present in wastewater treatment plants to remove nutrients from water at an incredibly high rate, through the use of natural processes for waste management (Antunes et al., 2019). The biological treatment processes can degrade a mixture of biodegradable organic compounds to specific end products such as methane, carbon dioxide, water, ethanol, and butanol among many other personalized chemicals. This depends on the type of microorganism used for fermentation and process conditions. Some, like those found in Geobacteracea microorganisms, are capable of transferring electrons from the bacteria to an anode which can lead to electricity generation (e.g., in MFCs) (Rulkens, 2008).

A novel method of photosynthetic microalgae-based wastewater treatment greatly reduces energy needs because the oxygen produced by these phototrophic microbes can be used instead of air, preventing typical aeration methods. The use of microalgae in water treatment reduces the production costs of activated sludge and also increases the value-added products obtained from activated sludge (Zhou & Yang, 2016). Microbial treatment technologies such as the combined anaerobic/anoxic/aerobic (A_2/O) and membrane bioreactor systems are key in ensuring stable system operation with the ability to remove problematic pollutants through degradation by activated sludge-fluidized either suspended or detached microorganisms. The concentration of dissolved oxygen and its pH have a major effect on both nitrogen metabolism and microbial interactions in activated sludge (Feng et al., 2022).

17.7 DRAWBACKS OF PHYSICAL AND CHEMICAL METHODS

Physical and chemical approaches of wastewater treatment can be costly to operate and maintain. These approaches may also have negative impacts on the environment and human health, as some may require chemicals that are not environmentally friendly and can damage ecosystems. For this reason, microbial technologies are often considered more ecologically beneficial than physical and chemical methods for removing PAHs from soils. Traditional methods also often fail to completely degrade newly emerging MPs, many of which are resistant to physical or chemical removal and pose potential ecological risks. Compared with biotreatment technologies, physical and chemical methods may require high energy consumption and large space to achieve effective removal of pollutants, making these techniques less sustainable and less economical (Kumar et al., 2024). Conventional physical and chemical methods for removing pollutants from wastewater are relatively expensive, especially when dealing with more undeveloped regions where alternatives to less economically viable methods are needed. These may not be as proficient in degrading the wide range of complex organic substances and cycling nutrients like nitrogen, phosphorus, and carbon (Dharani et al., 2023). The elimination of wastewater by using physical means such as filtration and sedimentation may have a lower capacity and fail to fully remove certain contaminants, especially those that are in dissolved form, resulting in poorly purified wastewater. Chemical methods may require strong chemicals that require tight control so as to avoid environmental or health risks, including chemical spills and long-term ecological harm. For example, polyaluminum chloride (PAC) is a common physical-chemical treatment agent, but it can be costly to implement and sustain, and requires careful dosage control to ensure optimum performance. In addition, chemical treatment can generate sludge that needs proper disposal in order to prevent possible contamination of water sources (Tuan et al., 2023).

Removal of mercury (Hg) in wastewater through conventional physical and chemical technologies, ion exchange, precipitation, adsorption, and membrane filtration has some disadvantages, such as low Hg removal efficiency. These methods also generate hazardous waste products during the treatment process, and these toxic residues could worsen environmental pollution and increase health risks.

Similarly, physical and chemical methods for MPs removal from industrial wastewaters can be expensive because of their requirements for specialized equipment, chemicals, and continuous monitoring. They may also be less economically viable than bioremediation approaches and may be less effective in removing MPs with diverse chemical properties and often require multiple treatment steps for complete removal. Traditional chemical agents used in treating wastewater may contain toxic compounds; in this respect, silver nanoparticles synthesized by bacteria are considered a more environmentally friendly alternative (Singh & Tiwari, 2021). The physical cleanup methods, which include techniques such as incineration and excavation of large quantities of material, raise issues regarding both cost and ecological disruption. Strong oxidizing agents or solvents are among the chemical techniques that can produce harmful by-products that might endanger human health or remain in the environment. Chemical treatments sometimes also call for careful handling and disposal practices to stop additional contamination (Sangwan & Dukare, 2018). As the study emphasizes, physical and chemical approaches for dissolved organic matter (DOM) control in wastewater treatment operations might not totally address the microbial-mediated modification of DOM. This restriction might cause inefficiencies in general treatment performance and DOM removal. The overreliance on physical and chemical techniques without recognizing the microbial involvement in DOM transformation can lead to the partial destruction of DOM compounds, resulting in the accumulation of resistant molecules which may be tough to remove from the wastewater (Wang et al., 2021). Physical and chemical approaches of textile wastewater treatment require a range of chemicals, which can be expensive and possibly detrimental to the environment. In particular, another disadvantage of these techniques is the creation of hazardous sludge as a side effect which needs appropriate disposal practices to avoid environmental contamination (Tusief et al., 2022).

Finally, physical and chemical approaches for bacterial identification can be costly and time-consuming, especially in large populations. Gene sequencing is a technique commonly used for identification; however, it becomes disadvantageous when there is need to rapidly evaluate large number of samples. Furthermore, techniques such as matrix-assisted laser desorption ionization-time-of-flight mass spectrometry (MALDI-TOF MS) do not serve the purpose as they require specialized equipment and trained personnels, which limits their use in many research and industrial settings (Antunes et al., 2019).

17.8 FUTURE RESEARCH

In view of high-throughput microbial biotechnology for wastewater treatment, potential future studies could focus on deciphering microbial distribution and evolutionary processes in relation to redox potentials involved in key reactions, shifts from dominant taxa to more diverse functional communities, and the control of niches that influence the removal of emerging contaminants. Another possible avenue for future work on improving current wastewater management practices is the incorporation of high-end technologies into innovative solutions, such as decentralized treatment systems and water–energy recovery nexus concepts. Further research may also explore alternative, more sustainable, and more efficient wastewater treatment methods based on recently discovered microbial enzymes. In addition, examining the application of state-of-the-art biotechnological solutions may enhance multiple enzyme capabilities for minimizing pollutants present in wastewater, thereby improving water quality (Yadav et al., 2024).

There is also significant potential to improve MFCs for wastewater treatment, so that they become more suitable for large-scale application (Dharani et al., 2023). Hybrid treatment technologies that combine microbial engineering with other sustainable methods should be developed to achieve more efficient wastewater treatment. Studies on new microbial consortia and their inter-microbial interactions in the biodegradation of a wide range of organic and inorganic contaminants offer opportunities for designing more eco-friendly treatment systems. Delineating the structure and function of biofilms, supported by molecular techniques, would help improve the performance of wastewater treatment plants. In biofilm-based sewage treatment systems, the inclusion of microalgae can further improve treatment efficiency. Studies aimed at increasing pollutant removal through nanotechnology also show promise for sustainable water treatment. Finally, research on smart sensors and real-time monitoring systems for wastewater treatment plants could help optimize effluent quality and enable rapid responses to varying pollution loads, thereby improving overall treatment performance and protecting the water environment.

Future research might consider the long-term performance and sustainability of upflow sludge blanket filtration (USBF) technology for domestic wastewater treatment, potential scalability of this approach as well as implementation cost-effectiveness on a larger scale. Further research should focus on combining PAC with biological processes at optimized dosages to increase removal rates of various pollutants in domestic wastewaters (Tuan et al., 2023). Further exploration of the development of end-step segregation strategies in metal nanoparticle-assisted, bacterial biofilm-based wastewater treatment is ongoing. Future studies may also investigate the role of bacteria in biofilms in the mineralization of wastewater toxins into simpler compounds and in immobilizing toxins to protect aquatic life. In addition, optimizing hydraulic retention time in anaerobic fluidized bed membrane bioreactors (AFBMR) may enhance biosorption efficiency in wastewater treatment. Integrating MFCs with AFBMRSs may also improve electricity generation and increase removal efficiency in low-strength wastewater. Future bioremediation studies may focus on examining the synergistic actions of various types of microbial enzymes, including laccases, peroxidases, and proteases, in degrading a wide range of organic pollutants

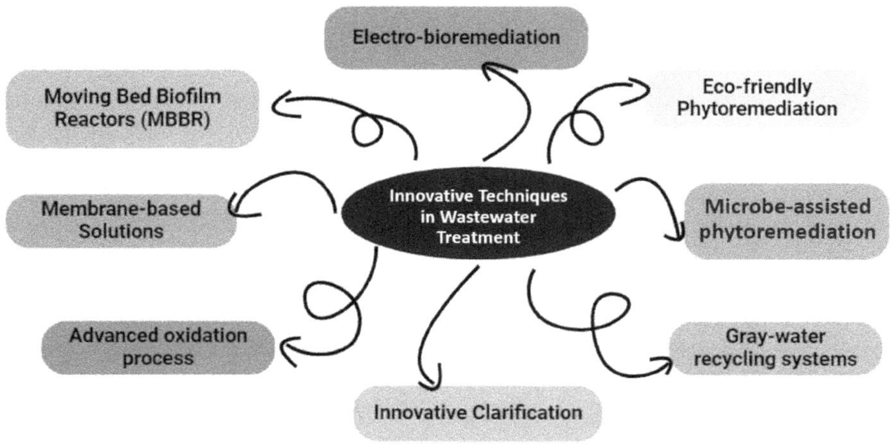

FIGURE 17.2 Innovative techniques in wastewater treatment.

at contaminated sites. Analysis of the impacts of various environmental parameters, e.g., temperature, pH, and co-contaminants on microbial degradation of xenobiotics in situ, may also be used to enhance bioremediation strategies at diverse polluted sites (Sangwan & Dukare, 2018).

Additional studies in microbial ecology using molecular techniques such as metagenomics, metatranscriptomics, and next-generation sequencing (NGS) will help better understand metabolic activity associated with different types of microbial consortia involved in activated sludge catabolic functionality. Thus, future research may examine how emerging contaminants such as microplastics, drugs, and heavy metals affect sewage ecosystems and help understaind their impact on the microbes involved in their removal and degradation, offering new directions for wastewater treatment research (Figure 17.2).

17.9 CONCLUSIONS

Optimizing the ability to digest microorganisms in activated sludge is one of the most important ways through which control and efficiency are improved when it comes to wastewater purification. Specifically, applying gas diffusion to return sludge and exposing it to ultrafine bubbles (UFB) in vitro with pure oxygen or, without any agent, while this is diluted using ozone at trace levels only will awaken inactive microorganisms to metabolize biodegradables through chemical processes different from those caused by their natural habitat (Ohki & Rich, 2020). With hundreds of operational taxonomic units and over 700 identified genera of microorganisms including bacteria, fungi, yeasts, and protozoa, microbial diversity within biological wastewater treatment plants (WWTPs) is essential for system stability and efficiency. Maintaining the functionality of WWTPs—including organic matter removal, flocs, bacterial population control, microbial diversity, and possibly survival advantages through encoding antibiotic synthesis genes—depends on the activity of diverse microbial groups including bacteria, protozoa, and viruses (phages) (Rezaei et al., 2024).

Given the importance of microbial community structure in wastewater treatment systems and the wide range of enzymes produced by microbiota to degrade pollutants, appropriate methods are needed to manipulate them adequately and control their bioremediation potential. The above mentioned methods require high energy, chemical inputs, long treatment time, and may also be limited by the presence of pathogens in wastewater, biological techniques remain attractive because they are relatively environmentally friendly and cost-effective (Dadrasnia et al., 2017).

The microbial consortium developed from *Pseudomonas putida*, *Candida utilis*, *Bacillus amyloliquefaciens*, *Clostridium papyrosolvens*, *Nitrosomonas europaea*, *Paracoccus denitrificans*, and auxiliary materials has proven effective in degrading phenylamine-resistant compounds in industrial wastewater treatment systems. In particular, the synergy among these microorganisms was remarkable, which improved the treated water quantity and quality, reduced the operating costs, and ensured that the discharged water quality was in compliance with the standards. These results were achieved when the microbial components remained compatible and non-antagonistic (Sun, 2016). Despite the advantages that physical and chemical methods may possess, they frequently do not measure up in terms of sustainability over the long term and the risks associated with the formation of harmful by-products. Consequently, the transition toward microbial wastewater treatment signifies a hopeful trajectory for realizing more sustainable and ecologically responsible methodologies for water purification. Ongoing research ought to persist in investigating and refining microbial treatment methods for heightened effectiveness and broader applicability within a variety of wastewater treatment contexts (Sen, 2015).

REFERENCES

Al-Jabri, H., Das, P., Khan, S., Thaher, M., & AbdulQuadir, M. (2020). Treatment of wastewaters by microalgae and the potential applications of the produced biomass: a review. *Water*, *13*(1), 27. https://doi.org/10.3390/W13010027

Antunes, T. C., Ballarini, A. E., & Sand, S. V. D. (2019). Temporal variation of bacterial population and response to physical and chemical parameters along a petrochemical industry wastewater treatment plant. *Anais da Academia Brasileira de Ciências*, *91*, e20180394. https://doi.org/10.1590/0001-3765201920180394

Bhtnagar, A., & Kalra, S. (2022). Autochthonous antimicrobial microorganisms: application in wastewater treatment. *Journal of Applied and Natural Science*, *14*(2), 660–676. https://doi.org/10.31018/jans.v14i2.3462

Chandel, N., Ahuja, V., Gurav, R., Kumar, V., Tyagi, V. K., Pugazhendhi, A., … & Bhatia, S. K. (2022). Progress in microalgal mediated bioremediation systems for the removal of antibiotics and pharmaceuticals from wastewater. *Science of the Total Environment*, *825*, 153895. https://doi.org/10.1016/j.scitotenv.2022.153895

Dadrasnia, A., Usman, M. M., Lim, K. T., Velappan, R. D., Shahsavari, N., Vejan, P., & Ismail, S. (2017). Microbial aspects in wastewater treatment: a technical. *Environmental Pollution and Protection*, *2*(2), 75–84. https://doi.org/10.22606/EPP.2017.22005

DelRio-Chanona, E. A., Cong, X., Bradford, E., Zhang, D., & Jing, K. (2019). Review of advanced physical and data-driven models for dynamic bioprocess simulation: case study of algae-bacteria consortium wastewater treatment. *Biotechnology and Bioengineering*, *116*(2), 342–353. https://doi.org/10.1002/bit.26881

Dharani, L., Umapriya, R., Rohan, J., Surendran, G., & Deepak, P. (2023). Chapter 1: Microbes and wastewater treatment. In P. Verma & M. P. Shah (Eds.), *Clean Technologies toward the Development of a Sustainable Environment and Future: Physicochemical, Biochemical, and Biotechnological Approaches* (pp. 1–17). IWA Publishing. https://doi.org/10.2166/9781789063783_0001

Feng, Z., Li, T., Lin, Y., & Wu, G. (2022). Microbial communities and interactions in full-scale A2/O and MBR wastewater treatment plants. *Journal of Water Process Engineering*, *46*, 102660. https://doi.org/10.1016/j.jwpe.2022.102660

Ghosh, S., Sarkar, B., Imboon, T., & Thongmee, S. (2024). Chapter 12 - Microbe-mediated mercury bioremediation for wastewater treatment. In M. P. Shah (Ed.), *Microbial Metagenomics in Effluent Treatment Plant Development in Wastewater Treatment Research and Processes* (1st ed., pp. 239–255). Elsevier. https://doi.org/10.1016/B978-0-443-13531-6.00004-5

Islam, M. D., & Mahdi, M. M. (2022). Chapter 1 - Evaluation of micro-pollutants removal from industrial wastewater using conventional and advanced biological treatment processes. In I. Haq, A. S. Kalamdhad, & M. P. Shah (Eds.), *Biodegradation and Detoxification of Micropollutants in Industrial Wastewater* (pp. 1–26). Elsevier. https://doi.org/10.1016/B978-0-323-88507-2.00011-7

Krzeminski, P., Tomei, M. C., Karaolia, P., Langenhoff, A., Almeida, C. M. R., Felis, E., Gritten, F., Andersen, H. R., Fernandes, T., Manaia, C. M., Rizzo, L., & Fatta-Kassinos, D. (2019). Performance of secondary wastewater treatment methods for the removal of contaminants of emerging concern implicated in crop uptake and antibiotic resistance spread: a review. *The Science of the Total Environment*, *648*, 1052–1081. https://doi.org/10.1016/j.scitotenv.2018.08.130

Kumar, V., Kumar, S., Verma, P., & Bhat, S. A. (Eds.). (2024). *Microbial Nexus for Sustainable Wastewater Treatment*. CRC Press eBooks. https://doi.org/10.1201/9781003441069

Kumar, V., Sehgal, R., & Gupta, R. (2023). Chapter 13 - Microbes and wastewater treatment. In M. P. Shah, S. Rodriguez-Couto, A. K. Nadda, & A. Daverey (Eds.), *Development in Wastewater Treatment Research and Processes Bioelectrochemical Systems for Wastewater Management* (pp. 239–255). Elsevier. https://doi.org/10.1016/B978-0-323-88505-8.00010-3

Li, M., Liu, Y., Zhou, X., Wang, N., & Yuan, B. (2023). A study on the carriers compound multi-stage mbbr biological treatment process for domestic sewage. *Sustainability*, *15*(10), 7922. https://doi.org/10.3390/su15107922

Mathew, M. M., Khatana, K., Vats, V., Dhanker, R., Kumar, R., Dahms, H. U., & Hwang, J. S. (2022). Biological approaches integrating algae and bacteria for the degradation of wastewater contaminants: a review. *Frontiers in Microbiology*, *12*, 801051. https://doi.org/10.3389/fmicb.2021.801051

Naz, A., Chowdhury, A., & Mishra, B. K. (2021). Applications of microbes in bioremediation of point source pollutants from wastewater. *IOP Conference Series: Earth and Environmental Science*, *796*, 012039. https://doi.org/10.1088/1755-1315/796/1/012039

Ohki, A., & Rich, W. (2020). System and method for wastewater treatment through microorganism biochemical pathway optimization. U.S. Patent No. US11046603B2. Washington, DC: U.S. Patent and Trademark Office.

Patel, H. K., Kalaria, R. K., More, B. S., Jokhakar, P. H., & Khimani, M. R. (2022). Chapter 13 - Microbial ecology of biofiltration. In M. Shah, S. Rodriguez-Couto, & J. Biswas (Eds.), *An Innovative Role of Biofiltration in Wastewater Treatment Plants* (pp. 235–266). Elsevier. https://doi.org/10.1016/B978-0-12-823946-9.00008-5

Renuka, N., Sood, A., Prasanna, R., & Ahluwalia, A. S. (2015). Phycoremediation of wastewaters: a synergistic approach using microalgae for bioremediation and biomass generation. *International Journal of Environmental Science and Technology*, *12*, 1443–1460. https://doi.org/10.1007/S13762-014-0700-2

Rezaei, Z., Giyahchi, M., & Moghimi, H. (2024). Enhancing microbial cooperation within the microbial nexus-based system. In V. Kumar, S. Kumar, P. Verma, & S. A. Bhat (Eds.), *Microbial Nexus for Sustainable Wastewater Treatment* (1st ed., pp. 1–21). CRC Press eBooks. https://doi.org/10.1201/9781003441069-1

Rulkens, W. (2008). Increasing significance of advanced physical/chemical processes in the development and application of sustainable wastewater treatment systems. *Frontiers of Environmental Science & Engineering in China, 2*, 385–396. https://doi.org/10.1007/S11783-008-0073-1

Sahal, S., Khaturia, S., & Joshi, N. (2023). Chapter 12 - Application of microbial enzymes in wastewater treatment. In V. Kumar, M. Bilal, L. F. R. Ferreira, & H. M. N. Iqbal (Eds.), *Genomics Approach to Bioremediation: Principles, Tools, and Emerging Technologies* (pp. 209–227). Wiley. https://doi.org/10.1002/9781119852131.ch12

Sangwan, S., & Dukare, A. (2018). Microbe-mediated bioremediation: an eco-friendly sustainable approach for environmental clean-up. In T. K. Adhya, B. Lal, B. Mohapatra, D. Paul, & S. Das (Eds.), *Advances in Soil Microbiology: Recent Trends and Future Prospects. Microorganisms for Sustainability* (1st ed., Vol. 3, pp. 145–163). Springer, Singapore. https://doi.org/10.1007/978-981-10-6178-3

Sen, T. K. (2015). *Physical, Chemical and Biological Treatment Processes for Water and Wastewater*. Hauppauge, NY: Nova Science Publishers, Incorporated.

Shahid, M. J., Al-Surhanee, A. A., Kouadri, F., Ali, S., Nawaz, N., Afzal, M., … & Soliman, M. H. (2020). Role of microorganisms in the remediation of wastewater in floating treatment wetlands: a review. *Sustainability, 12*(14), 5559. https://doi.org/10.3390/SU12145559

Sharma, A., & Dhiman, S. (2023). Microbial biofilms in wastewater remediation. In M. P. Shah (Ed.), *Microbial Technologies in Industrial Wastewater Treatment* (pp. 101–118). Springer. https://doi.org/10.1007/978-981-99-2435-6_7

Singh, S., & Shyu, D. J. H. (2024). Chapter 11 – Microbial technologies effectiveness and feasibility viewpoint. In V. Kumar, S. Kumar, P. Verma, & S. A. Bhat (Eds.), *Microbial Nexus for Sustainable Wastewater Treatment* (1st ed., pp 1–33). CRC Press. https://doi.org/10.1201/9781003441069

Singh, V., & Tiwari, A. (2021). Chapter 11 - Wastewater remediation through microbe-based nanoparticles. In M. Shah & S. Rodriguez-Couto (Eds.), *Microbial Ecology of Wastewater Treatment Plants* (pp. 237–248). Elsevier. https://doi.org/10.1016/B978-0-12-822503-5.00024-2

Sun, X. (2016). Microorganism bacterium agent for treating industrial wastewater pollution. Patent.

Suneja, G., & Sharma, R. (2020). Chapter 16 - Insights into the microbial processes prevalent in wastewater treatment systems using omics approaches. In M. P. Shah (Ed.), *Removal of Toxic Pollutants through Microbiological and Tertiary Treatment* (Vol. 1, pp. 409–430). Elsevier. https://doi.org/10.1016/B978-0-12-821014-7.00016-2

Taher, H. S., Wahdan, S. F. M., Sayed, R., Ouda, A., & Abdulla, H. (2024). Chapter 9 - The microbial nexus in effluent treatment plants. In V. Kumar, S. Kumar, P. Verma, & S. A. Bhat (Eds.), *Microbial Nexus for Sustainable Wastewater Treatment* (1st ed., pp. 1–27). CRC Press. https://doi.org/10.1201/9781003441069-9

Tuan, H. N. A., Hung, N. T. Q., Thi, C., Quyen, C. T., & Thanh, H. (2023). Combined biological and chemical-physical processes for domestic wastewater treatment. *Journal of Thu Dau Mot University*, 545–555. https://doi.org/10.37550/tdmu.ejs/2023.04.503

Tusief, Q., Mohsin, M., Malik, M. H., Asghar, H. N., & Sardar, S. (2022). Textile wastewater treatment through plant–microbe synergism in a floating treatment wetland system. *AATCC Journal of Research, 9*(4), 182–193. https://doi.org/10.1177/24723444221103679

Wang, L., Lin, Y., Ye, L., Qian, Y., Shi, Y., Xu, K., … & Geng, J. (2021). Microbial roles in dissolved organic matter transformation in full-scale wastewater treatment processes revealed by reactomics and comparative genomics. *Environmental Science & Technology*, *55*(16), 11294–11307. https://doi.org/10.1021/ACS.EST.1C02584

Yadav, K., Dwivedi, S., Gupta, S., Tanveer, A., & Yadav, D. (2024). Chapter 8 - The intervention of microbial enzymes in wastewater management an update. In V. Kumar, S. Kumar, P. Verma, & S. A. Bhat (Eds.), *Microbial Nexus for Sustainable Wastewater Treatment* (1st ed., pp. 1–19). CRC Press. https://doi.org/10.1201/9781003441069

Yalçın, Z. G., Dağ, M., & Aydoğmuş, E. (2023). Wastewater treatment using active microorganisms and evaluation of results. *International Journal of Advanced Natural Sciences and Engineering Researches*, *7*(4), 55–61. https://doi.org/10.59287/ijanser.550

Zhang, Y., Liu, F., Lin, Y., Sun, L., Guo, X., Yang, S., & He, J. (2022). Enhanced swine wastewater treatment by constructed wetland-microbial fuel cell systems. *Water*, *14*(23), 3930. https://doi.org/10.3390/w14233930

Zhou, W., & Yang, M. (2016). Photosynthetic organism-mediated wastewater comprehensive ecological treatment system and method. Patent.

18 Bioremediation as a Sustainable Management of Petroleum Hydrocarbon Pollutants

Smiti Kana Ray, Sudeshna Dey, Kiran Bala, Chhagan Lal, Chandra Mohan, Ipsita Dipamitra Behera, Krishna Kumar Yadav, and Alok Prasad Das

18.1 INTRODUCTION

The production, distribution, and use of petroleum hydrocarbons have increased due to the rapid advancement of human civilization, urbanization, population rise, economic expansion, and impetuous industrialization. Natural petroleum supplies gradually ran out as a result of this process, and demand for petroleum products increased. A complex chemical product prepared from a mixture of carbon and hydrogen found in crude oil or petroleum is termed petroleum hydrocarbons (Mohammadi et al., 2020). Dependence on fossil fuels in recent years has led to serious environmental issues global petroleum production in the year 2023. Global petroleum production has increased to 96,000,000 barrels/day. The major contributors to crude oil production include the countries of the US, Brazil, Iran, China, Norway, and Russia. They have been counted as rapidly increasing environmental pollutants of the 21st century. These hydrocarbons are primarily used as fuels and energy sources across various industries. Petroleum pollution has been a major concern for scientists across the globe because the chemicals present in PHCs act as a threat to both the ecosystem and humans if their concentration in the environment is high enough. Products derived from petroleum hydrocarbons can cause pollution that can have both direct and indirect consequences on ecotoxicology and human health. Due to the considerable amounts of specific petroleum hydrocarbon components, there are a variety of reasons for contaminants entering the environment (Mallick et al., 2024). Risks to human health and the environment in the long run can arise from spills, leaks, and discharges of petroleum hydrocarbon contamination (Roy et al., 2018). Sustainable solutions are necessary because traditional methods fail to address environmental impact. Modern greener technologies provide economical, ecologically friendly, and effective substitutes.

DOI: 10.1201/9781003508496-18

Crude oil is unprocessed petroleum, with its weight being made up of hydrogen and carbon. Small amounts of oxygen, nitrogen, and sulfur are other components. Additional trace elements include nichrome, silicon, iron, and aluminum. Petroleum hydrocarbons are organic molecules such as alkanes, cycloalkanes, aromatic hydrocarbons, and complex polycyclic aromatic hydrocarbons (PAHs) that are obtained from crude oil and natural gas. These hydrophobic compounds are persistent in the environment and may be carcinogenic, endangering both human and wildlife health (Kostka et al., 2014). Crude oil, a mixture of hydrocarbons, is the primary source of petroleum hydrocarbons. It is extracted from the earth's bottom and purified into petroleum products like liquefied petroleum gas, heating oil, kerosene, diesel fuel, and asphalt based at refineries. The fractional distillation method is employed to separate crude oil components based on their boiling points. Lower fractions of a fractionating column have higher viscosities and boiling temperatures. When released into the environment, petroleum hydrocarbon components experience a variety of weathering processes, such as adsorption, dissolution, oxidation, photolysis, biotransformation, and hydrolysis. By weathering, dissolving in water, volatilizing in the atmosphere, and attaching themselves to soil particles and organic molecules, hydrocarbon pollutants are discharged into the environment (Bala et al., 2024; Dey, Bissoyi, et al., 2024; Dey, Dash, Ray, et al., 2024; Mishra et al., 2024). Petroleum hydrocarbons have the potential to damage land, surface water, groundwater, and the atmosphere when they are discharged due to accidents, leaks, and inappropriate waste disposal. They have the power to destroy marine ecosystems, upend food networks, and make soil unfit for irrigation. Air pollution is caused by the volatilization of hydrocarbons, being especially dangerous because of their carcinogenic qualities.

The main causes of hydrocarbon pollution are leaks, oil spills, and incidents, including pipeline breaks and tanker accidents. Air pollution and deposition are caused by industrial operations such as fuel burning and refining, which release hydrocarbons into the atmosphere. The contamination of soil, groundwater, and surface water is exacerbated by the improper disposal of petroleum products, such as used motor oil and petrol. Risks to human health, biodiversity, and the integrity of the environment are presented by urban runoff and industrial discharges, which worsen the degradation of aquatic ecosystems. Freshwater, marine, and terrestrial biomes are all impacted by petroleum hydrocarbon contamination, which may coat the surfaces of aquatic ecosystems, suffocate marine life, and upset food chains (Maletić et al., 2019). Acute exposure to common hydrocarbon benzene can cause symptoms like elevated blood pressure, dizziness, and cardiovascular issues. Extended exposure can lead to neurological disorders, reproductive problems, liver and kidney damage, and respiratory ailments. It's also linked to malignancies, blood problems, and skin irritation. Extended exposure weakens immunity, making individuals more susceptible to infections and autoimmune diseases (Das & Singh, 2011; Paul et al., 2023).

Traditionally, in situ burning, chemical dispersants, mechanical recovery, and physical barriers like booms and skimmers have been used to control petroleum hydrocarbon contamination. Physically removing oil from the surface of water or land is known as mechanical recovery, yet this process is frequently laborious and ineffective, particularly in choppy waters or hard-to-reach locations. Oil is broken up into smaller droplets by chemical dispersants so that it can be broken down by the environment. Controlled burning of oil on the water's surface is known as "in situ burning," and while it can drastically cut down on oil spills, it also produces a lot of smoke and other air pollutants, including greenhouse gases. Weather and the availability of enough oil thickness to support combustion are other limitations. Although booms and skimmers are frequently utilized as physical barriers to collect and contain oil on the water's surface, their effectiveness is reduced in choppy conditions or in locations where the oil has already spread far (Zhang et al., 2019). They don't deal with oil that has seeped into the ground or that has reached shorelines.

Greener technologies are creative approaches used in waste management, energy, transportation, and industry to lower greenhouse gas emissions, resource consumption, and environmental impact. These technologies minimize pollution while placing a high priority on efficiency, sustainability, and the preservation of natural resources (Mallick et al., 2023; Nag et al., 2023). They combine environmental stewardship, circularity, and sustainability, offering benefits in the management of petroleum hydrocarbons (Bala et al., 2022). While modern oxidation techniques and surfactant-enhanced remediation can remediate hydrocarbon-contaminated areas effectively and economically, lowering hazards to human health and environmental quality, bioremediation and phytoremediation use natural processes to eradicate petroleum contaminants. More sophisticated and environmentally friendly solutions have been created to handle petroleum hydrocarbon emissions sustainably. These eco-friendly technologies break down, eliminate, or contain hydrocarbons by the use of biological, chemical, and physical processes. The sources, effects, and remediation of petroleum hydrocarbons are demonstrated. The creation of dispersants and absorbents using green chemistry, nanotechnology, bioremediation, and phytoremediation is an important strategy. Microorganisms are used in bioremediation to break down or purify contaminants, such as petroleum hydrocarbons (Mishra & Das, 2023; Pradhan et al., 2023). Depending on the kind and number of hydrocarbons, acceptable microorganisms, and the surrounding circumstances, it can be administered in situ or ex situ. Utilizing plants, phytoremediation purges break down, or confine contaminants found in soil and water. Rhizodegradation, phytoextraction, and phytodegradation are some of the methods. Nanotechnology provides novel approaches to hydrocarbon detection, degradation, and removal. By engineering nano sorbents, nanocatalysts, and nanosensors to specifically target particular hydrocarbon types, the adsorption process can be made more selective and efficient (Kebede et al., 2021). However, before widespread deployment, concerns regarding health and environmental consequences need to be carefully evaluated.

Green chemistry concepts can be used to create more ecologically friendly dispersants, absorbents, and remediation agents in the context of petroleum hydrocarbon pollution (Das et al., 2023; Dey, Das, et al., 2023; Singh et al., 2020). To minimize waste and provide the possibility of recovering and reusing absorbed oil, environmentally friendly dispersants, biodegradable absorbents, and green solvents and catalysts are crucial. This chapter describes various petroleum hydrocarbons, their effects, and outlines different greener technologies to remediate these toxic pollutants present in the environment.

18.2 CLASSIFICATION OF PETROLEUM HYDROCARBONS

Alkanes, cycloalkanes, aromatics, polyaromatic hydrocarbons, asphaltenes, resins, and heteroatomic molecules containing nitrogen, sulfur, and oxygen are the four groups of compounds that make up petroleum. Natural activities like migration, biodegradation, and water washing have an impact on its composition (Deng et al., 2025). Natural product metamorphosis brought about by localized chemical and physical changes results in petroleum. Aliphatic hydrocarbons, the most prevalent part of crude oil, are organic compounds found in natural gas and petroleum. These hydrocarbons consist of hydrogen and carbon atoms, typically joined in chains by single, double, or triple bonds. They can branch or take nonaromatic structures. Aliphatics are double-carbon-bonded saturated or unsaturated hydrocarbons with linear or branched open chains. Aliphatics are molecules with linear or branched open chains, including terpenes, steranes, n-alkanes, isoalkanes, and cycloalkanes. These structures are joined to each other by a single, double, or triple bond and are also known as paraffins. Due to their poor water solubility, biological availability, and structure, aliphatic hydrocarbons with longer chain lengths are more persistent in soil, slower to volatilize, and difficult to break down (Mishra, Dash, et al., 2023). Petroleum contains paraffin waxes at their higher boiling point fractions, including cyclopentane and cyclohexane. The cycloalkanes that are present in petroleum are alkanes in which the side chains form a ring-like structure. Larger chain-length aliphatic hydrocarbons (C20–40) are more persistent in soil, less volatilized, and difficult to degrade due to their low water solubility, biological availability, and structure.

Up to 50% of all petroleum hydrocarbons are classified as aromatic hydrocarbons because they include one or more benzene rings. This group is named for the intense and powerful aroma that these hydrocarbons emit. Petroleum contains a variety of aromatic hydrocarbons, such as lovely dancing xylene, 1,3,5-trimethyl benzene, etc. The most prevalent aromatic component found in petroleum is benzene, toluene, ethylbenzene and xylene (BTEX), which is added to combinations made by distilling or refining coal. These compounds are used to make a range of industrial goods, including paints, inks, cosmetics, and pharmaceuticals (Khatun et al., 2023; Mishra, Ghosh, et al., 2023). Organic molecules are classified as aromatic fractions if they have two to six aromatic rings in their structure, as well as alkyl substituents arranged in linear, cluster, or angular configurations. The extreme toxicity of polycyclic aromatic hydrocarbons and their potential to induce cancer and mutations make them a major concern for the environment and public health. The molecular weight and number of benzene rings in PAHs vary, influencing their physical–chemical characteristics. Their distribution, fate, transit, and ecological consequences are all determined by their physicochemical properties. High-molecular-weight compounds have low volatility, low water solubility, low chemical reactivity, and high carcinogenic potential. Common traits include resistance to heat, light, and corrosion. The toxicity and persistence of PAHs, especially those with higher molecular weights and more complexity, are linked to their long-term environmental impact (Das et al., 2011; Mishra et al., 2019, 2022). They are tough to decompose and resistant in soil due to their high hydrophobicity. The alkaline derivatives of PAHs, such as naphthalene, anthracene, and phenanthrene, are stable at room temperature and can be found in solutions. PAHs are composed of two or more benzene rings fused together. Asphaltene is a high-molecular-weight material that, according to confirmation, is composed of roughly 80% carbon, 10% hydrogen, and other components including oxygen and sulfur. Their molecular makeup may resemble that of naphthalene sheets with alkali side chains as well as polycyclic aromatic compounds.

Hetero atomic compounds, or NSAs, are composed of functional groups like oxygen, sulfur, nitrogen, and sulfur oxides. The oxides of sulfur and nitrogen are particularly dangerous to the environment because they can produce acid rain when they are discharged into the atmosphere after combustion. Petroleum contains oxygen in the forms of alcohols, phenols, ketones, and esters. Petroleum contains a variety of carboxylic acids, such as fatty acids, crude oils, and short-chain alkyl carboxylic acids. Nitrogen compounds are found in less than 1% of these oils and are primarily found in the form of pyridine and its homologs. Petroleum also includes several petrol additives, such as methyl tert-butyl ether, a gaseous additive used to increase the oxygen content and facilitate better petrol combustion. Hydrogen sulfide, or sulfur, is a common component in many crude oils. Petroleum contains sulfur compounds such as cyclic compounds, thiocycloalkanes, alkanethiols, and thioalkanes. When crude oil comes into contact with brines in the reservoir, it contains sodium chloride. Another form of inorganic component is metal purify rinse, which often comprises nickel and vanadium as the main metals. Asphaltenes are nonpolar, nonvolatile, brittle, glossy, solid compounds with a dark brown to black color that are precipitated by n-heptane. Alkanes do not dissolve them, but aromatic solvents do. Conversely, resins have a larger hydrogen-to-carbon ratio, a smaller molecular weight, and are less heteroatomized. They can be found in bitumen and asphalt materials that are used for roofing, road paving, and weatherproof coatings.

18.3 ENVIRONMENTAL EFFECT OF PETROLEUM HYDROCARBONS

Petroleum hydrocarbons have an impact on the environment that can be studied through a variety of processes, such as biological interactions with microbes and metabolic pathways, as well as physical abiotic reactions. In the environment, petroleum hydrocarbons can weather through a variety of mechanisms, including evaporation, photocatalysis, and photolytic degradation. The evaporation of light substances is the fastest physical process and is influenced by temperature, wind speed, water turbulence, and surface properties. The breakdown or movement of different chemicals and hydrocarbons included in petroleum products is facilitated by these processes. Hydrocarbons can lose their large proportions due to evaporation and photolytic breakdown, particularly of PAHs, is another method of petroleum weathering. Since petroleum hydrocarbons interfere with the soil/sediment matrix, they can exacerbate erosion. These hydrophobic pollutants can be transported via erosion, sedimentation, or soil mixing after they have absorbed. Chemical agents, such as dispersants and surfactants, are frequently employed to break down petroleum particle sizes and promote biodegradation. Due to their ability to repel water, petroleum hydrocarbons have a strong affinity for organic matter, making them capable of lingering for many years before degrading. Petroleum hydrocarbons' small surface area limits their availability for biodegradation.

Soil contamination by petroleum hydrocarbons is a major environmental problem that affects people's health and is a major worldwide concern. For soil biochemical reactions to occur, soil enzymes are essential, and petroleum hydrocarbons negatively impact these enzymes. Oil can affect plants either temporarily or continuously and disrupt their hydric relationships due to its higher inhibitory effect on nitrification (Maletić et al., 2019). Plant development and growth are restricted by the degradation of the physicochemical characteristics of soil induced by refinery products. Petroleum products lead to deficiencies in water and oxygen, limitations of nitrogen and phosphorus, and organic contamination of water, which restricts its use and results in losses and environmental issues. Petroleum hydrocarbons are toxic to humans, animals, plants, and microorganisms. The fate of petroleum hydrocarbons in terrestrial environments is influenced by soil composition, physical properties, and chemical composition, including air diffusion coefficient, solubility in water, and boiling point. Particle agglomeration and oxygen transfer are two effects of oil spills that might endanger coastal habitats. They can range from small-scale releases to large-scale mishaps. Oil spills can create large oil slicks that harm aquatic life, reducing their insulating ability and leading to hypothermia. Air pollution can result from the release of volatile organic compounds (VOCs), contributing to ground-level ozone and greenhouse gas emissions. Wildlife can bioaccumulate and biomagnify to the point where top predators—including humans—are exposed to toxic levels of it. Oil spills and leaks have the potential to destroy habitat, especially in delicate places like wetlands, mangroves, and coral reefs. Ecosystem disruption can result in loss of biodiversity and changes in species composition. PAHs, when present at low concentrations, can be poisonous and can be carcinogenic or mutagenic to both wildlife and humans.

Hydrocarbon exposure in symptomatic patients can occur unintentionally, through household products, occupational exposure, or intentional recreational inhalation. Techniques for purposeful inhaling include bagging, puffing, and sniffing. Seizures and death from respiratory failure, arrhythmias, or severe central nervous system CNS effects can result from exposure that affects the respiratory, cardiovascular, gastrointestinal, and kidney systems. Hydrocarbons have the potential to have systemic effects on the skin, heart, gastrointestinal tract, kidneys, and lungs, among other organ systems. Breathing difficulties and coughing might result from aspiration or inhalation that causes severe pneumonitis, inflammation, and fever. Petrogenic hydrocarbons accumulate in ecosystems, causing death or genetic mutations in animals and humans. Frequent exposure can cause physiological impairments, liver damage, impaired immune system, anemia, weight loss, reduced productivity, and gastrointestinal deterioration. Short-term CNS consequences include slurred speech and generalized depression. Although the precise mechanism underlying these effects is uncertain, some research indicates that they may impact the brain's dopamine, gamma aminobutyric acid (GABA), voltage-gated ion channels, glutamate receptors, serotonin, and nicotinic acid (N-Methyl-D-aspartic acid (NMDA)) pathways. Murmurs linked to pulmonary hypertension and pulmonary hypertension itself can be brought on by long-term usage. Hematemesis, nausea, vomiting, abdominal discomfort, and gastrointestinal tract irritation can all result from ingestion. Metabolic acidosis, urinary calculi, glomerulonephritis, hyperchloremia, and hypokalemia can all have an impact on the kidneys. Exposure to chemicals can burn the skin or produce minor irritation, which might result in systemic symptoms.

18.4 GREENER TECHNOLOGIES

Greener technologies aim to reduce environmental impact, promote sustainability, and enhance resource efficiency. They focus on pollution reduction, resource conservation, energy consumption reduction, and lowering greenhouse gas emissions. The goal is to create environmentally friendly processes, products, and systems. Sustainable management of petroleum hydrocarbons involves reducing their environmental impact through greener technology. These advancements aim to decrease pollution, enhance efficiency, and promote renewable energy use, as described in Table 18.1.

TABLE 18.1
Advantages and Disadvantages of Different Greener Technologies Involved in Petroleum Hydrocarbon Management

Sl. No.	Process	Advantages	Disadvantages	References
1.	Advanced oxidation process	• AOP helps in the breakdown of a wide range of complex aromatic hydrocarbons like benzene, xylene, etc. • It is a fast process and leads to complete mineralization of pollutants.	• It has a high operational cost, and the transport of oxidants is dangerous. • This process leads to the production of corrosive by-products and requires treatment to remove larger debris.	Kumari and Kumar (2023)
2.	Bioremediation and phytoremediation	• It is a natural, eco-friendly, in situ, and cost-effective processes. • A wide range of petroleum pollutants can be degraded through this process.	• This is a relatively slow process and can take years for complete degradation of pollutants. • The processes are highly site-specific and depend on environmental conditions.	Noor Zaliha et al. (2003)
3.	Nanotechnology	• Nanotechnology helps in enriched soil recovery and changes the characteristics of the soil. • Nanosensors can detect very less concentration of hydrocarbons in soil and degrade them.	• The process could lead to bioaccumulation of NP in the food chain and act as a threat to the ecosystem. • This is a highly expensive and complex process.	Kumari and Singh (2016)

(Continued)

TABLE 18.1 (*Continued*)
Advantages and Disadvantages of Different Greener Technologies Involved in Petroleum Hydrocarbon Management

Sl. No.	Process	Advantages	Disadvantages	References
4.	Electrocoagulation	• No additional chemicals are required for the process, making same to use. • It is highly effective and can be used to remove different types of pollutants other than petroleum hydrocarbons.	• The process requires a large amount of energy in the form of electricity. • It is a high-maintenance process where regular exchange of electrodes is necessary.	Mohammadi et al. (2020)
5.	Photocatalytic oxidation	• The source of energy used is renewable in the form of solar energy • It is adaptable and has no chemical additives.	• Photocatalytic oxidation (PCO) requires light for activation, which can be limited in low sunlight or poor lighting conditions, potentially increasing energy costs. • It also has slower reaction rates and is time-consuming.	Mohammadi et al. (2020)
6.	Electrokinetics	• It is highly efficient in low-permeability soils like clays, mobilizing contaminants and enhancing their mobility. • It is easier to extract pollutants like petroleum hydrocarbons, heavy metals, and organic compounds.	• There may be a limit to the depths at which electrokinetic remediation is effective. • Over time, electrodes may corrode or get fouled, which will diminish their efficiency.	Saini et al. (2021)
7.	Membrane filtration	• Membrane filtration offers high efficiency in removing contaminants like petroleum hydrocarbons and particulate matter from water. • Its flexibility allows for the use of various membrane types for specific pollution types.	• Systems can be costly due to clogging and degradation caused by particles, oils, and organic waste. • As high initial capital and ongoing operating costs.	Munirasu et al. (2016)

18.5 ADVANCED OXIDATION PROCESS

Aromatic compounds that undergo advanced oxidation process (AOP) can be oxidized chemically, photochemically, electrochemically, or sonochemically using chemical reagents, catalysts, solar energy, UV sources, electric sources, or ultrasound. Reactive oxygen species (ROS) are used in the AOP, a method that attempts to fully mineralize organic contaminants into safe by-products. ROS contain at least one or more unpaired electron, which is responsible for the process of oxidation. The first step in an AOP is to create ROS, such as hydroxyl radicals (OH), superoxide anions (O_2^-), or sulfate radicals (SO_4^-), by applying energy (such as UV radiation, electrical energy, or ultrasound) or by using chemical processes. After that, a reaction between the generated ROS and the organic contaminants occurs, forming intermediates through various response mechanisms. There are various possible response mechanisms which lead to different rates of degradation and efficiency. One way that hydroxyl radicals can break down complex molecules into smaller, more biodegradable intermediates is by interacting with other radicals to take hydrogen atoms from organic substances. Complete mineralization, which converts the pollutants into harmless by-products like water (H_2O) and carbon dioxide (CO_2), is the ultimate objective of the degradation process. After that, conventional treatment methods can be used to get rid of the produced inorganic by-products, which include salts, CO_2, and water. For aromatic hydrocarbons, the first step in oxidation involves hydroxylation of the rings and then reaction with ROS to give rise to open structures that are further changed to organic acids. These compounds undergo mineralization to produce biodegradable products. Strong oxidants, such as hydroxyl radical (OH), are produced by AOP at room temperature and pressure. Ozone and hydroxyl radicals have a high reduction potential and OH production, which makes them popular oxidants in AOP.

Chemical AOPs use chemical reagents and catalysts to generate ROS, such as Fenton's process, which involves the reaction of hydrogen peroxide with ferrous iron (Fe^{2+}) to produce hydroxyl radicals (OH) and includes a complex chain reaction as shown in equation (18.1). This method effectively degrades various organic pollutants, including phenols and dyes. The first ferrous ion is converted to ferric ion in the presence of hydrogen peroxide as an oxidizing agent (Kumari & Kumar, 2023). The ferric ion is returned to a ferrous state along with the formation of hydroperoxyl radical (equation 18.2). Other steps in the reaction cassette initiate the formation of hydroxyl radicals along with water (equation 18.3). OH, radical then reacts with the pollutants to form intermediates, and these intermediates undergo mineralization to produce CO_2 and water.

$$H_2O_2 + Fe^{2+} \rightarrow Fe^{3+} + OH^- + \cdot OH \tag{18.1}$$

$$Fe^{3+} + H_2O_2 \rightarrow \cdot OOH + Fe^{2+} + H^+ \tag{18.2}$$

$$2H_2O_2 \rightarrow \cdot OOH + \cdot OH + H_2O \tag{18.3}$$

In order to increase the amount of hydroxyl radical produced through ozonation, as the amount produced is not enough for the degradation of pollutants, a mixture of ozone and hydrogen peroxide is used as a substrate to enhance the degradation of ozone and produce hydroxyl ions (as shown in equation (18.4)). Hydrogen peroxide dissociates to form hydroperoxide radical and ozone reacts with this radical to release ozonide (as shown in equation (18.5)). The peroxide radical further breaks to form oxide and a proton (as shown in equation (18.6)). Oxide ion and ozonide react to release oxygen and ozonide radical (as shown in equation (18.7)). Ozonide adds with a proton to produce hydrogen ozonide (as shown in equation (18.8)). Later, hydroxyl radical and hydroperoxide radicals are produced (as shown in equations (18.9) and (18.10)). AOP containing ozone or peroxone is a better alternative for the removal of hydrocarbon pollutants from water sources.

$$H_2O_2 \rightarrow HO_2^- + H^+ \tag{18.4}$$

$$HO_2^- + O_3 \rightarrow \cdot OOH + O_3^- \tag{18.5}$$

$$\cdot OOH \rightarrow H^+ + O_2^- \tag{18.6}$$

$$O_2^- + O_3 \rightarrow O_2 + O_3^- \tag{18.7}$$

$$O_3^- + H^+ \rightarrow HO_3 \tag{18.8}$$

$$HO_3 \cdot \rightarrow \cdot OH + O_2 \tag{18.9}$$

$$\cdot OH + O_3 \rightarrow \cdot OOH + O_2 \tag{18.10}$$

Photochemical AOPs harness light energy to activate chemical reactions that produce ROS, such as photodecomposition and photocatalysis. In this process to increase the amount of hydroxyl radical, UV radiation along with oxidants like hydrogen peroxide and ozone, and catalysts like titanium oxide and ferric ion are used. Photodecomposition uses oxidants like H_2O_2 and ozone, which give hydroxyl radicals using UV radiation of wavelength 200–300 nm. Photocatalysis uses semiconductor catalysts like TiO_2 and ZnO, which enhance the process of degradation. Electrochemical AOPs use electric current to generate ROS at the anode or cathode, such as the Electro-Fenton process, which combines electrochemical oxidation with Fenton's chemistry. Sonochemical AOPs use ultrasound waves to generate cavitation bubbles in a liquid medium, which collapse to produce high temperatures and pressures, leading to the formation of hydroxyl radicals. An aqueous solution when treated with ultrasonic waves leads to the generation of oxidizing agent. Ultrasonic waves lead to the collapse of the bubble leading to the release of extra energy, which helps oxidize water to form hydroxyl radical. Sonophotochemical processes combine ultrasound with light to enhance degradation efficiency further.

18.6 BIOREMEDIATION AND PHYTOREMEDIATION

Bioremediation is a promising approach for managing petroleum hydrocarbons, using microorganisms like bacteria and fungi to degrade complex compounds into less harmful substances. Genetic engineering can enhance the process by modifying microorganisms to improve their degradation capabilities. Factors such as temperature, nutrient availability, and oxygen levels play a crucial role in the efficiency of bioremediation. *Bacillus, Pseudomonas, Sphingomonas, Flavobacterium, Nocardia, Rhodococcus*, and *Mycobacterium* are aerobic bacteria capable of degrading various complex organic compounds, including polyaromatic compounds (Oualha et al., 2019). Bioremediation includes the use of hydrocarbon-degrading microbes that naturally facilitate the process of degradation. These microbes degrade alkanes and PAHs using different enzymes like alkane monoxygenase and naphthalene dioxygenase through a sequential oxidation process. Fungal species like *Rhizopus* and *Penicillium* are seen to participate in the degradation process (Mishra et al., 2020, 2021). The process of biostimulation entails the addition of nutrients, biosurfactants, bulking agents, etc. to activate and support native hydrocarbon-degrading bacteria to facilitate the breakdown of toxic chemicals (Das & Ghosh, 2018). An increase in carbon stimulates the proliferation of these microbes following an oil spill, and an appropriate concentration of nutrients increases their propensity to break down hydrocarbons. Scientists have found that biostimulation of contaminated soil with nutrients and consortia can reduce the total environmental pollutants by 99%. Finding and holding onto the optimal nutrient content for microbe growth is the key to achieving optimum biostimulation (Das et al., 2015b). Bioaugmentation involves the addition of exogenous microbial growth, genetically modified microbes responsible for hydrocarbon degradation, or an increase in the rate of degradation to contaminated sources (Wu et al., 2017). However, because the environment naturally contains the microorganisms needed for hydrocarbon breakdown, bioaugmentation is not frequently used for this purpose. Bioattenuation uses different biological and physicochemical events without the intervention of man-made forces to naturally degrade and neutralize toxic petroleum hydrocarbons. The different methods used include degradation, stabilization, volatilization, dispersion, etc. The major benefits of this process involve minimal damage to the environment and can be used in addition to different mechanisms (Roy et al., 2018).

Bioslurry includes the use of bioreactors where contaminated soil is provided with nutrients at a continuous flow. This technique is used as an ex situ conservation method for petroleum hydrocarbons and VOCs. The bioreactors are designed in a way to control the different factors like temperature, pH, nutrients, etc. to obtain the best results. It was observed in a study that the use of Bacillus species along with the Pseudomonas strain can increase the degradation of PAHs by 70% (Ramadass et al., 2018). This is a faster and more developed method that offers more control over the process of degradation than any other. Bioventing is a specific bioremediation technique that injects air into contaminated soil, stimulating indigenous microbial populations to promote degradation (Mishra et al., 2009; Sahoo et al., 2022). It is particularly effective for treating soils contaminated with VOCs. The addition of air creates an aerobic environment suitable for the hydrocarbon-degrading microbes to perform (Noor Zaliha et al., 2003). The process can be used in both ex situ and in situ clearance of petroleum hydrocarbon contaminants. About 80%–90% of the total petroleum hydrocarbon (TPH) can be cleared through this method. Land farming leads to a method that includes plowing, tilling, and spreading of contaminated soil in a thin layer, and addition of nutrients and other products to improve microbial activity. VOCs, low-molecule-weight hydrocarbons, are degraded by this process. Biosparging adds oxygenated and nutrients back to the soil under high pressure to increase its concentration and improve the activity of microbes A 0 success rate of 60%–70% is seen in the degradation of compounds like BTEX, toluene, benzene, etc. Petroleum-contaminated soil has been bioremediated using mesophilic conditions (30°C–40°C) and low aeration rates for the removal of TPH. Biopile systems have been utilized for treating diesel-contaminated soil in the sub-Antarctic region, removing 93% of TPH within 1 year. The components of piling include piling of contaminated soil, followed by the addition of air, moisture, and nutrients to the soil to improve the activity of microbes. This is best used to clean large volumes of contaminants from various sources. This is a man-made and above-the-surface technique that limits the volatilization of low-molecular-weight compounds (Dey, Dash, & Das, 2024). Windrow treatment is a less common but notable method for managing petroleum hydrocarbons, promoting microbial activity, and enhancing biodegradation (Bal et al., 2018; Das et al., 2015a; Das & Mishra, 2008). However, it is generally less efficient than biopile systems and may produce greenhouse gases as a by-product. Windrow methods depend on timely tilling, plowing, and the addition of water to the contaminated source to increase the moisture and aeration of the soil. The major benefits of Windrow includes retaining soil fertility, maintaining soil pH, being eco-friendly, and requiring low capital investment. Bioslurping, which combines vacuum extraction with bioventing, is useful for recovering light nonaqueous phase liquids (LNAPLs) from soil and groundwater (Ossai et al., n.d.). Phytoremediation is an innovative approach that utilizes plants to manage petroleum hydrocarbons in contaminated soils. Certain plant species can absorb, sequester, and transform hydrocarbons through their root systems and metabolic processes, reducing pollutant concentrations and contributing to soil stabilization and habitat restoration (Koshlaf & Ball, 2017). During phytoremediation, plants produce different enzymes that are capable of sequestering and degrading the hydrocarbons. Phytotransformation, phytostabilization, and phytodegradation are different mechanisms of phytoremediation.

18.7 NANOTECHNOLOGY

Nanotechnology has been an up-and-coming technique for the management of petroleum hydrocarbons. Hydrophobicity of petroleum hydrocarbons makes it a limiting factor for degradation. Biosurfactants produced by microbes reduce the hydrophobicity of these compounds and make it available for the microbes. The amount of surfactants produced by microbes is limited due to the restrictions in metabolic pathways. Nanoparticles (NPs) can also modify their surface properties to achieve a desired affinity for hydrophobic contaminants, optimizing the interaction between NPs and hydrophobic compounds (Bal et al., 2018). Additionally, NPs can act as surfactants, reducing water's surface tension, facilitating the dispersion of hydrophobic compounds, and enhancing their bioavailability to microorganisms, crucial for effective bioremediation. It has been studied that the use of iron NPs in Nocardiopsis has increased biosurfactant production. NPs can enhance the solubility of hydrophobic compounds by forming stable suspensions, increasing their concentration in aqueous environments. This competition with nonaqueous phase liquids (NAPLs) enhances contaminants' mobility (Kumari & Singh, 2016). The utilization of NPs, specifically nano silica derived from rice husk, has the potential to greatly enhance soil microbial diversity and biomass, hence resulting in better soil fertility and ecosystem health. By encapsulating vital nutrients like potassium, phosphate, and nitrogen inside NPs, they also improve the efficiency of nutrient release by enabling controlled and prolonged release. By encouraging adequate nutrient levels, this delayed release technique encourages plant growth and lessens the need for chemical fertilizers.

As they increase the solubility and mobility of PAHs in contaminated settings, NPs are essential for PAHs adsorption. NPs that efficiently adsorb PAHs, including carbon nanotubes and organo-clay, greatly enhance the removal of PAHs from soil and water (Kumar & Das, 2017). In addition to lessening the toxicity of PAHs, this adsorption process encourages microbial development, which speeds up the breakdown of these dangerous substances. According to research, modified NPs can attain significant adsorption capabilities, which makes them useful instruments in bioremediation procedures meant to restore PAH-contaminated sites and enhance environmental health.

A nano-peroxide called calcium peroxide promotes the degradation of petroleum hydrocarbons by quickening their rate of solubility and reaction. In contaminated environments, NPs such as titanium dioxide (TiO_2) and nano zero-valent iron (NZVI) act as powerful catalysts in oxidative processes, causing a considerable breakdown of hydrocarbons. According to studies, NPs may remove up to 91% of pollutants within hours, demonstrating remarkable removal efficiency (Das, Dey, Bala, et al., 2024). For this reason, NPs are crucial in the fight against oil pollution and the enhancement of environmental recovery initiatives.

18.8 ELECTROCOAGULATION

One of the main sources of pollution in the environment is petroleum sludge, a complicated mixture of waste oil, wastewater, sand, and mineral materials. When treating petroleum-contaminated groundwater, the electrocoagulation process (ECP) is often assessed and measured based on the TPH eliminated by the electrode materials, which are primarily aluminum, iron, and steel. The anode–cathode electrode configuration of steel–iron has been shown to remove the maximum amount of TPH. Batch and continuous trials generally demonstrated that the ECP might be useful in removing TPH from water, indicating that it is a potential method for treating groundwater contaminated by petroleum (Mohammadi et al., 2020). Applications for ECP include purified terephthalic acid (PTA) wastewater treatment, gas refinery oily wastewater treatment utilizing aluminum electrodes, and turbidity removal from petrochemical wastewater.

18.9 PHOTOCATALYTIC OXIDATION

One technique that shows promise for breaking down resistant organic molecules is heterogeneous photocatalytic oxidation (AOP). One possible use of semiconductor heterogeneous photocatalysis is the removal of organic pollutants from air and water. Numerous semiconductors have been investigated for their ability to degrade petrochemical compounds, such as TiO_2, ZnO, Fe_2O_3, CdS, and ZnS. The degradation mechanisms of TiO_2 under visible and UV light irradiation have been extensively explored; the majority of these investigations have been carried out in laboratories under UV light. LEDs and other alternative light sources show promise in lowering power usage. However, catalytic activities only require 5% of solar light, and the end products or intermediates may be more dangerous than the original chemical (Mohammadi et al., 2020). Doping metals with other metals has been studied to increase the photo-response of TiO_2 toward the visible range, hence expanding the applicability of TiO_2. In one case, Ce-doped TiO_2 nanostructures (CeT) with varying concentrations of Ce were prepared using a sonochemical processing approach, and these nanostructures considerably enhanced photocatalytic activity when it came to the degradation of toluene and o-xylene. Additionally, compared to toluene, the produced CeT catalysts demonstrated higher photocatalytic degradation efficiency in o-xylene.

18.10 ELECTROKINETIC REMEDIATION

One technique that shows promise for treating soil and sediment contaminated with petroleum hydrocarbons is electrokinetic remediation. This technique is especially useful in low-permeability soils like clays and silts because it applies an electric field to transport pollutants and remove them. Electrodes (anodes and cathodes) are first installed into the polluted soil to start the process. Soil particles that are charged move when an electric field with a direct current is supplied between these electrodes. There are several electrokinetic mechanisms at play, including electroosmosis, electrophoresis, and electromigration. Petroleum hydrocarbon contamination can be effectively managed using electrokinetic remediation, especially in difficult situations. Particularly in low-permeability soils, these hydrocarbons, which include compounds like benzene, toluene, ethylbenzene, and xylene (often referred to as BTEX), can be challenging to repair using traditional techniques (Saini et al., 2021). The technology can be used in soil remediation, where it targets specific contaminated zones without the need for excavation, reducing environmental disturbance and costs. In contaminated groundwater treatment, electrokinetic processes can mobilize petroleum hydrocarbons toward electrodes, where they can be extracted and treated, particularly in cases where contaminants have migrated below the surface, making traditional extraction difficult (Munirasu et al., 2016).

18.11 CIRCULAR ECONOMY METHODS

The goal of circular economy methods is to establish a closed-loop system that maximizes resource efficiency and reduces waste, especially in petroleum hydrocarbons. These tactics emphasize cutting back on the use of nonrenewable resources, cutting down on waste production, and encouraging material recycling and reuse. Reducing the consumption of raw materials by increasing resource efficiency is one of the main strategies in a circular economy. Techniques like enhanced oil recovery (EOR), sophisticated refining methods, digitalization, and process optimization can help achieve this. EOR methods optimize hydrocarbon production from active oil fields while reducing environmental impact and preserving natural resources. Advanced refining methods maximize the conversion of heavier hydrocarbons into lighter, more valuable products like petrol and diesel through processes like hydrocracking and catalytic cracking.

Energy consumption and hydrocarbon waste during hydrocarbon extraction, processing, and transportation are decreased by more accurate monitoring and process optimization enabled by digitalization. Because they lessen the need for virgin petroleum resources and the related environmental effects, recycling and reusing petroleum-based products are essential to a circular economy. Petroleum waste can be chemically recycled, sometimes referred to as feedstock recycled, which reduces the demand for virgin petrochemical feedstocks by breaking down plastic waste into its original monomers or other valuable chemicals (Waly et al., 2023). Re-refining lubricants and oils entails clearing impurities from spent oil and returning it to its original state. Petrochemical by-products produced during refining can be recycled and utilized to make fertilizers, chemicals, building materials for roads, and fuel, among other industrial activities. Overall, circular economic strategies aim to ensure sustainable management of petroleum hydrocarbons.

Waste reduction and energy recovery are given top priority by circular economic strategies, especially in the petroleum sector. Waste-to-energy (WtE), I carbon capture and utilization (CCU), sustainable design and substitution, and market- and regulation-based incentives are some of these solutions. WtE technologies use techniques including gasification, anaerobic digestion, and incineration to transform waste that contains hydrocarbons into useful energy. This lowers the amount of waste produced and recovers energy for use in manufacturing or the production of power. Industrial symbiosis is the cooperative use of waste products and by-products from several sectors as resources. CO_2 emissions from hydrocarbon processing are captured by CCU and used as feedstock for synthesizing chemicals, fuels, and construction materials. For things to be properly recycled, reused, or disposed of at the end of their life cycle, sustainable design is essential. Goods ought to be made to be easily disassembled so that parts can be recycled (Ray et al., 2024). Emerging as petroleum-based product substitutes, bioplastics and biofuels have the potential to minimize environmental effects and lessen dependency on fossil fuels. The goal of green chemistry principles is to create chemical processes and products that use fewer or no harmful materials (Das, Ray, et al., 2024; Singh Jadaun et al., 2022).

The petroleum industry can benefit greatly from the promotion of circular economy principles through government regulations and market-based incentives. Policies pertaining to extended producer responsibility (EPR) hold manufacturers accountable for the complete lifecycle of their goods, encompassing disposal. By attaching a monetary value to carbon emissions, carbon pricing mechanisms like carbon taxes and cap-and-trade programs encourage businesses to lessen their carbon footprint. The development and application of circular technologies in the petroleum industry can be aided by grants and subsidies. The creation of bioplastics and research into chemical recycling technologies can receive funding, which can hasten the shift to a more circular economy (Kosiorowska et al., 2021). There are several obstacles and difficulties facing circular economy strategies in the petroleum sector. One technical problem is the complexity of recycling petroleum-based products, which calls for expensive development and cutting-edge technologies (Jagaba et al., 2022). Another major obstacle is economic viability because WtE, re-refining, and recycling can be expensive operations. Economic factors that may deter circular practices include low oil prices and other market conditions.

Adoption of circular economy methods is hampered by regulatory and legislative obstacles, such as uneven laws and policies in various areas. These obstacles breed uncertainty. The demand for sustainable products and consumer knowledge is also a major factor in the success of circular economy strategies. Prospects and developments for the petroleum industry in the future include the creation of circular business models, the integration of renewable energy sources, and improvements in recycling technologies (Das, Dey, & Das, 2024). Technological developments in recycling, like chemical recycling and bioconversion, have the potential to surmount obstacles and enhance the effectiveness and financial sustainability of circular activities. Table 18.1 shows the advantages and disadvantages of different greener technologies involved in petroleum hydrocarbon management. Petroleum processing and recycling processes may become more sustainable and have a smaller carbon footprint if renewable energy sources are incorporated (Jagatee et al., 2024). The adoption of circular practices in the petroleum sector could be facilitated by the creation of circular business models, such as leasing or product-as-a-service models, which move the focus from ownership to service supply and guarantee product recycling or reuse at the end of its life.

18.12 FUTURE PERSPECTIVE

There are various emerging trends that will drive major breakthroughs in the management of petroleum hydrocarbon emissions using advanced green technology in the future. By enabling real-time monitoring and optimization, the integration of digital technologies—such as artificial intelligence (AI), machine learning, and the Internet of Things (IoT)—will transform pollution control. AI will improve operational effectiveness and predictive maintenance, while IoT devices will supply vital data for dynamic modifications and early problem detection. Material science advancements will lower prices and increase performance, especially in membrane technologies like self-cleaning and graphene-based membranes. With the creation of sophisticated sensors for accurate contaminant detection and NPs for targeted degradation, nanotechnology will significantly revolutionize pollutant management. Adopting the ideas of the circular economy will be essential, since resource recovery technologies such as chemical recycling and closed-loop systems will minimize waste and reduce resource consumption. Green chemistry and sustainable design, which concentrate on developing biodegradable materials and safer chemical processes, will also be crucial. The adoption of these technologies will also be fueled by more financing for research and development as well as improved regulatory backing. Growing public awareness will drive demand for sustainable solutions, and successful implementation will depend on corporate, government, and academic collaboration. As a whole, these developments will improve the efficiency and sustainability of technology used to control pollution derived from petroleum hydrocarbons, opening the door to a future that uses resources more wisely and cleanly. Despite their potential, greener remediation technologies face challenges such as scalability, site-specific applicability, and regulatory acceptance. Integrating these technologies into comprehensive remediation strategies requires interdisciplinary collaboration among scientists, engineers, regulators, and stakeholders. Furthermore, public awareness and education are crucial for fostering acceptance and adoption of greener remediation practices.

18.13 CONCLUSION

Challenges to greener remediation solutions for petroleum hydrocarbon cleanup include regulatory approval, scalability, and site-specific applicability. Scientists, engineers, regulators, and stakeholders must work together in a transdisciplinary manner to integrate these technologies into comprehensive remediation methods. In order to promote acceptance and adoption of green remediation techniques, public education and awareness campaigns are also essential. A viable route toward sustainable environmental management is the shift from petroleum hydrocarbon treatment to greener technology. Progress toward more practical and financially sustainable solutions can be accelerated by recognizing these issues and investigating possible breakthroughs. To fully realize the potential of greener remediation approaches and to overcome technological, financial, and legal hurdles, it is imperative that research, innovation, and collaboration continue. We can lessen environmental impacts and encourage a healthier, more sustainable future by adopting eco-friendly alternatives. The management of petroleum hydrocarbon emissions requires a multifaceted approach involving financial removal, scalability enhancement, and integration of new technology with existing systems. Current developments and research offer ways to enhance the efficacy and sustainability of these technologies, aiming for a more sustainable future and reduced environmental impact.

BIBLIOGRAPHY

Bal, B., Ghosh, S., & Das, A. P. (2018). Microbial recovery and recycling of manganese waste and their future application: a review. *Geomicrobiological Journal*, *36*(1), 85–96. https://doi.org/10.1080/01490451.2018.1497731

Bala, K., Dogra, A., Lata, P., & Das, A. P. (2024). Discussing sources and biological-chemical remediation approaches for microplastic pollution. In *Sustainable Microbial Technology for Synthetic and Cellulosic Microfiber Bioremediation* (pp. 37–59). https://doi.org/10.1007/978-3-031-62600-5_3

Bala, S., Garg, D., Thirumalesh, B. V., Sharma, M., Sridhar, K., Inbaraj, B. S., & Tripathi, M. (2022). Recent strategies for bioremediation of emerging pollutants: a review for a green and sustainable environment. *Toxics*, 10(8). MDPI. https://doi.org/10.3390/toxics10080484

Das, A., Dey, S., Bala, K., Gnanasekaran, R., & Das, A. P. (2024). Microbial bioremediation technology for sustainable treatment and management of synthetic microfiber waste. In *Sustainable Microbial Technology for Synthetic and Cellulosic Microfiber Bioremediation* (pp. 221–238). https://doi.org/10.1007/978-3-031-62600-5_12

Das, A., Dey, S., & Das, A. P. (2024). Ecotoxicological impacts of synthetic microfiber pollutants and development of sustainable mitigation strategies. In *Environmental Chemistry and Ecotoxicology*. https://doi.org/10.1016/J.ENCECO.2024.12.003

Das, A., Ray, S. K., Mohanty, M., Mohanty, J., Dey, S., & Das, A. P. (2024). Ecotoxicity of microplastic wastes and their sustainable management: a review. *Environmental Chemistry and Ecotoxicology*, *6*, 144–152. https://doi.org/10.1016/J.ENCECO.2024.05.003

Das, A. P., & Ghosh, S. (2018). Bioleaching of manganese from mining waste materials. *Materials Today: Proceedings*, *5*(1), 2381–2390. https://doi.org/10.1016/J.MATPR.2017.11.459

Das, A. P., Ghosh, S., Mohanty, S., & Sukla, L. B. (2015a). Advances in manganese pollution and its bioremediation. In L. B. Sukla, N. Pradhan, S. Panda, & B. K. Mishra (Eds.), *Environmental Microbe Biotechnology* (1st ed., pp. 313–328). Springer, Cham. https://doi.org/10.1007/978-3-319-19018-1_16

Das, A. P., Ghosh, S., Mohanty, S., & Sukla, L. B. (2015b). Consequences of manganese compounds: a review. *Toxicological & Environmental Chemistry*, *96*(7), 981–997. https://doi.org/10.1080/02772248.2015.1005428

Das, A. P., Mallavarapu, M., & Ghosh, S. (2023). Ecotoxicity of mining pollutants on the environment and their remediation. *Environmental Chemistry and Ecotoxicology*, *5*, 165–167. https://doi.org/10.1016/J.ENCECO.2023.08.002

Das, A. P., & Mishra, S. (2008). Hexavalent chromium (VI): health hazards and environmental pollutant. *Journal of Environmental Research and Development*, *2*(3), 386–392.

Das, A. P., & Singh, S. (2011). Occupational health assessment of chromite toxicity among Indian miners. *Indian Journal of Occupational and Environmental Medicine*, *15*(1), 6. https://doi.org/10.4103/0019-5278.82998

Das, A. P., Sukla, L. B., Pradhan, N., & Nayak, S. (2011). Manganese biomining: a review. *Bioresource Technology*, *102*(16), 7381–7387. https://doi.org/10.1016/J.BIORTECH.2011.05.018

Deng, L., Peng, C., Li, S., Yang, R., Yan, C., Li, M., & Lu, L. (2025). Comparative effects of chemical dispersants and rhamnolipid biosurfactants on oil biodegradation and microbial community in coastal sediments. *International Biodeterioration & Biodegradation*, *196*, 105913. https://doi.org/10.1016/J.IBIOD.2024.105913

Dey, S., Bissoyi, A., Koteswara Reddy, G., & Das, A. P. (2024). Synthetic microfibers: an urban emerging pollutant demanding for sustainable environmental solutions. In *Sustainable Microbial Technology for Synthetic and Cellulosic Microfiber Bioremediation* (pp. 107–121). https://doi.org/10.1007/978-3-031-62600-5_7

Dey, S., Das, A., Mallick, K., Sahu, A., & Das, A. P. (2023). Environmental petroleum waste: pollution, toxicity, sustainable remediation. (pp. 159–175). https://doi.org/10.1007/978-3-031-48220-5_7

Dey, S., Dash, A., & Das, A. P. (2024). Bioremediation as a sustainable and circular solution for mining waste management and bioenergy generation. In *Sustainable Management of Mining Waste and Tailings: A Circular Economy Approach* (pp. 107–119). 10.1201/9781003442455-6.

Dey, S., Dash, N., Ray, S. K., Priyadarsini, S., Chhotaray, C., & Das, A. P. (2024). Occurrence and quantification of tire wear plastics pollutants from highway city dust of Bhubaneswar City of Odisha. In *Sustainable Microbial Technology for Synthetic and Cellulosic Microfiber Bioremediation* (pp. 77–89). https://doi.org/10.1007/978-3-031-62600-5_5

Dey, S., Tripathy, B., Kumar, M. S., & Das, A. P. (2023). Ecotoxicological consequences of manganese mining pollutants and their biological remediation. In *Environmental Chemistry and Ecotoxicology*. https://doi.org/10.1016/J.ENCECO.2023.01.001

Ghosh, S., Gandhi, M., van Hullebusch, E. D., & Das, A. P. (2020). Proteomic insights into Lysinibacillus sp.-mediated biosolubilization of manganese. *Environmental Science and Pollution Research*, *28*(30), 40249–40263. https://doi.org/10.1007/S11356-020-10863-4

Jagaba, A. H., Kutty, S. R. M., Lawal, I. M., Birniwa, A. H., Affam, A. C., Yaro, N. S. A., Usman, A. K., Umaru, I., Abubakar, S., Noor, A., Soja, U. B., & Yakubu, A. S. (2022). Circular economy potential and contributions of petroleum industry sludge utilization to environmental sustainability through engineered processes: a review. In *Cleaner and Circular Bioeconomy* (Vol. 3). Elsevier B.V. https://doi.org/10.1016/j.clcb.2022.100029

Jagatee, S., Priyadarshini, S., Rath, C. C., & Das, A. P. (2024). Synthetic microfiber: an enduring environmental problem linked to sustainable development. In *Environmental Science and Engineering, Part F2333* (pp. 93–112). https://doi.org/10.1007/978-3-031-51792-1_6

Kebede, G., Tafese, T., Abda, E. M., Kamaraj, M., & Assefa, F. (2021). Factors influencing the bacterial bioremediation of hydrocarbon contaminants in the soil: mechanisms and impacts. *Journal of Chemistry* (Vol. 2021). Hindawi Limited. https://doi.org/10.1155/2021/9823362

Khatun, R., Singh, S., Dubey, N. K., & Prasad Das, A. (2023). A review on marine-based phytochemicals and their application in biomedical research. In *Recent Frontiers of Phytochemicals: Applications in Food, Pharmacy, Cosmetics, and Biotechnology* (pp. 383–395). Elsevier. https://doi.org/10.1016/B978-0-443-19143-5.00013-X

Koshlaf, E., & Ball, A. S. (2017). Soil bioremediation approaches for petroleum hydrocarbon polluted environments. *AIMS Microbiology*, *3*(1), 25–49. AIMS Press. https://doi.org/10.3934/MICROBIOL.2017.1.25

Kosiorowska, K. E., Połomska, X., Wang, G., Borodina, I., & Mirończuk, A. M. (2021). Efficient biodegradation of aliphatic polyester by genetically engineered strains of the yeast Yarrowia lipolytica. *International Biodeterioration & Biodegradation*, *161*, 105232. https://doi.org/10.1016/J.IBIOD.2021.105232

Kostka, J. E., Teske, A. P., Joye, S. B., & Head, I. M. (2014). The metabolic pathways and environmental controls of hydrocarbon biodegradation in marine ecosystems. *Frontiers in Microbiology*, 5(SEP). Frontiers Research Foundation. https://doi.org/10.3389/fmicb.2014.00471

Kumar, M. S., & Das, A. P. (2017). Emerging nanotechnology based strategies for diagnosis and therapeutics of urinary tract infections: a review. *Advances in Colloid and Interface Science*, *249*, 53–65. https://doi.org/10.1016/J.CIS.2017.06.010

Kumari, B., & Singh, D. P. (2016). A review on multifaceted application of nanoparticles in the field of bioremediation of petroleum hydrocarbons. *Ecological Engineering*, *97*, 98–105. Elsevier B.V. https://doi.org/10.1016/j.ecoleng.2016.08.006

Kumari, P., & Kumar, A. (2023). Advanced oxidation process: a remediation technique for organic and non-biodegradable pollutant. *Results in Surfaces and Interfaces*, *11*. Elsevier B.V. https://doi.org/10.1016/j.rsurfi.2023.100122

Maletić, S. P., Beljin, J. M., Rončević, S. D., Grgić, M. G., & Dalmacija, B. D. (2019). State of the art and future challenges for polycyclic aromatic hydrocarbons is sediments: sources, fate, bioavailability and remediation techniques. *Journal of Hazardous Materials*, *365*, 467–482. Elsevier B.V. https://doi.org/10.1016/j.jhazmat.2018.11.020

Mallick, K., Sahu, A., Dubey, N. K., & Das, A. P. (2023). Harvesting marine plastic pollutants-derived renewable energy: a comprehensive review on applied energy and sustainable approach. *Journal of Environmental Management*, *348*, 119371. https://doi.org/10.1016/J.JENVMAN.2023.119371

Mallick, K., Sahu, S., Sahu, A., Dey, S., & Das, A. P. (2024). A critical review of marine microfiber pollution routes, toxicity, and its sustainable remediation. (pp. 189–211). https://doi.org/10.1007/978-3-031-51792-1_10

Mishra, S., & Das, A. P. (2023). Microbial remediation of synthetic microfiber contaminated wastewater. In M. P. Shah (Ed.), *Microbial Technologies in Industrial Wastewater Treatment* (pp. 337–350). Springer, Singapore. https://doi.org/10.1007/978-981-99-2435-6_17

Mishra, S., Das, A. P., & Seragadam, P. (2009). Microbial remediation of hexavalent chromium from chromite contaminated mines of Sukinda Valley, Orissa (India). *Journal of Environmental Research and Development*, (1), 1122–1127.

Mishra, S., Dash, D., Al-Tawaha, A. R. M. S., & Das, A. P. (2022). A review on heavy metal ion adsorption on synthetic microfiber surface in aquatic environments. *Applied Biochemistry and Biotechnology*, *194*(10), 4639–4654. https://doi.org/10.1007/s12010-022-04029-w

Mishra, S., Dash, D., & Das, A. P. (2023). Aquatic microbial diversity on plastisphere: colonization and potential role in microplastic biodegradation. *Geomicrobiology Journal*, 1–12. https://doi.org/10.1080/01490451.2023.2209750

Mishra, S., Dash, D., & Das, A. P. (2024). Isolation and molecular characterization of polyester-degrading bacterial strains from municipal wastewater and their application in microplastic degradation in contaminated wastewater. *Geomicrobiology Journal*. https://doi.org/10.1080/01490451.2024.2376101

Mishra, S., Ghosh, S., van Hullebusch, E. D., Singh, S., & Das, A. P. (2023). A critical review on the recovery of base and critical elements from electronic waste-contaminated streams using microbial biotechnology. *Applied Biochemistry and Biotechnology*, *2023*, 1–30. https://doi.org/10.1007/S12010-023-04440-X

Mishra, S., Rath, C. C., & Das, A. P. (2019). Marine microfiber pollution: a review on present status and future challenges. *Marine Pollution Bulletin*, *140*, 188–197. https://doi.org/10.1016/J.MARPOLBUL.2019.01.039

Mishra, S., Rout, P. K., & Das, A. P. (2021). Emerging microfiber pollution and its remediation. In R. Prasad (Ed.), *Environmental and Microbial Biotechnology* (pp. 247–266). Springer, Singapore. https://doi.org/10.1007/978-981-15-5499-5_9

Mishra, S., Singh, R. P., Rath, C. C., & Das, A. P. (2020). Synthetic microfibers: source, transport and their remediation. *Journal of Water Process Engineering*, *38*, 101612. https://doi.org/10.1016/J.JWPE.2020.101612

Mohammadi, L., Rahdar, A., Bazrafshan, E., Dahmardeh, H., Susan, M. A. B. H., & Kyzas, G. Z. (2020). Petroleum hydrocarbon removal fromwastewaters: a review. *Processes*, *8*(4). MDPI AG. https://doi.org/10.3390/PR8040447

Munirasu, S., Haija, M. A., & Banat, F. (2016). Use of membrane technology for oil field and refinery produced water treatment: a review. *Process Safety and Environmental Protection*, *100*, 183–202. Institution of Chemical Engineers. https://doi.org/10.1016/j.psep.2016.01.010

Nag, M., Lahiri, D., Ghosh, S., Sarkar, T., Pati, S., Das, A. P., Ram, D. K., Bhattacharya, D., & Ray, R. R. (2023). Application of microorganisms in biotransformation and bioremediation of environmental contaminant: a review. *Geomicrobiology Journal*. https://doi.org/10.1080/01490451.2023.2261443

NoorZaliha, R., Noor Zaliha Raja Abd Rahman, R., Basri, M., Bakar Salleh, A., Mohamad Ghazali, F., & Noor Zaliha Abd Rahman, R. (2003). Bioremediation of petroleum hydrocarbon pollution. *Article in Indian Journal of Biotechnology*, *2*. https://www.researchgate.net/publication/229000303

Ossai, C., Hamid, S., & Hassan, A. (n.d.). *Biological Treatments for Petroleum Hydrocarbon Pollutions: The Eco-Friendly Technologies*. www.intechopen.com

Oualha, M., Al-Kaabi, N., Al-Ghouti, M., & Zouari, N. (2019). Identification and overcome of limitations of weathered oil hydrocarbons bioremediation by an adapted Bacillus sorensis strain. *Journal of Environmental Management, 250*. https://doi.org/10.1016/j.jenvman.2019.109455

Paul, A., Dey, S., Ram, D. K., & Das, A. P. (2023). Hexavalent chromium pollution and its sustainable management through bioremediation. *Geomicrobiology Journal*, 1–11. https://doi.org/10.1080/01490451.2023.2218377

Pradhan, G., Tripathy, B., Ram, D. K., Digal, A. K., & Das, A. P. (2023). Bauxite mining waste pollution and its sustainable management through bioremediation. *Geomicrobiology Journal*, 1–10. https://doi.org/10.1080/01490451.2023.2235353

Ramadass, K., Megharaj, M., Venkateswarlu, K., & Naidu, R. (2018). Bioavailability of weathered hydrocarbons in engine oil-contaminated soil: impact of bioaugmentation mediated by Pseudomonas spp. on bioremediation. *Science of the Total Environment, 636*, 968–974. https://doi.org/10.1016/j.scitotenv.2018.04.379

Ray, S. K., Dey, S., Das, A. P., & John, A. (2024). Microplastic waste management through greener technologies: a review. In *Sustainable Management of Urban Plastic Waste through Circular Economic Approaches* (pp. 237–261). https://doi.org/10.1201/9781003465447-14

Roy, A., Dutta, A., Pal, S., Gupta, A., Sarkar, J., Chatterjee, A., Saha, A., Sarkar, P., Sar, P., & Kazy, S. K. (2018). Biostimulation and bioaugmentation of native microbial community accelerated bioremediation of oil refinery sludge. *Bioresource Technology, 253*, 22–32. https://doi.org/10.1016/j.biortech.2018.01.004

Sahoo, P. P., Singh, S., Rout, P. K., Mishra, S., & Das, A. P. (2022). Microbial remediation of plastic pollutants generated from discarded and abandoned marine fishing nets. *Biotechnology and Gnetic Engineering Reviews*. https://doi.org/10.1080/02648725.2022.2152629

Saini, A., Bekele, D. N., Chadalavada, S., Fang, C., & Naidu, R. (2021). Electrokinetic remediation of petroleum hydrocarbon contaminated soil (I). *Environmental Technology and Innovation, 23*. https://doi.org/10.1016/j.eti.2021.101585

Singh, R. P., Mishra, S., & Das, A. P. (2020). Synthetic microfibers: pollution toxicity and remediation. *Chemosphere, 257*, 127199. https://doi.org/10.1016/j.chemosphere.2020.127199

SinghJadaun, J., Bansal, S., Sonthalia, A., Rai, A. K., & Singh, S. P. (2022). Biodegradation of plastics for sustainable environment. *Bioresource Technology, 347*, 126697. https://doi.org/10.1016/J.BIORTECH.2022.126697

Waly, M. M., Mickovski, S. B., & Thomson, C. (2023). Application of circular economy in oil and gas produced water treatment. *Sustainability (Switzerland), 15*(3). https://doi.org/10.3390/su15032132

Wu, M., Ye, X., Chen, K., Li, W., Yuan, J., & Jiang, X. (2017). Bacterial community shift and hydrocarbon transformation during bioremediation of short-term petroleum-contaminated soil. *Environmental Pollution, 223*, 657–664. https://doi.org/10.1016/j.envpol.2017.01.079

Zhang, Y., Peng, C., Guo, Z., Xiao, X., & Xiao, R. (2019). Polycyclic aromatic hydrocarbons in urban soils of China: distribution, influencing factors, health risk and regression prediction. *Environmental Pollution, 254*. Elsevier Ltd. https://doi.org/10.1016/j.envpol.2019.07.098

Index

A

B

C

D

E

I

K

L

M

N

O

P

R

S

T

U

V

W

X

Y

Z

For Product Safety Concerns and Information please contact our EU
representative GPSR@taylorandfrancis.com
Taylor & Francis Verlag GmbH, Kaufingerstraße 24, 80331 München, Germany

www.ingramcontent.com/pod-product-compliance
Ingram Content Group UK Ltd.
Pitfield, Milton Keynes, MK11 3LW, UK
UKHW022214190926
475702UK00003B/347
* 9 7 8 1 0 3 2 8 2 4 3 2 1 *